The Correspondence of John Tyndall

VOLUME 5

THE CORRESPONDENCE OF JOHN TYNDALL

VOLUME 5

The Correspondence, January 1855–October 1856

Edited by
William H. Brock and Geoffrey Cantor

General Editors
James Elwick, Roland Jackson, Bernard Lightman,
and Michael S. Reidy

UNIVERSITY OF PITTSBURGH PRESS

Published by the University of Pittsburgh Press, Pittsburgh, Pa., 15260

Manufactured in the United States of America
Printed on acid-free paper
10 9 8 7 6 5 4 3 2 1

Cataloging-in-Publication data is available from the Library of Congress

ISBN 13: 978-0-8229-4532-1

CONTENTS

Acknowledgments	vii
List of Abbreviations	ix
Introduction to Volume 5	xiii
Editorial Principles	xxvii
Note on Money	xxxi
Timeline of John Tyndall's Life	xxxiii
Timeline of Events in John Tyndall's Life Specific to Volume 5	xxxvii
The Correspondence, January 1855–October 1856	1
Biographical Register	459
Index	477

ACKNOWLEDGMENTS

The editors of this volume would like to express their appreciation to the many people who generously assisted their editorial labors by providing information and advice on a wide range of issues, especially Anne Barrett, Gabriel Finkelstein, Erin Grosjean, Jon Hodge, John Hudson, Julie McDougall-Waters, Alison Morrison-Low, Sylvia Nickerson, Julie Orford, and Marek Petrik (Marburg); and, at the Royal Institution, Frank James, Jane Harrison, Roland Jackson, and Charlotte New. For invaluable help with the transcription and translation of Matteucci's and other French letters, we are particularly indebted to Juliette Reboul, and to James Braund and Nina Spieler for help with the German letters. Our reseach was supported by an Insight Grant from the Social Sciences and Humanities Research Council of Canada, which was helpfully and efficiently administered by Stephanie Collins in the Department of English at the University of Leicester. Finally, we acknowledge the help and support of the editors of other current volumes in the series; namely, Melina Baldwin, Ruth Barton, Janet Browne, Gowan Dawson, Ian Hesketh, and Efram Sera-Shriar; and especially the general editors of *The Correspondence of John Tyndall*—Bernard Lightman, James Elwick, Roland Jackson, and Michael Reidy—who have responded rapidly and efficiently to our numerous and diverse queries.

LIST OF ABBREVIATIONS

ADB	*Allgemeine Deutsche Biographie*
BAAS	British Association for the Advancement of Science
BL	British Library, London
Brit. Assoc. Rep.	*Report of the Meeting of the British Association for the Advancement of Science*, followed by the year
DSB	*Dictionary of Scientific Biography*
ERE	M. Faraday, *Experimental Researches in Electricity*, 3 vols. (London: J. Murray, 1839–55)
FRS	Fellow of the Royal Society
IC HP	Imperial College Archives, London, Huxley Papers
ICEP	*Proceedings of the Institution of Civil Engineers*
Faraday Correspondence	*The Correspondence of Michael Faraday*, ed. F. A. J. L. James, 6 vols. (London: Institution of Electrical Engineers / Institution of Engineering and Technology, 1991–2012)
Jackson	Roland Jackson, "John Tyndall and the Early History of Diamagnetism," *Annals of Science* 72 (2014), 439–85
JRL Frankland	John Rylands Library, University of Manchester, Manchester, Edward Frankland Papers
LI	London Institution
ODNB	*Oxford Dictionary of National Biography*
OED	*Oxford English Dictionary*
Phil. Mag.	*Philosophical Magazine*
Phil. Trans.	*Philosophical Transactions of the Royal Society of London*
Poggend. Annal.	*Annalen der Physik und Chemie*, edited by J. C. Poggendorff
RGO	Cambridge University Library, Royal Greenwich Observatory Archives
Roy. Soc. Proc.	*Proceedings of the Royal Society of London*
RI	Royal Institution of Great Britain
RI MS JT	Royal Institution, London, John Tyndall Papers, followed by manuscript number and page/folio number(s)
Roy. Inst. Proc.	*Proceedings of the Royal Institution of Great Britain*

RS RR	Royal Society of London, Referees Reports
Tyndall's Journal	RI MS JT/2/13b/ covers period October 17, 1850, to January 25, 1855; RI MS JT/2/13c/ covers period February 1, 1855, to July 1, 1857
Wellcome	Wellcome Library, London

Frontispiece: John Tyndall in a social setting with Michael Faraday, Mrs. Sarah Faraday (both seated), Jane Barnard (a niece of the Faradays; probably the woman seated on the ground), and Julia Moore. Julia's sister, Harriet Moore, took the photograph in the gardens of the Moores' Petersham home in June 1858. Reproduced with the permission of the Royal Institution, London, UK/Bridgeman Images.

INTRODUCTION TO VOLUME 5

This fifth volume of *The Correspondence of John Tyndall* contains 266 letters covering a period of twenty-two months, from 1 January 1855 to 31 October 1856. Tyndall was then in his mid-thirties and had been employed by the RI as professor of natural philosophy since September 1853. Many of the letters printed here concern the lectures he delivered at the RI and other institutions and his attempts to establish his reputation as a researcher. Although he published in several other areas—including the cleavage of rocks, colorblindness, and glaciers—the main focus of his research was the newly discovered and problematic phenomenon of diamagnetism. In several lectures and papers Tyndall reported his experimental results and theoretical views on this subject, including the Bakerian Lecture he delivered to the RS on 26 January 1855.[1] This prestigious lecture greatly enhanced his scientific standing, which was further extended by his contact with other scientists, not only in London (where he was patronized by several leading figures) but across the British Isles and on the Continent, especially in France and Germany. His French contacts were augmented through serving as a juror at the 1855 Exposition Universelle, held in Paris, and by attending a meeting of German scientists in Vienna in 1856. By the end of October 1856 Tyndall was a man of science with a European reputation. His scientific reputation was further confirmed and enhanced by his election to the council of the RS on 27 November 1856.

Family, Friendships, and Social Life

In contrast to the first two volumes in this series, few letters in the present volume engage with Tyndall's family. His father, to whom he had been very close despite their divergent views on various subjects, had died in 1847. His mother, still living in Ireland, was the recipient of only three letters that appear below, all of which are brief, factual, and distinctly impersonal.[2] Of his sister, Emma, there is not a single mention. Apart from a few references to another cousin (Thomas Payne), the only relative with whom Tyndall pursued a substantial correspondence, which includes some of the longest letters in this volume,[3] was his flamboyant American cousin Hector Tyndale, whom he had met in London in 1854.

The letters printed here can be related to different periods of Tyndall's life and to the friendships he had made before entering the RI's service. Barely represented are any of the people whom Tyndall had known when he was in Ireland (which he left in August 1842), the principal exception being George Wynne (now the government inspector of railways), who had commanded the 5th Division, C District of the Irish Ordnance Survey, for which Tyndall had worked as a civil assistant. From his time being employed as a surveyor in Halifax (1845) he remained on very close terms with Thomas Archer Hirst; correspondence with Hirst constitutes 30 percent of these letters. Hirst had replaced Tyndall when he left Queenwood College in 1853, and the Queenwood connection also was continued through Hirst, Edward Frankland (who moved to Owen's College, Manchester, in 1851), and Heinrich Debus. From his time studying in Germany (1848–51) he continued to correspond with several of his scientific friends, both British and German, including Hirst, Frankland, Debus, Wilhelm Weber, Heinrich Gustav Magnus, and Emil du Bois-Reymond, who lectured at the RI in 1855.

Having been appointed to the RI—one of the very few paid scientific positions in Britain—Tyndall was determined to make his career in science. To achieve this he not only pursued innovative research (see below) but also developed a network of scientific contacts. Several established men of science took Tyndall under their wings and helped cement his position in British science. Principal among these was Michael Faraday; the "father and son" intimacy between them is illustrated by Faraday's gift to Tyndall of the third volume of his *Experimental Researches in Electricity* in April 1855.[4] Tyndall also attracted the support and patronage of Edward Sabine of the Royal Military Academy, Woolwich, and the geologist Charles Lyell. Others, such as Lord Ashburton, Peter Barlow, Henry Bence Jones, James Coxe, John Peter Gassiot, and Charles Wheatstone, helped Tyndall by including him in their social circle, invited him to dinner, and thus enabled him to meet other scientists, potential patrons, and literary figures. At such events Tyndall encountered two of his literary and philosophical heroes—Thomas Carlyle and Alfred, Lord Tennyson.[5] Although he corresponded with several women, including Julia Moore, Juliet Pollock, Mary Anne Coxe, and Henrietta, Huxley's new wife, and was clearly attracted to some of the young ladies he met socially, he was not on particularly close terms with any eligible women.[6] He also took an interest in some of the children of friends, including Edward Wynne (George Wynne's son) and Olivia Jones (Bence Jones's daughter).

Outside London his circle of scientific acquaintances included William Thomson, with whom he had a tense relationship that evoked disagreements over diamagnetism,[7] and others whom he met at the BAAS meeting in

Glasgow, which ran from 12 to 19 September 1855. During his visit to Paris earlier in the same year, when he served as a juror at the Exposition Universelle, he had the opportunity of seeing the elite of French science during a visit to the Académie des Sciences.[8] As he told his mother, "the great advantage in being here for some time, [is] an opportunity of meeting with eminent men from all quarters of the world and whom I previously knew by their writings alone."[9] While in Paris he also made friends with César Despretz, the instrument maker Louis Jules Duboscq, and other jurors, including Elie-François Wartmann from Switzerland and Theodor Wertheim from Austria.

Tyndall appreciated the value of his position at the RI, which enabled him to work closely with Faraday, have access to a well-equipped laboratory, and enjoy contact with the many scientists whom he met through the RI and RS. Thus, when he was considered for a chair at the University of Toronto and the position of director of its observatory, he—assured by his friends—rapidly declined the invitation, as the position involved observation rather than innovative scientific experimentation.[10] By contrast with the Toronto post, which carried a basic salary of £450 per annum, his salary at the RI was £200 per annum, forcing him to seek other forms of remuneration, which included writing reviews for the *Phil. Mag.* and *Westminster Review* and contributing chapters on a variety of scientific topics to a series of school textbooks edited by Edward Hughes.[11] For setting and marking the twice-yearly examination papers sat by candidates for Royal Artillery and Royal Engineers commissions to the Royal Military Academy at Woolwich he received the princely sum of £50 per annum.

Although his position at the RI suited him well, Tyndall's vigor was frequently undermined by debilitating headaches, especially during the early months of 1855. Constipation was also a recurrent issue; as he informed Hirst during one period of good health in November 1855, "I have kept my eye upon the bowels for some time and without betaking myself to physic [i.e., to medicine] have brought them into a good condition."[12] His frequent visits to Queenwood and his enjoyment of country walks were, in part, antidotes to the more hectic life he led in London. His other main relaxation was reading. Thus during one of his severe headaches he recorded in his journal: "Spent most of this forenoon in bed reading and cogitating—a new way of spending time."[13] Among the many diverse books he consumed during 1855–56 were Henry Wadsworth Longfellow's *The Song of Hiawatha* (1855), Henry More's *Enthusiasmus Triumphatus* (1712 ed.), George Henry Lewes's *The Life and Works of Goethe* (1855; which he thoroughly disliked), Plutarch's life of Gaius Marcius Coriolanus (in *Plutarch's Lives*), and Walter Scott's *Rob Roy* (1817).[14] And, of course, he read profusely in the contemporary scientific literature.

The Queenwood Question

Although Tyndall had resigned from his teaching post at Queenwood College in 1853, having been appointed professor of natural philosophy at the RI, he remained a constant visitor during the years 1855–56. Apart from short weekend visits (often with friends such as William Francis and James Bevington), he borrowed accommodation at Queenwood during Easter vacations and the summer months. There he could walk, philosophize with teachers such as Hirst, Debus, and Wright, use the laboratory, or simply write, translate, and review. He was drawn there for a number of reasons: the presence of his closest friend, Hirst; the beauty of the neighborhood and the availability of pleasurable walks in the New Forest, which Debus urged as necessary for Tyndall's health;[15] and his genuine concern for the continuing success of the school as an educational institution. The latter was a constant source of worry because both he and Hirst had a very poor opinion of the headmaster, George Edmondson. We see only one side of the story through the correspondence between Tyndall and Hirst and the views they expressed in their journals. Edmondson was, in fact, always in a difficult position, as the school was only leased to him and his notional landlords, John Finch and William Pare and the cooperative movement that they represented, still owed a huge sum of money—some £14,000—to the real landowner, the London bullion merchant and philanthropist Isaac Lyon Goldsmid, following the collapse of Robert Owen's communitarian experiment of Harmony Hall in 1845.[16] This meant that Edmondson was continually worried that his socialist landlords might dispose of the school and its land. In December 1853 Tyndall had stepped in to prevent the sale of the estate to a community of American Shakers[17] and in 1855 Hirst and other teachers dissuaded Edmondson from being influenced by his wife and daughter to abandon Queenwood for a smaller school.[18]

Tyndall became concerned again for the school's future when Hirst had to resign in order to nurse his sick wife, Anna, in June 1856. Tyndall found himself torn by Hirst's (and other masters') requests that he abandon his position at the RI and return to Queenwood to save the school. Tyndall resolved to stay in London. There can be no doubt that, given Queenwood College's excellent laboratory facilities, Tyndall could have continued research in Hampshire,[19] but he relished his daily contact with Faraday and other London men of science as well as the rich social life that he found himself enjoying. Yet throughout 1855 and the first part of 1856 Tyndall and Hirst regularly exchanged information about pupils and their moral training.[20]

At the RI, Tyndall delivered eleven weekly lectures on magnetism and electricity in the early months of 1855 and a course of eight lectures on voltaic electricity after Easter. During the spring and summer of 1856 the subject of his weekly lectures—nineteen in all—was light. The preparation and delivery of such lectures clearly worried him and he usually deprecated his own delivery and performance in his journal, though simultaneously reporting that members of the audience seemed pleased with the result. Frankland captured the stage fright that Tyndall evidently experienced and hid so well in these public ordeals: "Those Royal Institution lectures will never agree with you. They keep you continually upon the rack for 6 months. Unless you can look a course of lectures in the face as coolly as an after dinner conversation you may depend upon it your health will suffer."[21]

Over the period under discussion Tyndall also delivered four Friday Evening Discourses at the RI. Two of these were on diamagnetism (26 January 1855 and 1 February 1856), the others on the Leyden battery (1 June 1855) and the cleavage of crystals and slaty rocks (6 June 1856). This last RI lecture reflected his increasing interest in glaciology, but also intimated a plan to gradually work his way over the years ahead through a complete course of experimental physics as defined by magnetism, electricity, light, heat, and sound. Tyndall was also often called upon to offer assistance to visiting speakers. For example, he assisted du Bois-Reymond when he gave a series of lectures on electrophysiology in the spring of 1855, and he helped Thomson prepare experiments for Thomson's RI discourse in February 1856. He also delivered several series of lectures at the LI (including six lectures on heat that he was to repeat later at the RI). The managers of the LI obviously hoped to tempt him to take a permanent position, even offering him double his RI salary.

At the RS he was honored by being invited to deliver the Bakerian Lecture on diamagnetism in January 1855. He lectured for two hours, and although the subject was a difficult one, the president and several fellows of the society congratulated him on his cogency. As a loyal fellow, Tyndall frequently attended meetings at the RS, and his growing expertise was recognized by the two secretaries, George Stokes and William Sharpey, who invited him to referee papers by Thomson, James Prescott Joule, Antoine Claudet, and Carlo Matteucci.[22] Forward-looking fellows clearly recognized him as a "modernizer," and (through the machinations of William Grove) he was elected to the exclusive Philosophical Club of the Society in May of the same year.

The only other society he attended sporadically was the Geological Society, usually at the invitation of Lyell. At such a meeting in January 1855 he expressed his disagreement with David Forbes's views on the causes of foliation of rocks in Scotland and Norway. As he wrote to Hirst a few days later, during the meeting he became increasingly critical of Forbes's assertion that heat had caused the foliation of rocks and he voiced his objections during the ensuing discussion, evoking "loud and general cries of 'hear hear' [that] shewed that my convictions had been contagious and had won the audience. The theory of Forbes was virtually declared 'not proven.'"[23] Yet, despite his apparent success at challenging Forbes's views, he soon "sank again into a state of intellectual meanness. My power over science is I think increasing greatly, but the power and vigour of what I call my soul is not commensurate."[24]

Tyndall remained one of the coeditors of the *Phil. Mag.* during the period—though he took pains to explain to Peter Riess that his role was purely advisory and that William Francis was the real editor. In practice, Tyndall continued to publish his own papers there, offer his translations from German to the journal,[25] and steer some of the letters he received toward the journal for printing.[26]

In 1855 Tyndall learned that the French were to hold an international exhibition of industry in Paris between May and November that was intended to rival the Great Exhibition masterminded by Prince Albert in Hyde Park four years earlier. He decided that he would exhibit a new instrument he had devised for his lectures on electricity and magnetism, the polymagnet. He was subsequently invited by Brewster to serve on a subcommittee of the international jury that would allocate prizes for instruments devoted to electricity and magnetism. He accepted, and, in order not to compromise himself, withdrew the polymagnet from the competition.

Tyndall had to be in Paris for the jury's deliberations in July and this gave him the opportunity for a protracted holiday tramp beginning in mid-June that took him and Debus across the seasick-inducing English Channel from Southampton to Jersey and from there to Saint Malo, on the tip of the Brittany coast. During the next fortnight they hiked to Dinan and on through the Normandy towns of Dol, Avranches, Mont-Saint-Michel, Vire, Caen, Havre, and Rouen before reaching Paris on 30 June.[27] His stay in Paris ended abruptly on 21 July, as he was required in London to mark the first examination of candidates for Royal Artillery and Royal Engineers commissions to the Royal Military Academy at Woolwich. These examinations were a direct result of reforms the War Department introduced to counter criticism of Britain's conduct of the Crimean War. Meritocracy was to replace the historic system of army commissions gained through wealth, birth, and patronage. The invitation to serve as an examiner came from Henry Moseley, a former professor of physics at King's College, London, who was working tirelessly to introduce scientific studies into the curriculum of English schools as one of Britain's first inspectors of education. He was an obvious choice to help in the reform of military education and had been asked by the government to organize the Woolwich entrance examinations.

The first and second examinations in August 1855 and January 1856 for which Tyndall set and marked a wide range of questions in elementary physics, chemistry, and natural history, passed uneventfully; in April 1856 Joseph Galbraith, a professor of natural philosophy at Trinity College Dublin, complained that the marking system favored candidates with a knowledge of Greek and Latin, modern languages, and English literature rather than mathematics and the physical sciences. Tyndall took this criticism seriously, complaining that it was absurd to allow the same number of marks for a candidate who answered questions across the whole range of sciences as for a candidate who concentrated on just one paper in a classical language. "The absurdity makes a man a partisan."[28] Rashly, he publicized Galbraith's complaint in a letter to the *Times* without first seeking permission from the War Department. Following a reprimand from Lord Panmure, the secretary for war, Tyndall penned a long apologetic letter, as he feared he would lose the examinership and its associated income.[29] Fortunately, it seems that Panmure (and Moseley) agreed with Galbraith and Tyndall's critique, and the marking scheme for future exams was adjusted to make it more equitable between subjects. However, Tyndall was to resign from the examinership in 1858 over rates of payment.

In early September 1855 Tyndall attended the BAAS meeting in Glasgow in his capacity as secretary to Section A (mathematics and physics). On his journey to Scotland he took the opportunity to have three days' holiday with Frankland in the Lake District before the meeting, which he found exhausting because the other secretary of the physical section was absent through illness.[30] When delivering his own paper on diamagnetism he succeeded in upsetting Whewell because he differed in opinion from Thomson.[31] On hearing about the dispute, Faraday wisely reflected:

> [L]et me as an old man, who ought by this time to have profited by experience, say, that when I was younger I found I often misinterpreted the intentions of people, and found that they did not mean what at the time I supposed they meant and further, that as a general rule it was better to be a little dull of apprehension when phrases seemed to imply pique and quick in perception when on the contrary they seemed to convey kindly feeling.—The real truth never fails ultimately to appear, and the opposing parties, if wrong, sooner convinced when replied to forebearingly than when overwhelmed.[32]

Despite being kept busy with administration, the friends were able to steal a day in the middle of the meeting to explore the Trossachs and Loch Katrine, where Tyndall seized the opportunity to revel in Scott's "Lady of the Lake." Frankland subsequently expressed his dissatisfaction with the BAAS, being "convinced of the deep injury it annually inflicts upon science." Apart from a convivial dinner with fellow scientists and their trip to Loch Katrine, Frankland regarded the Glasgow meeting with unmitigated regret.[33]

Tyndall continued to seize opportunities to get away from the smoke of

London. He took a week in June 1856 to stay with Richard Wright and his wife at Mudeford in Dorset and celebrated in his journal a memorable day at Alum Bay on the Isle of Wight with four hundred lines of doggerel largely in praise of Mrs. Wright.[34] Although Tyndall had made a brief vacation visit to Switzerland in September 1849 and climbed Mount Rigi in his slippers, he had not been overly impressed by the mountain scenery. By contrast, in 1856 he visited the Tyrol and became entranced by the alpine scenery. Prompted by Huxley he began to read John David Forbes's *Travels in the Alps* (1843), by which time his magnetic studies had led him to the investigation of crystal cleavage, from which he concluded that his own study of slaty cleavage had bearing upon the nature and structure of European glaciers. Moreover, he realized that Forbes's explanation of glacial structure as "a viscous body which is urged down slopes of a certain inclination by the mutual pressure of its parts"[35] was wanting. Always glad of an excuse to combine a fresh-air holiday with the scientific observation of nature, he accordingly planned to make a detailed study of Alpine glaciers, accompanied by Frankland, Huxley, and Hooker, while en route to the meeting of the Deutsche Naturforscher Versammlung in Vienna in September 1856. Tyndall's visit to the Tyrol from mid-August to early September 1856 provided him with the firsthand experience of the glaciers of Grindelwald, the Aar, and the Rhone that underpinned his later glaciological research.

Research and Publications

During the years 1855–56 magnetism, especially diamagnetism, was the focus of Tyndall's research and the subject through which he launched his "scientific career and gave him[self] the platform for his entry into elite Victorian Society."[36] A decade earlier Faraday had reported to the RS his discovery of diamagnetism, which was manifested by a weak repulsion between a block of "heavy glass" and a strong magnetic field. In contrast to the behavior of, say, iron filings, which align themselves parallel to the lines of magnetic force, the "heavy glass" bar became oriented at right angles to the lines of force, as if repelled.[37] Faraday and a number of other scientists, including William Thomson, Julius Plücker, Wilhelm Weber, and Tyndall, turned their attention to this new class of substances that behaved in such a surprising manner. They sought to clarify the phenomena through delicate experiments and to frame explanatory theories.

Tyndall's two earliest scientific publications, which were on the magnetic properties of crystals, appeared in the March and July 1850 issues of the *Phil. Mag.* and were based on experiments conducted in Marburg with his coauthor, Hermann Knoblauch. The authors refuted Plücker's claim that magnecrystallic action arose from the repulsion between the optical axes of a crystal and a strong magnet. They also challenged Faraday, who had proposed a separate magnecrystallic force. Instead, they showed that magnecrystallic action depends solely on the (para)magnetic and diamagnetic forces possessed by the crystals.[38] Tyndall also presented his research at the 1850 BAAS meeting in Edinburgh,[39] where he first crossed swords with Thomson.

The center of Tyndall's research then moved more firmly to the study of diamagnetism. Drawing on research on bismuth (a diamagnetic) carried out in Weber's laboratory in Berlin, Tyndall argued in a paper he presented at the 1851 BAAS meeting and published in a more extensive form in the *Phil. Mag.* in September 1851 that—contra Plücker—(para)magnetism and diamagnetism obey similar laws.[40] Two months later the *Phil. Mag.* carried Tyndall's sensitive investigations of the polarity of diamagnetism, in which he concluded that—just as with a bar of iron—a bar of bismuth possesses polarity.[41]

There followed a partial break from the subject of magnetism, during which time Tyndall returned to England, was hired by the RI, and was elected to the RS. Now ensconced in the RI, he recommenced his magnetic researches in the summer of 1854, using a powerful apparatus designed by Weber for measuring diamagnetic polarity. His initial findings were presented at the 1854 BAAS in Liverpool and then, in a far more detailed form, in his 1855 Bakerian Lecture titled "On the Nature of the Force by which Bodies Are Repelled from the Poles of a Magnet."[42] Several letters in this volume relate to this long and important paper,[43] in which Tyndall argued that the (para) magnetic or diamagnetic forces associated with a body depend on that body's structure; that is, "the manner in which the material particles are grouped together."[44]

One of the consequences of his Bakerian Lecture was an exchange with Faraday, Thomson, and Alexander Williamson on whether a magnetic medium existed in space, which Tyndall doubted.[45] During his Friday Evening Discourse on diamagnetism in January 1855 Tyndall publicly disagreed with Faraday's views on magnetic action, leading him initially to feel "saddened [but] I did not regret what I had done."[46] However, encouraged by the magnanimous Faraday, Tyndall expressed his criticisms in the *Phil. Mag.* in February 1855.[47] Keen to further Hirst's own career beyond that of a mere schoolmaster and translator, Tyndall also encouraged Hirst to publish his own critique of Faraday.[48] Another upshot of his Bakerian Lecture was a second paper on the polarity of the diamagnetic force that Tyndall read to the RS in December 1855.[49] The papers that Tyndall read before the RS and BAAS and published in the *Phil. Trans.*, *Phil. Mag.*, and BAAS reports provide informative accounts of his impressive skill as an experimenter, which helped him establish a number of new diamagnetic phenomena. For example, he was delighted when Matteucci, who had previously asserted the erroneousness of some of Tyndall's experiments with diamagnetism—and had publicized his doubts to both Grove and Faraday—subsequently confirmed those results.[50]

Although magnetism was the subject of eleven of the fifteen scientific papers that he published during 1855–56,[51] Tyndall also made contributions to several other topics during this period. One of these resulted from the invitation he received from Wynne to visit the railway works at Swindon in order to examine the damaged remains of a locomotive boiler that had exploded at Gloucester Station on 7 February 1855. In the report he wrote, which formed part of Wynne's published evidence, Tyndall attributed the corrosion of the boiler to electrochemical decomposition.[52]

While the letters in this volume shed light on some of his contemporary investigations, others do not feature prominently in his correspondence. For example, Tyndall possessed a long-term research interest in the production of sound by heating metal bars—known as "rockers." He had published a paper on the subject in the *Phil. Trans.* for 1854[53] but the subject barely entered his 1855–56 correspondence.[54] Likewise, these letters do not reflect Tyndall's interest in the phenomenon of colorblindness, on which he published a paper in the May 1856 issue of the *Phil. Mag.*, reporting the case of a man whose ability to differentiate colors had been greatly reduced after a riding accident.[55]

By contrast, the accompanying letters demonstrate Tyndall's increasing concern with geology—specifically the cleavage and foliation of rocks, "a question in which I take so deep an interest"[56]—and glaciology. While he had briefly considered the subject of cleavage in late September 1854,[57] Lyell's invitation to a meeting of the Geological Society on 31 January 1855 enabled him to hear—and publicly dispute—David Forbes's paper on the foliation of rocks.[58] In a Friday Evening Discourse at the RI on 6 June 1856, subsequently reprinted in the *Phil. Mag.*, Tyndall offered a forceful critique of those geologists who, in his opinion, failed to recognize that forces were involved in the cleavage of crystals and in the formation of slaty rocks. He argued that the two phenomena were due to similar causes and that slates had been produced by the exercise of very strong geological forces.[59] Subsequently he and the geologist Henry Sorby became embroiled in a controversy over the cause of slate rocks.[60] During his visit to the Tyrol in the late summer of 1856 Tyndall made extensive observations of glaciers, which he recorded in detail in his letters and journal. By the end of 1856 glaciology had become Tyndall's new (and enduring) field for research and mountaineering his vacation passion.

Tyndall's first public pronouncement on the structure and motion of glaciers was to be at the RI on 23 January 1857, when he boldly challenged James David Forbes's viscous theory of glacial motion.[61] Whereas his earlier disagreement with Sorby had been constrained and limited to the two of them, over the next few years his conflict with Forbes was to become very public and sufficiently bitter and *ad hominum* as to split the scientific community, leading Tyndall to acquire a reputation as a scientific controversialist. However, the question of the origins and continued existence of glaciers was also to lead to a new and exciting field of research: the interaction of the earth's atmosphere and its constituent gases with radiant heat. Mountaineering alone or with friends was also to lead to more fame when he and Hirst climbed Mont Blanc in August 1857 and Tyndall published his Alpine adventures in his first popular book, *Glaciers of the Alps*, in 1860.[62] Meanwhile, in his private life, he was to have a doomed love affair with one of the daughters of the Drummond family, who doggedly attended his lectures at the RI (although he was destined to remain a bachelor until 1876). These developments will be unveiled as the correspondence continues in the next volume.

Notes

1. J. Tyndall, "On the Nature of the Force by which Bodies Are Repelled from the Poles of a Magnet; to which Is Prefixed an Account of Some Experiments on Molecular Influences," *Phil. Trans.* 145 (1855): 1–52.
2. Letters 1103, 1136, and 1270.
3. Letters 1052, 1063, 1099, 1118, 1140, and 1274.
4. Letter 1069.
5. Letters 1089, 1187, and 1274.
6. Letters 1047 and 1089.
7. E.g., letters 1155, 1156, 1158, and 1199.
8. Letters 1105 and 1109.
9. Letter 1103.
10. Letters 1074 and 1075.
11. Letters 1141, 1146, and 1147. Tyndall contributed articles on natural philosophy to four volumes of Edward Hughes's *Reading Lessons* (London: Longman, Brown, Green & Longmans, 1855–58). A fifth volume was planned but appears not to have been published.
12. Letter 1152.
13. Journal, 2 March 1855, RI MS JT/2/13c/731.
14. Letter 1188; Journal, 9–10 March 1855, RI MS JT/2/13c/733–35; letters 1170, 1179, and 1187; letter 1241; Journal, 25 March 1855, RI MS JT/2/13c/738.
15. Letter 1065.
16. The financial details are given in R. G. Garnett, *Co-operation and the Owenite Socialist Communities in Britain* (Manchester: Manchester University Press, 1972), chapter 6.
17. Journal, 20 December 1853, RI MS JT/2/13b/647; vol. 4, letter 0856.
18. Letters 1086 and 1089.
19. As the chemical research output of Debus demonstrates.
20. Letters 1035, 1220, and 1221.
21. Letter 1073.
22. Letters 1082, 1116, 1181, 1263, and 1264.
23. Letter 1037.
24. Letter 1037.
25. E.g., letters 1139 and 1142.
26. E.g., letters 1053, 1060, 1158, 1257, and 1262.
27. Letters 1101 and 1102.

28. Journal, 3 August 1856, RI MS JT/2/13c/780.
29. Letter 1217.
30. Letters 1093 and 1138.
31. Letter 1137.
32. Letter 1144.
33. Letter 1149.
34. Journal, 30 June 1856, RI MS JT/2/13c/853–60.
35. James D. Forbes, *Travels through the Alps of Savoy and other parts of the Pennine Chain* (Edinburgh: Adam & Charles Black, 1843), 365.
36. Jackson, 435. Our discussion of Tyndall's diamagnetic researches draws heavily on Jackson's paper.
37. M. Faraday, "On New Magnetic Actions, and on the Magnetic Condition of All Matter," *Phil. Trans.* 136 (1846): 21–62.
38. J. Tyndall and H. Knoblauch, "On the Deportment of Crystalline Bodies between the Poles of a Magnet," *Phil. Mag.* 36 (1850): 178–83; J. Tyndall and H. Knoblauch, "On the Magneto-optic Properties of Crystals, and the Relation of Magnetism and Diamagnetism to Molecular Arrangement," *Phil. Mag.* 37 (1850): 1–33. This latter paper and later publications on related topics were reprinted in J. Tyndall, *Researches on Diamagnetism and Magne-crystallic Action, including the Question of Diamagnetic Polarity* (London: Longmans, Green, & Co., 1870).
39. J. Tyndall, "On the Magneto-Optical Properties of Crystals," *Brit. Assoc. Rep. 1850*, 23.
40. J. Tyndall, "On Diamagnetism and Magnecrystallic Action," *Brit. Assoc. Rep. 1851*, 15–18 and *Phil. Mag.* 2 (1851): 165–88.
41. J. Tyndall, "On the Polarity of Bismuth, including an Examination of the Magnetic Field," *Phil. Mag.* 2 (1851): 333–44.
42. J. Tyndall, "On the Diamagnetic Force," *Brit. Assoc. Rep. 1854*, 14–16; Tyndall, "On the Nature of the Force."
43. See especially letters 1029, 1031, 1076, and 1082.
44. Tyndall, "On the Nature of the Force," 3.
45. Letters 1042, 1053, 1057, and 1058.
46. Journal, 1 February 1855, RI MS JT/2/13c/724.
47. Letter 1042.
48. Letter 1058.
49. J. Tyndall, "Further Researches on the Polarity of the Diamagnetic Force," *Phil. Trans.* 146 (1856): 237–59.
50. Letters 1161, 1164, and 1168.
51. As listed in the *Catalogue of Scientific Papers. Compiled by the Royal Society of London* (1872), 6: 75–76.
52. See letter 1045. He subsequently exchanged letters with Magnus on boiler explosions; see letters 1216, 1230, and 1244..
53. J. Tyndall, "On the Vibrations and Tones Produced by the Contact of Bodies Having Different Temperatures," *Phil. Trans.* 144 (1854): 1–10.
54. Principally letters 1149 and 1153.
55. J. Tyndall, "On a Peculiar Case of Colour Blindness," *Phil. Mag.* 11 (1856): 329–33.
56. Letter 1034.
57. Journal, 30 September 1854, RI MS JT/2/12b/693.
58. D. Forbes, "On the Causes Producing Foliation in Rocks; and on Some Observed Cases of Foliated Structure in Norway and Scotland," *Quarterly Journal of the Geological Society of London* 11 (1855): 166–85.
59. J. Tyndall, "Comparative View of the Cleavage of Crystals and Slate Rocks," *Phil. Mag.* 12 (1856): 35–48.

60. Letter 1275; H. C. Sorby, "On the Theory of the Origin of Slaty Cleavage," *Phil. Mag.* 12 (1856): 127–29; J. Tyndall, "Observations on the Preceding Paper," *Phil. Mag.* 12 (1856): 129–35.
61. See especially letters 1274, 1275, and 1283; Journal, 12 August to 8 September 1856, RI MS JT/2/13c/866–903; J. Tyndall, "Observations on Glaciers," *Roy. Inst. Proc.*, 2 (1854–58), 320–27. J. S. Rowlinson, "The Theory of Glaciers," *Notes & Records Royal Society of London*, 26 (1971), 189–204.
62. J. Tyndall, *Glaciers of the Alps: Being a Narrative of Excursions and Ascents, an Account of the Origin and Phenomena of Glaciers and an Exposition of the Physical Principles to which They Are Related* (London: J. Murray, 1860).

EDITORIAL PRINCIPLES

Our aim is to include all extant letters to and from John Tyndall. Some of the letters are from published sources, such as letters written by Tyndall to newspapers or journals, but the majority are personal letters that are transcribed either from the original or from typescript copies produced by Louisa Tyndall from the originals, which are sometimes now missing. One central editorial principle has been to transcribe, wherever possible, from original letters rather than Louisa's typescripts. This is because of Louisa's conscious and unconscious editorial interventions, a point discussed in more detail below. We have also aimed to reproduce, as accurately as possible, the text of the letters as they were written. Idiosyncrasies of spelling and punctuation in the handwritten letters have therefore been preserved, but, because Louisa made many inadvertent errors, mistakes appearing in typescript sources have been silently corrected. But, in general, nonstandard spellings (e.g., the use of dialect) has been retained even when in a typescript-only letter (though there is the possibility that the spelling is Louisa's, not her husband's). Where the transcription is based on a typescript transcription by Louisa (or one of her assistants), this will be indicated by "LT Transcript Only." Some of the letters in this volume are transcripts made by Tyndall, indicated by "JT Transcript Only," while others are LT transcripts of letters copied from the originals by Tyndall, indicated by "LT Transcript of JT Transcript Only."

We have included illustrations that are part of the letter. In cases where it was possible to reprint the original handwritten image, we have done so. In other cases, volume editors have reconstructed drawings because it was not possible to reprint an image from the handwritten letter (due, for example, to deterioration of the original document).

The letters are presented in chronological order of writing. Where more than one letter has the same date, letters from Tyndall (in alphabetical order of addressee) come first, followed by letters to Tyndall (in alphabetical order of writer), with the exception that if the order of writing can be determined, that order takes precedence. Where letters cannot be dated precisely, they are placed at the latest likely date of writing. (In some cases, if it is possible, but unlikely, that they were written later, notes indicate such possibilities.) To indicate the beginning of a new year, the date of that year has been inserted in a large font before the first letter from that year. Readers should note that although volumes are arranged in chronological order, letters that are discovered from years already covered in published volumes will appear in one of the final volumes of the correspondence.

For non-English-language letters, both the transcription of the original and the English translation are presented. The transcription of the non-English letter appears first, followed by the English translation in a smaller font. Editorial notes are inserted in the English translation. Some foreign letters are in an original hand, while others exist only as transcriptions by Louisa or others who often did not have a full command of the language. In the latter case these letters contain frequent spelling, grammatical, and interpretation errors that do not seem to have been committed by the original author. In these typescript-only cases, and only in these cases, spelling and grammatical errors have been corrected, as have poor interpretations where a better word choice was obvious from the context.

There is a standard format for each letter. The first line lists the author, if it is to Tyndall, or the recipient, if it is from Tyndall; the date; and the letter number. The second line lists any information supplied by the writer in their salutation, such as the place, date, or time of day when the letter was written. (In some cases, though, where the writer put the date at the foot of the letter, we have left it there in order to reproduce the text accurately.) The opening salutation, the text of the letter, and closing then follow. Then come the source and editorial notes, the latter indicated in the text by arabic numerals.

Editorial notes are intended to provide the contextualization necessary to understand the contents of each letter. There are basically six types of editorial notes. Many of them are informative notes on persons mentioned in the letters. If a person referred to in a letter is not described in a note, they have been mentioned more than twice in the volume and they will have an entry in the biographical register at the back of the volume. For the rest—those mentioned once or twice—biographical information will appear in the editorial note appended to the letter in which he or she is first mentioned. When there is a second reference to that person, a note will refer back to the letter in which he or she was first mentioned. Biographical notes on well-known figures such as Faraday or Huxley stress what that person was doing at the time when the letter was written rather than providing a full overview of the person's life.

A second type of editorial note identifies allusions and quotations. In the case of quotations, where possible, the first edition of the work is cited unless a different edition is mentioned in the letter. If a quote is not in English, a translation is given. Glosses on obscure places, abstruse words, and words used in an unfamiliar sense make up the third type of editorial note.

The fourth type provides information about the context of the letter, drawing on such sources as contemporary publications, scholarly sources, Tyndall's personal journals and field notes, and other letters in the Tyndall Correspondence. The fifth type of editorial note is cross-references to other letters in the correspondence, referring to the letter number. The last type is information on significant textual changes that Tyndall made to the original that is thought to be relevant to the meaning of the letter.

There are a series of conventions for indicating when editors have given additional information about the text of a letter:

> When Louisa Tyndall has made an annotation or marginal insertion that the editors have deemed important enough to retain, we have included her annotation in an editorial footnote.
>
> When editors other than she have made insertions, they are indicated by unitalicized square brackets; for example, [word].
>
> When certain words are ambiguous and we have had to conjecture their meaning they are indicated by italicized square brackets; for example, *[*words*]*; when we have considered such words illegible the number of illegible words is indicated in full italicization and square brackets; for example, *[3 words illeg]*.
>
> If a word or series of words is not present or has been destroyed—by inkblots, holes, or mold, for example—they are indicated as missing, italicized and in angular brackets; for example, <*3 words missing*>.
>
> In extremely rare cases where it is clear that Louisa herself has intentionally destroyed a few words, they are indicated as excised, italicized and in angular brackets; for example, <*3 words excised*>.

Finally, a brief word about Louisa. She was responsible for collecting many of Tyndall's letters in order to write a biography of her late husband. This biography was not completed. It is important to recognize that Louisa was not always reliable as a transcriber. She often threw out the manuscript letter when she had made her transcription, but some original manuscripts remain, which indicate her editing processes. Like any transcriber, she was sometimes inaccurate, but corrections show that she proofread her transcriptions. However, she also edited Tyndall's writing to make his letters seem more consistent or grammatically correct. For example, she made punctuation more formal, corrected grammar, and, occasionally, corrected quotations. She consistently corrected Tyndall's usual (deliberate) lowercase for "god" to "God." Her typewriter was unable to insert the superscripts that Tyndall used in abbreviations, or the "&c" symbols, which stood for an address at the foot of the letter. (Modern typefaces are unable to deal with all the variant scrawls that the "etc." symbol took, and they are all formalized here to "&c.") Louisa also left out details that she considered too private or too salacious for public consumption. She equated postmark and date of writing and overlooked the time taken to write some letters. When she inferred dates from internal evidence, her reasoning was sometimes faulty. Similarly, her identifications of people are sometimes in error. Independent evidence has always been sought for her dates and identifications. Lastly, the patterns of punctuation differ between LT sources and manuscript sources. Tyndall himself used dashes more often than commas, semicolons, and stops, and he used dashes of many different lengths that are all standardized here to em-dashes. It is often difficult to distinguish between very short dashes and commas or very short dashes and stops. Similarly, capitalization is often ambiguous; distinguishing S/s for "science," for example, is a particular problem. Therefore, scholars for whom punctuation and capitalization are significant will need to consult the manuscripts. Users should also be aware that throughout this project the transcribers and editors have worked from scans and photographs rather than the original manuscripts.

NOTE ON MONEY

In the 1850s the currency used in both England and Ireland was the pound sterling, usually abbreviated to pound (£). A pound consisted of 20 shillings, and each shilling contained 12 pennies. In the notes to this volume the editors have used the following format: £3 4*s.* 5*d.* to express three pounds, four shillings, and five pence. The letter *L* (for livres) has sometimes been used in place of the £ sign and the *d.* (derived from *denarius*, Latin, a small coin) indicates pennies. A farthing was a quarter of a penny (¼*d.*). A crown was 5 shillings (5*s.*), and half a crown equaled two shillings and sixpence (2*s.* 6*d.*). A guinea was 21 shillings; that is, £1 1*s.*

Several forms of representation were used in these letters. Sometimes the numbers for pounds, shillings, and pence were separated by dots, forward slashes, or spaces; thus; £3 4*s.* 5*d.* could be written as £3.4.5, £3/4/5, or £3 4 5. A zero could be represented by a dash; thus, 7 shillings was often written as 7/- or just 7/. A half crown was often written as 2/6.

French currency is also occasionally mentioned in this volume. The franc was the national currency in France from 1795 until 1999, when France adopted the Euro. In the mid-nineteenth century the exchange rate between British and French currencies was approximately 25 francs per £. A centime was a coin worth one-hundredth of a franc.

TIMELINE OF JOHN TYNDALL'S LIFE

Year	Event
c. 1822	John Tyndall born at Leighlin Bridge, County Carlow, Ireland
c. 1836	Attends John Conwill's National School in Ballinabranna
1839	Begins employment as civil assistant with Ordnance Survey of Ireland in Carlow
1840	Becomes civil assistant in the Ordnance Survey Office at Youghal, Cork County
1841	First appearance in print (poem in *Carlow Sentinel* under pseudonym 'W[alter] S[nooks]')
1842	Joins English Ordnance Survey in Preston
1843	Leads written criticism of Ordnance Survey in *Liverpool Mercury*; dismissed November
1844	Home in Ireland, mostly unemployed until obtaining a position with the firm of Nevins and Lawton, Surveyors, of Manchester
1845–47	Works for Richard Carter, Surveyor, of Halifax
1845	Meets Thomas Archer Hirst in Halifax
1847	Begins teaching mathematics and surveying at Queenwood College, Hampshire
1848–50	At University of Marburg, working for PhD with Robert Bunsen and Friedrich Stegmann
1850	Returns to England: meets Michael Faraday and William Francis and attends his first British Association meeting before returning to Marburg
1851	Moves from Marburg to Berlin (April–June); then returns to Queenwood
1852	Elected Fellow of the RS
1852	Friendship developed with Thomas Henry Huxley
1853	First lecture at Royal Institution (RI)
1853	Meets Herbert Spencer
1853	Appointed professor of natural philosophy at RI
1853	Awarded Royal Medal for magnetic work; declines when controversy arises over award
1857	Climbs Mont Blanc for first time with Hirst
1859	Demonstrates existence of greenhouse gases and climatic implications
1859	Joins Government School of Mines as professor of natural philosophy
1860	*The Glaciers of the Alps* published
1860	Meets John Lubbock and his wife, Ellen Lubbock

Year	Event
1861	First ascent of the Weisshorn
1861	Delivers his first Christmas Lectures at RI
1862	*Mountaineering in 1861* published
1862	Begins to champion Mayer's priority in discovery of conservation of energy
1863	*Heat Considered as a Mode of Motion* published
1864	Awarded RS's Rumford Medal
1864	First meeting of X Club
1865	*On Radiation* published
1865	Engages in public controversy over the efficacy of prayer
1866	Succeeds Michael Faraday as scientific adviser to Trinity House
1867	Michael Faraday dies; Tyndall becomes superintendent of the RI
1867	*Sound: A Course of Eight Lectures* published
1868	Successfully climbs Matterhorn
1868	Discovers the cause of light scattering, to be known as 'Tyndall Effect'
1868	*Faraday as a Discoverer* published
1868	Lecture on 'Scientific Materialism' at the British Association
1868	Resigns from Royal School of Mines
1869	Joins Metaphysical Society
1870	'On the Scientific Use of the Imagination' lecture at British Association
1870	*Researches on Diamagnetism* published
1870	*Three Scientific Addresses* published
1870	*Researches on Diamagnetism and Magne-Crystallic Action* published
1871	Meets Louis Pasteur for first time while in Paris
1871	*Hours of Exercise in the Alps* published
1871	*Light and Electricity* published
1871	*Fragments of Science* published
1872	Debates over how to measure the efficacy of prayer, aka the 'Prayer-Gauge Debate'
1872	*Contributions to Molecular Physics in the Domain of Radiant Heat* published
1872	*The Forms of Water in Clouds and Rivers, Ice and Glaciers* published
1872–73	USA Lecture Tour
1873	*Six Lectures on Light* published
1874	As President of the BAAS, delivers the 'Belfast Address'
1876	Marries Louisa Charlotte Hamilton
1877	Develops 'Tyndallization' (discontinuous heating and cooling process to destroy heat-resistant spores)
1877	Tyndall and Louisa build summer cottage 'Alp Lusgen' at Bel Alp, northern side of Valais, above Brig
1877	*Fermentation and its Bearings on Phenomena of Disease* published
1879	*Fragments of Science*, which had gradually expanded, first published in two volumes
1881	*Essays on the Floating Matter of the Air* published
1885	Tyndall and Louisa build Hindhead retreat, Surrey Downs

Year	Event
1887	Resigns from RI
1890	Fight with Gladstone in *The Times* over Irish Home Rule
1892	Hirst dies
1892	*New Fragments* published
1893	Dies Haslemere, Surrey, accidentally poisoned

TIMELINE OF EVENTS IN JOHN TYNDALL'S LIFE SPECIFIC TO VOLUME 5

Month and year	Biographical details	Notable social, political, and cultural events
January 1855	Begins series of eleven weekly lectures at RI on electricity and magnetism Delivers Bakerian Lecture at RS on diamagnetism (25th) Gives RI Friday Evening Discourse on magnetic force (26th) Attends Geological Society with Lyell and criticizes David Forbes's view on foliation (31st) Experiences headaches throughout month	Crimean War, which began in October 1853, continues with alliance of Britain, France, and Turkey attacking Russia in Crimean peninsula Lord Aberdeen resigns as prime minister over conduct of the war
February 1855	Lecture on geysers at LI (14th) Investigates steam boiler explosions	Death of Gauss Lord Palmerston elected prime minister at age seventy
March 1855	Starts preparing apparatus to exhibit at Paris Exposition Universelle Publishes critique of Faraday's views on magnetic medium and encourages Hirst to write on that topic Writes critical review of new edition of *Hunt's Elementary Physics* for *Westminster Review* Visits Queenwood twice Reads Henry More and Rob Roy Headaches continue	Death of Napoleon I, Emperor of Russia Start of Great Reform Era under Alexander II Death of Charlotte Brontë Gustav Magnus awarded honorary membership of Chemical Society
April 1855	Declines directorship of Toronto observatory Begins weekly lectures on voltaic electricity at RI Invited to serve as juror at Exposition Universelle	France's Emperor Napoleon III on state visit to London Death of Henry de la Beche Robert Warrington develops Gosse's aquarium

Month and year	Biographical details	Notable social, political, and cultural events
May 1855	Helps du Bois-Reymond with lecture demonstrations at RI Sees Carlyle and Tennyson for first time Elected to Philosophical Club of RS Headaches continue	Exposition Universelle held in Paris (May–November) Mercury pump designed by Geissler Birth of Freud
June 1855	Hirst marries Anna Martin at Newry Delivers Friday Evening Discourse on Leyden battery Invited to examine candidates for Royal Artillery and Engineers Stays at Queenwood at end of month	Abolition of stamp duty on newspapers Siege of Sevastopol Alexander Williamson and Julius Plücker elected FRS Joseph Hooker appointed assistant director of Kew Gardens
July 1855	Walking holiday with Debus in Normandy Juror at Exposition Universelle Meets John Martin and leading French scientists in Paris Acts as examiner for War Department Returns to London (21st) Huxley appointed Fullerian Professor at RI Spends some time with du Bois-Reymond (24)	Walt Whitman publishes *Leaves of Grass* in New York Longfellow publishes *Hiawatha*
August 1855	Reports critically on paper by Joule Works on compression of bismuth Demonstrates diamagnetic experiments to Faraday and Augustus de la Rive Quits lodgings in Chadwell Street (18th) and stores possessions at RI Moves to Queenwood (18 August–5 September), where he continues research on diamagnetism	Thomas Cook leads his first continental tour to Exposition Universelle in Paris Liebig's final visit to England and Scotland

Month and year	Biographical details	Notable social, political, and cultural events
September 1855	Holiday in Lake District with Frankland Attends BAAS meeting in Glasgow as secretary of Section A Demonstrates polarity of diamagnetic bodies and offends Whewell by his remarks on Faraday Returns to Queenwood until end of October	Fall of Sevastopol Sylvester appointed professor of mathematics at Royal Military Academy Joseph Henry suggests that BAAS should compile catalogue of scientific papers since 1800
October 1855	Remains at Queenwood until moving to Mount Street (31st) to be close to RI Agrees to write further essays for *Hughes's Reader*	Bessemer patents the blast furnace Singer develops sewing machine
November 1855	Further experiments on the compression of materials Opens his first bank account Composes physics lessons for *Hughes's Reader* Meets Mrs. Huxley for first time	Foucault awarded Copley Medal by RS Livingstone discovers Victoria Falls British government awards £1,000 to RS to promote scientific research
December 1855	Taken by Faraday to shop where he was trained as a bookbinder Matteucci accepts validity of Tyndall's work on diamagnetism Criticizes Weber's and Thomson's views on diamagnetism Unhappy with Mount Street lodgings, as landlord is deceitful Hirst alludes to his wife's illness Hirst criticizes Tyndall's combative style	Metropolitan Board of Works established
January 1856	Breakfasts with Richard Owen Visits Lord Ashburton at the Grange (11th–13th) and converses with Carlyle Begins weekly series of lectures on light at RI Acts as examiner for War Department Knoblauch reports deaths of his wife and son	Victoria Cross established as a military honor

Month and year	Biographical details	Notable social, political, and cultural events
February 1856	RI Friday Evening Discourse on disposition of magnetic force (1st) Lectures at LI Helps Ashburton write report on scientific education Investigates effects of chemicals on water surfaces Investigates colorblindness Meets novelist Geraldine Jewsbury Arranges for Eddy Wynne to be taught by Hirst and accompanies him to Queenwood Dislikes Lewes's *Life of Goethe* Attends part of Thomson's Bakerian Lecture at RS (28th) Advises Thomson on his RI lecture and helps him with demonstrations (29th) Unwell throughout most of the month	Death of poet Heinrich Heine
March 1856	Experiments with Newton's rings Lectures on eye at RI Consolidates friendships with Huxley and Hooker Holiday at Queenwood (18th–26th) and discusses Kant with Debus Hears Jenny Lind sing	Treaty of Paris closes Crimean War University boat race between Oxford and Cambridge Society of Arts introduces examinations
April 1856	Delivers further eleven weekly lectures on light at RI Friendship with Moore and Pollock families Galbraith complains about artillery examinations and Tyndall publishes his concerns in the *Times* (28th)	

Month and year	Biographical details	Notable social, political, and cultural events
May 1856	War Department accepts Tyndall's criticism without reprimand Publishes paper on colorblindness Advises Hirst on outbreak of masturbation at Queenwood Meets Despretz in London Visits Brighton and Eastbourne over Whitsun Meets Emerson Tennent Looks at instruments with Wheatstone (23rd)	John Martin allowed to return to Ireland Death of Daniel Sharpe John Brown accused of murdering five Kansas homesteaders Senator Charles Sumner brutally assaulted for his antislavery views
June 1856	RI Friday Evening Discourse on slaty cleavage (6th) Huxley notes application to glaciation Stays at Queenwood (14th–22nd) Holidays with Wrights on Isle of Wight and writes account in verse Following Hirst's resignation from Queenwood in order to nurse Anna, considers whether to return to the school Acts as examiner for War Department	Mormons leave Iowa to found Salt Lake City Philip Goose, Henry Noad, and Antonio Secchi elected FRS
July 1856	Marks War Department examinations Debates on cleavage and glaciations and reads Forbes on glaciers for first time Takes Poggendorff to meet Richard Taylor at Richmond Referees papers by Claudet and Matteucci for RS	British government agrees to accommodate RS and Chemical Society in Burlington House
August 1856	Awarded RS government grant of £100 Writes for *Photographic Journal* Travels to Switzerland (12th) and explores glaciers with Hooker, Huxley, and Frankland Hirsts settle in Pau	School Grant Act extends British state education Perkin makes first aniline dyestuff Foundry casts Big Ben Discovery of Neanderthal skull in Germany

Month and year	Biographical details	Notable social, political, and cultural events
September 1856	Continues Alpine holiday Arrives in Vienna (16th) and attends Deutscher Naturforscher Versammlung Friendship with Josef Grailich Visits Reichenbach's castle and is skeptical of animal magnetism Asked by de la Rive to oversee translation of his text on electricity	
October 1854	Trip to Liverpool and then Wales Stayed with Edward Frankland in Richmond Grove near Manchester on 15–16	Commencement of Second Opium War between Britain and China

The Correspondence of John Tyndall

1855

From Hermine Ballin[1] 2 January 1855 1026

Stamford Hill[2] | the 2nd of January 1855

Dear Sir,

I feel very much obliged to you for the kind interest you take in my plans and the assistance you offer me in my pursuits in this country; it gives me encouragement and confidence to know that I have such kind friends here. It seems to me that you are not fully acquainted with the nature of my engagement with Miss Hornblower.[3] I undertook this situation with no other end in view but to acquire the knowledge of the English language and to have time and opportunity to improve in those branches of knowledge which then I did not feel competent to teach. Miss H. has given me every possible opportunity of carrying out this purpose and I have great reason to be thankful for it. Though I feel that I am far from knowing enough for the responsible career I have chosen and though I am convinced that another year under Miss H's kind instruction would do me much good, still there are other duties which it would be wrong of me to lose sight of—I am therefore determined not to prolong my stay at Miss H's under the present circumstances longer than a year. As Miss H. is satisfied with my small services and knows how gladly I would remain with her she is trying to find out means to make this possible to me by procuring me German lessons in the neighbourhood. But I am afraid it is not likely that this plan will succeed, but at all events Miss H's house will remain my home till I find another situation and I know that she will assist me in my endeavours to find a suitable one. However I feel that I must do my best myself and I have therefore asked Mrs. Frankland some months ago to write to me as soon as she heard of a situation[4] in her large circle of friends and acquaintances, but I am afraid it will be difficult for my friends to serve me as my not knowing music will always prove a great drawback in this country. But I will hope and trust that all will be well; it is a good thing that I am so very fond of teaching and that it is not a trouble to me as it is to many who like me are obliged to choose this career. I am very sorry not to be able to attend the lecture to-day;[5] Miss H. could not have given me a greater holiday pleasure than taking me to those interesting lectures.

Very truly yours | Hermine Ballin

RI MS JT/1/C/61

---

1. *Hermine Ballin*: For earlier correspondence re Hermine Ballin (b. c. 1831), see volume 4, letters 0879, 0884 and 1025. It would appear that Tyndall, who had known her in Marburg, was attracted to her and that Frankland hinted at their possible liaison: 'I understand that this [. . .] pretty little girl is an importation of yours, ostensibly to make a Governess of her, but in reality I know not what!' (volume 4, letter 0884). She had arrived in England about 20 February 1854 (Tyndall, Journal: RI MS JT/3/13b/660).
2. *Stamford Hill*: in the parish of Hackney, Middlesex, 4 miles north of St Paul's Cathedral. Owing its airy elevation, Stamford Hill attracted a number of successful Victorian businessmen who purchased large and spacious houses.
3. *Miss Hornblower*: Jemima Hanbury Hornblower (1800–61) was a deaconess in the Sandemanian sect and conducted an independent school for girls in Stamford Hill. Tyndall would have known her through Faraday, who was also a Sandemanian.
4. *Mrs. Frankland . . . heard of a situation*: see letter 1031.
5. *the lecture to-day*: On 2 January 2 Faraday delivered the third of his six RI Christmas lectures for juveniles on 'the Chemistry of Combustion'.

From Thomas Archer Hirst 2 January 1855 1027

January 2nd 1855 | Hotel Windsor, | Rue de Rivoli

My dear John,

At length I have time to write you a line to tell you that I am at length married[1] and as happy as I can well conceive. Anna and I have not quarreled yet nor do I see any cause to fear we shall do so. On the contrary I am more satisfied every day at the important step I have taken and more confident that its influence upon both of us will be a permanently good one. After spending a very pleasant week in Dublin and Newry,[2] I went with Anna on the day before our wedding to the residence of her married brother[3] at Kilbroney, near Rostrevor. I was very unwell all that day, and as I had reason to believe that my illness was caused by want of exercise and too many luxuries I fasted all day. I slept tranquilly all night and rose on the morning of my wedding day in better health and spirits than I ever enjoyed in my life before. The weather which for a long time had been dismal had suddenly changed to a bright clear frost. It would be difficult to imagine a prettier and more cheering view than the one I had from my bedroom window on that memorable morning. Carlingford

Bay[4] glittered like a silver mirror at the foot of the bold hills whose summits were white with frost and for the first time for a week the sun shone brilliantly on everything. I rose dressed myself in my wedding garments, threw a cloak around me and walked out with my cigar to ponder well what I was about to do that day. I returned encouraged by my walk and the change that had been wrought in my face since the previous day astonished everyone. I was a newborn man. I rejoiced in my own strength and my feelings no doubt radiated from my face. Soon after 12 we were married. The Presbyterian minister[5] gave us a short & excellent introductory address. I heard and weighed every syllable of it and with a few paltry exceptions I felt I could endorse all he had said as to our duties to one another to God & to society.[6] The wedding feast was gorgeous, my wife's and my own health were drunk of course and I stood on my feet without hesitation and with a great pride thanked them all. People told me afterwards that I spoke well and indeed I think I did for I never felt more power to do so. We left in the afternoon for Dublin[;] as we drove away with four horses and 2 postilions they cheered us in true Irish style and did not forget to throw after us the old slipper. The last cheer had died away in the distance before I became conscious that my wife, the soul I had promised to protect until God should separate us by death was at my side weeping bitterly, she knew not why. It was not very difficult to dry such tears and before we had reached the Railway station[7] about 12 miles distant she was as cheerful and trustful as I could wish and has continued so ever since and God willing will do so for some time to come. We made the rest of our way to Paris where we were heartily welcomed by John Martin.[8] I must speak to you about him another time. An hour ago he took her away from me for the first time & I have taken advantage of their absence to write this. In about a fortnight or so we shall be in London. I should like to stop with you best but I am not sure that we shall do so. I will let Anna do as she pleases for women have their own particular reasons for preferring one place of residence to another[;] she knows the people in Cecil Street[9] so well that it is probable she would prefer stopping there. We will see and tell you in time. Here we have two very cosy little rooms and since Anna went out I have been and hired a neat little piano to surprise her. I am just posting your letter to Moigno,[10] and I will call upon Despretz shortly. I hope you are well and in good spirits. Write to me here. Anna has got some Brides Cake[11] for you which she is going to deliver herself.

Thine affectionately | T. A. Hirst

RI MS JT/1/H/211

---

1. *I am . . . married*: Thomas Archer Hirst married Anna Martin, daughter of the late Samuel Martin (1749–1831), on 28 December 1854 at Kilbroney, Rostrevor, County

Down (*Dublin Evening Mail,* January 1, 1855). See also account in Hirst Journal, RI MS JT/2/32b/1155–59. Rostrevor is an attractively positioned village on the north side of Carlingford Lough (Bay); it stands on the east coast of Ireland about 52 miles north of Dublin and 37 miles south-west of Belfast. The 1851 census recorded a population of 764.

2. *Newry*: an historical city in County Down, about 7½ miles north-west of Rostrevor.
3. *her married brother*: Robert Martin (1814–58), who lived at Kilbroney.
4. *Carlingford Bay*: 'Carlingford Bay [. . .] is one of the most beautiful of the inland bays of Ireland. [. . .] It is about a mile and a half wide at its sea-gate, and runs up into the land rather more than seven Irish miles, of the same medium breadth, though in parts wider in parts narrower. It is closely bounded on the whole of its southern shore by the chain of Carlingford and Newry mountains, and, over a considerable portion of its northern border, by the still more magnificent range of the Mourne mountains and Slievebane' (J. Forbes, *Memorandums Made in Ireland in the Autumn of 1852,* 2 vols (London: Smith, Elder & Co., 1853), vol. 2, pp. 255–56).
5. *The Presbyterian minister*: probably the Rev. John Moran of the Rostrevor Presbyterian Church.
6. *our duties to one another to God & to society*: marriage vows, together with minister's own comments.
7. *the Railway station*: probably the railway station at Newry, the distance being about 12 miles by road.
8. *Paris . . . John Martin*: Anna's eldest brother, John, who was a prominent Irish nationalist, had been sentenced to transportation to Tasmania in 1848 but had been released in 1854 and granted a 'conditional pardon' that enabled him to travel to Paris. Under the terms of the pardon he was forbidden to return to Ireland. Hirst recorded in his Journal (RI MS JT/2/32b/1159) that John Martin was 'one of the gentlest, kindest souls I have met—one of "God's own gentlemen" [. . .] His sensitive, sympathetic, affectionate heart is the key to his political history. His intellect is kept too much under the sway of his feelings for him ever to be a practical and efficient politician; [. . .] His intellect is by no means strong, nor has he much decision of character. In fact. his character is more feminine than masculine in its nature.'
9. *the people in Cecil Street*: Cecil Street is to the south of the Strand. According to the 1852 directory of London, Mrs Mary Gray was the proprietor of the lodging house at 9 Cecil Street, where Anna stayed from 13 to 19 January. It would appear that Anna Hirst knew Mrs Gray or others living at that address.
10. *your letter to Moigno*: letter missing. The French Catholic priest François-Napoléon-Marie Moigno (1804–84) had published a number of scientific papers and popular books on such subjects as optics and the electric telegraph. Moigno later translated Tyndall's books into French.
11. *Brides Cake*: A rich, highly ornamented cake, eaten at a wedding (*OED*).

To Thomas Archer Hirst 4 January 1855 1028

Royal Institution London | 4th January 1855.

My dear Tom

For a man to write or to speak with vigour it is necessary that he should believe that he is not beating the air, or preaching to the dusty envelopes of the dead letter office—now how am I to get this faith on the present occasion? You write to me,[1] and on the NE[2] corner of your letter stand the simple words January 2nd 1855, in the body of the letter you inform me that you are in Paris, but in what part of Paris you leave me to guess—you tell me to write to you 'there'—well I will do so, and the sin falls upon your head if my letter, pinioned like a phoenix, and bursting from the fire of my rejoicing heart be ignobly quenched amid the lumber before referred to. I can truly say that I read your letter with the greatest satisfaction—it came just in time to gratify a longing which I had experienced for some days. Your cards came to me,[3] but they were directed by a strange hand, and I yearned once more for those familiar pothooks & hangers[4] which, like magicians wands, or like the witch of Endor when she called the ghost of Samuel from the dead,[5] place your image bodily before me. I have copied that letter of yours into my journal:[6] not at all on your account, but with the prospective view that when I get married myself one of these days, my inexpressible darling, that is to be, will be amused by the perusal of it; I know that ladies feel an interest in such matters and therefore on <u>her</u> account I have copied the letter. I am far from being in an impressionable mood at present, but I did feel a stirring of my dry bones as you referred to your speech when Anna's & your healths were drunk. 'With great pride I thanked them all'—I should like to have been present to witness the outgush of that pride, and had Anna been near me I should have kissed her with the pride of a proud father glorying in his son. The whole matter was to all appearance very beautifully arranged, and I am thankful to the gods for granting you that gush of sunshine, and that diamond frost-sparkle to beautify the day of your nuptials. I am loaded with the dross of science, otherwise my heart would burst forth into singing and I would send you an epithalamium of which Edmund Spenser[7] need not be ashamed. I am also thankful to the minister[8] for having talked so very little nonsense, and if ever I meet him he may reckon upon the best glass of whiskey punch that I can procure. I owe him this, as he has taken a burden off my shoulders, for if he had not acquitted himself properly I should have felt it incumbent on me to make my own experience in matrimonial matters the subject of an exhortation to you and Anna on commencing life. There is something to my mind highly philosophical in this tending of human hearts to draw towards each other; and I doubt not, were there a man wise enough to do it, that a treatise on a kind of moral mechanics might be written in which love should be the force, and which would possess strict scientific precision. I can imagine myself for example beside a fair girl, looking steadfastly at the spirit in her eyes, feeling a force, symbolised by magnetic induction, rising within me until finally yielding to the resistless influence I clasp her strongly to my heart while our lips flash together like the union of opposite electricities![9] —What an old fool I am you will say to imagine such a thing! But remember I only imagine it as an object of philosophic interest. I should like extremely to make an investigation in connexion with this subject, and there is very great probability of my taking it up at some future day. If ladies could be purchased by the pound like bismuth, and affection excited by means of a copper helix,[10] the Royal Society should not be long without a communication on this important subject. You will deem me crazy Tom, for I believe I never wrote to you in this foolish way before; but what can a man say when his heart is soft iron, and his brain an assemblage of molecular currents! I could write to you very gravely on magnetic philosophy but on nothing else. My cousin[11] is here still and is not altogether satisfied with his progress. He is very quiet, calculated to make an excellent husband; and although not at all possessing the intellectual power of Hector,[12] is endowed with a large common sense—it is this latter, combined with a hankering after beauty, that renders him difficult to please. He meets the qualities he seeks separate in separate individuals, but has not yet met them combined.

My first lecture is given on the 16th.[13] I expect to give an evening lecture at the London Institution on the 17th[14] and shall be happy to see you and Anna there. If Anna knows the people in Cecil Street,[15] and thinks the place comfortable I would advise her to go there—as indeed I doubt my own powers to make her as comfortable here as I could wish.

affectionately hers and yours | John Tyndall

My eye has just caught the stamp of your paper: I will direct there at a venture.[16]

RI MS JT/1/T/591

1. *You write to me*: letter 1027.
2. *NE*: north-east; i.e. top right-hand corner of the letter.
3. *Your cards came to me*: cards missing.
4. *familiar pothooks & hangers*: Hirst's idiosyncratic handwriting. Pot-hook refers to the 'curved or hooked stroke made with the pen'; hanger is a nursery name for 'the stroke with a double curve' (*OED*). Thus the phrase 'pothooks and hangers' refers to a person's handwriting.

5. *the witch of Endor . . . Samuel from the dead*: Hirst's handwriting evoked Hirst's 'bodily image' before Tyndall, just as the Witch of Endor conjured up, for Saul, the ghost of Samuel from the dead (1 Samuel 28:3–25).
6. *copied that letter . . . my journal*: letter 1027; Tyndall's Journal, 3 January 1855: RI MS JT/2/13b/413–14.
7. *an epithalamium . . . Edmund Spenser*: An epithalamium is a nuptial song or poem in praise of the bride and bridegroom, and praying for their prosperity (*OED*). To mark his wedding in 1594 the poet Edmund Spenser (1652–99) wrote an *Epithalamion,* which was published the following year.
8. *the minister*: see letter 1027, n. 5.
9. *the union of opposite electricities*: The view that there are two opposite types of electricity—often called vitreous and resinous, and conceived as two separate fluids—which, when the come together, neutralize each other producing a spark.
10. *affection excited by means of a copper helix*: Tyndall likens the attraction of the sexes to the magnetic field produced in his experiments.
11. *My cousin*: Tom Payne, who had recently visited Tyndall in London.
12. *Hector*: Hector Tyndale.
13. *My first lecture . . . 16th*: The first of Tyndall's 11 lectures on 'Electricity and Magnetism' was delivered in the RI at 3pm on Tuesday, 16 January.
14. *an evening lecture at the London Institution on the 17th*: Tyndall subsequently recorded in his Journal for 17 January: 'My lecture pleased my audience better than it pleased myself. It is said that they spoke highly of it' (RI MS JT/2/13b/419). The title of the lecture has not been ascertained, but it was probably on magnetism.
15. *the people in Cecil Street*: see letter 1027, n. 9.
16. *at a venture*: in realization of the risk involved.

To Thomas Archer Hirst [4?] January 1855 1029

My dear Tom

Try to find a flaw in this

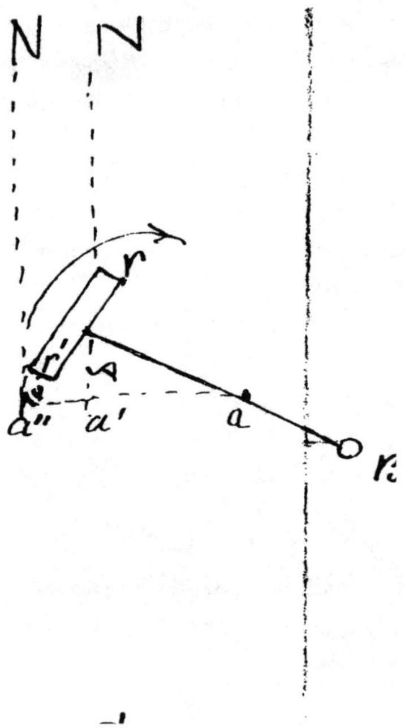

AB is a lever suspended horizontally by an axis passing through a round which point it is free to turn. A repulsive force acts upon the rectangle rr′ fixed at the end of the lever—The direction of the force being from NN, it is mechanically impossible that such a force could cause the lever to rotate in the direction of the arrow. Take for example the forces on the two ends of the rectangle—their movements are respectively measured by the intensity of the force multiplied into the distance of its line of direction from the axis of suspension and in the case before us both these movements tend to turn the bar the same way—that is to make it rotate in a direction opposed to the arrow.

If then the lever rotate with the arrow it is an irresistable proof that the force cannot be purely repulsive.

Supposing the force f at r to be equal or very nearly so to that f′ at r′ and that the former force is repulsive while the latter is attractive. Then the product f x a′a is less than the product f′ x a″a—the moment of force tending to turn the lever in the direction of the arrow would be therefore in the ascendant and an <u>attraction</u> of the mass towards NN would be the consequence.

I have so arranged matters as to be able to produce this result with a mass of bismuth. The artifice consists of making the field of force such that the repulsion at r is very nearly = that at r′ both forces thus neutralize each other as regards a motion of translation (as the force of the earth on a magnetic needle) but they are still effective in producing a motion of rotation which evidences itself in the manner described—All this is hastily written but you will see what I mean, and that by it the polarity of Diamagnetic bodies,[1] or in other words the simultaneous development of repulsion & attraction is reduced to a physical certainty.

affectionately John

I wish you would call at No 20 Rue Cassette and enquire of Mr Despretz whether the books which I sent him[2] have arrived safely. If you wish I will give you a note of introduction to him.

Direct the enclosed[3] to Anna.

RI MS JT/1/T/1008[4]

1. *polarity of Diamagnetic bodies*: cf. Tyndall's Bakerian lecture, 'On the Nature of the Force by which Bodies Are Repelled from the Poles of a Magnet; to which Is Prefixed an Account of Some Experiments on Molecular Influences', *Phil. Trans.*, 145 (1855), pp. 1–52, esp. pp. 45–46.
2. *the books which I sent him*: see volume 4, letter 0980.
3. *the enclosed*: enclosure missing.
4. This letter may have been sent with letter 1028 or shortly thereafter. Hirst was in Paris from 30 December 1854 until 12 January 1855.

From Michael Faraday 16 January 1855 1030

Various philosophical notes of experimental investigation on foolscap paper, paged in series & partly bound in five volumes,[1]—a quarto MS. book of Philosophical notes,[2]—a second larger quarto of similar notes,[3]—some of my printed papers collected in two bound volumes & illustrated by letters &c,—the one quarto the other 8[vo],[4] and a bound copy of Davy's chemical elements,[5] being a copy of that which, whilst abroad in 1814–5, he prepared for a second edition, (which was never published);—these I offer for the Library of the Royal Institution, if the Managers should think them worth a place:—if not to remain at the disposal of my executors

Whatever my wife may think fit to give, from among my personal things, to relatives or friends as remembrances, will I hope be accepted as from both of us. We were always as one, and whatever she may do is what I should do.

I desire that, at my death, all cases in which I have become security should be put an end to, that they may not be continued over my wife. I only remember one, namely, for Mr Geo Whitelaw to Messrs Spottiswoods.[6]

16 January 1855 M Faraday.

Dr Tyndall | Dr Frankland | Mr Barlow | Dr Bence Jones[7]

Newcastle University Library, MS Album 38
Faraday Correspondence, 4:2932

---

1. *Various philosophical notes . . . in five volumes*: Frank James (*Faraday Correspondence,* 4:2932) references this as 'RI MS F2 C–G. That is Faraday, *Diary* [ed. by T. Martin, 7 vols. (London, G. Bell, 1932)], 1828–1832, vol. 1, pp. 327–430.'
2. *a quarto MS. book of Philosophical notes*: James (ibid.) references this as 'RI MS F2A. That is Faraday, *Diary,* 1820–1823, vol. 1, pp. 1–117.'
3. *a second larger quarto of similar notes*: James (ibid.) references this as 'RI MS F2B. That is Faraday, *Diary,* 1820–1823, vol. 1, pp. 121–323.'
4. *the one quarto the other* 8vo: RI MS F3B and RI MS F3A.
5. *Davy's chemical elements*: Humphry Davy, *Elements of Chemical Philosophy* (London: J. Johnson, 1812); annotated copy located at RI MS HD 29.
6. *Geo Whitelaw to Messrs Spottiswoods*: Faraday had stood surety for George Whitelaw (1804–72), a fellow member of the London Sandemanian community, who had been employed as a manager in a printing company; probably Eyre & Spottiswoode of East Harding St, off Fleet St.
7. *Dr Tyndall | Dr Frankland | Mr Barlow | Dr Bence Jones*: These names are not in Faraday's hand.

To Edward Frankland 20 January 1855 1031

20th Jan. 1855

My dear Frankland

I ought to have written to you two days ago—but I have been so hard pushed at the commencement of our session that I have deferred writing till the close of the week—Tuesday last[1] my lectures commenced—Tuesday next of course I have another—Thursday I am to give the Bakerian Lecture at the Royal Society[2]—on the following evening (Friday) I am to give a lecture at the Royal Institution[3] so that a load of work is crammed into the commencement of the year.

I purchased Duboscq's lamp[4] in Liverpool for Dr Bence Jones. I have worked some with it and found it a beautiful instrument—I have heard it complained of as being complicated and liable to be deranged, but as yet

I have met no obstacle in the use of it. The cost of the lamp alone is 200 francs[5]—other appurtenances for optical experiments make it a dear piece of apparatus—I am not acquainted with the merits of the lamp to which you allude. I saw one of them at the great Exhibition in 1851,[6] but did not examine it minutely—It is said that a Dr Watson has invented a much simpler lamp[7] than that of Duboscq but I have never seen the instrument. The light continues as long as the coal points continue—it could readily be kept constant for half an hour or more. The points are of retort charcoal—cut into thin prisms—were the lamp my own I would offer you the use of it for the occasion but I have no right to it whatever—nor has the Institution.

I hope you will be successful in your efforts for Miss Ballin[8]—she has a good friend in the lady with whom she is at present staying.[9]

Kind remembrance to Mrs Frankland | & Believe me | Most faithfully yours | John Tyndall

Faraday gave us a remarkable lecture last night[10]—poor man he was greatly smitten down by the death of a friend[11] which had occurred during the day, but he managed to speak in the extraordinary vigour notwithstanding.

JRL Frankland, 190

---

1. *Tuesday last*: 16 January; see letter 1028, n. 13.
2. *the Bakerian Lecture at the Royal Society*: The Bakerian Lectures were established in 1775 through a bequest of Henry Baker (1698–1774) to the RS. Each year a lecturer was chosen to deliver a lecture on a topic in the physical sciences. Tyndall's lecture was due to be delivered on 18 January 1855 (see minutes of RS Council meeting on 14 December 1854, RS CMO/13) but was instead given on 25 January; see letter 1029, n. 1 for title and publication details.
3. *a lecture at the Royal Institution*: On January 26 Tyndall delivered a Friday Evening Discourse at the RI entitled 'On the Nature of the Force by which Bodies are Repelled from the Poles of a Magnet', which was subsequently published in *Roy. Inst. Proc.*, 2 (1854–58), pp. 13–17.
4. *Duboscq's lamp*: a self-regulating arc lamp developed by César Duboscq in 1851.
5. *200 francs*: approximately £8.
6. *I saw one of them at the great Exhibition in 1851*: For Duboscq's exhibits (France, 1197), see *Official Descriptive and Illustrated Catalogue of the Great Exhibition of the Works of All Nations,* 3 vols (London: Spicer Bros, 1851), vol. 3, p. 1235.
7. *Dr Watson has invented a much simpler lamp*: Joseph John William Watson (c. 1831–86), described as a 'civil engineer' in the 1871 census (Ancestry), was the author of *A Few Remarks on the Present State and Prospects of Electrical Illumination . . . With a Description*

of the Author's Patented Inventions in Galvanic Batteries and Electric Lamps (London: Saunders & Stanford, 1853).

8. *Miss Ballin*: see letter 1026, n. 1.
9. *the lady with whom she is at present staying*: Jemima Hornblower; see letter 1026, n. 3.
10. *a remarkable lecture last night*: Faraday's Friday evening discourse on 19 January was entitled 'On Some Points of Magnetic Philosophy', which was subsequently printed in *Roy. Inst. Proc.*, 2 (1854–58), pp. 6–13.
11. *the death of a friend*: In his Journal entry for 19 January (RI MS JT/2/13b/715) Tyndall noted that Faraday 'was smitten down by the death of some of his friends but spoke with uncommon vigour notwithstanding', which suggests that more than one of his friends had recently died. Among these would have been his father-in-law, Edward Barnard (b. 1767), who had died in 4 January 1855.

From Juliet Pollock 23 January 1855 1032

59 Montague Square | Jan. 23rd

My dear Mr Tyndall,

I hope you may be able to come to us on Tuesday evening[1] the 30th between 9 and 10 o'clock. The Barlows[2] and I hope some other of your friends will be here.

The weather[3] has spited me with regard to your lectures. I was actually driven back by rain when I had made some part of my way towards you on the last occasion and to day I don't feel strong enough to face the snow—It is a real disappointment to me.

Yours very truly | Juliet Pollock

RI MS JT/1/6/1882
LT Typescript Only

---

1. *come to us on Tuesday evening*: Tyndall recorded in his Journal, 'On Tuesday evening [30 January] went to Pollock's' (RI MS JT/2/13c/724).
2. *The Barlows*: John Barlow and his wife Cecilia, née Law (1795–1868).
3. *The weather*: On 16 January, when Tyndall delivered his first lecture on 'Electricity and Magnetism' at the RI, 'Snow fell early in the morning, which was followed by rain. About four p.m. the wind suddenly changed to E., which immediately introduced frost and snow' (*Morning Chronicle,* 23 January 1855, p. 8). On 23 January ('today'), the weather in London was 'Snow falling from 8a.m. till noon. Snow now 4 inches deep on Blackheath, without drifting' (*Morning Chronicle,* 31 January 1855, p. 3).

From Charles Lyell 24 January 1855 1033

53 Harley St. | Jany 24. 1855

My dear Sir

I find that Mr David Forbes is to read a really important paper[1] on the foliation of rocks in Scotland & Norway the gist of which is to show the gneiss, mica schist & crystalline limestone have their foliation, as a general rule parallel to the planes of stratification, what I have always thought to be the rule tho' not without exception.

I expect Mr Darwin[2] to be present & should be very glad if you could give me the favour of your company at dinner at the Geological Club[3] at Clunns Hotel[4] Covent Garden at ½ past 5 o^{ck} & go with us to the Soc^{y}[5] Wed^{y} 31s^{t} Jan^{y}.

Believe me | sincerely yrs | Chas Lyell

Profr Tyndal

RI MS JT/1/L/46

1. *Mr David Forbes is to read a really important paper*: Forbes's paper, 'On the Causes Producing Foliation in Rocks; and on Some Observed Cases of Foliated Structure in Norway and Scotland' was read at the Geological Society on 31 January 1855 and published in the *Quarterly Journal of the Geological Society of London,* 11 (1855), pp. 166–85.
2. *Mr Darwin*: the eminent naturalist and geologist Charles Darwin (1809–82).
3. *Geological Club*: A club consisting of a select group of Fellows of the Geological Society of London who dined together before attending meetings of the society.
4. *Clunns Hotel*: Thomas Clunn was the proprietor of Richardson's Hotel in Covent Garden.
5. *the Soc^{y}*: the Geological of London, located in Somerset House on the Strand.

To Charles Lyell 25 January 1855 1034

25th Nov. 1855.[1]

Dear Sir Charles.

I am truly grateful to you for granting me the opportunity of instructing myself on a question in which I take so deep an interest as the cleavage and foliation of rocks.[2] It will give me the greatest pleasure to join you at the Geological Club[3] on Wednesday next.

If you should happen to be at the Royal Society tonight,[4] or at the Royal Institution tomorrow night[5] I might shew you an experiment or two on the influence of structure upon the magnetic force, which would probably interest you. Magnetism may possibly become a valuable aid in the investigation of this important question.

Most faithfully yours | John Tyndall

Sir Charles Lyell | &c. &c. &c.

RI MS JT/1/T/1050

---

1. *25th Nov. 1855*: This letter was incorrectly dated, apparently by Tyndall. However, the Geological Society meeting was held on Wednesday, 31 January 1855 and Tyndall's Bakerian Lecture was delivered on Thursday, 25 January 1855.
2. *the cleavage and foliation of rocks*: see letter 1033, n. 1.
3. *to join you at the Geological Club*: see letter 1033. Tyndall subsequently gave Hirst an account of this meeting; see letter 1037, especially n. 7.
4. *the Royal Society tonight*: Tyndall delivered his Bakerian Lecture at the RS on 25 January; see letter 1031, n. 2.
5. *the Royal Institution tomorrow night*: Tyndall delivered a Friday evening discourse at the RI on 26 January; see letter 1031, n. 3.

From Thomas Archer Hirst 28 January 1855 1035

Broughton | Jan 28th 1855

My dear John,

I half expected before this to have a line from you though I am not disappointed at not doing so for I know well how busy you must be. Anna and I have regretted several times that we saw so little of you in London and we are looking forward with pleasure to our papa's next visit to Broughton.[1] We have got peacefully settled here and there is an atmosphere of domestic happiness about Vine Cottage[2] that I am sure would please you well and would make your lecture all the better if some Friday night you would run down and return on Monday morning. Every body here received us kindly; of welcomes and good wishes we have plenty and as far as I can calculate the latter are likely to be fulfilled. To tell most people that I am satisfied with my wife would be absurd for of course they would know that if I was not I should never admit it, but to you the assertion may have a different meaning. Yes, John, so far my hopes and anticipations are all fulfilled. I love my wife to-day as warmly as I did a month ago. Already I begin to feel a better man and I feel sure that we shall influence one another for our good.

The boys are slowly assembling at Queenwood[3] and as far as we can judge at present our prospects this half year are in every respect greatly improved.

Mr E.[4] has consulted me more this time than ever before. I hope for his sake, or rather for Queenwood's that he may continue to be as communicative. He shewed me a letter from Mr Gale,[5] who writes to Mr E. for more particular reasons why his son is not to return to Queenwood. (Edmondson in informing Gale's father of his intention not to readmit Gale did so in a very smooth, quaker-like and unsatisfactory way). It is an affecting and truly fatherly letter which I requested Edmondson to allow me to answer. This answer you shall see soon.

Let me hear from you soon John and tell me can I do anything for the Phil. Mag. just now if so give me orders as soon as you please.

I have had a kind note from Bevington, he seems to be more than ever satisfied with Henry Hill[6] and writes very cheeringly about him.

Anna sends her love and so do I.

Thine affectionately | T. A. Hirst

RI MS JT/1/H/212

---

1. *our papa's next visit to Broughton*: Tyndall was humorously being described as their 'papa', presumably in response to his claim in letter 1028 that he was Hirst's 'proud father glorying in his son'.
2. *Vine Cottage*: Thomas and Anna Hirst lived in Vine Cottage, in the nearby village of Broughton.
3. *assembling at Queenwood*: after the Christmas vacation.
4. *Mr E.*: George Edmondson.
5. *a letter from Mr Gale*: Mr Gale was the father of one of the pupils at Queenwood. A pupil with this surname had clashed with a schoolmaster, Haas, but refused to apologise. Although Edmondson recommended expulsion, Hirst persuaded the boy to apologise. There was subsequent correspondence with the father including the letter referred to here (Hirst's Journal, RI MS JT/2/32b/1152–54 and 1157–58).
6. *Henry Hill*: Henry Hill (c. 1830–86) had been a member of the Halifax Mechanics Society where he had become a friend of Hirst's. Tyndall and Hirst had subsequently recommended him for employment in James Bevington's leather factory.

To Thomas Archer Hirst 29 January 1855 1036

29th January 1855

My dear Tom.

Your letter[1] came to me this morning as I was about to commence preparations for my lecture tomorrow.[2] I read it with great gratification and at intervals during the day it dawned in upon my thoughts and imparted a pleasant warmth to my feelings. It assured me among other things that love for your wife did not necessarily undermine your love for your friend, indeed that the one may brighten and enhance the other.

It is now night and I am about to step from the world of action into the world of dreams. I shall go to bed all the more contented, with sweeter and more placid thoughts for having read your simple statement of your present feelings. God grant that these feelings may flourish like the amaranth[3] in the gardens of heaven and never know what it is to fade. It was impossible for me to contemplate the change which you proposed to yourself[4] without some degree of anxiety, but the consciousness that you alone were the competent judge in this matter held me back from trying to influence you in the slightest degree. Your statement gratifies me therefore for a double reason—It proves that you were wise in the choice you have made and that I was wise in believing that your choice would be the right one.

I have had a hard week of it. On Tuesday I think I gave them a good lecture at the Institution[5]—On Thursday I gave the Bakerian Lecture[6] at the Royal Society—it lasted nearly two hours, but even this time was far too short to do justice to the subject. I by no means filled the measure of my own ideal, but Lord Wrottesley called the lecture an able one—Lord Ashburton I am told expressed himself greatly delighted with it—Grove did not see how the arguments brought forward could be overcome—so that on the whole it cannot be called a failure.

On Friday evening I lectured at the Institution[7]—I fear I made a slight mistake—I said once that I was compelled to dissent from the views put forward by Faraday in his lecture of the foregoing week.[8] Faraday's own feelings I do not know. He shook hands with me at the conclusion of the lecture, but I know the thing has been commented on by people in the house. If I had known that the matter would have caused such a stir I should probably not have alluded to the difference between us. As it is I am content—I have a clearer appreciation of Faraday's merits than those who would condemn me. I love and reverence the man, but it is with the heart of a freeman who will ever maintain his right to differ from him. The world will yet see that I do not differ from him on insufficient grounds.

I am glad Edmondson is commencing the half year wisely—I told the old man when he was here with me[9] that your opinions influenced me—and that I consulted you on almost all matter of importance. I should not wonder if this simple remark had some influence upon him. It is past bedtime—goodbye—my love to Anna—did you get the portrait from John Martin?[10]—

Your affectionate | John[11]

RI MS JT/1/T/592

1. *Your letter*: letter 1035.
2. *my lecture tomorrow*: Tyndall's third lecture in his 'Electricity and Magnetism' course at the RI on 30 January.
3. *amaranth*: Amaranth is an imaginary flower reputed never to fade (*OED*) which has often been evoked by in poetical contexts (as in the present instance).
4. *the change which you proposed to yourself*: marriage to Anna Martin.
5. *a good lecture at the Institution*: Tyndall gave the second lecture in his 'Electricity and Magnetism' course on 23 January.
6. *the Bakerian Lecture*: see letter 1029, n. 1.
7. *On Friday evening I lectured at the Institution*: see letter 1031, n. 3.
8. *Faraday . . . the foregoing week*: see letter 1031, n. 10. See also Tyndall's Journal entries for late January 1855: RI MS JT/2/13c/723.
9. *when he was here with me*: Tyndall had met Edmondson (together with Hirst and Debus) in London on the evening of 16 January (RI MS JT/2/13c/714).
10. *the portrait from John Martin*: portrait missing.
11. Signature is followed by deleted text: 'I have lately made the acquaintance of Sir Charles Lyell the geologist—he is very kind to me—I dine with him on Wednesday [31 January].'

To Thomas Archer Hirst 4 February 1855 1037

4th Feb. 1855

My dear Tom

Would you kindly glance over the enclosed[1] and tell me if you think the argument of any value. This is Sunday and I have been within all day—never once shaved nor drew clean linen over my shoulders. I walked out tonight in the mud and mist and felt all the better for the draught of adulterated oxygen. On my return I encountered two young men on the footway distributing tracts to a rabble of big and little boys which surrounded them. One as he handed the tract to each pronounced in a low voice 'the Sabbath'—the other an older hand or perhaps of a warmer temperament exclaimed as he distributed his ware 'Christ the only Saviour'—I stood and watched them for some time with no small interest, for I could discern beneath the wildness of the enthusiasts eye the working of that spirit which keeps the world out of the mud. As far as intellect and recognition of intellect go I was never better than I am at present, but there was a something at the heart of these methodist fanatics which I lacked and which I longed for—Not of course to make use of it as they did, but to apply it to my own private purposes. I was invited to dine with Sir Charles Lyell on Wednesday last at the geological club[2] and we went afterwards to the geological society together—There was a paper to be read on cleavage by a brother of the late Edward Forbes,[3] and Lyell wrote to me[4] informing me of the fact. Well the paper was read—the results of experiments were described & shewn and the whole question of cleavage and foliation was deemed settled for ever. Geologists of great eminence[5] rose and eulogized the memoir, and all seemed to admit that the important question with which it professed to deal was completely solved—During this time the conviction was rising more clearly in my mind that the experiments were just as much in need of explanation as the subject which they professed to clear up. The president[6] had mentioned my name in connexion with one or two others when he asked for remarks—I waited for a fitting opportunity, it came and I rose and stated my views[7] of the experiments and of their value with reference to the point in question. At first I was heard with silent attention, it became more earnest as I proceeded and as I drew towards the conclusion of what I had to say the loud and general cries of 'hear hear' shewed that my convictions had been contagious and had won the audience. The theory of Forbes was virtually declared 'not proven'[8]—He replied to me afterwards but evidently made no way in removing the impression which my remarks had created.

Now here was an exhibition of intellectual vigour which was pleasant while it lasted—but it was founded on an external impulse and when this was withdrawn I sank again into a state of intellectual meanness. My power over science is I think increasing greatly, but the power and vigour of what I call my soul is not commensurate. But why do I tell you so? Not for your edification alone but through the selfish prompting that by dwelling upon the subject I may clear away the moss which chokes those inner fountains and allow the abundant spirit to gush forth once more. On Friday night Airy gave us an excellent lecture[9] on his recent pendulum experiments—Next morning I breakfasted with him and Faraday at Barlow's.[10] He is a proud man they say and I was led to deem him a dry crust—but the man is full of humour and he and Barlow spouted parodies during half of breakfast time. He has had a discussion with Scoresby[11] lately in the Athenaeum[12] on the magnetism of iron ships—I see a gap to be filled in the science of the subject and tomorrow I intend to make some experiments for the purpose of supplying the needful facts.

But all this is vapid to me. I think I have been too comfortable all day and thus set my spirit to sleep—these English fires are not so good for the students as the dingy German stove—the latter does not woo you with its gleam and sparkle and you can bear to sit far away from it—here on the contrary I feel an unconscious tendency to draw towards the fire—if I obey this tendency I become warm and comfortable, and dozy, and finally all manful vigour seems squeezed out of me. I am sometimes inclined to make a vow against fires altogether; and I sympathise strongly with the habit of Socrates of going barefoot and using his hair for shelter. I will not contend for an uneasy fretful activity which breeds disease of brain and body—but a wakeful and vigorous habit of thought--a prompt realization of that which duty dictates are conducive to health both mental and bodily. With regard to my lectures I feel far less anxious than I used to be for I feel far more power—The lecture that you heard[13] was certainly by a good deal the worst I have given this year and I hope it will continue so. It is now nearly bed time and I will soon bid the world goodnight—send me back that letter quickly—Now I have given you an intellectual beggar's dish[14] which will improve the flavour of your breakfast. Good bye and believe me

Yours as ever | John

RI MS JT/1/T/593

---

1. *the enclosed*: enclosure missing. This enclosure was a copy of the Tyndall's draft response to Faraday, the revised version being letter 1042.
2. *Sir Charles Lyell . . . at the geological club*: see letter 1033.
3. *brother of the late Edward Forbes*: David Forbes; see letter 1033, n. 1. Edward Forbes, who had held the chair of natural history at the University of Edinburgh, had died on 18 November 1854.
4. *Lyell wrote to me*: letter 1033.
5. *Geologists of great eminence*: In his Journal for 30 January Tyndall mentioned that Charles Darwin, Charles Lyell and Samuel Sharp (1814–82) complimented David Forbes (RI MS JT/2/13c/724).
6. *The president*: Lord Wrottesley.
7. *stated my views*: 'It seemed to me evident that the hypothesis that the rock was perfectly homogeneous before heating—an hypothesis which nobody thought of disputing—was wholly baseless. The heat merely demonstrated the non homogeneous nature of the rock, and <*words missing*> by removing the dark colour which masked the true structure of the rock had merely exhibited the latter: there appeared to be no grounds for supposing that any such movement of solid matter through solid matter across sensible distances as inferred by Forbes and admitted by others had ever taken place' (Tyndall's Journal, 31 January 1855: RI MS JT/2/13c/724).
8. *not proven*: The verdict of 'not proven' was one of the three outcomes available in Scottish courts (the other two being 'guilty' and 'not guilty'). Other courts in the UK did not have this third option.
9. *Airy gave us an excellent lecture*: Airy's Friday Evening Discourse at the RI was entitled

'On the Pendulum Experiments Lately Made in the Harton Colliery, for Ascertaining the Mean Density of the Earth' and was published in *Roy. Inst. Proc.*, 2 (1854–58), pp. 17–23.

10. *Barlow's*: John Barlow lived on Berkeley St, Piccadilly.
11. *Scoresby*: William Scoresby (1789–1857), the curate of Bessingby, Yorkshire, had been an arctic explorer and made significant contributions to the study of terrestrial magnetism.
12. *a discussion . . . lately in the Athenaeum*: Airy, 'Correction of the Compass in Iron Ships', *Athenæum,* 3 February 1855, pp. 146–48.
13. *The lecture that you heard*: Hirst had attended the first of Tyndall's 'Electricity and Magnetism' lectures, which had been held on 16 January. In his Journal entry for that day Tyndall complained of a bad headache which 'continued bad throughout the lecture' (RI MS JT/2/13c/714).
14. *beggar's dish*: The 'beggar's dish' was 'a regular and proper part of his [the mendicant's] professional accoutrements' and was used by 'itinerant professed beggars and [. . .] wandering tribes of gypsies' to collect offerings. However, according to the article 'Beggar's Dish' in *Chamber's Edinburgh Journal* (25 February 1832, p. 27) it had by the 1830s largely gone out of fashion.

From Edward Sabine — 5 February 1855 — 1038

Feb. 5th | 13. Ashley Place

My dear Sir,

I enclose a note from Mr Welsh,[1] by which you will see that the Unifilar[2] will be with you on Tuesday morning.

also I enclose a volume of instructions with the pages marked describing its use.

also, a second copy of my report at Liverpool,[3] as you will probably mark up the 1st copy.

also, a note from Dr Lloyd,[4] that you will like to see, as showing the amiable & candid disposition of the writer, as well as the satisfaction I must have derived from such a concession on the part certainly of my best informed opponents.

In regard to the question whether we shall ultimately learn more of the nature of Terrestrial Magnetism from the small variations occasioned by the Sun & Moon, or from the study of magnetism as developped to us on the largest scale in our own planet, I remain of the same opinion as formerly—viz that the distribution of magnetism over the globe, and its secular changes, are by much the most important to study.

Sincerely yours | Edward Sabine

Your lecture[5] is I suppose at 3.PM—can you send me a ticket for myself and Mrs. Sabine?[6]

RI MS JT/1/S/30

1. *Mr Welsh*: John Welsh (1824–59), meteorologist and superintendent at Kew Observatory.
2. *the Unifilar*: a magnetometer in which the magnet is suspended by a single thread or filament.
3. *my report at Liverpool*: At the 1854 meeting of the BAAS, which was held in Liverpool, Sabine delivered a report 'On Some of the Results Obtained at the British Colonial Magnetic Observatories', *Brit. Assoc. Rep. 1854,* pp. 355–69.
4. *a note from Dr Lloyd*: Humphrey Lloyd (1800–81), the professor of natural and experimental philosophy at Trinity College, Dublin, pursued research primarily on optics and terrestrial magnetism. The 'note' is missing.
5. *Your lecture*: Tyndall delivered the fourth lecture in his series on 'Electricity and Magnetism' at 3pm on 6 February at the RI. Sabine attended the lecture (Tyndall's Journal: RI MS JT/2/13c/726).
6. *Mrs. Sabine*: Elizabeth Juliana, née Leeves (1807–79), wife of Edward Sabine.

From Thomas Archer Hirst 6 February 1855 1039

Broughton | Feb 6th 1855

My dear John,

Since I read Faraday's paper in last month's Phil. Mag.[1] I have been wishful to ask you how you would treat his argument about the deportment of the several solutions of protosulphate of Iron between the poles of a magnet.[2] It appeared to me a very ingenious one and though from instinct I could not admit that it at all proved the identity of Magnetic and Diamagnetic polarity yet I confess it shook my faith in their distinctive character and rather confounded my notions on the subject. This morning brought me your reply[3] to the very point in question and after reading it with great care I am inclined to believe that as an indirect proof of the opposite natures of magnetism and diamagnetism it will not be easily overcome[.] I see no escape from the contradiction between experiment and the consequence you have legitimately deduced from Faraday's notions of a mere differential action. As to other parts of your letter not involving the argument I like most of it though as a general remark I would say do not allude too much to the value of his researches and his opinions but you will feel better than I can tell how much is due to him. As to your letter to me it is all interesting although the general tone has not a cheering influence on me[.] I notice in it the old consciousness of the effects which ever result from a too exclusive devotion to intellectual pursuits the only satisfactory fact to me is that such self-consciousness usually indicates a wish to remedy the evil which is felt is in fact a first step towards the reaction which in a properly balanced man always follows the disturbing action. I knew the good effect of writing to me at such times; it renders your cause for grievance more visible and strengthens you to undertake once more its remedy. For your sake I am glad you have me to write to; for my sake I am glad to be written to for apart from my brotherly affection toward you it has a salutary effect on me. For now that I have a wife whose harmonizing influence upon me is certain enough and as it may be carried too far and abused I see quite well the advantages of and even necessity of a Tyndall to shew me occasionally the other side of the question. So you see my dear John your selfishness accords with my selfishness and in following both we may do one another service.

My days pass pleasantly now. Until 6 P.M daily (from 7½A.M) I devote myself strictly to my several duties after that time I am my wife's husband. No more at present write again soon and so will I.

Yours affectionately | T A Hirst

RI MS JT/1/H/213

---

1. *Faraday's paper in last month's Phil. Mag.*: Faraday's 'On some Points of Magnetic Philosophy' was published in the February issue of the *Phil. Mag.*, 9 (1855), pp. 81–113.
2. *protosulphate of Iron . . . a magnet*: ibid.: 89–91. Protosulphate of iron is ferrous sulphate, $FeSO_4$.
3. *your reply*: For Tyndall's reply to Faraday see letter 1037, n. 1.

To Thomas Archer Hirst 7 February 1855 1040

Wednesday, written in a scamper | 1855

My dear Tom

Thank you many times for your capital letter[1]—dont let mine cast the slightest shadow on your soul—I know how to get back my force, and the point from which I wrote to you was as you rightly estimated past that turning point when the pendulum having swung is about to return. As I read your letter today I felt a stir of vigour which could I but communicate it would dispel all dismay on my account. I had a strange dream about you last night—a dream probably brought on by the mixture of Rhine wine with port during a dinner which I had with Carter yesterday.[2] I thought you had shaved off your barb and looked quite a boy as regards years—You gave me I thought some childish excuse for this act, the real reason of it being however that Anna disliked the thing—I thought of the arguments which you once brought forward in favour of it and I felt angry to see you turning recreant to your own creed.

I looked at you and seeing you so youthful I said 'the boy is a child—he is soft & flexible and I have totally miscalculated him'—I turned away half in anger and met Anna smiling kindly into my eyes—she spoke to me and I know not what necromancy caused me to exclaim 'By the Lord Tom is right—I would cut off half a dozen beards for her sake!' —

I gave them a lecture yesterday[3] which though difficult appeared to interest every body very much. It was amusing to see the eager attention with which they listened to me, and when the lecture was ended a great number of them with earnest eye, gathered round the lecture table—It cost me an additional half hour to answer all their questions. I have just the temptation you lay before me on your envelope—depend upon it I shall invade Anna's larder some day soon. I am mightily busy at present having to join a soirée lecture next week at the London Institution[4]—I shall modify the letter to Faraday.[5] Had I an opportunity of talking the whole subject over to you I think little doubt would remain upon your mind as to diamagnetic polarity.

RI MS JT/1/T/867

1. *your capital letter*: letter 1039.
2. *with Carter yesterday*: On 6 February Tyndall 'Dined with Carter at the Gray's Inn Coffee House and revived some old recollections' (Journal: RI MS LT 2/13c/726). Richard Carter (1819–95) was the surveyor and land agent from Halifax, Yorkshire, who had earlier employed both Tyndall and Hirst.
3. *a lecture yesterday*: see letter 1038, n. 5. 'The lecture [of 6 February] appeared to interest the audience very much—with earnest eyes numbers of them came down to the table when the lecture was over, and half an hour further was spent answering questions' (Journal: RI MS LT 2/13c/726).
4. *soirée lecture next week at the London Institution*: Tyndall delivered a lecture on 'On Boiling and Boiling Springs' at a conversazione at the LI on 14 February. See also Tyndall's notes for the lecture (RI MS JT/4/7a/51–53) and his Journal account (RI MS JT/2/13c/728).
5. *I shall modify the letter to Faraday*: Tyndall had sent Hirst a draft of his reply to Faraday (see letter 1037, n. 1), but subsequently revised the text (letter 1042).

From Viktor Hueter[1] 7 February 1855 1041

Mein lieber John!

Wenn auch schon beinahe mehr als ein lustrum zwischen unserem Abschiede auf den hehren Philippinen und dem jetzigen Augenblicke liegt, so hat mir dieser Zeitraum doch noch genug der Gefühle von Liebe und Freundschaft zu Ihnen gelassen, um mich vor mir zu rechtfertigen, wenn ich Sie auch jetzt noch nach so langer Trennung und aus so weiter Ferne mit den

Worten anrede, die ich im Eingange gebrauchte. Vielleicht hat die Zeit, das Leben, der gelehrte Umgang, der Ihnen zu Gebote steht, der Gewinn eines in der Wissenschaft bedeutenden Namens die Vergangenheit bei Ihnen in den Schatten gestellt, und Sie denken vielleicht nur selten noch der Marburger Tage mit ihrer Idylle; doch weiß ich gewiß, daß Sie nicht Alles vergessen, daß auch Ihnen bisweilen die Vergangenheit noch freundlich winkt, daß Sie dann vielleicht auch meiner gedenken, der er manchmal in später Abendstunde neben Ihnen saß, Thee trank, Englisch paukte und <*word illeg*> las. Deshalb hege ich keine Besorgnis, daß Sie diese Zeilen unfreundlich aufnehmen möchten.

Ihnen geht es gut, das habe ich zu meiner Freude von unserm gemeinsamen Freunde Dèbus vernommen; doch hatte ich nie daran gezweifelt. Mir geht es auch in zufriedenstellender Weise; eine erträgliche Einnahme, eine gute Frau, ein lively boy von 3½ Jahren lassen mich nur eine Stellung vermissen, die auch etwas mehr wissenschaftlichen Gewinn brächte. Damit sieht es aber verzweifelt flau aus, und es bleibt mir auch daher weiter nichts übrig als mich mit Horazischer Weisheit zu trösten. Aequam memento etc.

Weshalb ich nun eigentlich schreibe? Dessen ist eine englische Erbschaft Schuld, bei der ich indirect interessirt bin, um so mehr als sich vielleicht bei günstiger Conjunctur eine Reise zum fröhlichen England damit verbindet. Zu dessen Ende nehme ich meine Zuflucht zu Ihrer Güte und Freundschaft, die mir zunächst einiges Licht über den Thatbestand der Erbschaft zuführen sollen. Die Erbschaft ist schon alt, in London angeblich unter dem Namen der Ifflandschen Erbschaft bekannt. Der Erblasser war ein gewisser Stuecker aus Kurhessen, der zu Gunsten eines gewissen Johannes Iffland und dessen Ehefrau Anna Christina Stuken testirte. Die Fonds der Erbschaft sollen in der Londoner Bank liegen.

Vielleicht haben Sie Connexionen, die es Ihrer Spürkraft möglich machen, zu erkunden, wo das Testament liegt, was, wenn die Fonds wirklich in der Londoner Bank liegen, dieser bekannt sein muß; ferner, ob man eine Abschrift des Testaments, die gar nicht beglaubigt zu sein braucht, erhalten kann; wie hoch der Betrag der Erbschaft ist; und ob es wohl den Erben des Joh. Iffland sehr schwer gemacht werden müsste, die Erbschaft zu heben.

Es wäre doch eine erzfidele Geschichte, wenn ich eines Tages bei Ihnen einträte, um durch Ihre Vermittlung den Schatz zu heben.

Wenn es Ihnen nun nicht zu mühevoll ist, so verschaffen Sie mir Licht in dieser Angelegenheit und ich werde Ihnen dankbar sein. Um Ihnen nicht allzu viel Last zu machen, bitte ich, die fragliche Kunde via Dèbus an mich gelangen zu lassen.

Ich hoffe, Sie deuten meine Bitte nicht übel, ich bin zu jedem möglichen Gegendienste bereit.

Leben Sie wohl, lieber John. Ich umarme Sie im Geiste.
Ihr | Hueter
Niederaula bei Hersfeld | d[en] 7 Februar 1855.

My dear John!

Even though nearly more than five years[2] lie between our farewell in the noble Philippines[3] and the present moment, this period has nevertheless still left me with enough feelings of love and friendship for you to justify myself addressing you, even now after such a long separation and from such a great distance, with the words that I used in the opening. Perhaps time, life, the learned company that is at your command or the benefit of an important name in science, have overshadowed the past with you, and perhaps you think only seldom still about the idyllic days at Marburg.[4] But I know for certain that you have not forgotten everything, that from time to time the past still gives you a friendly wave also, that you then perhaps think of me also, who sometimes sat beside you at a late evening hour, drinking tea, cramming English and reading <*word illeg*>. I therefore have no concern that you might receive these lines unkindly.

You are doing well, this I have heard to my delight from our mutual friend Debus; but I had never doubted this. I am doing satisfactorily also; a tolerable income, a good wife, a lively boy of 3½[5] leave me only feeling the need of a position that would also bring somewhat more scientific gain. Things are looking desperately dull with this, however, and there therefore remains nothing further for me but to console myself with Horatian wisdom. Aequam memento,[6] etc.

Why am I really writing now? This is due to an English inheritance,[7] in which I have an indirect interest, all the more so as, with a favourable economic outcome, a trip to merry England will perhaps be connected with it. To this end, I take my refuge in your kindness and friendship, which should first of all supply me with some light about the facts of the inheritance. The inheritance is already old and known in London supposedly under the name of the Iffland inheritance.[8] The testator was a certain Stuecker from the Electorate of Hesse,[9] who made a will in favour of a certain Johannes Iffland[10] and his wife Anna Christina Stuken.[11] The funds of the inheritance are said to be deposited in the Bank of England.

Perhaps you have connections that will make it possible for your investigative powers to ascertain where the will is deposited, which, if the funds really are deposited in the Bank of England, must be known to the latter; further, whether one can obtain a copy of the will, which need not be authenticated at all; how great the amount of the inheritance is; and whether things might possibly be made very difficult for the heirs of Joh. Iffland to collect the inheritance.

It would be a thoroughly jolly story indeed, if I were to turn up at your place one day in order to collect the treasure through your good offices.[12]

If it is not too troublesome for you, then please provide some light for me in this affair and I shall be grateful to you. In order not to create all too much of burden for you, I ask that you send the information in question to me via Debus.

I hope you will not construe my request unkindly, I am ready to render any possible service in return.

Fare well, dear John. I embrace you in spirit.

Your | Hueter

Niederaula, Hersfeld[13] | 7 February 1855.

RI MS JT/1/H/511

---

1. *Hueter*: Viktor Hueter (1832–97) who studied medicine at the University of Marburg (1851–55) and was a son of Carl Christoph Hueter, a professor of medicine at the University. He became a gynocologist and philanthropist.
2. *five years*: The German refers to 'ein lustrum', a lustrum being the period of five years separating censuses in ancient Rome.
3. *farewell in the noble Philippines*: Both Tyndall and Hueter had studied at the (Philipps) University of Marburg—Philipps-Universität Marburg—which had been founded in 1527 by the Landgrave Philip I of Hesse.
4. *the idyllic days at Marburg*: where they had both been students.
5. *a good wife, a lively boy of 3½*: not identified.
6. *Horatian wisdom. Aequam memento*: **'Aequam memento rebus in arduis servare mentem'** (Latin): 'Remember when life's path is steep to hold your mind steady'. From the Roman poet Horace (Quintus Horatius Flaccus, 65–68 BC), *Odes*, II, iii, 1–2.
7. *an English inheritance*: not identified.
8. *Iffland inheritance*: not identified.
9. *Electorate of Hesse*: The state created by Napoleon that existed from 1803 to 1866, when it was annexed by Prussia.
10. *Johannes Iffland*: not identified.
11. *Anna Christina Stuken*: possibly Stuecker, not identified.
12. *through your good offices*: There is no evidence that Tyndall responded to this request.
13. *Niederaula, Hersfeld*: a market town in the Grand Duchy of Hesse, Germany, about 5½ miles south-west of spa at Bad Hersfeld.

To Michael Faraday [10?] February 1855 1042

Royal Institution, | February 1855.

MY DEAR MR. FARADAY,

FEW, I imagine, who read your memoir[1] in the last Number of the Philosophical Magazine, will escape the necessity of reconsidering their views of magnetic action. We are so accustomed to regard the phænomena of this portion of science through the imagery with which hypothesis has invested them, that it is extremely difficult to detach symbols from facts, and to view the latter in their purity. This duty, however, is now forced upon us; for the more we reflect upon the results of recent scientific research, the more deeply must we be convinced of the impossibility of reconciling these results with our present theories.[2] In the downfall of hypotheses thus pending, the great question of a universal magnetic medium has presented itself to your mind. Your researches incline you to believe in the existence of such a medium,[3] and lead you, at the same time, to infer the perfect identity of magnetism and diamagnetism.[4]

In support and illustration of your views, you appeal to the following beautiful experiments:[5]–Three solutions of protosulphate of iron are taken; the first, 1, contains 4 grains; the second, *m,* 8 grains; and the third, *n,* 16 grains of the salt to a cubic inch of water. Enclosed in hollow globules of glass, all these solutions, when suspended in the air before the pole of a magnet, are attracted by the pole. You then place a quantity of the medium solution, *m,* in a proper vessel, immerse in it the globule containing the strong solution *n,* and find that the latter is still attracted; but that when the globule containing the solution 1is immersed, the latter is *repelled* by the magnetic pole. Substituting elongated tubes for spheres, you find that when a tube containing a solution of a certain strength is suspended in a weaker solution, between the two poles of a magnet, the tube sets from pole to pole; but that when the solution *without* the tube is stronger than the *within* it, the tube recedes from the pole and sets equatorially.

Here then, you state, are the phænomena of diamagnetism. It is maintained by some, that, to account for these phænomena, it is necessary to assume, in the case of diamagnetic bodies, the existence of a polarity the reverse of that of iron. But nobody will affirm that the mere fact of its being suspended in a stronger solution reverses the polarity of a magnetic liquid: —to account for the repulsion of the weak solution, when submersed in a stronger one, no such hypothesis is needed; why then should it be thought necessary in the case of so-called diamagnetic bodies? It is only by denying that space presents a medium which bears the same relation to diamagnetic bodies that the stronger magnetic solution bears to the weaker one, that the hypothesis of a distinct diamagnetic polarity is at all rendered necessary.

The effects upon which the foregoing striking argument is based are differential ones, and are embraced, as already observed by M. E. Becquerel,[6] by the so-called principle of Archimedes.[7] This principle, in reference to the case before us, affirms that the body immersed in the liquid is attracted by a force equal to the difference of the attractions exerted upon the liquid and the body immersed in it. Hence, if the attraction of the liquid be less than that of the immersed body, the latter will approach the pole; if the former attraction be the greater, the immersed body recedes from the pole, and is apparently repelled. The action is the same as that of gravity upon a body immersed in water: if the body be more forcibly attracted, bulk for bulk, than the water, it sinks; if less forcibly attracted, it rises; the mechanical effect being the same as if it were repelled by the earth.

The question then is, are all magnetic phænomena the result of a differential action of this kind? Does space present a medium less strongly attracted than soft iron, and more strongly attracted than bismuth, thus permitting of the approach of the former, but causing the latter to recede from the pole of a magnet? If such a medium exists, then diamagnetism, as you incline to believe, merges into ordinary magnetism, and 'the polarity of the magnetic force,' in iron and in bismuth, is one and the same.

Pondering upon this subject a few evenings ago, and almost despairing of seeing it ever brought to an experimental test, a thought occurred to me, which, when it first presented itself, seemed to illuminate the matter. Such illuminations vanish in nine cases out of ten before the test of subsequent criticism; but the thought referred to, having thus far withstood the criticism brought to bear upon it, I am emboldened to submit it to you for consideration.

I shall best explain myself by assuming that a medium of the nature described exists in space, and pursuing this assumption to its necessary consequences.

Let a cube, formed from the impalpable dust of carbonate of iron,[8] which has been compressed forcibly in one direction, be placed upon the end of a torsion beam, and first let the line in which the pressure has been exerted be in the direction of the beam. Let a magnet, with its axis at right angles to the beam, and hence also at right angles to the line of pressure, be brought to bear upon the cube. The cube will be attracted, and the amount of this attraction, at any assigned distance, may be accurately measured by the torsion of the wire from which the beam depends.[9] Let this attraction, expressed in degrees of torsion, be called *a*. Let the cube now be turned round 90°, so that the line of pressure shall coincide with the direction of the axis of the magnet, and let the attraction *a′* in this new position be determined as in the former instance. On comparison it will be found that *a′* exceeds *a;* or, in other words, that the attraction of the cube is strongest when the force acts parallel to the line of compression.[10]

Instead of carbonate of iron we might choose other substances of a much feebler magnetic capacity, with precisely the same result. Let us now conceive the magnetic capacity of the compressed cube to diminish gradually, and thus to approach the capacity of the medium in which, according to our assumption, the carbonate of iron is supposed to be immersed. If it were a perfectly homogeneous cube, and attracted with the same force in all directions, we should at length arrive at a point, when the *magnetic weight* of the cube, if I may use the term, would be equal to that of the medium, and we should then have a substance which, as regards magnetism, would be in a condition similar to that of a body withdrawn from the action of gravity in Plateau's experiments.[11] Such a body would be neither attracted nor repelled by the magnet. In the compressed cube, however, the magnetic weight varies with the direction of the force; supposing the magnetic weight, when the force acts along the line of compression, to be equal to that of the medium, then if the force acted across the line of compression, the magnetic weight of the cube would be less than that of the medium. Acted upon in the former direction, the cube would be a neutral body; acted upon in the latter direction, it would be a diamagnetic body. If the magnetic capacity of the cube diminished still further, it will, according to your hypothesis, become wholly diamagnetic. Now it is evident, supposing the true magnetic excitement to continue, that the cube, when acted on by the magnet in the direction of compression, *will approach nearer to the magnetic weight of the medium* in which we suppose it immersed, than when the action is across the said line; and, hence, the repulsion of the cube, when the force acts along the line of compression, must be *less* than when the force acts across it.

Reasoning thus from the assumption of a magnetic medium in space, we arrive at a conclusion which can be brought to the test of experiment. So far as I can see at present, the assumption is negative by this test; for in diamagnetic bodies the repulsion along the line in which the pressure is exerted is proved by experiment to be a *maximum.*[12] An ordinary magnetic excitement could not, it appears to me, be accompanied by this effect.

The subject finds further, and perhaps clearer, elucidation in the case of isomorphous crystals. It is not, I think, questioned at present, that the deportment of crystals in the magnetic field depends upon their molecular structure; nor will it, I imagine, be doubted, that the molecular structure of a complete crystal of carbonate of iron is the same as that of an isomorphous crystal of carbonate of lime. In the architecture of the latter crystal, calcium simply takes the place which iron occupies in the former. Now a crystal of carbonate of iron is attracted most forcibly when the attracting force acts parallel to the crystallographic axis.[13] Let such a crystal be supposed to diminish gradually in magnetic capacity, until finally it attains a magnetic weight, *in a direction parallel to its axis,* equal to that of the medium in which we assume it to be immersed. Such a crystal would be indifferent, if the force acted parallel to its axis, but would be repelled, if the force acted in any other direction. If the magnetic weight of the crystal diminish a little further, it will be repelled in all directions, or, in other words, will become diamagnetic; but it will then follow, that the repulsion in the direction of the axis, if the nature of the excitement remain unchanged, will be less than in any other direction. In other words, a diamagnetic crystal of the form of carbonate of iron will, supposing magnetism and diamagnetism to be the same, be repelled with a *minimum* force when the repulsion acts parallel to the axis. Here, as before, we arrive at a conclusion which is controverted by experiment; for the repulsion of a crystal of carbonate of lime is a *maximum* when the repelling force acts along the axis of the crystal. Hence I would infer that the excitement of carbonate of iron cannot be the same as that of carbonate of lime.

Such are the reflections which presented themselves to my mind on the evening to which I have referred. I now submit them to you as a fraction of that thought which your last memoir upon this great question will assuredly awaken.

Believe me, | Dear Mr. Faraday, | Yours very faithfully, | JOHN TYNDALL.

Royal Institution, February 1855.[14]

Phil. Mag., 9 (1855), pp. 205–9[15]

---

1. *your memoir*: M. Faraday, 'On Some Points of Magnetic Philosophy', *Phil. Mag.*, 9 (1855), pp. 81–113.
2. Footnote in published text: 'Some of the reasons which induce the writer to hold this opinion are given in the Bakerian Lecture of the Royal Society for the present year.' For Bakerian Lecture see letter 1031, n. 2.
3. *the existence of such a medium*: see Faraday, 'On Some Points'.
4. *the perfect identity of magnetism and diamagnetism*: see Faraday, 'On Some Points'.
5. *the following beautiful experiments*: see Faraday, 'On Some Points', 89–90.
6. *observed by M. E. Becquerel*: The French physicist Edmond Becquerel (1820–91), who was a professor at the Conservatoire Impérial des Arts et Métiers, opposed Faraday's view that diamagnetic phenomena were essentially different from those of regular magnetism. Instead, he postulated a law of magnetic action based on Archimedes' principle (next note) to account for diamagnetism. Becquerel, 'De l'Action du Magnétisme sur Tous les Corps', *Annales de Chimie et de Physique,* 28 (1850), pp. 283–350; 32 (1851), pp. 68–112.
7. *principle of Archimedes*: 'that a body immersed in a liquid loses a portion of its weight equal to the weight of the quantity of liquid which it displaces' (J. Tyndall, 'Natural Philosophy', in E. Hughes (ed.), *Reading Lessons, Advanced Series. Second Book* (London: Longman, Brown, Green & Longmans, 1855), pp. 289–322, on pp. 301–2).
8. Footnote in published text: 'For an ample supply of this most useful mineral, I am indebted to the kindness of J. Kenyon Blackwell, Esq. F.G.S.' John Kenyon Blackwell (1812–69) was a mining engineer and inspector of mines.

9. *depends*: possibly a misprint for 'descends'.
10. Footnote in published text: 'Phil. Mag. Sept. 1851; Pogg. *Ann.* 1851.' See J. Tyndall, 'On Diamagnetism and Magnecrystallic Action', *Phil. Mag.*, 2 (1851), pp. 165–88; *Poggend. Annal.*, 83 (1851), pp. 1–37.
11. *Plateau's experiments*: Joseph Antoine Ferdinand Plateau (1801–83) was a professor of physics at the University of Ghent. His work on the statics of liquids was originally published in a series of papers 'Sur les Phénomènes que Présente une Masse Liquide Libre et Soustraite à l'Action de la Pesanteur', *Bulletins de l'Académie Royale des Sciences et Belles-lettres de Bruxelles*, 9 (1842), pp. 17–19 and 298–301; 14 (1847), pp. 429–32.
12. Footnote in published text: 'Phil Mag. Sept. 1851. Pogg. *Ann* 1851.' See n. 10.
13. Footnote in published text: 'Phil. Mag. Sept. 1851. Pogg. *Ann.* 1851.' See n. 10.
14. *February 1855*: Tyndall's Journal entry for 4 February 1855: 'Wrote letter to Mr Faraday on the question whether there is a magnetic medium in space' (RI MS JT/2/13c/725). However, this letter predates letter 1040 of 7 February, in which Tyndall stated 'I shall modify the letter to Faraday'.
15. Published in the *Phil. Mag.* with the title 'On the Existence of a Magnetic Medium in Space'.

From Thomas Henry Huxley 13 February 1855 1043

14 Waverley Place | St Johns Wood

My dear Tyndall,

Many thanks for your note[1]—The kind of praise your friend[2] gave me is just what I would desire & I confess my vanity is pleasantly tickled thereby

Had I imagined you needed a ticket[3] I would have sent you one—But I imagined that as one of their instructors[4] you were free to come and go—I am sorry I can't send you one now but they only gave me three & I don't like to ask for any more—being a paid servant of their worships.[5] Little Brayley[6] (who knows a thing or two more than God almighty) says its the most heretical Syllabus[7] he has seen—so I send you one to judge.

I wish I could come & hear you tomorrow[8] but its impossible as I dined out on Monday & have to go to the Royal[9] on Thursday and to dine out on Saturday—

When will you come up here? I have a hundred things to talk to you about[.] Will you come & breakfast on Sunday morning[10] & then we can have a long yarn.

If any body asked me to do such a thing I should consider it a personal insult but you know you confessed to a habit of walking for an hour before breakfast.

Fix your own hour—9. 10 or 11.

Ever yours faithfully | T. H. Huxley.

Prof. Tyndall

IC HP 9.20

1. *your note*: letter missing.
2. *your friend*: not identified.
3. *a ticket*: to attend Huxley's course at the London Institution on 'The General Laws of Life'; see n. 7, below.
4. *one of their instructors*: Huxley was presumably referring to Tyndall's lecture at the LI on the following day; see letter 1040, n.4 .
5. *a paid servant of their worships*: the Committee of Managers of the LI.
6. *Little Brayley*: the science writer and lecturer Edward William Brayley (1802–70), who served as joint librarian at the LI. The epithet 'Little' may have been described here to distinguish him from his father, the late Edward Wedlake Brayley, who wrote on topography and archaeology.
7. *the most heretical Syllabus*: Beginning on 8 February 1855, Huxley delivered a series of six weekly lectures on 'The General Laws of Life' at the LI: see 'London Institution. Syllabus for a Course of Six Lectures on "The General Laws of Life," February to March 1855' (T. H. Huxley Manuscripts, Imperial College, London: HM 3:122.7). See also the *Athenæum* (14 October 1854), p. 1221.
8. *hear you tomorrow*: Tyndall's lecture at the LI; see letter 1040, n. 4.
9. *the Royal*: the RS.
10. *come and breakfast on Sunday morning*: It would appear that Tyndall did not breakfast with Huxley on Sunday, 18 February, although he met Huxley the previous evening at the home of the geologist Andrew Crombie Ramsay (Journal, RI MS JT/2/13c/729).

To Thomas Henry Huxley 14 February 1855 1044

Wednesday night | 11½ o'c

My dear Huxley

I have just finished my lecture at the London[1] and have been lucky enough to please them all—I am spirited off tomorrow to Swindon[2] by the Government Inspector of Railways[3] to see a boiler which has exploded. Friday I have two engagements but Saturday God willing I shall meet you at Ramsay's.[4] You mean to go there of course. I interpret your dining out on Saturday to mean so much. We will talk of the breakfast then.

Thank you for the Syllabus[5]—it is a deep one—I breakfasted with Owen[6] a few days ago and ventured at the time upon one or two of your 'heretical' remarks—

I dare say the people would not shut the door against me if I presented myself.[7] So if no other matters interfere I shall contrive to be present at your next lecture.

Goodbye till Saturday | Ever yours | John Tyndall

IC HP 1.333

---

1. *my lecture at the London*: see letter 1040, n. 4.
2. *Swindon*: Located 70 miles to the west of London, Swindon was a railway town with the major locomotive workshops of the Great Western Railway, which Isambard Kingdom Brunel had designed.
3. *the Government Inspector of Railways*: Colonel George Wynne of the Royal Engineers.
4. *Ramsay's*: The Scottish geologist Andrew Ramsay (1814–91), was professor of geology at the Royal School of Mines. 'Spent the evening [Saturday, 17 February] at Ramsay's. Huxley was there, and we had much philosophic talk' (Journal, RI MS JT/2/13c/729).
5. *the Syllabus*: see letter 1043, n. 7.
6. *Owen*: On 10 February Tyndall recorded in his Journal (RI MS JT/2/13c/727) that he had breakfasted at the home of the surgeon William White Cooper, where he met Richard Owen.
7. *if I presented myself*: if Tyndall attended Huxley's lecture at the LI; see letter 1043, n. 7.

To George Wynne 16 February 1855[1] 1045

MY DEAR SIR,

In placing before you the reflections suggested by our joint examination of the boiler[2] which exploded some time ago[3] at the Swindon Station of the Great Western Railway, I have to express my regret that the absence of safe experimental data on the question of steam boiler explosions generally, renders every opinion upon the subject in a great degree conjectural and uncertain. Our two hours' examination,[4] instead of enabling us to assign the exact source of the actions which led to the destruction of the boiler, merely suggests the mode of attacking the question. While venturing, therefore, upon the following remarks, I cannot help feeling how much more satisfactory it would be to be able to substitute for supposition, however probable, the surer evidence of experimental fact.

The appearances presented by the boiler were as follows:—The iron plates[5] as far as the water usually extended, were deeply pitted and furrowed by some corrosive agency. At the centre portion of the boiler, the pits, I think, were deepest and best defined; towards the end at which the cold water entered to feed the boiler, they were less pronounced; towards the end nearest to the fire-box[6] they were more diffuse and general. While pondering over these curious appearances, my attention was attracted by the tubes which passed through the boiler; these were of brass, and were attached at one end to the fire-box, and at the other end to the smoke-box of the engine. In the normal condition of the boiler, these tubes were separated from the bottom and sides by an interval of an inch or two, this interval being usually filled with hot water.

Without pretending to a degree of confidence unwarranted by the circumstances of the case, I think it may be affirmed that such a boiler constitutes a veritable voltaic couple,[7] consisting of brass, iron, and the exciting liquid water. Feeble currents of electricity will be established, the direction of which, through the liquid, will be from iron to brass, the iron being what is called the positive element of the couple. A decomposition of the water will take place; hydrogen will be liberated against the brass tubes; oxygen (and acid, if a salt be dissolved in the water,) will be liberated against the iron plates. The quantity thus evolved may be very small, but, acting incessantly for a number of years, it would, I imagine, be sufficient to produce the observed corrosion.

In the course of our examination you directed my attention to the singular preservative influence which the fire-box seemed to exercise on some iron bars which passed through it. There were, I believe, four such bars; solid cylinders, which passed, as longitudinal stays, from the front to the rear of the engine. Up to the point where these bars entered the steam dome they were deeply corroded; in the steam dome they were comparatively unaltered. This seems to be in harmony with the foregoing explanation. Up to the steam dome the iron bars ran side by side with the brass tubes which traversed the boiler; at the steam dome the brass tubes ceased, and from this to the end of the boiler, the iron bars, instead of being *electro moters,*[8] were merely the *conductors* of the currents generated where the bars and the tubes were in close proximity. Other explanations may possibly be suggested, the exact merits of which experiment alone can decide.

Parallel to the row of rivets which united the bottom plate of the boiler to the smoke-box, a deep furrow had been eaten into the plate. A similar corrosive action was not observed parallel to the corresponding row of rivets at the fire-box end. Parallel to one longitudinal row of rivets a furrow had been eaten so deep as to reduce the thickness of the plate from three eighths to less than one eighth of an inch. Along this furrow the boiler is supposed to have given way; the plates were torn asunder, and the total destruction of the boiler was the immediate consequence.

Now, whatever be the cause of the corrosion, whether we refer it to the oxygen held in solution by the water, or to the decomposition of the water itself, science probably furnishes the means for its prevention. The experiments of Pepys, Davy, Van Beck,[9] and others show that by suitably connecting a more positive metal with iron, the corrosion is diverted from the latter, so that it is thus possible to preserve iron uninjured in a liquid which, under ordinary circumstances, would rapidly attack it.

If the explanation which refers the corrosion to electro chemical decomposition be the correct one, then in boilers where iron tubes are used, instead of brass ones, the pitting and oxidation ought not to take place to the same extent. I say 'to the same extent', because a difference in the manipulation of two pieces of iron is sufficient to destroy their electro homogeneity and to produce feeble currents when the pieces are immersed in an exciting liquid. An action of this kind may perhaps be established between the iron rivets and the plates they unite, and, possibly, to this cause may be referred the furrows which are sometimes observed to follow the lines of junction. It would, however, be hasty to infer from a single instance that this is a general result; the whole subject demands a thorough examination, and a far wider acquaintanceship with facts than two hours' inspection of a single boiler could possibly furnish.

The longer we reflect upon this subject, more deeply must we be impressed with the necessity of associating, in the construction and management of steam boilers, the knowledge of the natural philosopher with that of the engineer. The consciousness of our deficiency in this respect was present with me when, on a recent occasion, I ventured to state, that 'there are agencies at work in a locomotive of which the maker of it never dreamed, and which may nevertheless convert it into an engine of destruction'.[10] At the present moment it might be difficult to say, how many boilers are on the verge of explosion without any single engineer being aware of the danger. Another point deserving of attention is the following:—Common water always contains a quantity of atmospheric air in solution. This air may be expelled from the liquid by continued boiling; and it is an experimental fact, that water thus freed from air possesses mechanical properties widely different from ordinary water. The cohesion of the liquid is enormously increased, in consequence of which it may be heated to a temperature far beyond its ordinary boiling point without boiling; but when it does boil, it is not with the quiet ebullition of common water, the liquid particles snap suddenly asunder like a broken spring, and ebullition is converted into explosion. It is, I believe, a fact that boilers, after standing for a time, have often exploded at the precise moment when the engineer turned on the steam. The question presents itself, whether the action just referred to may not here come into play? If the water has been sufficiently purged of air and then left at rest, its augmented cohesion may permit of its being heated far above the boiling point due to the pressure upon its surface; the mechanical disturbance produced by turning on the steam would destroy this cohesion, and the superheated liquid would develop a force resembling that of gunpowder on the application of a lighted match. I would wish it to be borne in mind, that we are in total ignorance as to the practical value of these suggestions; but the investigation which the subject so pressingly demands must take cognizance of them all.

I am, &c.

To Lieut.-Col. Wynne, R.E. (Signed) JOHN TYNDALL.

House of Commons Parliamentary Papers, 1854–55 [1951], *Two Reports of an Inspecting-Officer of the Railway Department to the Lords of the Committee of Privy Council for Trade, on the Explosion of Boilers of Locomotive Engines, One of which Occurred at the Gloucester Station of the Great Western Railway, On the 7th February 1855, and the Other at the Greenock Station of the Caledonian Railway, on the 5th April 1855, Together with a Copy of a Circular Issued by their Lordships to the Railway Companies on the Subject* (London: HMSO, 1855), 63–64.

---

1. *16 February 1855*: After visiting Swindon to inspect the damaged boiler on 15 February Tyndall 'Wrote report on boiler', which he showed to Faraday (Journal, 16 February 1855: RI MS JT/2/13c/729).
2. *the boiler*: of a locomotive.
3. *exploded some time ago*: A 'fearful explosion' had occurred on 7 February at Gloucester Station. The engine-driver 'had brought the carriage to within a few yards of the platform, when an awful explosion rent the air. The boiler of the engine had burst into a thousand fragments, and yet there stood the driver and stoker unharmed! [. . .] The explosion was heard for miles round, and a great many windows were broken by the concussion' (*London Standard,* 8 February 1855, p. [1]).
4. *Our two hours' examination*: with George Wynne, on 15 February.
5. *The iron plates*: forming the boiler.
6. *fire-box*: The chamber of a steam-boiler in which the fuel is burnt (*OED*).
7. *voltaic couple*: Two connected plates of different metals that produce an electric current.
8. *electro moters*: Electromoters are sources of electromotive force; in particular, metallic plates (of different metals) which, on contact, produce an electromotive force (*OED*).
9. *The experiments of Pepys, Davy, Van Beck*: The English scientists William Haseldine Pepys (1775–1856) and Humphry Davy (1778–1829) and the Utrecht physicist Albert van Beek (1787–1856) were among those who had contributed to research on the corrosion of metals through the study of electrochemistry.
10. Footnote in text reads: 'On the Study of Physics, a lecture; John Parker and Son, West Strand.' On 27 May 1854 Tyndall had delivered the lecture at the RI 'On the Importance of the Study of Physics as a Branch of Education for All Classes' (*Morning Post,* 27 May 1854, p. [1]). This lecture was subsequently published in *Lectures on Education Delivered at the Royal Institution of Great Britain* (London: J. W. Parker & Son, 1854), pp. 171–211.

From Edward Frankland 19 February 1855 1046

Owens College | Manchester Feb. 19/55

My dear Tyndall

Many thanks for your hint about the chlorine experiment.[1] I certainly intend to mention the precaution you name as I have a strong expression[2] that it was at one time in my manuscript so that I cannot exactly account for its omission. In the next edition[3] I shall alter the experiment altogether as I now know of a very beautiful and perfectly safe way of performing it, viz. by putting the gaseous mixture into a collodion balloon and then allowing it to ascend into the sunlight, or simply exposing it in the hand to the sun's rays. The explosion is immediate & perfectly free from danger. This experiment could probably be made by the electric light & if so should form a nice illustration in the Royal Institution Theatre. In fact these collodion balloons[4] are invaluable for lecture illustrations, for instance the explosibility of O&H is beautifully shewn by filling a balloon with them & then allowing it to ascend after tying to its neck a short ignited thread impregnated with nitre. But you probably knew all about this long ago & I am merely wasting time & paper should you however require to know how to make these balloons I will give you every information with pleasure as I frequently make them myself. I was much pleased with your article in Hughes' book,[5] & also congratulate you on being honoured by a call to the Bakerian Lecture.[6]

Ever yours sincerely | E. Frankland

RI MS JT/1/F/42

---

1. *your hint about the chlorine experiment*: if given in a letter, that letter is missing.
2. *expression*: Frankland presumably intended to write 'impression'.
3. *the next edition*: The present edition being Edward Frankland's article 'Chemistry', in E. Hughes (ed.), *Reading Lessons, First Book* (London: Longman, Brown, Green & Longmans, 1855), pp. 221–45. The experiment (p. 242) concerned the action of light, especially sunlight, on a mixture of chlorine and hydrogen.
4. *collodion balloons*: Collodion is a solution of gun-cotton in ether, forming a colourless gummy liquid, which dries rapidly in the air, owing to evaporation of the ether (*OED*). On 25 March 1855 Tyndall and Debus attempted to make collodion balloons but 'thus far it has been a failure' (Journal: RI MS JT/2/13c/830).
5. *your article in Hughes' book*: Tyndall's article 'Natural Philosophy', in Hughes, *Reading Lessons, First Book,* pp. 296–338.
6. *the Bakerian Lecture*: see letter 1029, n. 1.

To Thomas Archer Hirst 21 February 1855 1047

Feb. 21s[t] 1855

My dear Tom.

I said to Francis today that I was thinking of taking a trip to Broughton on Saturday. He said to me that if I would wait till the beginning of next month he would accompany me—this I agreed to do—so that if nothing inopportune occurs you may expect to see us both on Saturday week.[1]

Your last note[2] is at Chadwell Street[3] and I am at the R.I. and have but a dim recollection of its contents. I remember a leg of mutton, and must ask you to thank Anna for her hospitable intentions towards me—Your account of Queenwood too made a pleasant impression on me which still continues—The Saturday evening meeting—provided it does not degenerate (as there is always a danger) into boiled onions and apple patties cannot but be conducive to the welfare of the establishment. I hope its meeting will be conducted with calmness and patience to the exclusion of the feeling that it is in any way irksome. The duties which your new position[4] imposes on you are far from being heavy ones—I should enact them with great content under the circumstances. —I appreciate Bevington's generous disposition and sweet kind heart—nevertheless that 10 guinea affair[5] does not suit me—It is all right on his part, but I should have been perfectly content had you left the choice of a suitable gift to himself—If however both you and he feel that it has no effect whatever upon the relationship between you the matter is allright.

Since I wrote to you[6] I accompanied Col. Wynne to Swindon to examine a boiler[7] which exploded there some time ago. According to my explanation the whole boiler formed a galvanic element, against the positive electrode (the plates of the boiler) of which oxygen was liberated, causing corrosion so deep as to eat the plate quite thin and render it unable to sustain the inward pressure.

I also gave a lecture at the London Institution[8] at the conclusion of which one of the managers told me that he had never seen the audience so delighted before—My lectures at the R.I. appear to give great satisfaction and my audience at present is a very large one—Yesterday, was about the poorest lecture[9] I delivered except the first and the cause of it was that my brains were parboiled by a hot fire in the vicinity of which I was compelled to stand.

I think I told you that I had made Professor Owen's acquaintance[10]—not by my pushing towards him but by his drawing towards me.

On Monday night I was in the midst of a large assemblage in Hyde Park Gardens[11]—Among the throng were one or two maidens shining resplendent[12] as the young May moon—The majority however were things of clay, with here and there the sparkle of a parasol to irradicate its dullness.—

Frances was about sending you some translations, but he will now probably let them stand over until he comes to see you—My paper in the Phil Trans[13] is not yet in the Phil. Trans. but I expect it soon will be—As soon as I lay hold of it I will send you a copy—Kind remembrance to Anna

as usual yours | John

RI MS JT/1/T/594

1. *Saturday week*: Tyndall and Francis went to Queenwood on Saturday, 3 March 1855 (Journal: RI MS JT/2/13c/731).
2. *Your last note*: letter missing.
3. *Chadwell Street*: Tyndall was lodging at 9 Chadwell Street, his landlady being Mrs Whitehouse.
4. *your new position*: as husband to Anna.
5. *that 10 guinea affair*: Hirst recorded in his Journal (18 February 1855: RI MS JT/2/32c/1161) that he had received 10 guineas as a wedding present from James Bevington. This gift had been delayed because, although Bevington had written a letter enclosing the cheque on 20 November 1854, he had forgotten to post it.
6. *I wrote to you*: letter missing.
7. *to examine a boiler*: see letter 1045.
8. *a lecture at the London Institution*: on 14 February; see letter 1040, n. 4.
9. *the poorest lecture*: On 20 February Tyndall delivered the sixth of his series of lectures on 'Electricity and Magnetism'. He recorded his dissatisfaction in his Journal: 'Too much work in the presence of a strong fire muddled me at length and I could not shake off the muddle during lecture. The consequence was that whatever satisfaction it may have given the others it was far from satisfactory to myself' (RI MS JT/2/13c/730).
10. *Professor Owen's acquaintance*: see letter 1044, n. 6.
11. *a large assemblage in Hyde Park Gardens*: presumably at the home of Maria Drummond, 18 Hyde Park Gardens, widow of Thomas Drummond (1797–1840), the Scottish engineer and surveyor.
12. *one or two maidens shining resplendent*: one or both of the Drummond's younger daughters—Mary ('May', b. 1837) and Emily ('Pussey', 1838–1930).
13. *My paper in the Phil Trans*: Tyndall was waiting for pre-prints of his Bakerian Lecture (see letter 1029, n. 1) which would later be published in the *Phil. Trans.*

From George Wynne 27 February 1855 1048

Board of Trade | 27 Feb. 1855

My dear Tyndall,

I send you one of our returns containing a report of mine upon a boiler explosion, it occurs at page 79,[1] I send it because it contains a short resumé of some of the explosions that have been enquired into by this office. The Gloucester explosion[2] is another instance in favour of my view that these explosions take place under the ordinary working pressure of the engine, the safety valves were furnished with ferrules which are stops that prevent the valves being screwed tight down so as to prevent them from acting.

I also send you Fairbairn's report[3] on the boiler explosion on which I have reported at page 79—the gent' high in authority[4] is I myself—I have to go to Ireland this week which will prevent me from making my report until my return

Most truly yours | Geo. Wynne

RI MS JT/1/5/1853
LT Typescript Only

---

1. *a report ... page 79*: Wynne contributed a report, dated 9 March 1854, on the explosion of an engine's boiler in the goods shed of Longsight Station on the London and North Western Railway on 9 March 1853: House of Commons Parliamentary Papers, 1854 [1747], *Reports to the Lords of the Committee of Privy Council for Trade upon the Principal Accidents which have Occurred on Railways, during the Year 1853* (London: HMSO, 1854), pp. 79–83.
2. *The Gloucester explosion*: see letter 1045, n. 3.
3. *Fairbairn's report*: The Scottish engineer William Fairbairn (1789–1874) presented his report entitled 'Experimental Researches to Determine the Strength of Locomotive Boilers which Lead to Explosion' to the 1853 meeting of the BAAS: *Brit. Assoc. Rep. 1853*, pp. 53–55.
4. *the gent' high in authority*: Fairbairn's report began: 'A difference of opinion having arisen between a gentleman in high authority and myself concerning the causes of an accident' (ibid., p. 53).

To Edward Frankland 28 February 1855 1049

28th Feb. 1855

My dear Frankland.

I am much obliged to you for your valuable hints regarding the application of collodion balloons.[1] I never saw the experiment you describe, but shall make it on the first suitable occasion—Meanwhile I shall thank you to let me know how the balloons are made.

Hughes has not yet struck off the final sheets of his first books[2]—My article is wrong in various places, the grouping of the crystals[3] was wrong—the printer having contrived to make my description of rock crystal refer to a group of alum crystals—I have altered this and other matters which will be corrected before the book is published.

When may we expect to have your paper at the Royal Society? —I must contrive to be present when it is read.[4] We have given a translation of a good paper of Bunsen[5] in the last & present number of the Phil. Mag.

I often wish to have you or some other first rate chemist near me—I would persecute you with questions on these and other matters—they interest me much—Kind regards to Mrs Frankland—

Ever yours | J Tyndall

I sent you a scrap[6] on a curious subject which I hope you will have patience to read

JRL Frankland, 191

1. *hints . . . collodion balloons*: see letter 1046, n. 4.
2. *his first books*: see volume 4, letter 1002, n. 4.
3. *the grouping of the crystals*: Tyndall does not refer to the grouping of crystals in the published volume.
4. *when it is read*: Tyndall was present when Frankland's paper, 'Researches on Organo-metallic Bodies' was read at the RS on 15 March 1855. A summary was subsequently published in *Roy. Soc. Proc.*, 7 (1854–55), pp. 303–6, and the full version in *Phil. Trans.*, 145 (1855), pp. 259–75.
5. *a good paper of Bunsen*: Robert Bunsen, 'On the Law of Absorption of Gases', *Phil. Mag.*, 9 (1855): pp. 116–30 and 181–201.
6. *a scrap*: not located.

To Thomas Archer Hirst 1 March 1855 1050

Thursday

My dear Tom

If the weather be at all decent Francis and myself will take a run down to you on Saturday.[1] With regard to your absence on Saturday night, I can only say that you will be present where I should best like to see you—and where I shall undoubtedly be present myself if I go. Instead therefore of finding your attendance at the mutual improvement Society[2] a cause of interference with our plans, I for my part shall enjoy the meeting all the more by finding myself at your side. The little meeting is one of Queenwood's greatest attractions to me. We shall sup with you afterwards if Mr Edmondson has sufficient hospitality to give us a supper—If he has not I know that Mrs Edmondson has, and I shall therefore calculate upon a supper at Queenwood for both of us.

You may, if you like, shew this to Mr Edmondson. We will, if necessary, leave you to your counsels afterwards and adjourn to Debus's room until you are ready to accompany us to Broughton—By the way you may not be able to afford lodging to two of us. So if you cannot the Dog[3] will be a pis aller[4] for one of us—I am glad that you are turning your thoughts to the subject of the magnetic field, but learn from your last letter[5] that you have now for the first time apparently experienced a feeling which I thought was known to you long ago—You are no mathematician I am bold to tell you—you know nothing of the sweets which your science can yield as long as you merely confine yourself to what others have done before you. The profit & intellectual strength derivable from learning what others have found is nothing in comparison with that which results from our own personal achievement—In the one case we look through a glass darkly,[6] in the other we see nature face to face. For this high privilege we have to pay the bounty of hard and perseverant thought, and the laziness & incapacity which induce men to shirk this sufficiently explain[s] the prevalence of intellectual mediocrity.

Ever your John

RI MS JT/1/T/895

1. *If the weather . . . on Saturday*: This letter appears to have been written on Thursday, 1 March, as (despite the poor weather) Tyndall and Francis went to Queenwood on Saturday, 3 March, and they were 'In time for the end of the Mutual Improvement Society's meeting' (Journal: RI MS JT/2/13c/321).
2. *the mutual improvement Society*: The Mutual Improvement Society at Queenwood College, which met on Saturday evenings, was a discussion group consisting of the more advanced students and some of the teachers.
3. *the Dog*: an inn, possibly the Greyhound at Broughton (*Post Office Directory of Hampshire, Wiltshire and Dorsetshire* (London: Kelly, 1855), p. 29).
4. *pis aller*: the less or least desirable alternative (*OED*).
5. *your last letter*: letter missing.
6. *through a glass darkly*: This phrase, which is taken from 1 Corinthians 13:12, means to fail to possess a firm grasp on reality.

From George Wynne 9 March 1855 1051

Whitehall | 9 March /55.

My dear Tyndall,

I have just returned from Ireland. I find Mr Gooch[1] has sent the iron stays[2] viz one that covered the crown of the fire box and one of the adjoining longitudinal stays, as they are rather heavy articles to send about perhaps you could make time to call here—to-morrow I have to attend the funeral of a little girl[3] a cousin that was very dear to me, but on Monday unless something unforeseen should occur I shall be at my office all day from 11 a.m. in case I should be out, I will leave word with Mr Hardy the office keeper[4] to shew you into my office and lay before you the specimens. Or if to-morrow would be more convenient come—in fact in every way consult what will be most convenient to yourself and I shall feel obliged.

Most truly yours | Geo. Wynne.

RI MS JT/1/5/1854
LT Typescript Only

---

1. *Mr Gooch*: Daniel Gooch (1816–89), an acolyte of Isambard Kingdom Brunel, was the chief engineer of the Great Western Railway,
2. *the iron stays*: 'The ends of the boiler are strengthened by wrought-iron double ribs, which are riveted upon them, reaching from side to side, and to which the wrought-iron stays are attached that extend the whole length of the boiler and brace the two ends to each other' (R. Armstrong, *A Rudimentary Treatise on Steam Boilers; Their Construction and Practical Management* (London: J. Weale, 1850), p. 117).
3. *the funeral of a little girl*: Possibly Catherine Beevor Wynne, aged about 15, who was buried on 10 March 1855 in the parish of St Thomas, Charlton, Greenwich (Ancestry).
4. *Mr Hardy the office keeper*: not otherwise identified.

To Hector Tyndale 10 March 1855 1052

Royal Institution | Albemarle street | London | 10th March 1855

My dear Hector

Nine long months have flown by since we saw each other[1] and no word has passed between us. This is not as it should be, for I believe neither one nor the other of us has, sincerely speaking, any better excuse than mere laziness to assign in extenuation of his omission. On the whole I think you are more to blame than I am—for in the first place you have more force, and secondly you have fewer objects on which to expend it. Therefore you ought to have written, and I blame, censure, and condemn you for not having done so, whereas my present act evokes a corresponding amount of self gratulation on finding myself your pioneer in the path of duty. But speaking seriously the rugged fact which is the cause of my writing to you just now is my inability to do any thing which draws heavily upon my brain—I have been smitten down during the last week by influenza or something of the kind. My bodily fortress was already in a state of dilapidation when I was thus assailed, so that the shock has made some impression upon me. I have been compelled to dismiss science from my head for the last two days, and when I do this thoughts of friends enter, and I become more humanized, so that a fit of illness has in some measure its compensations. Under the circumstances I thought I might occupy myself pleasantly for an hour in writing to you, and this is the simple genesis of my present letter.

Oh this miserable pulpy brain of mine! would that it were made of some tougher material—even the simple effort of writing to you sets it simmering, and I dare say I shall have to desist before my task is complete. You left me early in August—I continued to work until the 20th of September when I went to Liverpool to attend the meeting of the British Association.[2] It was a pleasant & successful meeting—one night it fell to my lot to explain Foucault's experiments proving the rotation of the earth,[3] and I nearly cracked my voice in my endeavour to fill the vast hall where the multitude was assembled. From Liverpool Francis and myself dived into Wales[4] and had a most glorious trip of about 10 days duration. It is a beautiful country, and as gratifying to the soul as the more stupendous scenes of Switzerland. There is such a harmony in the mountain grouping, the colours are so beautiful, and then the dinners of trout, plucked bleeding from their nature streams give a completeness to the country and enables it to appeal to the sum total of man animal and angelical. After my Welsh trip I took a rush over to Ireland, and my last impression of the country was that the people there were extremely dirty.

On my return I attacked my work once more and got a long investigation safely finished. This I handed in to the Royal Society in November—it was afterwards chosen by the Council for the so called Bakerian Lecture,[5] which is delivered annually before the Fellows. I send you an abstract of the Lecture[6] and when the paper is presented in full, if the postage offer no great barrier, I will send you a copy.

During the Autumn I had another cousin of mine from Melbourne with me here—not a Tyndall, but a Payne;[7]—he went out to Melbourne 15 years ago and has made an immense fortune there. He is about the age of ourselves and came over to this country, taking Palestine in his route, to seek out a wife. He spent a good while in Dublin; the rumour of his wealth having been spread abroad, and his object being pretty well known, he was literally besieged—His strong common sense however enabled him to see all the little artifices that were laid for him, and he finally left Dublin & came over to London. He was just as unsuccessful here, so he went-back to Ireland in January and I have not heard from him since. He is very different from you in disposition, and although so successful as a man of the world possesses nothing like your force of intellect. He is however steady, sincere & punctual and enjoys a large portion of practical common sense.

A little discussion has recently arisen between Faraday & myself regarding a very interesting point in magnetism.[8] Many people no doubt thought it rash, and perhaps wrong in me to differ from him—But Faraday's own life is a practical illustration of independent thought, and in expressing an honest, and I believe well founded dissent from him I do no more than follow the example which he has set. The man is too great and too good to allow this to interfere with the cordial personal relationship which subsists between us; but such a man is always surrounded by lesser men, and if a shade should ever cross his feelings towards me (which I do not in the least anticipate) it will proceed from them. He supposes a universal magnetic medium to exist in space, and entertains peculiar notions regarding the nature of the magnetic force, to which I cannot subscribe—Some of my reasons I have given in the last number of the Philosophical Magazine[9]—Many, & among the number the Astronomer Royal,[10] think them sound, and if you should chance to see this magazine you will I doubt not be able to judge of them yourself as well as any body else.

Since your departure I have heard frequently from Dr and Mrs Coxe. He sent me once a newspaper containing something concerning 'types'[11]—with the request to send it to you—I did so and hope you have received it safely. He wrote to me sometime ago saying that he thought the state of his health would soon induce the gentleman who now fills the chair of Natural Philosophy at Edinburgh[12] to resign his post, and asking me whether he might name me as one willing to accept the chair supposing it to be offered to me—My election to it he said would be next to a certainty—The post would be worth from 800 to £1000 a year, and I should have 6 months for private researches. I replied saying that I would not put myself forward as a candidate but if the post were offered to me I did not see any insuperable obstacle in the way of my accepting it—Thus the matter rests; I think however the contemplated resignation is not likely to take place soon—not perhaps for years.

I took a run down to Hampshire last Saturday accompanied by Francis and remained there till Monday morning—That huge young fellow with the black barb[13] whom I took you to see, has recently got married to a sister of John Martin's, who was sent out to Sidney with Smith Obrien, but who is now in Paris.[14] I have had many acquaintances and friends during my brief career, but they have all slipped away from me, all slid into matrimony and left me alone. Under ordinary circumstances, when I can pursue my speculations and experiments without any miserable physiological drawback, my life is so full of satisfaction that the thought of a wife, or the need of a wife never reveals itself to me; but when smitten with debility, when I am forced to sit by the fire and to think as little as possible then I feel that 'a rare and radiant maiden'[15] would be of inestimable value to me. Would that I could charm one from the clouds to keep me company at these special seasons always provided that I might send her back to the clouds as soon as her services were no longer necessary. —Dont shew this heretical stuff to any of your lady friends.

I have been reading this evening, and from time to time during the week, with infinite disgust, the remarks of certain patrons of the English press upon the death of the Emperor Nicholas.[16] Surely there is something very ignoble in the joyful yelping of these curs. He was a noble foe, and for my part I could not contemplate his extinction without a feeling of solemnity almost approaching to sorrow. Would to God that we had a few men of his calibre in this country at present. As far as I can see he was a real ruler of men and no mockery. Instead of meanly rejoicing at his death our prayer and effort ought to be to find men so fit to cope with him as to neutralise him of the fair dignified opposition of force to force. These newspapers I must say sometimes sicken me—there one lies before me in which the writers talk about literature, the dignity of literature, and rejoice in the fact that government has given some literary man a post—evidently making themselves not only members but exponents of this guild of literature, and yet the fellows who so talk are fellows whom God intended to be tailors and cheesemen, and into whose dull souls his celestial light has never shone. We have scarcely a paper worth reading except the Times[17]—It is changeful and inconsistent people say, but I find blood and life in its sentences—Its very errors offend the intellectual nostrils less than that flat putrescence which is the characteristic of more than half what emanates from the English newspaper press.

It is drawing towards bed-time and simultaneously towards the close of my sheet of paper. We have had an awfully severe winter here—last week the snow & frost had cleared away and the blood of spring appeared to be tingling in the world's cheeks—but winter has come back again, coy and stern and today there is a pretty deep layer of snow upon the ground. Now I think for a man with a broken brain I have managed to fill my sheet pretty well; I need not of course ask you to write to me, for a cohort of voices will doubtless rise within you on the perusal of this, crying aloud 'you must write to John'. I expect therefore a speedy outrush from your heart, for I know you have got a heart, and that there are forces within it which can act when called upon. Tell me all that has happened to you since our separation—present my regards to your wife, to your sister, to your brother Sharon,[18] to all my unseen friends—Your picture is behind me as I write—and I have once heard you called 'a handsome fellow'—This you are not, so dont suffer yourself to be elated. And now my dear Hector I stretch my thoughts across the thousands of miles that separate us; squeeze you by the hand—give one brief warm hug and bid you goodbye.

Affectionately yours | John Tyndall

RI MS JT/1/T/1442

1. *we saw each other*: Hector Tyndale had left London on or about 5 August 1854, a little over seven months before this letter was written.
2. *the meeting of the British Association*: The twenty-fourth meeting of the BAAS had been held in Liverpool between 22 and 27 September 1854.
3. *Foucault's experiments . . . earth*: In 1851 the French physicist Léon Foucault (1819–68) had performed an experiment that demonstrated the earth's rotation. Using a long and heavy pendulum suspended from the roof of the Panthéon in Paris, Foucault showed that the pendulum's plane of motion rotated in a clockwise direction in accord with the earth's rotation.
4. *dived into Wales*: Tyndall and Francis toured Wales from 27 September to 5 October 1854.
5. *Bakerian Lecture*: see letter 1029, n. 1.
6. *an abstract of the Lecture*: possibly one of the newspaper reports of his lecture.
7. *Payne*: Thomas Payne.
8. *discussion . . . in magnetism*: Letter 1042 was Tyndall's response to the views Faraday had expressed in 'On Some Points of Magnetic Philosophy', *Phil. Mag.*, 9 (1855), pp. 81–113.
9. *the last number of the Philosophical Magazine*: letter 1042.
10. *the Astronomer Royal*: According to Tyndall's Journal entry for 9 March, Faraday had shown him a letter that Barlow had received from George Biddell Airy (the Astronomer Royal) in which, 'if I recollect right, [Airy] expressed his approval of the strictness and mathematical conception of' Tyndall's letter to Faraday—letter 1042—published in the *Phil. Mag.* (RI MS JT/2/13c/681).
11. *newspaper containing something concerning 'types'*: see letter 1099, n. 3.
12. *He wrote to me . . . chair of Natural Philosophy at Edinburgh*: 'I suppose you know that James [David] Forbes has been for some time in very delicate health, and it seems to me very probable that before long he will feel himself obliged to resign his Chair. Should this happen, what would you think of being a Candidate?' (J. Coxe to Tyndall, 20 November 1854: volume 4, letter 1004). See also volume 4, letters 1006, 1008, 1012, 1013 and 1014. Forbes retained the chair until 1859.
13. *That huge young fellow with the black barb*: Thomas Archer Hirst.
14. *John Martin's, . . . now in Paris*: see letter 1027, n. 8. Together with John Martin and others accused of the 1848 Young Irelander Rebellion, William Smith O'Brien (1803–64) was transported to Tasmania.
15. *a rare and radiant maiden*: quotation from E. A. Poe, "The Raven," ii.5; see also letter 1063, n. 3; 1099, n. 8; 1118, n. 15; 1140, n. 28.
16. *the death of the Emperor Nicholas*: Nicholas I of Russia (b. 1796) had died on 2 March 1855. 'No single event could have happened in Europe of such momentous importance at the present time to the whole family of civilized nations; no event could have occurred more startling between the pride and power of a ruler who sent forth but yesterday his myriads to battle [against the English and French in Crimea . . .], but who is now less than

the least of his serfs, and lower than the dust of that empire that was lately his own' (*Times,* 3 March 1855, p. 8).

17. *the Times*: The *Times,* which began publication as a daily newspaper in 1785, was edited by John Delane. In 1855 it had a circulation of about 60,000 and was priced at 5d. Generally aligned with free trade and the reformist middle classes, it had supported the repeal of the Corn Laws in 1846, but retained a significant degree of political independence. It was widely respected, especially for its foreign reportage.
18. *your wife, to your sister, to your brother Sharon*: In 1842 Hector had married Julia Nowlen (1823–97), the daughter of Carlow immigrants; his sister was Elisabeth (dates unknown); his brother Sharon (1816–71) worked as a civil engineer and later served as the Secretary of State of Illinois.

From William Thomson 12 March 1855[1] 1053

2 College, Glasgow, | March 12, 1855.

MY DEAR SIR,

Allow me to thank you for the abstract of your lecture on magnetism,[2] and the copy of your letter to Mr. Faraday,[3] which I have recently received from you,[4] and have read with much interest. I am still strongly disposed to believe in the magnetic character of the medium occupying space,[5] and I am not sure but that your last argument in favour of the reverse bodily polarity of diamagnetics[6] may be turned to support the theory of universally direct polarity. There is no doubt but that the medium occupying interplanetary space,[7] and the best approximations to vacuum which we can make, have perfectly decided mechanical qualities, and among others, that of being able to transmit mechanical energy in enormous quantities (a platinum wire, for instance, kept incandescent by a galvanic current in the receiver of an air-pump, emits to the glass and external bodies the whole mechanical value of the energy of current spent in overcoming its galvanic resistance). Some of these properties differ but little from those of air or oxygen at an ordinary barometric pressure. Why not, then, the magnetic property? (of which we know so little that we have no right to pronounce a negative). Displace the interplanetary medium by oxygen,[8] and you have a slight increase of magnetic polarity in the locality with a drawing in of the lines of force. Displace it with a piece of bismuth or a piece of wood,[9] and a slight decrease of magnetic polarity through the locality takes place, accompanied by a pushing out of the lines of force. A state of strain by compression may enhance, in the direction of the strain, that quality of the substance by which it lessens the magnetizability of the space from which it displaces air or 'æther';[10] just as a similar state may enhance, in the direction of compression, the augmenting power of a paramagnetic substance.

By the bye, a long time ago (rather more than a year after the Edinburgh meeting of the British Association)[11] I repeated with much pleasure some of your compression experiments, and found a piece of fresh bread instantly affected by pressure, so as always to turn the compressed line perpendicular to the lines of force, to whatever form the fragment was reduced. A very slight squeeze between the fingers was quite enough to produce this property, or again to alter it so as to make a new line of compression set equatorially. I repeated it a few days ago with the same results, and got a ball of bismuth, too, to act similarly. I remember formerly finding the bread *attracted* as a whole, instead of being repelled, as I expected from your results. I suppose, however, this must have resulted from some ferruginous impurities,[12] which it may readily have got either in the course of the experiments with it, or in the baking. I mean to try this again.[13]

I do not quite admit the argument you draw from your compression experiments[14] regarding the effect of contiguity of particles, because in fact we know nothing of the actual state of the molecules of a strained solid. You have made out a most interesting fact regarding their magnetic bearings; but experiments are neither wanted, nor can be made, to show any sensible effect whatever of the mutual influence of a row of small pieces of bismuth placed near one another, or touching one another. It is perfectly easy to demonstrate that it *must* be such as to impair the 'diamagnetization' of each piece when the line of the row is parallel to the lines of force, and to enhance it when that line is perpendicular to the lines of force, but in each case to so infinitesimally minute a degree, as to be wholly inappreciable to the most refined tests that have ever been applied. For let the lines of force be parallel to the line shown in the figure,

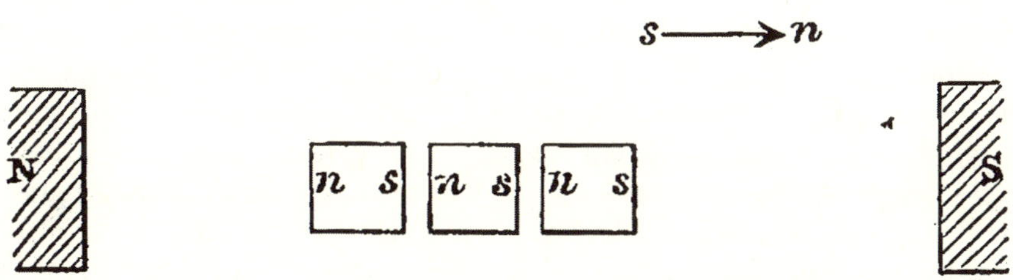

and act on a steel needle in the manner there represented. Then, whatever hypothesis be true for diamagnetism, there is not a doubt but that each piece is acted on, and consequently reacts, precisely as a piece of steel very feeble magnetized, with its magnetic axis reverse to that of a steel needle free to turn, substituted for it, would do. Each piece of bismuth therefore acts as a little magnet, having its polarity as marked in the diagram, would do. Hence the magnetizing force by which the middle fragment is influenced is less than if the two others were away (this being such a force as would be produced by a north pole on the right-hand side of the diagram, and a south pole on the left). It is easily seen, similarly, that if the line joining the centres be perpendicular to the lines of force, the magnetizing force on the space occupied by the middle fragment is increased. Corresponding assertions are true for the terminal fragments, although the disturbing effect will be less on them in each case than in the middle one. Hence the diamagnetization of each will be enfeebled in the former case and

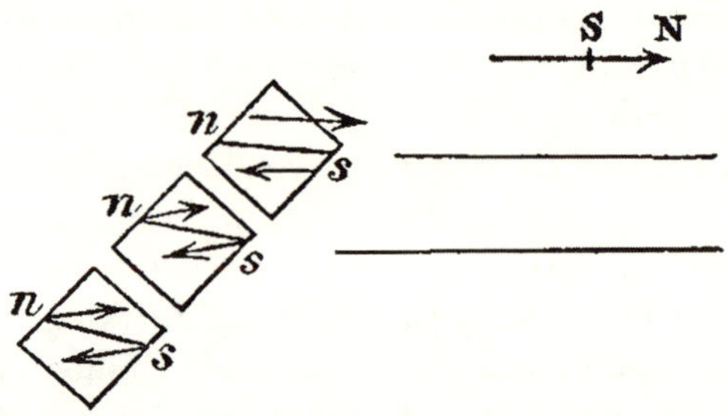

enhanced in the latter, by the presence of the others. It follows, according to the principles of super-position of magnetizations,[15] that if the line of the row be placed obliquely across the lines of force, the magnetic axis of each particle, instead of being exactly parallel to the lines of force, will be a little inclined to them, in the angle between their direction and the direction transverse to the bar. The magnets causing the force of the field must act on the little diamagnets, each with its axis thus rendered somewhat oblique, so as to produce on it a statical couple (as shown by the arrow heads), and the resultant of the couples thus acting on the fragments will, when all these are placed on a frame, or rigidly connected, tend to turn the whole mass in such a direction as to place the length of the bar along the lines of force. Still, I repeat, this action, although demonstrated with as much certainty as the parallelogram of forces, is so excessively feeble as to be absolutely inappreciable. A fragment of bismuth, of any shape whatever, held in any position whatever in any kind of magnetic field, uniform or varying most intensely, only exhibits the resultant action of couples on all its small parts if crystalline, and of forces acting always according to Faraday's law[16] on them if the field in which it is placed be non-uniform. Some phænomena that have been observed are to be explained by the resultant of forces from places of stronger to places of weaker intensity in the field, others by the resultant of couples depending on crystalline structure, and others by the resultant of such forces and couples coexisting; and none observed depend at all on any other cause.

I gave a very brief summary of these views (which I had explained somewhat fully and illustrated by experiments on paramagnetics of sufficient inductive capacity to manifest the effects of mutual influence, at the meeting at Belfast) as an abstract of my communication, for publication in the Report of the Belfast meeting of the British Association,[17] where you may see them stated, I hope intelligibly. The experiments on the paramagnetics are very easy, and certainly exhibit some very curious phænomena, illustrative of the resultant effects due to the attractions experienced by the parts in virtue of a variation of the intensity of the field, and to the couples they experience when their axes are diverted from parallelism to the lines of force by mutual influence of the magnetized parts.

I had no intention of entering on this long disquisition when I commenced, but merely wished to try and briefly point out, that the assertions I have made regarding mutual influence are demonstrable in every case without special experiment, are confirmed amply by experiment for paramagnetics, and are absolutely incontrovertible, as well as incapable of verification, by experiment or observation on diamagnetics.

Believe me, yours very truly, | WILLIAM THOMSON.

Phil. Mag., 9 (1855), pp. 290–3[18]

---

1. *12 March 1855*: This date may be incorrect as Tyndall reported in his Journal for 9 March that he had received 'two papers [from Thomson], one for myself and the other for Faraday'. Tyndall then added a comment about Thomson's style: 'There is a certain strain and haste in Thomson's writing which wearies me amazingly. He writes as if his object was to cover paper and make a noise' (RI MS JT/2/13c/733).
2. *the abstract of your lecture on magnetism*: Probably the abstract of his Bakerian lecture as sent to Hector Tyndale, letter 1052, n. 6.
3. *your letter to Mr. Faraday*: letter 1042.
4. *received from you*: letter (containing the above enclosures) missing.
5. *the magnetic character of the medium occupying space*: see n. 10, below.
6. *your last argument . . . diamagnetics*: see letter 1042.
7. *the medium occupying interplanetary space*: An ethereal medium was required for the transmission of light across interplanetary space; light being a vibration of that ether.
8. *oxygen*: a paramagnetic substance.
9. *piece of bismuth or a piece of wood*: both diamagnetic substances.
10. *air or 'æther'*: '[B]etween 1851 and 1854 Thomson developed a firm attachment to his air-ether theory as forming a "continuous material communication" throughout space' (C. Smith and M. N. Wise, *Energy and Empire: A Biographical Study of Lord Kelvin,* (Cambridge: Cambridge University Press, 1989), p. 400).
11. *Edinburgh meeting of the British Association*: in 1850.

12. *ferruginous impurities*: contaminated with iron rust.
13. Footnote in published text: 'Prof. Thomson's supposition is correct; pure bread is *repelled* by a magnetic pole. I may remark that I am at present engaged in the further examination of the influence of compression, and have already obtained numerous instructive results.—J.T.'
14. *the argument you draw from your compression experiments*: Probably the experiments discussed in Tyndall's RI lecture 'On the Influence of Material Aggregation upon the Manifestations of Force', *Roy. Inst. Proc.*, 1 (1851–54), pp. 254–59.
15. *principles of super-position of magnetizations*: see W. Thomson, 'On the Theory of Magnetic Induction in Crystalline and Non-crystalline Substances', *Phil. Mag.*, 1 (1851), pp. 177–86, on pp. 180–81.
16. *Faraday's law*: 'The *law* of action appears to be, that, *the line or axis* of MAGNECRYSTALLIC *force* (being the resultant of the action of all the molecules), *tends to place itself parallel, or as a tangent, to the magnetic curve or line of magnetic force, passing through the place where the crystal is situated*' (*ERE,* vol. 3, p. 90).
17. *Report of the Belfast meeting of the British Association*: W. Thomson, 'On the Equilibrium of Elongated Masses of Ferromagnetic Substance in Uniform and Varied Fields of Force', *Brit. Assoc. Rep. 1852,* pp. 18–20.
18. This letter was published in the April 1855 issue of the *Phil. Mag.* under the title 'Observations on the 'Magnetic Medium' and on the Effects of Compression'. It was preceded by a note by Tyndall, dated 24 March 1855: 'The following letter was received a few days ago. It was not written for publication, but the subject to which it refers being of general interest at present, I ventured to suggest to Professor Thomson the desirableness of having the letter printed. This he at once agreed to. With the exception of a paragraph relating to matters of a purely private nature, the letter appears as I received it.'

From Michael Faraday 14 March 1855 1054

Royal Institution, | March 14, 1855.

MY DEAR TYNDALL,

IN relation to your letter of last month,[1] I write, not for the purpose of giving what might be taken as an answer, but to say that it seems to me expedient and proper to wait and allow the thoughts that my papers may raise, to be considered and judged of at their leisure by those who are inclined to review and advance the subject. Perhaps, after a respectful interval, I may be induced to put forth such explanations, acknowledgements, or conclusions, as the state of the subject may then seem to render necessary or useful.

In the mean time, the more we can enlarge the number of anomalous facts and consequences the better it will be for the subject; for they can only remain anomalous to us whilst we continue in error. I may say, however, that the idea you suggest presents no difficulty to me; for having on former occasions (Exp. Res. 2501)[2] had to consider the magnecrystallic phænomena presented by the same body in different media, and having found the magnecrystallic difference unchanged in the media, I have no difficulty in conceiving that a body (as bismuth), which in the amorphous state is of the same magnetic character as the medium around it, shall, when employed as a crystal, be paramagnetic in one direction and diamagnetic in another (3157).[3] What happens in a medium may, according to my knowledge of the facts, happen in space; and is in full accordance with Thomson's clear paper on the theory of magnetic induction in crystalline bodies.[4]

In respect of the effects of pressure, to which you refer in your letter, we cannot easily draw conclusions on either side until we know better what pressure does. I am not aware whether you consider that pressure on bismuth, whilst it makes the metal more diamagnetic in one direction than another, also makes it more diamagnetic as a whole than before; or whether you suppose it *less* diamagnetic in the transverse direction of the pressure than at first. Gmelin says, on the authority of Marchand and Scheerer (vol. iv. p. 428),[5] that the density of bismuth is diminished as pressure upon it is increased, and extraordinary as the fact seems, gives densities of the following degree for increasing pressures, 9.783, 9.779, 9.655, 9.556; a change in texture at the same time occurring. If the statement be true, then the line of pressure in your beautiful experiments may be the line of *least density* or of *least approximation,* though I hardly know how to think so; still it becomes difficult for us to draw reasons from the constitution of a compressed body, until we know what happens during the compression, although no difficulty arises in considering it, after compression in one direction, like to a magnecrystallic substance.

You are aware (and I hope others will remember) that I give the lines of force[6] only as *representations* of the magnetic power, and do not profess to say to what physical idea they may hereafter point, or into what they will resolve themselves. Advancing no principle, I say, that the hypothetical fundamental ideas already advanced, when taken in relation to the body of facts now known, are self-contradictory and inapplicable.[7] The following points, namely,—that the *direction* and *polarity* of lines of magnetic force are always shown truly by the electric current induced in metal moving within their influence;—that the dualities of electricity and magnetism are always respectively and essentially related;—that the dualities of an isolated magnet are not related back in straight lines through the magnet;—are to my mind not hypothetical in character, but easily proveable by experiment:—and they, with the considerations arising from the principle of the conservation of force,[8] seem to me to be left unexplained by, and in opposition to, the usual hypotheses. No difference arises about the laws of magnetic action and their mathematical development; and that, simply because they are as yet applied only partially, and thus far are in accordance with *all* the views taken, including mine. When the attempt is made to apply them so as to include at once *paramagnetic, diamagnetic,* and *electro-magnetic* phænomena, and at the same time to deduce them from *one* hypothetical cause, then they may become so large and yet precise as to enable us to distinguish between true and false assumptions. On my part I endeavour not to assume anything, but only to draw such conclusions from the assumptions already made, and the phænomena now discovered, as seem subject to experiment and tangible by facts.

Some persons may feel surprised that I dwell upon points which are perfectly and mathematically explained by the hypothesis of two magnetic fluids, as, for instance, places of little or no action (3341 &c.[9]). My reason is, that being satisfied by the phænomena of diamagnetism, &c. that that hypothesis cannot be true, all these and such like phænomena acquire a new character and a high importance which they had not before, and amongst other philosophical uses, point most emphatically to the essential relation of the dualities and their equivalency in power. They do not contradict the old hypothesis when that is partially applied, but they are not the less strong and striking as evidence in favour of the view of lines of force.

I am, my dear Tyndall, | Yours very faithfully, | M. FARADAY.

Phil. Mag., 9 (1855), pp. 253–5[10]

---

1. Footnote in published text: 'Phil. Mag. 1855, vol. ix. p. 205.' which is reproduced here as letter 1042.
2. *Exp. Res. 2501*: paragraph 2501 of *ERE,* vol. 3, p. 97.
3. *3157*: paragraph 3157 of *ERE,* vol. 3, pp. 361–62.
4. Footnote in published text: 'Phil. Mag. 1851, vol. i. p. 177.' William Thomson, 'On the Theory of Magnetic Induction in Crystalline and Non-Crystalline Substances', *Phil. Mag.,* 1 (1851), pp. 177–86.
5. *Gmelin says, on the authority of Marchand and Scheerer (vol. iv. p. 428)*: Leopold Gmelin, trans. H. Watts, *Hand-book of Chemistry,* 18 vols (London: Cavendish Society, 1848–72), vol. 4, p. 428. Leopold Gmelin (1788–1853), who had been professor of chemistry at the University of Heidelberg, cited R. F. Marchand and T. Scheerer, 'Untersuchungen über einige Gegenstände aus dem Gebiete der Atomentheorie', *Journal für Praktische Chemie,* 27 (1842), pp. 129–210, on p. 209. See also note to *ERE,* vol. 3, p. v.
6. Footnote in published text: 'It is nearly twenty-four years since I first called attention to these lines; Exp. Res. 114, note.' This note was appended to Faraday's paper read at the RS on 24 November 1831 (*ERE,* vol. 1, p. 32).
7. *the hypothetical . . . are self-contradictory and inapplicable*: Faraday had discussed the competing hypotheses in his 'On Some Points of Magnetic Philosophy' (*Phil. Mag.*, 9 (1855), pp. 81–113) and found them inadequate.

8. *the principle of the conservation of force*: Faraday argued that force is conserved and could only be created or destroyed by God. This principle underpinned Faraday's science.
9. *3341 &c.*: paragraphs 3341–50 of Faraday, *ERE*, vol. 3, pp. 553–58.
10. Published in *Phil. Mag.* for April 1855 with the title 'Magnetic Remarks'.

From Thomas Archer Hirst 14 March 1855 1055

Broughton | March 14th/55

My dear John,

I was sorry to hear from Wright that you were suffering a great deal from your headaches on Saturday last.[1] I expected a letter from yourself before this to tell me for I remember that when you can no longer work you can find employment and occupation in writing to me. I wish you would do so for your silence makes me anxious when I know you are not well. I wish you would come down and see us again and stop a day or longer so that the good effects of your visit may be more durable.

Since you left nothing new has taken place here, except what you already know from Wright. We expect a new master (Foreman)[2] daily. I am not yet quite well but I can do my work. In my spare time I work away at the magnetic problem which makes a little but only slow progress. Francis has not yet sent me Reiss' paper.[3] In your lectures on Heat last session[4] I believe you froze water in red hot crucibles.[5] I wish you would describe to me how you did this.

I often think of your suffering there alone from your headaches and wish I could be near you. Rest is absolutely necessary for you and yet for you it is difficult to procure, for it is wearisome when you are alone. I wish I was nearer you. Write to me immediately if you please and tell me exactly how you are. With love from Anna and myself, believe me

Yours affectionately | T A Hirst

RI MS JT/1/H/214

1. *your headaches on Saturday last*: In almost every Journal entry between 6 and 17 March Tyndall complained about his headaches (RI MS JT/2/13c/732–36).
2. *Foreman*: not identified.
3. *Reiss' paper*: P. Riess, 'Ueber die Wirkung nicht-leitender Körper bei der electrischen Influenz', *Poggend. Annal.*, 92 (1854), pp. 337–54; see letter 1068, n. 5.
4. *In your lectures on Heat last session*: Tyndall delivered a series of twelve weekly lectures 'On Heat' at the RI, beginning at 3pm on Tuesday, 17 January 1854.
5. *you froze water in red hot crucibles*: Tyndall reported a similar experiment in a later series

of published lectures on heat at the RI: see, J. Tyndall, *Heat Considered as a Mode of Motion: Being a Course of Twelve Lectures Delivered at the Royal Institution of Great Britain in the Season of 1862* (London: Longman, Green, Longman, Roberts & Green, 1863), p. 165.

To Thomas Archer Hirst 15 March 1855[1] 1056

My dear Tom,

It is some time since I have been so sorely smitten as lately[2]—I am now gradually getting over it & trust when I next write to you to be able to report my complete restoration. Kind remembrance to Anna.

as ever yours | John

My head is weary just at the moment & therefore I write but little.

RI MS JT/1/T/598

1. *15 March 1855*: The letter includes the date 'March 16th 1855', which is not in Tyndall's hand. However, Tyndall left London on the afternoon of the 16th and met Hirst at Broughton. March 15 seems more likely date as he recorded the following in his Journal: 'Head heavy. [. . .] My head was so bad that I felt unable to enter upon the discussion [at the RS] of the subject [of magnetism]' (RI MS JT/2/13c/736).
2. *so sorely smitten as lately*: see letter 1055, n. 1.

To William Thomson 16 March 1855 1057

Royal Institution | 16th Mar. 1855.

My dear Sir,

A letter addressed as follows to Mr. Weber will be sure to find him:— Herrn Prof. Mr. Weber | Göttingen | Hanover. Germany.

The address of Dr Bence Jones | 30 Grosvenor Street | Grosvenor Square. London.

I am much obliged to you for your long and interesting letter,[1] and have also to thank you for your memoir on Thermo-Electric currents.[2] I handed his copy to Mr Faraday on the day it reached me.

Far be it for me to deny dogmatically the existence of a magnetic medium[3]—I long for more light on the subject, and I have no doubt that the agitation of the question will greatly contribute to advance our knowledge of magnetism.

At the present time I regret to say I am obliged to banish the matter as far as possible from my thoughts, as too much pondering upon it and kindred subjects has thrown my brain into a state of confusion from which I must escape as soon as possible. Prof. Williamson of University Coll. has sent a paper to the Royal Society upon the subject of a medium.[4] I heard it read last night, but did not feel able to enter upon the discussion.[5] You will find that pure lead is diamagnetic, but it must be carefully handled. The fingers if not very clean contaminate the lead very soon. I always use an ivory blade to cut lead wax and such substances for diamagnetic experiments.

I hope to be able to send you an account of a great number of facts in connection with the subject of compression as soon as my lectures here are concluded.[6] There is still a good deal to be done here and I have purposely avoided mingling the matter too much with my paper on polarity[7] which is now in the hands of the Royal Society—The desire to keep this latter question as much as possible to itself, and not the want of material, has prevented me from entering more fully upon the subject of compression.

A short communication on the 'medium' from Mr Faraday will appear in the next number of the Phil. Mag.[8]

With best wishes | Believe me | Sincerely yours | John Tyndall

RI MS JT/1/5/1538–9[9]

---

1. *your long and interesting letter*: letter 1053.
2. *your memoir on Thermo-Electric currents*: probably W. Thomson, 'Account of Researches in Thermo-Electricity', *Roy. Soc. Proc.*, 7 (1854–55), pp. 49–58.
3. *the existence of a magnetic medium*: which Thomson had supported in letter 1053.
4. *Prof. Williamson . . . a medium*: see letter 1057, n. 1.
5. *the discussion*: After hearing Williamson's paper at the RS on 15 March, Tyndall wrote in his Journal: 'My head was so bad that I felt unable to enter upon the discussion of the subject' (RI MS JT/2/13c/736).
6. *my lectures here are concluded*: On 27 March 1855 Tyndall delivered the last RI lecture in the series 'Electricity and Magnetism'.
7. *my paper on polarity*: Tyndall's Bakerian Lecture (see letter 1029, n. 1) had yet to be refereed before publication in the *Phil. Trans.* Unknown to Tyndall, Thomson was one of his referees; see letter 1076, n. 4.
8. *A short communication . . . Phil. Mag.*: letter 1054.
9. Transcript Only; Part of original letter at RI MS JT/1/T/1532 and part of the original manuscript exists (RI MS JT/1/T/1532) and has been used for transcription of 'an ivory blade' to end of letter.

To Thomas Archer Hirst [c.20] March 1855 1058

My dear Tom.

The following thought has just occurred to me:—

The 'magnetic medium' is a subject which excites real interest at the present time. A sound remark upon the subject will command attention and bring the maker of the remark into notice.

Williamson you know has made some statements upon my letter to Mr Faraday.[1] I send you his paper—it is ingenious, and apparently convincing. In my opinion however it is altogether wide of the mark. The paragraph marked with the cross[2] and which contains the marrow of the whole paper, will not I think bear strict analysis. If you are of the same opinion then I thought it might be of service to you to write a short paper upon the subject[3] which I will be glad to bring before the Royal Society.

For my part I intend to avoid skirmishing at present. I shall deal comprehensively with the subject at some future day.

affectionately yours | John

RI MS JT/1/T/1007

1. *Williamson . . . my letter to Mr Faraday*: A. W. Williamson, 'Note on the Magnetic Medium', *Roy. Soc. Proc.,* 6 (1856), pp. 306–8, which was read at the RS on 15 March 1855. Having studied in Heidelberg, Giessen and Paris, Alexander William Williamson (1824–1904) was appointed in 1849 to the chair of practical chemistry at University College, London.
2. *The paragraph marked with the cross*: Probably the paragraph (in Williamson, 'Note', p. 307) in which Williamson suggested that 'all repulsion by the magnet might be owing to the attraction exercised on the medium being stronger than that on the body repelled [. . .] And as phenomena of diamagnetism are observed in a so-called vacuum, it was thought that some "magnetic medium" might be present there.'
3. *a short paper upon the subject*: This became Hirst's paper, 'On the Existence of a Magnetic Medium', which Tyndall read at the RS on 21 June 1855 and was subsequently published in *Roy. Soc. Proc.,* 7 (1854–55), pp. 448–54.

To Margaret Ginty[1] 21 March 1855 1059

9 Chadwell Street | Myddleton Square | Pentonville | London |
21st. March 1855.

My dear Mrs Ginty,

In the midst of my present occupations, which usually keep my brain busy, there are times when I think of the long, kind, and pleasant letters[2] which you used to send to me long ago, and often my head has been unable to do its work, and I have been obliged to sit at home and feed myself on memories, I have experienced a feeling of regret that a correspondence which I once so highly esteemed should have so effectually ceased. The cause which prompts me to write to you at present is that William[3] has been kind enough to send me a long letter,[4] giving me an account of his doings in Brazil,[5] and informing me that you intend to start to Rio on the 9th of April.[6] He wishes, if possible, that I should see you and your children[7] safely on board, and I can only say that if it be at all possible I shall feel a true pleasure <*words missing*> 9th, and will do all in my power to help you.

Believe me | most sincerely yours | John Tyndall.

RI MS JT/1/11/3642
LT Typescript Only

1. *Margaret Ginty*: wife of William Ginty; see n. 3, below and volume 2, letter 0335.
2. *the long, kind, and pleasant letters*: including volume 3, letter 0517.
3. *William*: William Ginty (1821–66) had worked closely with Tyndall on the Ordnance Survey, both in Ireland and later in England. Like Tyndall, he was dismissed in November 1843 for signing a letter complaining about the ineffective management of the Survey. He then worked as a surveyor for the Tithe Commission, before being appointed engineer and manager of the Hydro-Carbon Gas Company in Manchester. In 1854 he was appointed general manager of the Imperial Gas Works in Rio de Janeiro and moved to Brazil. To mark his contributions to public works he was knighted by the Emperor of Brazil in 1860.
4. *a long letter*: letter missing.
5. *his doings in Brazil*: see n. 3, above.
6. *9th of April*: Margaret Ginty and her children sailed on *The Great Western* from Southampton on 9 April 1855.
7. *your children*: Tyndall and the Hirsts travelled to Southampton on 9 April to bid farewell to Margaret Ginty and her five young children (Hirst's Journal: RI MS JT/2/32b/1156). Four of the children were named Kathleen (b. c. 1848), William Roberts (b. 1849), Fanny (b. 1850) and George (b. 1851) (Ancestry).

To William Thomson 21 March 1855 1060

21s[t] March 1855

My dear Sir,

I have read your letter[1] a second time this morning. It appears to me to supply a want in the writings you have hitherto published on the subject of molecular induction in paramagnetic and diamagnetic bodies, on this account if you thought well of it I should be glad to have the portion of the letter which refers to this subject (or the whole letter if you prefer it) published in the next number of the Philosophical Magazine.[2]

Would you kindly let me know your mind[3] on this point as soon as possible, as the month is drawing to a close.

Believe me, | most truly yours | John Tyndall.

RI MS JT/1/5/1537
Transcript Only

---

1. *your letter*: letter 1053.
2. *next number of the Philosophical Magazine*: The April 1855 number of *Phil. Mag.* included Thomson's letter of 12 March (letter 1053), which was published together with a short note by Tyndall.
3. *let me know your mind*: letter 1061.

From William Thomson 22 March 1855 1061

2 College, Glasgow | March 22, 1855

My dear Sir

I have no objection to the publication of any part of my last letter[1] to you in the Philosophical Magazine, which you may think suitable. I wrote it very hastily, and I should be obliged by your making any remarks or corrections that may be required or that may occur to you; or allowing me to see a proof which I could return without delay if I receive it any day next week.

Since I wrote to you I have found that either iron filings or little pieces of very thin iron wire, mixed up with soft wax or dough, give a body which always places the line of compression perpendicular to the lines of force; doing so, & with great energy when suspended 4 or 5 inches from the centre of the field in a Ruhmkorf double electro-magnet,[2] immediately on being pressed between the hands, over and over again in different directions. I was quite prepared for this in the case of the little pieces of iron wire; but I rather expected the opposite bearing when iron filings were used. The explanation in each case must be that the little fragments of iron get pushed into such positions by compression that they act upon one another inductively like the parts of a bar or row of fragments perpendicular to the line of compression, or like the parts of a set of bars or rows at small distances from one another but on the whole more nearly continuous perpendicular to than along the line of compression. Why this should be is quite obvious in the case of the little pieces of iron wire; but, though not unintelligible, it is not to me so obvious in the case of the iron filings. Both these cases exhibit a phenomenon opposed to that which you have I believe ascertained for paramagnetic substances in general, in a state of strain,[3] and it follows that, although the effect of a state of strain by compression of the actual substance of a paramagnetic is, you find it to be, similar, as regards inductive capacity, to that which would exist in a compound solid composed of a number of small pieces of soft iron on the whole nearer one another along lines parallel to a certain direction than along lines perpendicular to that direction; yet as actual compression of a plastic medium containing fragments of soft iron mixed up with it does not produce the same effect on the resultant inductive capacity in the direction of compression, but in lines perpendicular to that of the compression.

I sent off to Professor Stokes[4] a short time ago a short paper[5] containing elementary demonstrations of some of the propositions regarding magnetic forces & the tendency of paramagnetics and diamagnetics as regards places of stronger and weaker force which have been already published as results of experiment, or as conclusions from mathematical analysis. It will probably be published before long, but I do not know in what journal.

Believe me, My dear Sir, | Yours very truly | William Thomson

RI MS JT/1/T/12

1. *my last letter*: Letter 1053, which was published in the April 1855 issue of the *Phil. Mag.* The present letter was a reply to letter 1060.
2. *field in a Ruhmkorf double electro-magnet*: The very strong magnetic field produced by high voltage electricity from a Ruhmkorff's apparatus with two windings. The high voltage induction coil bears the name of Heinrich Daniel Ruhmkorff (1803–77), who appears not to have been the original inventor.
3. *opposed to . . . paramagnetic substances in general, in a state of strain*: Thomson was probably challenging the results that Tyndall had announced in his paper 'On Diamagnetism and Magnecrystallic Action', *Phil. Mag.*, 2 (1851), pp. 165–88, esp. pp. 180–85.
4. *Professor Stokes*: William Thomson to G. G. Stokes, March 14, 1855; see letter 1076, n. 4.
5. *a short paper*: W. Thomson, 'Elementary Demonstrations of Propositions in the Theory of Magnetic Force', *Phil. Mag.*, 9 (1855), pp. 241–48.

To Edward Frankland 31 March 1855 1062

31st March 1855

My dear Frankland.

If you should be writing to LeNoir[1] would you be kind enough to order a Pachytrop[2] for me. I think you said it was £2–10—I lecture on Voltaic

Electricity after Easter[3] & if I could have the Instrument within the coming month I might make use of it my course.

Most sincerely yours | John Tyndall

JRL Frankland, 192

1. *LeNoir*: George André Lenoir (1825–1909), chemist and instrument maker in Vienna who in 1847 had studied with Bunsen in Marburg, where he met Frankland and Tyndall. See also letter 1053.
2. *Pachytrop*: A pachytrop(e) is an electromagnetic device through which an induced current is produced by spiral coils. By using different settings of the pachytrop—involving different resistances of the circuits—the output can be altered; for example, a small quality of electricity (current) at a high intensity (voltage), or a large current at low intensity.
3. *I lecture on Voltaic Electricity after Easter*: Tyndall delivered a course of lectures at the RI on 'Voltaic Electricity' starting on 17 April and ending on 5 June 1855.

From Hector Tyndale 31 March and 2 April 1855 1063

Philadelphia March 31st 1855

Dear John

Lies are much like curses, with the same domestic propensities or, otherwise, I might make use of one now, and with a 'violent, indignant cry' demand why you have not replied to my last,[1] announcing the safe arrival of your humble servant at home? This would be a 'safe dodge' too, for mails <u>do</u> miscarry, and I could thus have an <u>object</u> and a <u>subject</u> at the same time. But—Honour! Honour! Above all things Honour! But in good truth I had intended to write—and so contented myself with the intention. Besides this I knew you to be much occupied and blessed withal with a high sense of duty, which would perhaps impel you to write whether pleasant & convenient or not. And so I lapsed a little deeper into my lazy-pride mood and did <u>not</u> write. But I can't afford, my dear John, to have you <u>quite</u> forget me and so I hope you won't, like a good fellow as you are—always 'barrin' the <u>choleric grumpus</u>.[2] And thus humbly I <u>force</u> myself between you and the image of that 'lone and lovely maiden. Whom the angels call Lenore'[3] (!) soliciting a slight token of remembrance. But away with <u>stilts</u> cousin and let us stand up and shake hands. How are you? And how have you been this <u>thundering</u> long time? Let us hear all about it and you.

As I have said above I have had many a strong <u>desire</u> to write to you, but I refrained from two motives, one being that I knew you really very much occupied and the other—Oh shame—the longest and strongest fit of laziness I ever had. I have written to no one except 'upon compulsion' since my return.

I have received two or three papers from you[4] and they served me as a broomstick serves a witch to ride off upon into air. But a good stout horse in the shape of a letter would have been still more pleasant to me to amble on. Now mind I don't object to the sticks at all, on the contrary I like them much, but once in a while send me also a good cob! But don't 'cob'[5] me too hard though for you know my tender points and can lay it on heavy when you try. What a lot of trash I have scribbled—but, I don't know why, the Isle of Wight always comes to me when I think of you and off I go following your wild antics on the downs![6] And after all it is no wonder for sick as I was, after my arrival from Paris,[7] that short excursion began a change which has been going on ever since. In truth I was very sick, more sick than I seemed and have suffered more or less ever since. I don't know what it was, but it seemed as a general prostration of cheerfulness and a setting up of gloominess instead. I am not over sweet at any time, but I got a thorough acidulation[8] in France—how I can't tell. And so perhaps was more unpleasant than common. But let that go—much of it did go upon the free winds we met on that dear little island.

And now having thrown off a good deal of the egotism with which one should always welcome a friend (!) I come to some serious matters.

What are you doing, firstly as John? That is, how is your health and how are you enjoying yourself and in what state are you generally? Do let me know something of you, the man of you, the good hearty man, the real poetical man of you. Pray is there nothing you ought to have told me before this? What are your dreams made of? What about that 'rare and radiant maiden'—I hope not dead and buried maiden—'whom the angels call Lenore'? Have you no other life than your chrystal life there in Albermarle Street?[9] Is there no other thing than Bismuth with its labyrinthine forms in which you might lose yourself?[10] Is there no other axis than the magnetic? Is there no other current than that of electricity? Are there no other roads in which you delight to roam than electrodes? Finally is there no other counter magnetism than diamagnetism and no other affinity than chemical affinities? Lastly are you in no other way dissolved than as a salt is—with so many equivalents? And conclusively is there no other efflorescence for you than that of exuding whatever the surrounding atmosphere may ask for? My dear fellow rouse up, fly up, right up through the skylight in that workshop of yours. Unfold your wings, ruffle your feathers, shake off the dust of accumulated facts and soar straight up to the centre of yourself. And when you are there let me hear your voice coming down to me cheerfully and cheeringly.

And now secondly what are your professional labours, towards what inclining now? You have sent me, as I have said, two or three papers which

I read with interest, but have you no papers of your own, or any product or result of your labour to send me. You may, perhaps, think that I care little or nothing for such things, but if so you err. I may not understand thoroughly as a scientific man would do, but nevertheless all such things have an interest for me and had I not been otherwise engaged might perhaps have turned a constant attention to them. At any rate whenever you may have any publication of your own or any other in which you take an interest you may remember, if you please, that I shall always receive them not only for their worth but as evidences of your remembrance and esteem. Come now that last sentence is pretty good 'genteel begging'. Hey?

And so far I have rattled away without saying anything and by the way I suppose it will be the same to the end hereof. I found all our family, the whole of the noble race of Tyndale, very hearty, as hearty as though they were common folks! And so they remain until this day.

Your daguerreotype[11] has been seen by every one of the family. I was about to say it was seen by 'admiring thousands'—but as I started out to tell the truth, and nothing else, I can't say that. I leave to your vanity and insight the settling of the word, the adjective or the noun (or both) as the word I cannot utter truly! My dear John the Spring is coming on us slowly slowly it is true, but I sniff the green grass, yet under the ground, I feel in the rude winds once in a while the warm kiss of the summer air. Through the rapid flying grey clouds I catch a point of the deep blue Summer sky. In the rushing of the March winds I hear anticipated the whisperings of the green velvet leaves, and sometimes a liquid note from a coming spring bird sweetens the air. In short with something of the feelings of a gay colt in young clover I have run on through this letter and perhaps from the same cause, and that cause unknown to both—but I will guess at it—the green-ness of the pasture!

April 2[d]

I left off writing from the fear of offending you. On reading it over I find it so foolishly absurd that I would tear it up, but that I would not have time to write again very soon, and bad as it is I leave it to your good nature to pardon and—reply. It is very strange how sometimes one is struck with an accidental word and the whole course of life is thereby influenced from that time. Like a ship on the great wide ocean meeting a lone iceberg, on all the seas perhaps the only one afloat, she strikes and sinks, or perhaps more fortunately may only make for the nearest port no matter where that may be. I have thought often of this and some day, it may happen, I will tell to you how sailing one fair summer's day I met with such a berg and had in consequence to put up my helm for another haven! Only when that haven may be reached, in all probability never, I will tax you with the tale! Have you no inclination to visit us (U.S.) here? It is true to a small extent, that remark of your home-sick friend,[12] 'the men are all sallow and the women all tallow'[13] (or yallow I forget which).

But still life is somewhat rosy-hued even here; and we try to content ourselves in this small country place. Come over and see us once, not expecting great things, but simply new ones, of combinations of elements not yet classified, that is all. This latter part of my letter you may think as absurd as the first, but you will still pardon if you do not read.

All our family desire their regards to 'Cousin John'. Make my kindest remembrances to Dr Francis, to old Mr Taylor & his daughter,[14] to Mr Debus & my thanks & remembrances to all your good friends in Hampshire.[15]

I am John very truly & affectionately yours | Hector Tyndale

RI MS JT/1/T/46

---

1. *my last*: letter missing; the last extant letter from Hector is volume 4, letter 0957, dated 21 July 1854. Tyndall's letter 1052 of 4 March 1855 had not yet reached Tyndale.
2. *choleric grumpus*: bad-tempered.
3. *lone and lovely maiden. Whom the angels call Lenore*: Cf. 'the rare and radiant maiden. Whom the angels call Lenore': E. A. Poe, 'The Raven', ii:5. It is not clear whether there was at that time a maiden in Tyndall's life, but if there was one she has not been identified.
4. *two or three papers from you*: In a letter to James Coxe of 4 December 1854, Tyndall mentioned posting a copy of the *Scotsman* to Hector: volume 4, letter 1016.
5. *cob*: strike (*OED*). Tyndale was punning on the word 'cob', as the phrase 'good cob' in the previous sentence means a good horse.
6. *the Isle of Wight . . . the downs*: The Isle of Wight is located off the south coast of England; the South Downs are a range of chalk hills close to the south-east coast of England. Soon after Hector arrived in England, he and Tyndall visited these areas (Journal: RI MS JT/2/13b/680).
7. *my arrival from Paris*: probably in late July 1854; he left London in early August 1854.
8. *acidulation*: the action or process of becoming or being made more acidic (*OED*). Tyndale appears to have suffered gastric problems during his visit to France, possibly owing to an overzealous sampling of local food and wine.
9. *in Albermarle Street*: at the RI.
10. *Is there no other thing . . . lose yourself?*: In this and the following sentences Tyndale humorously characterized Tyndall's scientific interests.
11. *Your daguerreotype*: A daguerreotype is a photograph produced by the process, first published by Daguerre of Paris in 1839, in which the impression was taken upon a silver plate sensitized by iodine, and then developed by exposure to the vapour of mercury (*OED*). Tyndall and Tyndale presumably exchanged photographs when they met in London in August 1854; see Journal, August 5, 1854 (RI MS JT/2/13b/681).
12. *your home-sick friend*: possibly Tyndall's childhood friend William Wright (1818/19–92), who had emigrated to Cincinnati in March 1841.
13. *the men are all sallow and the women all tallow*: A variant on lines from the poem

'Duhallow' (the name of a barony in County Cork) by James Clarence Mangan: 'But the men are all short | And the women all sallow' (*Romances and Ballads of Ireland,* ed. H. Ellis (Dublin: J. Duffy, 1850), pp. 154–56, on p. 156).

14. *old Mr Taylor & his daughter*: Richard Taylor was often accompanied by his daughter Sarah (1804–84), who lived with him.
15. *all your good friends in Hampshire*: at Queenwood College.

To Thomas Archer Hirst 3 April 1855 1064

April 3. 1855

My dear Tom

I greatly fear that it will not be in my power to go down on Wednesday[1]—I have an apparatus[2] in hand for the Paris Exhibition[3] and must remain here until it is complete & ready for packing—This I trust will be tomorrow but too late to permit of my going to Broughton—Next day or Friday I will start—The gods know I am anxious to get out of this and to shake the scientific dust & lethargy from my veins amid the breezes of the country.

I send you a note from Mrs Ginty.[4]

Your paper on the medium[5] pleases me in many respects—you have doubtless thought closely on the matter—You will I imagine be disposed to modify a point or two on reconsideration. For my part your whole reasoning passed through my mind before ever I put my pen to the letter to Faraday.[6] But it was not for me to make suppositions—Let this be done by the advocates of the theory[7] and when all is done we shall be the better judges of its value as theory. If these suppositions are unnatural, they will diminish in proportion as they are so they need the credibility of the hypothesis whose life they are intended to save—I will hunt up Thomson's papers[8] and bring them to you when I come.

Affectionately Yours | John Tyndall

RI MS JT/1/T/596

1. *to go down on Wednesday*: Tyndall had intended to visit Hirst at Queenwood College (Broughton) on the following day, Wednesday, 4 April.
2. *an apparatus*: Tyndall contributed 'Instruments to illustrate magnetic and diamagnetic phenomena' to Class VIII ('Arts relating to the Exact Sciences, and to Instruction'), Section 4 ('Instruments employed in the Study of Physics, Chemistry, and Meteorology, or for Common Use') at the Exposition Universelle in Paris (*Paris Universal Exhibition, 1855. Catalogue of the Works Exhibited in the British Section of the Exhibition* (London: Chapman & Hall, 1855), p. 24).

3. *the Paris Exhibition*: The Exposition Universelle was held in Paris from 15 May to 15 November 1855.
4. *a note from Mrs Ginty*: letter missing.
5. *Your paper on the medium*: see letters 1058, n. 3; 1080, n. 8.
6. *the letter to Faraday*: letter 1042.
7. *the advocates of the theory*: As well as Faraday, Williamson and Thomson advocated a medium in space to account for magnetism.
8. *Thomson's papers*: including Thomson's 'Observations on the 'Magnetic Medium' and on the Effects of Compression' (letter 1053) which was published in the April 1855 issue of the *Phil. Mag.* See also letter 1053, n. 10.

From Heinrich Debus 15 April 1855 1065

Queenwood College | 15 April 1855.

Lieber John!

Körperliche Arbeit macht den Körper stark; Zuviel Arbeit ruinirt und reibt den Körper bald auf. Eine unserem Geiste angemessene Thätigkeit übt und vermehrt gleichsam unsere intellectuellen Kräfte, macht dieselben biegsamer und das Gedächtniss hält manche schätzbare Thatsache zurück. Aber wie der Körper so kann auch der Geist zu viel angestrengt werden, und in beiden Fällen Krankheit ist die Folge.

Ich erscheine vielleicht zudringlich oder es ist vielleicht unnöthig weil Du selbst weis[s]t was Du ja zu thun hast, wenn ich Dich in Diesem Brief mit Deinem Körperzustandt beschäftige. Wenn ich es recht verstehe dann kommt Dein Unwohlsein von einem Übermass von Arbeit. Daher möchte ich Dich bitten von jetzt bis Mitte Juni, ausser Deinen Vorlesungen Nichts zu thun; und von Juni bis zum December Alle wissenschaftliche Beschäftigung ein für allemal aufzugeben. Sehr wohl weiss ich dass ich viel verlange aber ich glaube nicht zu viel. Wenn Du es abschlägst, wenn Du im nächsten halben Jahr angestrengt arbeitest dann wirst Du immer schlechter werden und zuletzt möchtest Du das Capital mit den Zinsen verlieren. Deine Constitution ist sehr gut, aber auch Eisen kann in Staub verwandelt werden.

Komme am Freitag oder Sonnabend recht oft hierher und bleibe über Sonntag hier und Du weisst dass Du immer willkommen bist. Mische Dich ein Bischen mehr mit der Gesellschaft, gehe mehr aus, mache eine Reise zu Fuss über den Continent wie Sir William Thornhill und bald wirst Du wieder blühend und kräftig wie in den besten Tagen Deiner Jugend einherschreiten.

Mit herzlichen Grüssen | Dein | H Debus

Queenwood College | 15 April 1855.

Dear John!

Manual work makes the body strong; too much work ruins and soon wears down the body. An activity that is appropriate for our mind exercises and multiplies our intellectual powers as it were, makes them more supple and the memory retains many an estimable fact. But like the body, the mind too can be strained too much, and in both cases illness is the consequence.

I shall, perhaps, appear intrusive, or perhaps it is unnecessary because you yourself know what you have to do, if I bother you in this letter with your physical condition. If I understand correctly, then your feeling unwell[1] comes from an excess of work. I would therefore ask you to do nothing, apart from your lectures, from now until mid-June, and from June to December to give up all scientific work once and for all. I know very well that I am asking much, but not, I believe, too much. If you refuse to do so, if you work intensely over the next half year, then you will get worse and worse, and in the end you might lose the capital along with the interest. Your constitution is very good, but even iron can be turned into dust.

Come here fairly often on Fridays or Saturdays and stay here over Sundays, for you know that you are always welcome. Mix a bit more with people, go out more, go on a journey on foot across the continent, like Sir William Thornhill,[2] and soon you will be striding along again, lively and vigorously, as in the best days of your youth.

With warm regards | Your | H Debus

RI MS JT/1/D/18

1. *your feeling unwell*: the severe headaches that Tyndall frequently reported in his Journal.
2. *Sir William Thornhill*: Sir William Thornhill, a character in Oliver Goldsmith's novel *The Vicar of Wakefield: A Tale,* 2 vols (Salisbury: B. Collins, 1766), is said to have 'travelled through Europe on foot' 'in his own whimsical manner' (vol. 1, p. 29).

To Mary Anne Coxe 16 April 1855 1066

16th April 1855

My dear Mrs Coxe

Your last kind letter[1] lies before me, as yet unanswered. I have been waiting until I should be able to grasp my pen with the vigour of a man, and make my friend feel through the very energy of my penmanship that my heart is still beating, and that the 'struck eagle'[2] can yet clap his dilapidated wings.

At all events my ailments are useful, though stern teachers—they let me know the needs & infirmities of human nature—I believe every man must be able at times to withdraw himself from the world and hold pleasant communion with his own soul, but he cannot permanently sever himself from his fellows without injury to the best part of him. The shock of personal contact is necessary to stir up the forces resident within him. I once thought it my duty to fast and afflict my unfortunate body in a variety of ways—Now fasting and other acts of self denial are very useful sometimes, but their range of usefulness is I believe very limited, and I am inclined to think that a good song over a good glass of whiskey toddy, might, under certain circumstances be made conducive to religion and morality. Luther you know once recommended a young clergyman who tortured himself on the question of predestination & election to get well drunk as the best method of surmounting his difficulty.[3]

If Dr Coxe were in town at present he would feel an interest in hearing one of our courses of lectures which commences on Saturday next. It is by DuBois Reymond[4]—the subject being electro-physiology.[5] I admired your husband's caution in dealing with this point in his edition of his uncle's works[6]—Many would have treated it more boldly and talked eloquent nonsense about it—Dr Coxes caution was never more prudently exercised than in connexion with this subject—but I must stop.

Goodbye & believe me ever | Most faithfully yours J Tyndall

RI MS JT/1/T/247

---

1. *Your last kind letter*: letter missing.
2. *struck eagle*: 'So the struck Eagle stretched upon the plain, / No more through rolling clouds to soar again, / Viewed his own feather on the fatal dart, / And winged the shaft that quivered in his heart' (G. G. Byron, 'English Bards, and Scotch Reviewers', ll. 823–26).
3. *Luther . . . his difficulty*: The probably source is Ralph Waldo Emerson's lecture on Montaigne, in which he related Luther's advice to 'a young scholar perplexed with foreordination and free-will, to get well drunk' (Emerson, *Representative Men. Seven Lectures* (London: J. Chapman, 1850), pp. 112–13). See also Tyndall's Journal entry for 21 September 1850 where he recounted having enjoyed a 'supper of black-pudding and some fine ale and porter. The change [in fare] roused me and did my system good. [That reminded him of] Martin Luther's advice may under some circumstances be a sound one—"Get well drunk!"' (RI MS JT/2/13b/510).
4. *courses of lectures . . . by DuBois Reymond*: Beginning on 21 April, Emil du Bois-Reymond delivered a course of eight weekly lectures on 'Electro-Physiology' at the RI.
5. *electro-physiology*: A branch of physiology that deals with the electrical phenomena associated with physiological processes, esp. in nerve and muscle; also the electrophysiological characteristics of a process, tissue, disorder, etc. (*OED*).
6. *this point in his edition of his uncle's works*: Coxe was the nephew of the phrenologist George Combe (1788–1858) and the physiologist Andrew Combe (1797–1847). Tyndall was probably referring to the fourteenth edition of Andrew Combe's *The Principles of*

Physiology Applied to the Preservation of Health, and the Improvement of Physical and Mental Education (Edinburgh: Maclachlan & Stewart, 1852), which was revised and enlarged by Coxe.

To Emil du Bois Reymond 16 April 1855 1067

Monday

My dear DuBois

If you are not better engaged will you dine with me on Thursday à la Luther & Wegner[1] (but not half so good) at Anderton's Hotel[2] Fleet Street at a quarter before 6 o/c. It is quite near Red Lion Court. We can afterwards walk to the Royal Society[3] together if you like. I hope Mrs Dubois[4] will pardon this attempt to bring you back to your bachelor habits. On Wednesday I trust you will allow me the pleasure of paying my respects to her—Till then I shall be dabbling in all sorts of chemical pollution.

Ever Yours | John Tyndall

RI MS JT/1/T/427

1. *à la Luther & Wegner*: Lutter & Wegner (f. 1811) is a famous restaurant in Berlin that would have been familiar to du Bois-Reymond.
2. *Anderton's Hotel*: Tyndall often took visitors to dine at Anderton's Hotel, 164 Fleet Street; proprietor Frank Clemow.
3. *the Royal Society*: At the RS meeting on Thursday, 19 April, Henry Moseley read a paper entitled 'On the Descent of Glaciers' and the RS presented its response to the Lords of the Committee of Privy Council for Trade on the subject of marine meteorological observations (*Roy. Soc. Proc.*, 7 (1855–56), pp. 333–61).
4. *Mrs Dubois*: Jeannette, née Claude (1833–1911).

From Thomas Archer Hirst 17 April 1855 1068

Broughton | April 17th 1855

My dear John.

You did not tell me in your letter this morning[1] whether you were better or worse for your trip.[2] You left us in such a precarious state of health that I should like to know. I have packed up 20 cells[3] complete and will send them tomorrow. I have only one pair of zincs suitable for poles so that I was obliged to send you others. Send all back as soon as you have done with them. I am sending the translation of Verdet[4] to Francis by this post.

Riess' paper[5] is also on the point of being finished. We have had delightful weather since you left us. I was not at all surprised to find you had given Anna your watch,[6] the idea suggested itself suddenly to me one day when she was wearing it and you were present. One day indeed I thought of telling her not to wear it as it looked as if she wanted it but I did not for I saw she was perfectly innocent of any such intention and that if she had not been you would not have given her it for that reason. It is a gift she will prize. I will have it repaired and in future I am sure it will be better cared for than it has been. I intend coming to hear Du Bois on Saturday next[7] and stopping Sunday with you. I shall be alone.

Yours affectionately | T. A. Hirst

RI MS JT/1/H/215

---

1. *your letter this morning*: letter missing.
2. *your trip*: Tyndall visited the Hirsts at Queenwood, arriving on 4 April, when they attended an Easter concert at the school. On 9 April they went to Southampton to see Margaret Ginty and her children off to Rio de Janeiro; see letter 1059. Tyndall returned to London on 12 April 'with health improved but slightly' (Hirst's Journal: RI MS JT/2/32c/1165).
3. *20 cells*: a battery consisting of 20 voltaic cells, which Tyndall was borrowing from the laboratory at Queenwood.
4. *the translation of Verdet*: Hirst translated the first part of 'Recherches sur les Propriétés Optiques Développées dans les Corps Transparents par l'Action du Magnétisme' (*Annal. de Chimie,* 41 (1854), pp. 370–412) by the French physicist Emile Verdet (1824–66) as 'On the Optical Properties Developed in Transparent Bodies by the Action of Magnetism', *Phil. Mag.,* 9 (1855), pp. 481–509.
5. *Riess' paper*: Hirst translated Peter Riess's paper 'Ueber die Wirkung nicht-leitender Körper bei der electrischen Influenz' (*Poggend. Annal.,* 92 (1854), pp. 337–54) as 'On the Action of Non-Conducting Bodies in Electrical Induction', *Phil. Mag.,* 9 (1855), pp. 401–13.
6. *given Anna your watch*: see letter 1070.
7. *Du Bois on Saturday next*: Hirst attended the first of du Bois-Reymond's lectures on 'Electro-Physiology' which was delivered at the RI on Saturday, 21 April.

To Michael Faraday 18 April 1855 1069

Royal Institution | 18th April 1855

My dear Mr Faraday

You can hardly estimate how I prize your gift.[1] My highest ambition at the present moment is to work until I shall be able to present you with a volume worthy of your acceptance. If the head holds out[2] I hope to accomplish this—if not it will be my sweetest consolation to reflect that you have not thought me unworthy of your kindness.

Ever yours affectionately | John Tyndall

RI MS Conybeare Album, f. 3
Faraday Correspondence, 4:2974

1. *your gift*: The third volume of Faraday's *ERE,* which had just been published by R. & J. E. Taylor.
2. *If the head holds out*: Tyndall had repeatedly suffered severe headaches.

To Thomas Archer Hirst 18 April 1855 1070

18th April 1855.

My dear Tom

Very many thanks to you for the battery[1] which I doubt not will soon reach me. My health when I left you was very precarious and since that time it has not mended much—This is saying too much for I feel myself better & more hopeful today than I have been for a long time. To what do you think I attribute this favourable change? To giving up tea, coffee, wine & beer. But is only a proof of the abnormal state of my nerves when such trifles can effect me so seriously—I have felt very low at intervals since I left you—almost low enough to clench my teeth, bid good bye to the Royal Institution and sally forth to the mountains in search of the treasure I had lost. But I still expect to get through my course[2] if not with glory at least with satisfaction. It is almost certain that I shall go to Paris[3] & while there I shall spend a holiday time, and recruit myself. Today as I have said I feel better than I have felt for some time. With regard to the watch[4] you cannot be more convinced of Anna's innocence than I am—I am perfectly sure she never entertained a thought of my presenting her with the watch. But the thought of presenting it when it occurred to me gladdened me. I hate your formal presents, but this watch I regard as an appropriate gift—it is a part of myself and bears about it the marks and tokens of our companionship. It is fit that it should fall into Anna's hands—Nobody knows so much of its history as you do and hence nobody would be likely to treat it with more tenderness and affection.

I saw the Emperor & Empress in their passage through London,[5] and heard hundreds of thousands give the man the cheer which I gave him when all England was denouncing him.[6] I saw them twice once in Pall Mall. The Empress seemed pale and anxious then; I saw them again at the end of

Albermarle Street. The anxiety had fled and some passing accident called the bloom to her cheeks and she looked radiant—I am rejoiced to find that you are coming up on Saturday. Du Bois dines with me at Andertons tomorrow.[7] I wish you were there; but if he is not engaged I will try and secure him for Saturday evening.[8]

Kiss Anna for me | affectionately yours | John Tyndall

RI MS JT/1/T/595

---

1. *the battery*: see letter 1068, n. 3.
2. *my course*: On the previous day, 17 April, Tyndall had delivered the first lecture of his course on 'Voltaic Electricity' at the RI.
3. *go to Paris*: Tyndall was to serve as 'Supplementary Juror' in Class VIII ('Arts relating to the Exact Sciences, and to Instruction') at the Exposition Universelle in Paris; see letter 1064, nn. 2 and 3.
4. *the watch*: see letter 1068, n. 6.
5. *Emperor & Empress . . . London*: The Emperor of France, Napoleon III (1808–73), and the Empress Eugenie (1826–1920) visited London in April 1855. They traveled from Windsor, arriving at the Bricklayers' Arms Station (a railway terminus a mile south of London Bridge) late in the afternoon of Monday, 16 April, and then toured London by carriage, their route including Pall Mall. 'A numerous and highly-respectable class of persons crowded Pall-mall and the approaches leading thereto' (*London Standard,* 16 April 1855, p. 4).
6. *all England was denouncing him*: Some sections of the press (e.g. the radical *Reynolds's Newspaper,* 15 April 1855, p. 16) denounced Napoleon because of the bloody coup d'état he initiated in December 1851 when 3000–4000 of his opponents were killed and many republicans were subsequently arrested and exiled. Another unsuccessful attempt was made on the Emperor's life in Paris on 28 April 1855.
7. *Du Bois . . . tomorrow*: see letter 1067.
8. *Saturday evening*: After du Bois-Reymond's lecture at the RI on the afternoon of Saturday, 21 April, Tyndall and Faraday dined at Gassiot's house, but Hirst went to Tyndall's lodgings and 'had to [. . .] spend the evening alone' (Hirst's Journal: RI MS JT/2/32c/1166).

To Thomas Henry Huxley 18 April 1855 1071

My dear Huxley

I am more pleased with the fact of your bearing me in mind than thankful for the ticket[1]—No man wishes more sincerely that I do that your lectures may fill your own wishes to overflowing. I was sorry to hit Hunt so hard[2]—and today walking through Cheapside the thought occurred to me—'What if you meet Hunt?—Will you shake hands with him & smile as if nothing had happened—You must act no such damned hypocritical part—You must tell the man that you have smitten him & if he takes it in good temper you may then give him your hand.'—

Poor Sir Henry,[3] I was grieved to hear of his death.

I thought as I read that review of yours of Ehrenbergs works[4] that on many points there is a great similarity of thought between us—Do not damn my impertinence for I verily believe I state the truth. —When I stand alone in the woods hear the birds chirping & see the trees sprouting I feel like a puzzled infant amid things which baffle my comprehension—I like to hear a man who instead of turning my stomach with dry theories of this universe is able to appreciate the difficulty of the problem and recognize the fluxional character of our knowledge regarding it. The dogmatist however gets astride his little arc and swears it is the whole 360 degrees—We know better—

But I wont bore you further | Goodbye | Ever yours | John Tyndall
18th April 1855

IC HP 1.17–18

1. *thankful for the ticket*: see letter 1043, n. 3.
2. *I was sorry to hit Hunt so hard*: Tyndall's critical review of *Elementary Physics: An Introduction to the Study of Natural Philosophy,* new edn (London: H. G. Bohn, 1855) by Robert Hunt (keeper of the Mining Record Office and professor of mechanical science at the Royal School of Mines; 1807–87) had appeared in the *Westminster Review,* 63 (1855), pp. 551–58. Tyndall had reviewed an earlier edition of Hunt's *Elementary Physics* in 'Notices Respecting New Books', *Phil. Mag.,* 3 (1852), pp. 57–60.
3. *Sir Henry*: The geologist and paleontologist Henry de la Beche (b. 1796) who died on 13 April 1855. He had been the first director of the Geological Survey of Great Britain and also served as director of London's Museum of Practical Geology.
4. *review of yours of Ehrenbergs works*: Huxley had reviewed *Mikro-Geologie,* 2 vols (Leipzig: Voss: 1854) by the German naturalist, paleontologist and microbiologist Christian Gottfried Ehrenberg (1795–1876) in the *Westminster Review,* 63 (1855), pp. 558–63.

To Thomas Archer Hirst 19 April 1855 1072

My dear Tom

I forgot when I expressed my gladness at your coming that I had already engaged to Pinner to Col. Wynne[1] on Sunday next—I cannot get off the engagement as Mrs Wynne came here to see me & to her I have passed my word. Du Bois also tells me that Saturday week the lecture will be more important. He enquired cordially after big Tom Hirst this morning. I propose then that you defer your journey for a week.[2] What say you?

Yours affectionately | John
Thursday

RI MS JT/1/T/597

1. *to Pinner to Col. Wynne*: The Wynne family lived at Pinner, in the Borough of Harrow, 12 miles north-west of Charing Cross.
2. *defer your journey for a week*: Hirst did not postpone his journey for a week but instead attended du Bois-Reymond's first RI lecture, on 21 April.

From Edward Frankland 21 April 1855 1073

Owens College | Manchester April 21/55

My dear Tyndall

Your letter[1] has been very unfortunate having had to make the trip to Windermere[2] & back before it reached me this afternoon. We returned from Windermere on Wednesday where we had spent the Easter vacation most pleasantly. I have founded a laboratory there[3] & have become so far naturalised as to be the owner of two boats upon the Lake which are intended for fishing, shooting, & general rusticating & brain-clearing purposes. I hope to see you there this Summer. I cannot offer you a bed as our little cottage is quite full when wife & bairns[4] are there but you shall have a share of the prog[5] & I promise you that the fine mountain breezes will soon clear out the diamagnetic scum from your noddle[6] & put you in thorough working order again. I have just made you 2 balloons with the last of my collodion[7] & send them with two others the only ones I have remaining in stock otherwise I would send you more. You will find that one of them shews the prismatic colours, but it is rather too thick & is not to compare with one I had a short time ago which was probably less than half the thickness. I fear the colours are too weak to be seen by a large audience, but you can try how it looks when illuminated by the electric light which should be concealed from the spectators & thrown upon the inflated balloon with a reflector.

When I see you again I must initiate you into the mysteries of balloon making, for I am sure that such collodion films would be capable of endless applications at the Royal Institution. To explode the balloon with the mixed gases you must tie its neck with a short piece of thread impregnated with nitre,[8] light the end of the thread & let the balloon ascend to the ceiling.

I am sorry to hear of your indisposition. Those Royal Institution lectures will never agree with you. They keep you continually upon the rack for 6 months. Unless you can look a course of lectures in the face as coolly as an after dinner conversation you may depend upon it your health will suffer. I have not even yet completely shaken off the effects of my course in 1853[9] & for a whole year I suffered severely from indigestion brought on in a great measure I believe by the continued strain & anxiety attending those lectures.

Hoping that the balloons are still in time & that they will answer your purpose

Believe me | ever yours sincerely | E. Frankland

RI MS JT/1/F/43

1. *Your letter*: letter missing.
2. *Windermere*: the town in the Lake District where Frankland had purchased a 3-bedroom cottage.
3. *a laboratory there*: 'I have facilities there for performing any chemical operation of moderate character'; see letter 1149.
4. *bairns*: children.
5. *prog*: food (*OED*).
6. *your noddle*: your head.
7. *collodion*: see letter 1046, n. 4.
8. *nitre*: potassium nitrate.
9. *my course in 1853*: Commencing 7 April 1853, Frankland had delivered ten weekly lectures at the RI on 'Technological Chemistry'.

From Edward Sabine 24 April 1855 1074

13 Ashley Place | April 24

My dear Sir,

Some time since I may, perhaps, (little as it seemed to be in my way to do so, and little as it now seems possible that it should have been in my way to do so), have had some slight influence[1] upon a wish of yours to change your scene of action to a Canadian University.[2] I have never felt a moment's regret at the counsel which I then permitted myself to offer to you; on the contrary, the position which in the short time that has since elapsed you have established for yourself has been attentively observed by me since as a full justification of the freedom which I used to you when we were yet personally strangers.

Nevertheless the offering counsel to another is so little in my way, & I have so little confidence in my own judgment to advise, that I rejoice in an opportunity of making known to you, that if you have ever doubted of the expediency of the step you then took, you may have it in your power now to obtain a situation at Toronto, more eligible I believe in a pecuniary point of view than the one you had then in view, and far more important in scientific respects, inasmuch as the directorship of the physical observatory[3] attached to it, in hands such as yours, would secure opportunities scarcely inferior to that of any European or British establishment, for obtaining a present and an enduring scientific fame. It is, it is true, in the line of <u>observation</u> rather than that of <u>experiment</u>, that the enquiry into nature could be carried on; and your course hitherto has been rather in the field of experiment, but both are alike handmaids to knowledge; the principles of both are the same; and the aptitudes which you have so manifested in so remarkable a degree for experiment would soon lead you to distinction in the kindred field of observation.

Whatever may be your decision,—(and there must be much unknown to me to influence it)—I cannot but feel a great satisfaction in the opportunity it affords me of assuring you how cordially I join in the general feeling amongst your friends of admiration for the devotion to science which has hitherto so strongly marked your progress in life.

Very sincerely yours | Edward Sabine

RI MS JT/1/S/35
[Surviving extract from enclosure]

I am happy to inform you that his Excellency the Governor General has at last taken steps which will secure the permanence & efficiency of this Institution[4] & has been pleased to approve of the plan submitted by me for that purpose. A chair of meteorology is to be founded in University College to which the directorship of the observatory will be attached, the salary is £450 currency a year, together with lecture fees, and the house and grounds—on the whole worth about 600L[5] a year. There is also to be a permanent sub-director and three assistant observers, who shall be graduate scholars of the University of Toronto, and one mechanic (Sergt. Menzies)[6] with an attendant—thus forming an effective staff of seven persons. The expenses will be defrayed partly by the University, partly from the public funds, and I believe there is every disposition on all sides to treat the matter liberally. I have reason to expect that the new chair with the directorship will in the first place be offered to me—if so I shall decline it, so that if you are aware of any one who is worthy and desirous of filling it might be as well if he <u>were to make application at</u>

once to the Honble G. Et. Cartier[7] Provincial Secretary, Quebec Believe me my dear Sir |Yours very faithfully | W. Cherriman[8]

Col. Ed. Sabine, RA

RI MS JT/2/7/48–9
LT Typescript Only

1. *some slight influence*: Edward Sabine had directed the worldwide project to study the changes in magnetic declination. This had involved the construction of a number of magnetic observatories including the Toronto Magnetic and Meteorological Observatory, which was completed in 1840. Although the project at the observatory had ended in 1853, Sabine's earlier connection with the Toronto observatory would have given him some influence over appointments.
2. *to change . . . Canadian University*: With Sabine's encouragement in the autumn of 1851 Tyndall had applied for the chair of natural philosophy at the University of Toronto; see volume 3, letter 0559. The University of Toronto had been founded in 1827 as King's College, an Anglican institution. However in 1850 it was renamed the University of Toronto and severed its connection with the Anglican Church.
3. *the directorship of the physical observatory*: John Henry Lefoy had served as director of the Magnetic and Meteorological Observatory from 1842 to 1853, after which John Cherriman (see n. 8, below) became the acting directory. Cherriman wanted to ensure the observatory's continued existence and to put it on a firm financial footing. In 1855 his brother-in-law, G. T. Kingston, became the director (as professor of meteorology), a position he held until 1880.
4. *this Institution*: the University of Toronto.
5. *600L*: The character after the number 600 is unclear but could be an L (perhaps Livres?) or an incomplete £ sign.
6. *Sergt. Menzies*: not identified.
7. *Honble G. Et. Cartier*: George-Étienne Cartier (1814–73) was a Canadian lawyer and politician who served in the Legislative Assembly of the Province of Canada.
8. *W. Cherriman*: The Cambridge-educated John Bradford Cherriman (1823–1908) was the professor of mathematics and natural philosophy at the University of Toronto. In 1855 a chair of meteorology was established, which included the directorship of the magnetic and meteorological observatory. Cherriman was briefly appointed to the new position, which was soon transferred to G. T. Kingston. The initial 'W' in LT's transcription is an error.

To Edward Sabine 26 April 1855 1075

My dear Sir

I have reflected on the subject of your note[1] and the wish to do this before writing to you is the cause of the tardiness of my reply. The best proof that I can give of my estimation of the counsel which you have already given to me is that I should, in the present instance also, permit myself to be influenced in a great degree by your advice. I will now state the two opposing aspects of the question as they present themselves to my mind.

It is certainly an object with me to obtain a position which would make my mind easy on matters of personal finance:[2] this object would be attained at Toronto, and besides this, as you remark, an honourable sphere of action would open to me. On the other hand there would be the withdrawal of those stimulating influences to which you referred in one of your earliest letters[3] to me, and which habitually spur a man to action. There would be the separation from England and from the valued friends which now surround me. But as regards science the gravest point appears to be the change from the field of experiment to that of observation. Both are, as you say, alike handmaids of knowledge, but my adherence to the former has opened possibilities to me now which I could not relinquish without regret. There is I believe a wide field opening before the experimentalist of the present day, and it is not perhaps a blameable ambition to desire to contribute something in this field to the maintenance of the scientific fame of this country among the nations of Europe. The thought may be a dream—the dream may be exploded by the failure of working power with which I am sometimes threatened; but still the hope is sufficiently strong to enable me cheerfully to forego for the sake of it many of the agremens[4] which in the opinion of most render life enjoyable.

A man is not a constant quantity—his power and courage ebb and flow, and there are times when I confess the desire of a more solid and less anxious position than that which I now occupy lays hold of me. But I believe my healthiest moments are those when I can look the trials of life fearlessly in the face, and make up my mind to accept even poverty, if necessary, sooner than swerve from the vocation in which I believe I can labour with most effect—namely the advancement of exact scientific knowledge by the method of experiment.

Had time permitted I should gladly have consulted Bunsen and Magnus upon this subject. I have but one object in view, the attainment of that position in which I should have most opportunity to pursue my own researches. If Canada afforded me such an opportunity I would cheerfully go there. But in the present instance I fear the advantages offered, which are doubtless considerable, would have to be purchased by the sacrifice of many of the hopes and aims which the hard study of the last 7 years has engendered.

These are my feelings with regard to the point in question. But there are others associated with them which are a source of the purest pleasure to me. I trust that you will never have occasion to think worse of me than you now think—I trust you will never find me doing any thing to discredit the good opinion which you have so kindly expressed regarding me. When I cast my thoughts back on the rather unusual incidents of my life, the part you have acted in helping me to carve my way is not absent from my remembrance; and my hope is that the kindness then manifested, and again so strikingly evidenced in this your last letter to me, may see reason to perpetuate itself, and that I shall long have occasion to rejoice in the possession of your good will.

Believe me | most faithfully yours | John Tyndall

RI MS JT/2/7/50–2
LT Typescript Only

1. *the subject of your note*: the offer of the directorship of the Magnetic and Meteorological Observatory at the University of Toronto; see letter 1074. Tyndall subsequently asked the advice of Hirst (see letters 1080 and 1081) and others before refusing the offer.
2. *matters of personal finance*: Tyndall was offered a salary of £450 per annum at Toronto, with the possibility of earning a further £150 per annum. By contrast, his salary at the RI was £200 per annum in addition to which he would have received payment for his lectures at the LI and also for his contributions to the *Phil. Mag.,* the *Westminster Review* and to the reading books edited by Edward Hughes. A few months later he received (at least) £50 for his examining on behalf of the Board of Ordnance (see letter 1098). In total his income in 1855 was about £400 per annum.
3. *one of your earliest letters*: volume 3, letter 0559.
4. *agremens*: delights (French).

From George Gabriel Stokes 26 April 1855 1076

Royal Society | Somerset House | April 26 1855

My dear Sir,

At the last meeting of the Council[1] it was voted, on the recommendation of the referees,[2] that your paper[3] should be printed in the Transactions. Both the referees have made remarks in side papers on points here and there in your paper. These I submit to your consideration. Should you be disposed to make any material alteration in the paper (which I send you) please to send back the paper in the original state with fresh leaves containing the substitutions in order that we may have the evidence in case of any future dispute relating to priority.

Thomson in his report[4] seemed to think that you have been contending in part against an imaginary adversary, for with the exception of Van Feilitch[5] whose opinion he conceived originated in an obvious mistake, nearly all who had attended to the subject adopted polarity. For my own part (and I conversed with Thomson on the subject almost from the first) I don't think I ever had any other idea about it than that of a polarity (relative or absolute) of a contrary sign to that in a paramagnetic substance. I mention myself merely as a specimen of the public who had paid no particular attention to the subject; but I am certain that was Thomson's and it seems also it was Faraday's view from the first.

Please write me a line to Cambridge to let me know approximately when your paper will be ready that I may arrange about the printing.

Yours very truly | G. G. Stokes.

RI MS JT/1/S/217

1. *the last meeting of the Council*: The RS Council met on 21 April 1855, with some business held over until 23 April.
2. *the referees*: William Thomson and William Hallowes Miller (1801–80), who was the professor of mineralogy at the University of Cambridge, submitted reports (RS RR 2/252 and RR 2/253).
3. *your paper*: Tyndall's Bakerian Lecture; see letter 1029, n. 1.
4. *Thomson in his report*: 'In fact except for Von Feilitsch whose "theory" is founded on a mistake most obvious from the beginning and in my opinion not worthy of more notice than very shortly to point out that mistake, all Mr Tyndall's experiments and views are in perfect accordance with those indicated by Faraday from the beginning, and advocated by myself' (Thomson to Stokes, [14 March 1855], in *The Correspondence between Sir George Gabriel Stokes and Sir William Thomson, Baron Kelvin of Largs,* ed. D. B. Wilson, 2 vols (Cambridge: Cambridge University Press, 1990), vol. 1, p. 124).
5. *Van Feilitch*: O. von Feilitzsch, 'On the Physical Distinction of Magnetic and Diamagnetic Bodies', *Phil. Mag.,* 1 (1851), pp. 46–51.

To Mary Anne Coxe 27 April 1855 1077

27th. April 1855.

My dear Mrs. Coxe

Those who have chosen Dr. Coxe for his present eminence have in my opinion done more credit to themselves than to him, for in making such a choice they prove that they know what they are about which is more than many people know in the present age of the world. I wish him pleasure, profit and success in his new Vocation.[1] I spent last Sunday with the family of Col. Wynne at Pinner—I asked Mrs. Wynne had she heard of Miss Dix[2]—'Oh do' said she 'let me hear as much as possible about her for I reverence her in my heart', so I gave her your letter and let her enjoy it. As regards 'the son'[3]—I have sometimes found myself on the side of the young man in the apocrypha who when asked what was the strongest thing in the world answered 'Woman is the strongest'.[4] For my own part though naturally stubborn and insubordinate I am very dutiful when woman brings a womanly influence to bear upon me. But a lady[5] once made a demand upon my courtesy and I rather astounded her by the manner in which I replied to her demand. I told her plainly that I was courteous because I willed it, and not because she demanded it. From that forward she had the good sense to permit my courtesy to be my own free act and we have continued the best of friends. I am tempted to go out to Toronto as the accompanying note from Col. Sabine[6] will show. But the temptation is hardly strong enough—the post is worth about £600 a year but the sacrifices required are too great to be made. It is not therefore at all probable that I shall close with the proposal.[7] Be good enough to send the note back to me when you next write.

I long to see the leaves and blossoms appear once more, yet each succeeding day brings with it a keen north east wind which keeps them back—Thus cradled in adversity they will perhaps be all the more vigorous when they do appear. And then the news from the Crimea[8] adds itself to the wind to make one's spirits low. If we felt that all was done that might be done we should be content; but the feeling that disaster results from the absence of a head large enough and a will strong enough for the occasion is not a pleasant one. I wish Miss Dix could transfuse some of that dynamic power which reigns in her breast in her own sphere into the hearts of our Statesmen and Generals. It is that force which underlies mere insight, and which converts the latter into action that we need, but alas it is not forthcoming.

Good bye for the present | Always yours | John Tyndall.

RI MS JT/1/1/271
LT Typescript Only

---

1. *his new Vocation*: The *London Gazette* (10 April 1855, p. 1424) carried an announcement of James Coxe's appointment as one of 'Her Majesty's Commissioners for the purpose of enquiring of the state of the Lunatic Asylums in Scotland, and also into the present state of the law respecting Lunatics and Lunatic Asylums in that part of the United Kingdom'.
2. *Miss Dix*: Dorothea Lynde Dix (1802–87), an American advocate of mental health reform. In 1855 she was in close contact with James Coxe and visited Scotland to lobby for a Royal Commission to examine the lunatic asylums in Scotland.
3. *the son*: probably Edward, one of the sons of George and Anne Wynne.
4. *Woman is the strongest*: 1 Esdras 3:12 and 4:32–37. 1 Esdras is based on an early version of the

Book of Ezra. These verses are to be found in some bibles, such as the Wycliffe Bible, but not the King James Version. This quotation raises interesting questions about Tyndall's source.

5. *a lady*: Anne Edmondson; see volume 3, letter 0579.
6. *the accompanying note from Col. Sabine*: letter 1074.
7. *the proposal*: In his Journal for 30 April, Tyndall recorded that Anne Coxe had advised him to refuse the job offer from Toronto; but it is not clear whether she communicated with him verbally or by letter (RI MS JT/2/13c/744).
8. *the news from the Crimea*: Although the press frequently issued positive reports of the campaign against the Russian army in Crimea, the extensive lists of the dead and wounded indicated the extensive losses sustained by the British army. The view that British 'Statesmen and Generals' were incompetent received strong support from events at the infamous Battle of Balaklava; see letter 1096, n. 4.

To Thomas Archer Hirst 27 April 1855 1078

27th April 1855

My dear Tom.

I have an offer of a post in Toronto[1] (Canada)—the chair of Meteorology, & the directorship of the Physical Observatory. It is worth about £600 a year—but I should be obliged to change my career from one of experiment to one of observation—What say you to it? I am disposed to say no.

Yours as ever | John

RI MS JT/1/T/599

---

1. *post in Toronto*: see letter 1074

From Juliet Pollock 27 April 1855 1079

59 Montague Square | April 27th.

My dear Mr Tyndall,

I have promised to forward the enclosed[1] to you. I believe it contains an invention for making paper and that the inventor is very anxious to have your opinion upon it. I am not personally acquainted with Mr Schachner[2] and it is for the sake of my friend Mrs Brookfield[3] who is a friend of his, that I venture to trouble you on his behalf.

I was sorry not to see you when you so kindly called the other day, but the death of Mr Herries[4] is a very great affliction to me, and on that day I really felt too unwell to see any body.

Mr Herries was my Uncle, but he took the place of a Father to me from the time I was a year old, and I loved and revered him as such.

Little Walter[5] begs me to inform you that he has just spent his last sixpence on a box of turks,[6] but that he thinks he shall get more sixpences in the course of time

Yours most truly | Juliet Pollock

RI MS JT/1/6/1882
LT Typescript Only

1. *the enclosed*: enclosure missing.
2. *Mr Schachner*: not identified; name probably incorrectly transcribed by Louisa.
3. *Mrs Brookfield*: possibly Jane Brookfield of Cadogan Place, wife of William Brookfield, a clergyman and school inspector. See also letter 1187.
4. *the death of Mr Herries*: The statesman and financier John Charles Herries (b. 1778) who died on 24 April 1855 at Sevenoaks, Kent (*Morning Post,* 25 April 1855, p. 8). He was Juliet Pollock's uncle.
5. *Little Walter*: Walter Herries Pollock (1850–1926), who later became a novelist, poet and journalist, was the son of Juliet and William Fredrick Pollock.
6. *a box of turks*: a box of Turkish delight, a sweetmeat consisting of gelatine boiled, cubed, and dusted with sugar (*OED*).

From Thomas Archer Hirst 29 April 1855 1080

Broughton | April 29th 1855

My dear John,

With respect to the offer made to you of a chair in Toronto[1] my advice is quite decided. Refuse it. Whoever offered you it does not in my opinion appreciate the importance of your present researches and of your presence in England in the centre of the scientific world. Three years ago I spoke against your going to bury yourself in the backwoods[2] and if the step was unadvisable then it is far more so now. Refuse it.

On Friday last I exhibited my experiments[3] in Broughton and pleased the people pretty well I think. The experiments with Ruhmkorff's coil were more brilliant than usual. The concert also passed off well. At Queenwood everything goes on well. Pare (?) and Finch are there at present. They attended the Imp. Soc.[4] last night. It was a very good meeting and Pare was evidently pleased. I had some talk to him, and it is just possible that I may have removed the rather poor impression of me which he expressed to you.[5]

How is du Bois succeeding? He managed quite as well as I expected,[6] sometimes indeed I thought he spoke well, but on the whole I did not think he made the most of his experiments he was not sufficiently lucid in his explanations and descriptions of what his experiments were intended to prove. I remember one instance, the convulsions of the frog when a spark was taken from the machine. He explained it too loosely. I begin to fear that I shall not come and hear him again. I do not like leaving Queenwood on a Saturday evening, and again it is very expensive. We cannot get a return ticket for 14/6.[7] Such tickets are only granted at London. Give my kind regards to du Bois and tell him I wish very much he would come and spend a day or two with us, you might come with him. But I will write to him soon.

Anna is in good health and good spirits. We have never been quiet since before Easter but now that concerts and visitors are past I shall be left once more undisturbed with my work. How about my paper to the R.S.[8]—have you heard any thing of it?

Yours as ever | T A Hirst

RI MS JT/1/H/216

---

1. *chair in Toronto*: see letter 1074.
2. *Three years ago . . . the backwoods*: Hirst had advised Tyndall against going to Canada when he applied for the professorship of natural philosophy at the University of Toronto in the autumn of 1851; see letter 1074, n. 2.
3. *exhibited my experiments*: 'On Friday [April] the 27th I exhibited some experiments on Electricity, etc. at the Lecture Hall in Broughton, and afterwards took part in a concert conducted by [John] Cornwall', the music master (Hirst's Journal: RI MS JT/2/32c/1166).
4. *Imp. Soc.*: Mutual Improvement Society at Queenwood College: see letter 1050, n. 2.
5. *poor impression of me which he expressed to you*: The reason for Pare's 'poor impression' of Hirst has not been ascertained. See also letter 1081.
6. *He managed . . . expected*: Hirst had attended du Bois-Reymond's first lecture at the RI but not the second (which was delivered on 28 April 1855).
7. *We cannot get a return ticket for 14/6*: To travel from Waterloo Station, in London, to Dunbridge Station, which was the closest to Queenwood College, was a distance of 84½ miles. *Bradshaw's General Railway and Steam Navigation Guide* for February 1855 indicates (p. 21) that express trains, which carried only 1st class passengers, completed the journey in 2 hrs, 47 mins, while the fastest journey for 2nd and 3rd class travellers was 3hrs, 27 mins, although some trains took much longer. The cost of a single 1st class ticket was 18/6; 2nd class 12/6; 3rd class 7/-.
8. *my paper to the R.S.*: see letter 1058, n. 3.

To Thomas Archer Hirst 30 April 1855 1081

My dear Tom

Bravo!—there can be no mistake about your opinion[1]—and mine is just as decided. Before I wrote to you I had made up my mind—indeed before I finished the reading of the letter (which I will send you in a day or two) my resolution was formed. I am glad that your experiments passed off well, glad that you had an opportunity of speaking to Pare who I believe is a well-meaning man & very clever. But I do not remember his expressing a poor opinion to me concerning you[2]—He said something about you—something about some addition to your vigour—but never any thing poor regarding you.

I know nothing of your paper to the R.S.[3] I know nothing of a paper of my own which was sent into them last November—Both I trust will appear in due season. Du Bois is getting on pretty well—his last lecture[4] I should not call quite so successful as his first—But in both cases he has been far more fortunate than in his lecture at the London Institution. I should like to see you here—nevertheless as you say the trip is expensive—and if you now give yourself the habit of keeping down expense it will cause you no trouble by & by when you must do it.

Do you think you could lend me your coil for my Friday evening 1st June?[5]—If you could lend it to me and describe one or two pretty experiments it would be of service to me. I might get one from Gassiot—indeed there is one in the house here,[6] but Faraday is using it so that I think on the whole if yours could be sent to me it would be most pleasant. Tell Anna I will soon write to her.

Affectionately | John

Monday

RI MS JT/1/T/600

---

1. *your opinion*: on the offer of a post at the University of Toronto; see letter 1080.
2. *a poor opinion to me concerning you*: see letter 1080, n. 5.
3. *your paper to the R.S.*: see letter 1058, n. 3. Hirst's paper 'On the Existence of a Magnetic Medium' was received by the RS on 14 April 1855 (*Roy. Soc. Proc.,* 7 (1854–55), p. 448).
4. *his last lecture*: on 28 April.
5. *coil for my Friday evening 1st June*: Tyndall was to deliver the Friday Evening Discourse at the RI on 1 June 1855, entitled 'On the Currents of the Leyden Battery'. His account of the lecture refers to a 'wooden cylinder [. . .] round which two [insulated] copper wires, each 75 feet in length, were wound' (*Roy. Inst. Proc.,* 2 (1854–58), pp. 132–35, on p. 133).
6. *the house here*: the RI.

To George Gabriel Stokes 6 May 1855 1082

6th May 1855.

My dear Sir,

Two causes have prevented me from writing to you sooner. The first is that a wrong paper was in the first instance sent to me, and the correction of this mistake occasioned a delay of a couple of days. The second is that a bad head and many other engagements conspired to prevent me from making a few remarks on Mr Thomson's memoranda[1]—I will now swiftly refer to his notes taking them in the order in which they come.

It is perfectly certain that not only is another explanation conceivable but that such has been conceived and supported for years by writers on diamagnetism. Plücker himself made some experiments[2] for the express purpose of proving the explanation referred to [as] false—but he now just as expressly admits the error of his experiments and the truth of mine. With regard to the question of priority mooted in the commencing paragraph a reference to both our papers will show what each has done—no doubt the superior repulsion of bismuth along the planes of cleavage was established before my results were published; but I think I did something to generalise this result. If Mr Thomson however thinks otherwise I am content.

The experiments of Joule[3] and others on magnetic saturation are very familiar ones to me, but I leave you to judge whether the following fact does not justify the caution which I have observed. I have found many feeble diamagnetic bodies which, operated on by the self-same magnetic forces, do not follow the law of repulsion observed in the case of bismuth, but whose repulsion increases much less rapidly. This of course can only be explained by reference to the condition of the body operated on. The whole matter however affects the general aim of my paper so little that to avoid all discussion I will strike out the passage altogether.

A similar wish to remove all possible grounds of dispute induces me to strike out what has been said regarding places of stronger and weaker force. Its omission, though deprecated by some, does not affect the paper. At some future time[4] I shall take occasion to say a few words upon the subject.

With regard to Mr Thomson's views as to the influence of diamagnetic particles upon each other, I can only say that they do not coincide with mine. When I spoke of the 'promised proof' I was well acquainted with the paper in the British Association Report.[5] But if you look at the paper I am sure you will agree with me that it does not go beyond simple affirmation without any proof what ever.

Not long ago I had a letter from Mr v. Feilitsch accompanying his paper[6] and drawing my attention to his results. He is far from believing that his views originate in an obvious mistake. I hardly think the mistake was obvious until my paper proved it to be so. Even at the present moment there are experiments by v. Feilitsch which I know influence the minds of some who have thought deeply upon these matters and which still remain unanswered. In fact, however clear Mr Thomson's views may be, this subject is really in a state of great confusion and if I have succeeded in stating the truth with clearness it is all that I desire.

From many of his published papers[7] I should have inferred that Mr Thomson assumed the existence of diamagnetic polarity. Now even supposing this polarity never to have been doubted it is certainly worth while to convert the assumption into a fact and this I think has been done in the paper under consideration. But from what he mentioned to me himself in private, from his reply to me in Belfast,[8] from Mr Faraday's reference to his supposed refutation of the doctrine of diamagnetic polarity in the lecture upon education[9]—from the recent repetition of Thomson's argument[10] that a perpetual motion must result if diamagnetic polarity be conceded, I think I was warranted in concluding that Mr Thomson had changed his views. I should be the last to endeavour to hold him to an opinion once expressed, but I should also like to show that in supposing him to be the opponent of diamagnetic polarity I had grounds for believing him to be more than 'an imaginary adversary'. However all that involves his name in connexion with the question of polarity has been struck out and this makes an end of the discussion.

I am not aware that I have been anticipated by 'various writers' on the point referred to in Mr Thomson's last paragraph,[11] but am rather of opinion that I am the first who pointed it out. Nevertheless in this case also I have struck out the passages referred to.

I trust the paper in its present state will meet the wishes of all concerned. I have also attended to remarks of the second referee as far as my time permitted me.

The paper is now quite ready.

Believe me | Yours very truly | John Tyndall.

If you wish I will take the paper to the printers. I shall only add a short note or two which will bear the present date and which indeed contain nothing to contend about. I shall have to make some arrangement about the plates | J.T.

RI MS JT/1/4/1464–6
LT Typescript Only

1. *Mr Thomson's memoranda*: Thomson's letter to Stokes of 14[?] March 1855 included 'some memoranda hastily put down in reading' Tyndall's RS paper 'On the Nature of the Force

by which Bodies Are Repelled from the Poles of a Magnet' (RS RR 2/252; Thomson to Stokes, [14 March 1855], cited in letter 1076, n. 4.

2. *Plücker himself made some experiments*: Plücker published several papers on diamagnetic properties of crystals including 'Ueber die Abstossung der Optischen Axen der Krystalle durch die Pole der Magnete', *Poggend. Annal.,* 72 (1847), pp. 315–43; 'Ueber die Theorie des Diamagnetismus, die Erklärung des Ueberganges Magnetischen Verhaltens in Diamagnetisches und Mathematische Begründung der bei Krystallen beobachteten Erscheinungen', *Poggend. Annal.,* 86 (1852), pp. 1–34. After reading Tyndall's Bakerian Lecture, Plücker expressed his annoyance to Faraday: see *Faraday Correspondence,* 5:3109.
3. *The experiments of Joule*: J. Joule, 'Account of Experiments Demonstrating a Limit to the Magnetizability of Iron', *Phil. Mag.*, 2 (1851), pp. 306–15 and 447–56.
4. *At some future time*: Tyndall subsequently addressed this issue in his 'On the Relation of Diamagnetic Polarity to Magnecrystallic Action', *Phil. Mag.,* 11 (1856), pp. 125–37.
5. *the paper in the British Association Report*: probably W. Thomson, 'On the Equilibrium of Elongated Masses of Ferromagnetic Substance in Uniform and Varied Fields of Force', *Brit. Assoc. Rep. 1852,* pp. 18–20.
6. *a letter . . . accompanying his paper*: letter missing. The paper was probably F. C. O. von Feilitzsch, 'Erklärung der diamagnetischen Wirkungsweise durch die Ampère'sche Theorie', *Poggend. Annal.,* 92 (1854), pp. 366–401 and 536–76.
7. *many of his published papers*: see Jackson, pp. 453–55.
8. *his reply to me in Belfast*: presumably in the discussion at the BAAS meeting on 1 September 1852, following Thomson's paper 'On Certain Magnetic Curves'; see Jackson, p. 458.
9. *Mr Faraday's . . . upon education*: Faraday used the following example to caution against drawing conclusions prematurely. To account for the diagmagnetic behaviour of bismuth or phosphorus between the poles of a strong magnet 'I at one time suggested the idea that a polarity was induced in the phosphorus or bismuth the reverse of the polarity induced in iron, and that opinion is still sustained by eminent philosophers.' Owing to contrary evidence, Faraday subsequently withheld judgment on that theory ('Observations on Mental Education', in M. Faraday, *Experimental Researches in Chemistry and Physics* (London: R. Taylor & W. Francis, 1859), pp. 453–91, on p. 482).
10. *the recent repetition of Thomson's argument*: In a note to his Bakerian Lecture (*Phil. Trans.,* 145 (1855), pp. 50–51), Tyndall drew attention to the report in the *Athenæum* of Thomson's comments on a paper ('On the Diamagnetic Force') that Tyndall had delivered at the 1854 meeting of the BAAS. There Thomson had admitted 'that he had not perfect confidence in the truth of the conclusion, as one of the assumptions on which the reasoning was founded admitted of doubt' (*Athenæum* (7 October 1854), pp. 1203–4).
11. *Mr Thomson's last paragraph*: The last paragraph of Thomson's report held in the archive of the RS (RS RR 2/252) makes no mention of 'various writers'.

From Thomas Archer Hirst [10?] May 1855 1083

Broughton | May 6th 1855

My dear John,

You see by the above date that I commenced writing to you on Sunday but did not conclude. Anna asks me every day if I have heard from you and when I say no she says she fears you are not well. I have promised to write to-day and enquire. Almost every Saturday she prophecies you will come before evening, at first I believed her, but now I begin to think her a false prophetess. The Excursion trains[1] have commenced n'est-ce pas[2]? I will send you Ruhmkorff's apparatus[3] in time. My experiments are very little different to the one Faraday shewed us.

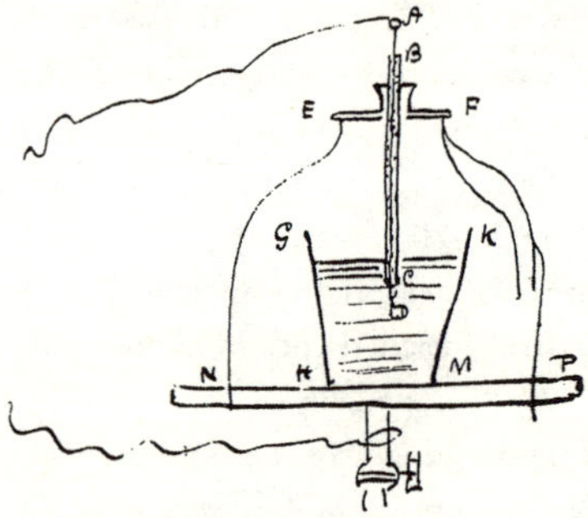

On the plate of the air pump NP a common receiver open at the top is placed, through the cap EF of which a glass tube BC is passed. Through this tube BC a wire AD passes and the rest of the tube is filled with sealing wax, so as to have the wire from B to C perfectly insulated. To secure this I closed the end C with the blow pipe and having made the tube hot I dropped in powdered sealing wax which melted as fast as I threw it in, and thus the wire from B to C was imbedded in a stick of sealing wax. This would not be necessary if a thick glass tube were used & simply closed at C with the blow pipe. Well the extremity C just dips below the surface of some mercury in a tumbler GHMK. The plate of the air pump and the knob A are connected with the two poles of Ruhmkorff's apparatus and—you know what takes place.[4]

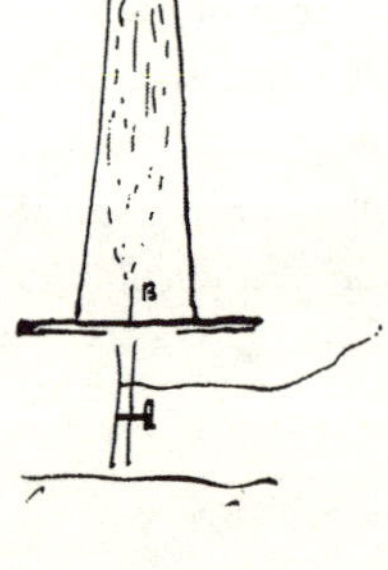

Another very pretty experiment is to make the electricity jump from one of the points A to another B in a tall receiver.

You will find that the coil of my Ruhmkorff has got a little displaced (by carriage) and will have to adjust it.

What is the title of your Friday evening lecture?[5] We are just beginning to prepare our examination papers for Midsummer.

Write soon to | Yours affectionately | T. A. Hirst.

RI MS JT/1/H/217

1. *Excursion trains*: trains intended to convey persons making a pleasure excursion, usually at reduced fares (*OED*); see letter 1080, n. 7.
2. *n'est-ce pas*: is it not (French).
3. *Ruhmkorff's apparatus*: a Ruhmkorff coil for producing pulses of high voltage electricity.
4. *what takes place*: a discharge.
5. *the title of your Friday evening lecture*: On 1 June Tyndall lectured 'On the Currents of the Leyden Battery' at the RI.

From Heinrich Gustav Magnus 11 May 1855 1084

Berlin 11 Mai 1855

Theurer Freund

Endlich bin ich im Stande Ihnen ein Exemplar meiner Abhandl. zu senden, dieselbe hat einen etwas umfassenden Titel „Hydraulische Untersuchungen" erhalten. Der Verleger der Poggendorffschen Annalen hat dieselbe besonders abdrucken lassen und will sie so in den Handel bringen. Sie ist etwas umfangreich und zerfällt eigentlich in 3 Abhandl. Die erste „über die Einwirkung zweier Strahlen auf einander die unter einem Winkel zusammentreffen" Die 2te „über die Gestalt von Strahlen welche aus Oeffnungen in dünner Wand ausfliessen" Die 3te. „Uber das Eindringen von Luftblasen in eine Flüssigkeit." Sie erinnern sich vielleicht noch daß ich bei Ihrer letzten Anwesenheit in Berlin Ihnen eine Abhandl vorlaß die über den 3ten Gegenstand handelte; diese bildet zum Theil die letzte Abhandl. Das übrige habe ich unterdrückt. Ich wäre begierig zu wissen was Sie über die ganze Arbeit denken, nicht sowohl über die einzelnen Erklärungen und Facta als über den Werth des ganzen. Mir geht es immer eigenthümlich mit meinen Arbeiten. Ich beginne sie mit großem Eifer und voll von Interesse, aber wenn sie vollendet sind und namentlich wenn sie redigirt sind so sage ich mir, wozu all die Mühe, was ist nun dadurch herausgekommen?—Doch der Mensch ist nun einmal so! Man muß arbeiten und das Seinige thun und mehr kann man nicht. Die Vorrede die Sie in der

Abhandl finden, fehlt natürlich in den Annalen, ich wollte die Abhandl dort in mehrere Hefte getrennt erscheinen lassen, aber Poggendorff zog es vor sie auf einmal abzudrucken.

Für Ihren freundlichen Brief vom 15 v.M. meinen verbindlichsten Dank. Ich kann mir denken wie beschäftigt Sie jetzt sind. Ihr Streit mit Faraday thut mir leid. Zwar kann ich Faraday's Ansichten nicht überall theilen, aber ich liebe ihn persönlich so sehr, daß mich alles trübt was ihn unangenehm berühren könnte.

Durch Asher habe ich 4g 20th = 14 Schill. erhalten. Ist denn das kleine Instrument zu Ihrer Zufriedenheit ausgefallen?

Wie geht es Du Bois, sind seine Vorträge recht besucht? Bitte grüßen Sie ihn.

Jetzt kann ich Ihnen auch über meinen Besuch in Paris sagen. Ich werde Anfang oder Mitte August dort eintreffen, Anfangs wollte ich zu Pfingsten hin, allein das Ministerium hat mich zum Jury bestimmt, und da die Jury im August zusammentreten wird, so werde ich meine Reise bis dahin verschieben. Sehr schön wäre es wenn Sie dann auch nach Paris kämen. Sie würden noch Dove finden, der auch zur Jury gehört, und außerdem geht Dr. Rose hin, auch hoffe ich noch Rieß und Poggendorff zu bestimmen, so daß wir dann in großer Zahl zusammenträfen.

Wenn Sie Faraday sehn bitte grüßen Sie ihn und sagen Sie ihm recht viel Freundliches von mir.

Setzen Sie Ihre Versuche fort und lassen Sie nicht nach, es ist doch eigentlich ein Scandal daß man nicht soll ermitteln können ob die diamagnetischen Körper polarisch sind oder nicht. Wenn die Sache auch schwierig ist, so haben Sie doch den Vortheil daß Sie in einem Gebiete arbeiten, in dem jetzt sehr Viele thätig sind, daß dadurch Ihre Resultate sowohl für Sie wie für Ihre Freunde, und auch für Ihre Gegner von Interesse sind, und Sie dadurch zu immer neuen Untersuchungen anregen.

Grüßen Sie Francis, Mr. Gassiot und alle die sich meiner erinnern. Meine Frau und Kinder senden Ihnen die freundlichsten Grüße. Ihre Schülerin Anna versteht jetzt ganz gut wenn man englisch mit ihr spricht, und treibt es mit Leidenschaft.

von Herzen | Ihr | G Magnus.

Berlin 11 Mai 1855

Dear friend

I am finally in a position to send you a copy of my treatise;[1] it has received a somewhat broad title, 'Hydraulic Investigations' ['Hydraulische Untersuchungen']. The publisher of Poggendorff's Annalen[2] has had it printed separately and wants to put it on sale like that. It is somewhat extensive and really falls into 3 papers. The first one, 'on the effect of two jets on one another that meet at an angle', the 2nd 'on the shape of jets which flow out of openings in a thin wall', the 3rd 'on the penetration of air bubbles into a liquid'. You perhaps still remember that I read you a paper that dealt with the 3rd subject the last time you were in Berlin;[3] this partly forms the last paper. I have suppressed the remainder. I would be curious to know what you think about the whole work, both about the individual explanations and facts as well as about the value of the whole. I always have a curious feeling about my works. I begin them with great enthusiasm and full of interest, but once they are completed and especially once they are edited, then I say to myself, why all the effort, what has now eventuated through it all?—But Man is just like that! One has to work and do one's bit, and one cannot do anything more. The preface that you find in the paper is, of course, missing in the Annalen, I wanted to have the paper appear there separated into several issues, but Poggendorff preferred to print it all in one lot.

My most profound thanks for your kind letter of the 15th of last month.[4] I can imagine how busy you are now. I am sorry about your quarrel with Faraday. Admittedly I cannot share Faraday's views on every matter, but I like him personally so much that everything that could affect him in an unpleasant way saddens me.

Through Asher[5] I have received 4g 20th = 14 Schill.[6] Did the small instrument[7] turn out to your satisfaction then?

How is Du Bois doing? Are his lectures well-attended? Please give him my regards.

I can now tell you about my visit to Paris also. I shall arrive there at the beginning or in the middle of August. Initially I wanted to go at Whitsun, but the ministry has appointed me to the jury,[8] and as the jury will meet in August, I shall postpone my trip until then. It would be very nice if you were to come to Paris then also. You would still see Dove, who is also a member of the jury, and Dr. Rose[9] is going there as well, I also hope to appoint Riess and Poggendorff[10] still, so that we can then meet in a large number.

If you see Faraday, please give him my regards and give him very many kind wishes from me.

Continue your experiments and do not ease off, it is really a scandal that one should not be able to determine whether diamagnetic bodies are polar or not. Even if the matter is difficult, you nevertheless have the advantage of working in a field in which very many are active now, and that, through this, your results are of interest both for you as for your friends, and also for your adversaries, and through this they encourage you to ever-new investigations.

Give my regards to Francis, Mr. Gassiot, and everyone who remembers me.

My wife and children[11] send you their kindest greetings. Your pupil Anna[12] understands very well now when one speaks English with her, and she pursues it with passion.

sincerely | Your | G Magnus.

RI MS JT/1/M/16

1. *my treatise*: G. Magnus, 'Hydraulische Untersuchungen', *Poggend. Annal.*, 95 (1855), pp. 1–59.
2. *The publisher of Poggendorff's Annalen*: The physicist Johann Christian Poggendorff (1796–1877), of the Humboldt University, Berlin, edited the *Annalen der Physik und Chemie* (cited here as *Poggend. Annal.*) from 1824 to 1876.
3. *the last time you were in Berlin*: Tyndall's Journal entry for 25 July 1851 reads: 'Saw the experiments of Magnus on the deviation of projectiles yesterday evening' (RI MS JT/2/13b/613).
4. *your kind letter . . . month*: letter missing.
5. *Asher*: In 1830 Adolphus Asher and Isaac Simon founded a bookshop in Linden Walk, Berlin that attracted many foreign visitors, specialised in British publications and often supplied the British Museum Library with German language books. Following Asher's death in 1853 the firm continued trading under the name A. Asher & Co.
6. *4g 20th = 14 Schill*: 4 guilders and 20 Thaler = 14 Schilling.
7. *the small instrument*: not located.
8. *jury*: Magnus served on one of the juries at the Exposition Universelle, which was held in Paris from 15 May to 15 November 1855.
9. *Rose*: The mineralogist Gustav Rose (1798–1873), who worked principally in the areas of petrology and mineralogy.
10. *Riess and Poggendorff*: neither appears on the list of jurors.
11. *My wife and children*: Bertha Magnus (née Humblot; 1820–1910) was his wife and their children were Anna (1841–75), Christine (1842–1936) and Paul Gustav (1845–1930).
12. *Your pupil Anna*: Magnus's elder daughter.

To Thomas Archer Hirst 13 May 1855 1085

13th May 1855

My dear Tom.

This is Sunday afternoon—almost evening—the bells are ringing for Church and the rain is drizzling heavily down and keeping me within doors—Thus imprisoned I will give my thoughts at last a quarter of an hour's holiday by writing to you. I have nothing to say to you still the highest evidence of creative power theologians say is to produce something from nothing[1] and I don't see why I should not by my creative faculty—Still I can gossip at least—I conveyed to Du Bois your wish regarding a visit to you. The expense of the journey was evidently a barrier with him, but since he received your note he evinced symptoms of taking you at your word—I may possibly go along with him when he goes. Francis also talks of going down. So that you are threatened with a Wolkenbruch of visitors[2] one of these fine days. I think the lecture of yesterday[3] was one of the best which Du Bois has given to us—his utterance was better than hitherto. Last Wednesday I dined with an old gentleman named Goldsmid,[4] some relative probably of the proprietor of Queenwood.[5] He is a hearty good old fellow and not at all like a screw[6]—his board was sumptuous, and the good old man had the kindness to give me a passable girl[7] to take to dinner with whom I had a very affable & sparkling conversation. Since then the said lady's father has invited me to dine with him so that thus I am dragged away from the magnets & molecules. There is a club here in connexion with the Royal Society which seems to be rather exclusive in its character—it is called the Philosophical Club[8]—and last Friday I received an intimation that I had been elected a member of it. Stokes and I think Huxley was elected at the same time[9]—but what surprised me greatly considering how much influence the man possesses Warren DelaRue was blackballed. Percy[10] was also blackballed some time ago. My health is getting slowly and steadily better—I had the gravest apprehensions about it a few weeks ago and sometimes used to smile as I looked at my whole cheeks and aching bones in the looking glass—a kind of hopeless smile expressive simply of my resignation to the worst. I watched the effects of various aliments upon my system and I discerned a fortnight ago at Windsor that bitter beer seemed to do me harm, I thought it might do good. I tried it copiously—took a small bottle at 11 o/c in the forenoon, a glass at dinner, a bottle after dinner, and found my downward progress arrested, and a sure & steady upward motion established. Some time ago I had fallen into the habit of having a chop[11] at a confectioner's and a cup of coffee afterwards instead of dinner—This I have soon abandoned, and will the most beneficial results—for I had an application[12] from the association of Schoolmasters[13] yesterday to deliver an address to the Society on the first of June. I have a lecture that day at the Royal Institution. Of course this alone if no other obstacle intervened (which then does) would prevent me from undertaking the thing—Say to Anna that if she has not received a certain article already it will come to her very soon.

John | Dont send me the coil until you hear from me again.

RI MS JT/1/T/601

1. *to produce something from nothing*: The prevalent, but much disputed, view that God created the universe *ex nihilo* (out of nothing) based on such texts as Genesis 1:1: 'In the beginning God created the heaven and the earth'.
2. *a Wolkenbruch of visitors*: a deluge of visitors; literally a cloudburst.
3. *the lecture of yesterday*: Du Bois-Reymond's fourth RI lecture 'On Electro-Physiology' was delivered on 12 May 1855.
4. *Goldsmid*: The Jewish financier Aaron Asher Goldsmid (1785–1860), who served as a manager of the RI.
5. *the proprietor of Queenwood*: In 1839 the socialist Robert Owen had leased the land on which Queenwood College was built from the financier Isaac Lyon Goldsmid (1778–1859), who owned the Tytherley estate and was the brother of Aaron Asher Goldsmid.
6. *a screw*: one who forces down (prices) by haggling; a stingy, miserly person (*OED*); an oblique reference to Shylock, the Jew, in Shakespeare's *Merchant of Venice.*
7. *a passable girl*: not identified.
8. *the Philosophical Club*: Founded in 1847, the Philosophical Club of the Royal Society of London was a select dining club limited to 47 FRSs. The club's aim was 'to promote as much as possible the scientific objects of the Royal Society, to facilitate intercourse between those Fellows who are actively engaged in cultivating the various branches of Natural Science and who have contributed to its progress, to increase the attendance at the evening meetings, and to encourage the contribution and discussion of papers' (T. G. Bonney, *Annals of the Philosophical Club of the Royal Society* (London: Macmillan, 1919), p. 1). Tyndall was elected to the Club on 10 May 1855.
9. *Huxley was elected at the same time*: Stokes and Huxley were also elected at the meeting of the Philosophical Club on 10 May 1855.
10. *Percy*: The Birmingham metallurgist John Percy (1817–89), who served as professor of metallurgy at the Government School of Mines.
11. *a chop*: a slice of meat, usually mutton or pork, including generally a rib, intended to be cooked and served by itself (*OED*). Chops were widely available in London coffee shops and other outlets.
12. *an application*: letter missing.
13. *the association of Schoolmasters*: A meeting of the United Association of Schoolmasters of Great Britain was held in St Martin's Hall, Longacre, London on 1 June (*London Standard,* 2 June 1855, p. [4]).

From Thomas Archer Hirst 20 May 1855 1086

Broughton, | Sunday May 20th /55

My dear John.

Your last letter to me,[1] for which I am told to thank last Sunday's rain, pleased me as you anticipated it would. I was truly glad to hear you were strong again. I hope your strength is not as temporary as usual but I have my apprehensions. I am glad that you are able and willing to take a good dinner and bitter ale your willingness to do so is more pleasing to me than the act itself. Your lectures you say this morning to Anna cease in three weeks,[2] what do you do then? When do you go to Paris? Our Queenwood concert will be on the 11th of course you will be here then. And after that I would advise you to start for the continent, to remain there as long as you possibly can to give up your scientific labours for a time, and to seek the occupation you need in walking and in literature. Try a change in your habits of study, quit molecules (to return to them) and write us a novel, or a description of your tour, or some poetry, original or translated. Let your occupation be what you like but let it be entirely changed. We shall not spend our vacation at Broughton, we shall be away most of the six weeks. I have some thoughts of spending ten days or a fortnight in London, I want to obtain the perusal of some mathematical books which I do not possess and cannot buy. I have been thinking (merely thinking) that if you were absent from London we might occupy your lodgings. Could you obtain permission for me to use the Royal Institution library? In pursuing my physical studies I find it necessary to be acquainted with many of the late expansions in methods of calculation among the rest: Determinants,[3] Quaternions,[4] and the Calculus of Forms and Operations.[5] I begin to think it would be wisest for me to leave original investigation for a short time and devote myself to the study of what has been done by others bearing on the subject. I could use a fortnight in London well. Affairs at Queenwood as far as the boys are concerned were never more prosperous. As far as Edmondson is concerned the prospect was not cheering a short time ago but is better now. He has had another meeting with his partners[6] and they have not yet agreed about his salary. Ten days ago he was determined (you know how much his determination is worth) to dissolve the partnership and take a situation (!) or commence a smaller school. I believe the women were persuading him to this, but thanks to us, Wright, Haas and myself, I think he has abandoned the foolish notion. Take a situation! Who would have him? Commence a school! He is not worth 2 pence when his debts are paid. Wright has proved clearly that with a salary of £300 for himself, he & his family considering their style of living, etc. would be receiving virtually an income of about £900. He would find his position miserably worse were he to give up Queenwood, not to mention the actual sin of doing so in its present state of prosperity. But these women, these women! They hate Edmonsons partners, they have but a very imperfect knowledge of Edmondsons financial position, they are hankering after more ease, they hint suspiciously and foolishly at the pleasant kind of life led by Mrs Hirst and Mrs Wright, they dislike the thought of George William[7] succeeding his father at Queenwood and inheriting his present anomalous position, &c &c. In short with their likes and dislikes they had nearly turned the poor weak man's head. But thanks to Haas and Wright to whom he has confided his affairs I think he is almost past the perilous crisis. Haas has been let into his confidence on account of the transactions he has lately had with Pare and Finch about a new house for himself. Through Wright & Haas I am perfectly well acquainted with the state of affairs, but Edmonson never speaks to me about them, perhaps because he does not think I know as much as I do. On school matters he is far more communicative than he was but about his own affairs he has not yet said a word to me. God knows were it not for Queenwood and the lads in it whom I love I would be sorry to soil my hands with his petty troubles that have all arisen from his incurable incompetence. But last week I was on the point of breaking through all reserve, to try and rescue him from drowning; luckily matters took a turn through Wright and Haas to whom I can talk and there was no necessity for my interference. For his own sake however I wish he would break down this reserve between us. Anna thanks you for your letter this morning. As usual as she prophesied yesterday that you would come and even made your bedroom ready but as usual she was mistaken. Let us hear again soon and believe me

As ever | Yours | T A Hirst

RI MS JT/1/H/218

---

1. *Your last letter to me*: letter 1085.
2. *Your lectures . . . three weeks*: Tyndall's last RI lecture was on 6 June 1855.
3. *Determinants*: A determinant is the sum of the products of a square block or 'matrix' of quantities, each product containing one factor from each row and column, and having the plus or minus sign according to the arrangement of its factors in the block (*OED*).
4. *Quaternions*: a non-commutative form of numbers involving complex numbers. They were devised by the Irish mathematician William Rowan Hamilton in 1843.
5. *Calculus of Forms and Operations*: 'THE Calculus of Operations, in the greatest extension of the phrase, may be regarded as that science which treats of the combinations of symbols of operation, conformably to certain given laws, and of the relations by which these symbols are connected with the subjects on which they operate' (R. Carmichael, *A Treatise on the Calculus of Operations* (London: Longman, Brown, Green and Longmans, 1850), p. 1.
6. *his partners*: Hirst was referring to John Finch and John Pare who (together with Robert Owen) were the governors of Queenwood College.
7. *George William*: George William Edmondson (1837–1913), the son of George and Anne Edmondson, wanted to be an artist and defied his father, who intended that he become an

engineer. He emigrated to Canada, participated in an expedition promoted by the RS and later settled in Norwalk, Ohio, where he established himself as a photographer.

From Friedrich Ludwig Stegmann[1] 20 May 1855 1087

Marburg, 20 Mai 1855

Theurer Freund!

Es ist länger als ein halbes Jahr verstrichen, seitdem ich für Sie die zweite Hälfte meines Lehrbuchs über Variationsrechnung zur Absendung bereit legte. Ein Bekannter von mir, welcher aus England gekommen war und bei seiner Durchreise durch Marburg mich besuchte, wollte das Packetchen mitnehmen, derselbe hat aber einen andern Rückweg eingeschlagen und so kam es in Vergessenheit. Da Sie indeßen die erste Hälfte freundlich acceptirt haben, so wird es Ihnen hoffentlich, wenn auch nur der Completirung wegen, nicht unangenehm sein, wenn ich die Reise meines Freundes Dunker benutze um Ihnen den Schluss zu senden. Hoffentlich haben Sie sich seither immer recht wohl befunden. Ihre vielen vortrefflichen Arbeiten, denen wir auch hier mit grossem Interesse folgen, lassen wenigstens auf die rüstigste Gesundheit schliessen.

Herr Prof. Dunker, als Nachfolger Girard's unser Mineraloge und Geologe, zugleich ein grosser Conchylien-Kenner u. eine anerkannte Autorität in diesem Fache, ist als Gast seines Freundes Mr. Cumming, im Begriffe nach London zu reisen. Sie werden, wenn er so glücklich ist Sie zu treffen, in ihm einen höchst liebenswürdigen Gelehrten kennenlernen; und wenn Sie ihm, da er das Englische nicht geläufig spricht, bei einem Besuche in der Royal Institution mit ihren Sehenswürdigkeiten vielleicht einige Unterstützung zu gewähren vermögen, so dürfen Sie einer aufrichtigen Würdigung derselben sich versichert halten.

Gar sehr habe ich mich über Ihren letzten lieben Brief gefreut, welcher so unverkennbar ausdrückte, dass Sie mir ein freundliches Andenken bewahren, so wie unserer kleinen Universitäts-Stadt überhaupt. Ihre herzliche Einladung, Sie demnächst einmal zu besuchen, ist sehr lockend für mich, und wird auch sicher Früchte tragen. Ich habe anfangs, als ich mit Prof. D. die von ihm beabsichtigte Reise besprach, grosse Lust zum Mitreisen gehabt; indeßen hat die Erwägung, dass die Zeitdauer unserer Pfingstferien zu kurz ist, um ausser London auch wenigstens noch meinen Freund Dreßner in Sandhurst zu besuchen, bei welchem ich doch mindestens 5 bis 6 Tage verweilen müsste, allzu stark abgerathen. Für nächsten Herbst habe ich mich bereits bei einigen Freunden versprochen nach Wien zur Versammlung der Aerzte und Naturforscher zu reisen. Allein im nächsten Jahre hoffe ich endlich einmal Ihnen meinen lange gewünschten Besuch machen zu können. Vielleicht habe ich vorher noch einmal das Vergnügen Sie hier zu sehen.

Kürzlich wäre ich beinahe veranlasst worden meine hiesige Stellung aufzugeben und nach Heidelberg überzusiedeln. Prof. Bunsen hat sich sehr dafür interessirt. Indessen ist eine Stelle dort eigentlich nicht vakant, sondern ein sehr alter Herr Schweins fungirt als Mathematikus, und es scheint, dass dieser nachdem er von der Sache Kenntnis erhalten, eine Contre-Operation eingeleitet hat, wodurch in Verbindung mit der Finanz-Frage ein Stillstand herbeigeführt wurde.

Ich habe meinen Freund Dunker gebeten, wenn sich ihm Gelegenheit darbietet, zwei Bücher

a) eine vierstellige Logarithmentafel
b) Green's Essay on the Application of Mathematical Analysis to the Theories of Electricity and Magnetism.

für mich mitzubringen. Dieser Essay ist bereits im Jahr 1828 in Nottingham (?) erschienen, er scheint aber kaum aufzutreiben zu sein. Ich habe bereits früher durch eine hiesige so wie durch eine Leipziger Buchhandlung Versuche gemacht, dasselbe zu erhalten, aber vergebens. —Was die Logarithmentafel betrifft, so sind ausser den 5-stelligen und 7-stelligen in England auch solche, worin die trigonometrischen Functionen blos mit vier Decimalstellen aufgeführt sind, viel gebräuchlich, ich weiss aber nicht, welche durch typographische Anordnung und Correctheit sich am meisten empfiehlt. Im Vertrauen auf Ihre freundschaftlichen Gesinnungen habe ich Prof. Dunker gebeten, Ihren Rath darüber einzuziehen; wenn Sie ihm den Titel einer guten kleinen Logarithmustafel, und die Addreße einer Antiquariats-Handlung; wo das Gewünschte [voraussichtlich] zu [kaufen] wäre <*words illeg*> so werden Sie mich recht sehr zu Dank verpflichten.

Mit aufrichtiger Hochachtung | Ihr | Treuer Freund | Stegmann.

Marburg, 20 May 1855

Dear friend!

More than half a year has passed since I prepared the second half of my textbook on the calculus of variations[2] to be sent to you. An acquaintance of mine, who had come from England and was visiting me on his way through Marburg, was going to take the little package with him, but he took a different return route and thus the matter was forgotten. As you have kindly accepted the first half in the meantime, then hopefully it will not be disagreeable to you, if only for the sake of completion, if I take advantage of my friend Dunker's trip[3] to send you the conclusion. Hopefully you have been very well all the time since then. Your many excellent works, which we follow even here with great interest, suggest at least the sprightliest health.

Prof. Dunker, as Girard's successor[4] our mineralogist and geologist, and at the same time a great expert in conchology and a recognised authority in this subject, is about to travel to London as a guest of his friend, Mr Cumming.[5] You will, if he is so fortunate to meet you,[6] make the acquaintance of a highly amiable scholar in him; and if, as he does not speak English fluently, you can perhaps give him some assistance during a visit to the Royal Institution with its points of interest, then you may consider yourself assured of a sincere acknowledgement of this.

I was very pleased indeed about your last kind letter,[7] which so unmistakably expressed the fact that you retain a friendly memory of me, as well as of our small university town generally. Your cordial invitation to visit you soon is very tempting for me and will certainly bear fruit too. Initially, when I discussed with Prof. D.[8] the trip that he was intending to do, I was greatly inclined to travel with him; however, the consideration that the duration of our Whitsun holidays is too short to visit, in addition to London, at the very least even my friend Dressner[9] in Sandhurst as well, with whom I would have to stay at least 5 to 6 days, has advised all too strongly against this. For next autumn, I have already promised some friends that I shall travel to Vienna to the meeting of physicians and natural scientists.[10] Only the following year do I hope to finally be able to pay my long wished-for visit to you. Perhaps I shall have the pleasure of seeing you here once again before then.

Recently, I would almost have been prompted to give up my position here and move to Heidelberg. Prof Bunsen was very interested in it. However, a position is not really vacant there, but a very old Herr Schweins[11] functions as a mathematician, and it seems that after he got knowledge of the matter, he initiated a counter-operation, by which means a standstill was brought about in connection with the question of finances.

I have asked my friend Dunker, if the opportunity presents itself to him, to bring two books with him for me

a) a four-digit table of logarithms
b) Green's Essay on the Application of Mathematical Analysis to the Theories of Electricity and Magnetism.[12]

This essay was published in Nottingham (?) in 1828, but it seems to be hardly to be found. I have already made attempts earlier to obtain it through a bookseller here as well as through one in Leipzig, but in vain. —As far as the table of logarithms is concerned, in addition to the 5-digit and 7-digit ones, some in which the trigonometric functions are listed with four decimal places only are also widely used in England, but I do not know which one recommends itself most of all through its typographic organisation and correctness. Trusting in your kind views, I have asked Prof. Dunker to seek your advice about it; if you could [recommend] <*words illeg*> him the title of a good small table of logarithms and the address of an antiquarian bookstore where the desired [item might probably be bought] <*words illeg*>, then you will oblige me very much indeed.

With sincere respect | Your | loyal friend | Stegmann.

RI MS JT/1/S/210

---

1. *Friedrich Ludwig Stegmann*: Stegmann (1813–91) was the professor of mathematics at the University of Marburg who had supervised the dissertations of both Tyndall and Hirst.
2. *my textbook on the calculus of variations*: L. Stegmann, *Lehrbuch der Variationsrechnung und ihrer Anwendung bei Untersuchungen über das Maximum und Minimum* (Kassel: Luckhardt, 1854).
3. *my friend Dunker's trip*: Also a professor at Marburg, Wilhelm Dunker (1809–85) pursued research in the areas of palaeontology and zoology.
4. *Girard's successor*: Prior to his move to the University of Halle in 1854, Heinrich Girard (1814–78) had been an associate professor of mineralogy and geology at the University of Marburg. He was succeeded by Dunker.
5. *Mr Cumming*: probably Mary Anne Coxe's brother, William Fullerton Cumming (1804–92), who graduated in medicine from Edinburgh University in 1827 and was then employed by the Bengal Medical Establishment. He visited Germany—where he might have met Dunker—to prepare his *Notes on Lunatic Asylums in Germany and Other Parts of Europe* (London: J. Churchill and Edinburgh: Maclachlan and Stewart, 1852).
6. *meet you*: Dunker was the 'Professor from Marburg [who] called [on 1 June] with a letter of introduction from Professor Stegmann' (Journal, RI MS JT/2/13c/749).
7. *your last kind letter*: letter missing; see also volume 4, letter 0823.
8. *Prof. D.*: Dunker.
9. *my friend Dressner*: Carl Dressner was a teacher of German at the Royal Military Academy, Sandhurst.
10. *the meeting of physicians and natural scientists*: Like the BAAS the Gesellschaft Deutscher Naturforscher und Ärzte (Society of German Natural Scientists and Physicians) held an annual meeting at which scientists liaised and communicated their research. Meetings were held in Vienna.
11. *Herr Schweins*: The mathematician Ferdinand Schweins (1780–1856) served as professor of mathematics at the University of Heidelberg for 46 years.
12. *Green's Essay . . . Magnetism*: *Essay on the Application of Mathematical Analysis to the Theories of Electricity and Magnetism* (Nottingham: printed by T. Wheelhouse, 1828) by the largely self-educated mathematician and miller George Green (1793–1841).

From James Joseph Sylvester[1] 21 May 1855 1088

26 Lincoln Inn Fields | 21s^t^ May 1855

My dear Sir,

I expect the profound Cayley,[2] the Sagacious Latham[3] and the Myriad-minded Buckle[4] to dine with me on Monday next the 28th Inst at 7 o'clock will you do me the favor of enabling me to add the name of the acute and philosophic Tyndall to the list?[5]

Yours very truly | J J Sylvester

RI MS JT/1/S/290

---

1. *Sylvester*: James Joseph Sylvester (1814–97), a creative mathematician, had held posts at University College, London, and the University of Virginia and then worked as an actuary. Later in 1855 he was appointed professor of mathematics at the Royal Military Academy, Woolwich.
2. *Cayley*: Arthur Cayley (1821–95), an innovative mathematician and close friend of Sylvester's, worked for many years as a lawyer, only gaining a professorship at Cambridge in 1860.
3. *Latham*: The physician Robert Gordon Latham (1812–88), who wrote several books on ethnology.
4. *Buckle*: Thomas Henry Buckle (1821–62), the author of *History of Civilization in England.* On May 9, 1855, Tyndall had met Buckle, whom he described as 'no common man. We soon forgot the rest of the table and plunged into the philosophy of history' (Journal: RI MS JT/2/13c/475).
5. *add . . . Tyndall to the list*: Due to headaches, Tyndall did not dine with Sylvester on 28 May (Journal: RI MS JT/2/13c/749).

To Thomas Archer Hirst 27 May 1855 1089

My dear Tom,

A dim impression of your last letter[1] rests upon my mind and from this I must endeavour to reply to it; for the letter itself lies in the deep recess of one of the pockets of my laboratory coat. You spoke of coming to London—If I am master of these lodgings I shall have them well prepared for your reception, but this is not probable—In fact I have for some time thought of moving nearer to the Institution, and as I shall be leaving London some time in June I thought of separating from my present lodgings at the same time. However of this we may talk again.

A new element of confusion appears to have made itself manifest at

Queenwood. The thought occurred to me that if Edmondson[2] is so absurd as to do what he talks of you ought yourself to try and make some arrangement with Pare & Finch, and undertake the direction of the establishment—I would of course help you to the utmost extent of my power and you might use my name as that of your cooperator. But I hope Edmondson is not going to dash out his own unfortunate little brains—poor man, he has taught me more, notwithstanding his profound shallowness, than any other man in this world. He is to me a key to ten thousand larger phenomena of social and political life.

As for his 'confidence' I hope you can afford to be indifferent about it—his 'confidence'!—there is no such thing. He is incapable of feeling it, and merely blabs about his concerns when he must to do so, to screen or extenuate his own folly. If you thought it well I would write to him upon the subject of his proposed move.

I wrote to Debus some time ago[3] asking him to accompany me to Normandy, he has not yet replied to me. Could not you & Anna come? The place is so lovely, and you could spend your vacation there just as cheaply as in England. I hope Debus at all events will come; I promise myself a glorious fortnight or three weeks there.

Last Sunday I spent at Richmond with Francis and felt the benefit of the holiday throughout the week. It is now Sunday again and the glorious sun is shining and beckoning me outward—I'll go immediately great Sol![4]—Just let me say two words to Tom & then I'm with you.

Faraday gave us a most pleasant lecture on Friday night[5]—he lectured instead of Hofmann who ought to have lectured,[6] but who is so absorbed in money grubbing that he could not come. By robbery & intrigue I am told this fellow manages to suck an income of 12 or 14 hundred a year[7] out of England. I come on next Friday and hope to give them at least a pleasant lecture.[8]

DuBois lectured yesterday[9]—the day was warm, the people drowsy and the experiments all unsuccessful. His two last lectures were not so good as his earlier ones, and he has evidently had enough of the work to satisfy him. The subject in fact is hardly sufficient to fill 8 lectures and illustrate them so as to keep the interest of our multifarious audience alive.

I cling to the notion of animal electricity which I ventured to express to you some time ago. A few easy experiments have recently occurred to me which will probably throw much light upon some portions of the subject. I dare not make them while DuBois is here but will do so afterwards, and communicate to him the result—It may be in his favour, but this I at present doubt.

I was at Lord Ashburton's a few evenings ago[10] & experienced much friendliness from many of those present. Once while talking to a gentleman whom I know not but who was introduced to me by Lord Ashburton, Cardwell,[11] the expresident of the Board of Trade came near me—He looked kindly in my face, took my hand and squeezed it cordially—So that although I refused his offer of Jermyn Street[12] he has not lost his respect for me. But the man who interested me most was a hairy faced individual, who was pointed out to me just before we separated—he talked sometimes with those present—and sometimes walked up & down looking at the splendid paintings alone—It was Thomas Carlyle.[13] I looked deeply into his countenance as he was putting on his coat to go away. His eyes appeared sad and beautiful to me—not a low complaining sadness, but the sadness of one who while he mourned the follies of men had his own high trust and consolations. There were many other notabilities present, Sir Robert Peel[14]—a fine young fellow—and his brother Frederick,[15] Lord Elgin,[16] the Duc D'Aumale,[17] &c &c.

Next evening I was at a soirée given by Weld at Somerset House,[18] and saw Tennyson there—I did not speak to him. He entered the room, walked straight forward through his arm out and seeing the hand of a friend greeted him with a rough how do you do!—He has a fine head; and there is something uncompromising and thorough about the fellows appearance that makes a man long to exchange a shock with him.

Now I will end my rigmarole—Sylvester sends me the following note[19] it is characteristic of the man:—

'I expect the profound Carlyle, the Sagacious Latham and the myriad minded Buckle to dine with me on Monday next the 28th. Inst. at 7 o'clock. Will you do me the favour of enabling me to add the name of the acute & philosophic Tyndall to the list?'

J. T.

I have lain on my back on the sofa this glorious morn and read Emerson's Essay on the poet.[20] But the Church Bells are ringing and my friend Sol is impatient—give my love to Anna—she asks me have I seen Beta[21] lately—alas! alas! It was ever thus—

'The fairest things the shortest stay
Peep at J. T. then start away'

Like 50 others (not one less) the thought of her has been swept off by the tide of other thoughts which ebbs & flows everlastingly in this doomed existence of mine. Good bye. | Affectionately | John

27th May 1855

RI MS JT/1/T/602

1. *your last letter*: letter 1086.
2. *Edmondson*: Hirst had devoted much of letter 1086 to Edmondson's predicament.

3. *I wrote to Debus some time ago*: letter missing.
4. *Sol*: sun (Latin).
5. *Faraday . . . Friday night*: On 25 May Faraday lectured at the RI 'On Electrical Conduction', later published in *Roy. Inst. Proc.,* 2 (1854–58), pp. 123–32.
6. *Hofmann who ought to have lectured*: Hofmann had recently been appointed one of the jurors at the Exposition Universelle and had presumably already travelled to Paris.
7. *money grubbing . . . 12 or 14 hundred a year*: Although Hofmann was involved in a number of remunerative activities and was attracted to a luxurious lifestyle, there appears to be no obvious trigger for Tyndall's comment, except that Hofmann's income, which may well have been £1200 or £1400 per annum, far exceeded Tyndall's (of about £400 per annum).
8. *next Friday . . . lecture*: Tyndall's Friday Evening Discourse at the RI 'On the Currents of the Leyden Battery', later published in *Roy. Inst. Proc.,* 2 (1854–58), pp. 132–35.
9. *DuBois lectured yesterday*: the sixth (of eight) of Du Bois-Reymond's RI lectures 'On Electro-Physiology'.
10. *a few evenings ago*: Tuesday 22 or Wednesday 23 May.
11. *Cardwell*: The Peelite politician Edward Cardwell (1813–86) had served under Lord Aberdeen as President of the Board of Trade from 28 December 1852 until 31 March 1855.
12. *his offer of Jermyn Street*: Tyndall had been offered (but declined) a position at the Government School of Mines (in Jermyn St.), which was under the Department of Science and Art, which, in turn, was under the control of the Board of Trade. On 17 October 1854, Playfair told Tyndall 'that Mr Cardwell intended after the next estimates to offer me a post in the School of Science' (Journal: RI MS JT/2/13b/628).
13. *Thomas Carlyle*: Although this was the first occasion when Tyndall had seen Carlyle, he had enthusiastically read his works and had written to him from Marburg on 6 June 1849; see volume 2, letter 0376.
14. *Sir Robert Peel*: The Liberal politician and MP for Tamworth, Peel (1822–95) had recently been appointed a junior Lord of the Admiralty.
15. *his brother Frederick*: Frederick Peel (1823–1906) was the MP for Bury and was serving as the under-secretary of state for war under Lord Palmerston.
16. *Lord Elgin*: James Bruce, 23rd Earl of Elgin (1811–63) was a colonial administrator and diplomat who had recently served as Governor General of the Province of Canada, 1847–54.
17. *Duc D'Aumale*: Henri Eugène Philippe Louis d'Orléans, duc d'Aumale (1822–97) was the fifth son of King Louis-Philippe of France.
18. *a soirée given by Weld at Somerset House*: Charles Weld (1813–69) was secretary of the Royal Society, which met in Somerset House on the Strand.
19. *the following note*: Tyndall copied letter 1088.
20. *Emerson's Essay on the poet*: Emerson's 'The Poet' was first published in his *Essays: Second Series* (London: J. Chapman, 1844).
21. *Beta*: Emily ('Pussey', 1838–1930), second daughter of the late Thomas Drummond; see letter 1047, n. 12.

To Thomas Archer Hirst [May?] 1855 1090

<*text missing*> the last week I have felt nothing that I could dignify with the name of a headache although I have worked sufficiently at times to bring it on if I were so disposed. I feel greatly changed indeed—from a mere child in strength I am becoming a man, and a man's hopes fill my heart as a man's vigour accumulates there. I can even smoke without feeling the slightest bad effect. Wine I cannot yet take—it soon clouds my brain, but the beer is not too strong for me.[1] I know you will like to hear this & therefore I thought refusing myself the aforesaid brief holiday I should be giving you pleasure also

affectionately John

RI MS JT/1/T/601a

1. *the beer is not too strong for me*: cf. letter 1085.

From William Thomson 1 June 1855 1091

Malvernbury House | Malvern, June 1, 1855

My dear Sir

I have asked Mr Weld[1] to set aside for you four copies of the joint paper by Mr Joule and myself recently published in the Transactions,[2] of which the authors will be gratified by your accepting one. If you will kindly take charge of one for Faraday also, and, in case of Du Bois Reymond being at present lecturing at the Royal Institution, if you will take two copies and give them to him, one for himself and the other for Helmholz,[3] I should be much obliged by your taking some opportunity of being at the Royal Society to ask Mr Weld for the four copies which are lying there for you. If you do not chance to be at the Royal Society before the end of the session do not be at the trouble of sending for the papers, as after that time I shall have them all forwarded to myself, and should you not have received a copy I shall have the pleasure of sending you one by post.

Can you tell me whether Helmholz has yet moved to Bonn[4] where I understand he is, if not there already, shortly to be settled? I am anxious to know the post office address both for him and Du Bois Reymond, and I should be very much obliged if you could inform me.

I hope you intend to be present yourself at the Glasgow meeting of the British Association[5] and to bring any apparatus you may have for illustrating your investigations in magnetism or other branches of experimental science, which may be suitable for bringing before our Section. I shall have much pleasure in hearing from you regarding the instrument of which you promised me a description, or still more in seeing it if you can bring it with you to Glasgow.

I think the phenomena presented by pieces of soft bread containing iron filings or fragments of fine iron wire, regarding which I wrote to you in March,[6] must be entirely due to the tendency to form into little rows perpendicular to the line of compression, and to place the lengths of little bars perpendicular to the same line, which such a compound may experience when compressed in one direction.

I am quite out of the world here and have not seen either Philosophical Magazine or Proceedings of the Royal Society for five weeks so that I am in the dark as to recent progress in some of these branches of science to which so much thought is now directed.

Do you know Werner Siemens' plan for measuring the times occupied by military projectiles in passing over short distances in various parts of their paths?[7] I only know it by an allusion to it in the short article on Helmholtz's observations regarding the velocity of transmission of nervous excitement, w[h] appeared in the Philosophical Magazine.[8] I thought in writing to you to suggest that, if you, or some other competent investigator, could be induced to take it up, it would be an excellent object for an application of part of the Government Grant Fund[9] in the present state of national affairs. I see however by a newspaper paragraph that Wheatstone's plan,[10] which he is said to have proposed some years ago, is at last about to be taken up by government, with a view to testing & improving the practice of fire arms. I sh[d] think Wheatstone's must be something like (whether possessing priority or not) that of Siemens—but I am sure it cannot be better.

Believe me, Dear Sir, Yours very truly | William Thomson

Do you by chance to know in what place (some part of the North of Italy I presume), Maggi,[11] (who discovered the unequal thermal conductivity of magnetized iron in different directions) is to be found?

RI MS JT/1/T/13

1. *Mr Weld*: see letter 1089, n. 18.
2. *of the joint paper by Mr Joule . . . Transactions*: J. P. Joule and W. Thomson, 'On the Thermal Effects of Fluids in Motion. Part II', *Phil. Trans.*, 144 (1854), pp. 321–64. As the *Phil. Trans.* was an annual publication, these copies would have been pre-prints distributed by the authors.
3. *Helmholz*: The eminent German physicist and physiologist Hermann von Helmholtz (1821–94).
4. *whether Helmholz has yet moved to Bonn*: In 1855 Helmholtz, who had been lecturing on physiology at the University of Königsberg since 1849, was appointed professor of

anatomy and physiology at the University of Bonn. 'On July 29 [1855], Helmholtz left Königsberg, and went, after a short visit to his relations in Dahlem and Potsdam, to Bonn' (L. Koenigsberger, trans F. A. Welby, *Hermann von Helmholtz* (Oxford: Clarendon Press, 1906), p. 144).

5. *at the Glasgow meeting of the British Association*: Tyndall did attend the BAAS meeting in Glasgow, 12–19 September 1855 and presented some of his experiments; see later letters.
6. *I wrote to you in March*: see letter 1058.
7. *Werner Siemens' plan . . . paths*: The German industrialist and inventor Werner Siemens (1816–92) devised a method for measuring the speed of projectiles by means of an electric spark: see 'Ueber die Anwendung des electrischen Funkens zu Geschwindigkeitsmessungen', *Poggend. Annal.,* 64 (1845), pp. 435–44.
8. *Helmholtz's observations . . . Philosophical Magazine*: H. von Helmholtz, 'On the Methods of Measuring Very Small Portions of Time, and their Application to Physiological Purposes', *Phil. Mag.,* 6 (1853), pp. 313–25. Siemens' work is discussed on pp. 315–16.
9. *of the Government Grant Fund*: In 1849 the British government had instigated an annual grant of £1000 to the Royal Society to aid the promotion of science.
10. *newspaper paragraph . . . Wheatstone's plan*: '"THE TORPID HANDS" OF THE ENGLISH GOVERNMENT.—Some time ago Professor Wheatstone invented an instrument or means, by an electric wire attached to a gun and a target, of measuring with a degree of accuracy never before attained the time occupied in the flight of projectiles. For the improvement of ordnance practice it is acknowledged to be an invention of great value. [. . .] It is now, we are informed, about to be adopted by the Board of Ordnance' (*Dundee Courier,* 30 May 1855, p. [2]).
11. *Maggi*: Pietro Giuseppe Maggi (1817–73), who may have been a doctor living in Milan, published on a wide range of scientific and non-scientific subjects, including poetry. The paper referred to by Thomson is 'De l'Influence du Magnétisme sur la Conductibilité pour la Chaleur du Fer Doux', *Bibl. Univ. Archives,* 14 (1850), pp. 132–34.

To Thomas Archer Hirst 3 June 1855 1092

3r[d] June 1855

My dear Tom,

I got your note[1] which Fox was kind enough to bring me. By the way I am not sure that it was our friend the Dr.[2]—for when I asked the porter[3] what he was like he replied that he was a fine young man about 30 years of age[4]—Can it be that the father of such a family was mistaken for a man younger than I am that have still my family to produce!

Well my lectures are nearly over. I gave them a lecture on Friday evening[5] which seemed to please them mightily—I had steadied my thoughts beforehand and resolved to make the lecture an instructive one. As I sat pondering in my little room Barlow came to me saying that 'The Duke'[6] was there and asking me would I see him—I went and was very cordially received by his grace. At the appointed moment I found myself behind the lecture table. I introduced the subject by a few general remarks on the laws of mind in relation to scientific symbols. Faraday I saw was caught by them—and led on the rumble of applause which followed. I then went at the experiments, taking time to clear the way for each, and to extract from it all the instructions which it was capable of giving. The bursts of applause were frequent and warm, and the experiments succeeded every one to perfection. I had my doubts of some of them, but no doubt was justified by the result. The applause at the conclusion was loud long and enthusiastic, the Duke of Northumberland & many others came and shook me by the hand so that on the whole the lecture was as successful as the most exorbitant egotist could desire it to be.

I have now cast it[7] behind me and its benefit to me is this—it tranquilizes me by the assurance that I have done my work successfully and this tranquility enables me to attack other work with advantage—God knows I do not feed upon the memory of my success. It simply relieves me of that dissatisfaction which accompanies failure and thus enables me to apply my thoughts without any disturbing influence. I shall be down with you on Saturday.[8] How long do you purpose remaining in Broughton—I <u>must</u> remain there for a few days to get some writing finished for the Westminster Review[9]—You <u>must</u> give me lodgings for a week & remain with me until the week is spent. Love to Anna.

Your affectionate | <u>John</u>

RI MS JT/1/T/603

---

1. *your note*: possibly letter 1089.
2. *the Dr*: the local surgeon Luther Owen Fox (c. 1812–79).
3. *the porter*: The porter, probably Henry Lacey (c. 1818–62), would have received letters delivered to the RI.
4. *a fine young man about 30 years of age*: unidentified.
5. *a lecture on Friday evening*: 'On the Currents of the Leyden Battery'; see letter 1081, n. 5.
6. *The Duke*: Algernon Percy, 4th Duke of Northumberland (1792–1865), was a landowner and Conservative politician with a strong interest in science. He collected Egyptian antiquities, helped to finance John Herschel's expedition to the Cape of Good Hope to observe the stars in the southern hemisphere and was the current president of the RI. In his Journal entry for 1 June 1855, Tyndall recorded his meeting with the Duke (RI MS JT/2/13c/750–51).
7. *it*: the Friday Evening Discourse at the RI.

8. *down with you on Saturday*: Tyndall and Debus travelled to Broughton on Saturday, 9 June 1855.
9. *some writing finished for the Westminster Review*: Tyndall reviewed four works on geology, under the title 'The Earth and Man', in the July 1855 issue of the *Westminster Review*, 64 (1855), pp. 151–70.

To Edward Frankland 6 June 1855 1093

My dear Frankland

If you tell me how I am to send the instrument[1] to Mr Lenoir[2] I will avail myself of his offer.[3] in its present state it will not work. It was certainly too great a risque to send it off without previous trial—The workmanship is superior, and expected to be able to use the instrument in my last lecture but one when I spoke of the law of Ohm,[4] but when I came to buy it I found it would not help me.

It would give me great pleasure indeed to spend a week with you among the lakes[5]—so much pleasure indeed that I think I should make a great heave to accomplish it. Whether it is possible or not I do not yet know—I am going over to Normandy in a few days when the memory of our trip to Havre[6] will be revived.

Ever Yours | John Tyndall

JRL Frankland, 202

---

1. *the instrument*: probably the Pachytrop(e), which could be used to demonstrate Ohm's Law; see letter 1062, n. 2.
2. *Mr Lenoir*: see letter 1062, n. 1.
3. *his offer*: presumably to repair the instrument.
4. *my last lecture but one when I spoke of the law of Ohm*: Tyndall's lecture on 29 May was concerned with the electric telegraph and the use of an electrical signalling device that could track the progress of a railway train. The short account of Tyndall's lecture published in the *London Daily News*, 30 May 1855, p. 2, makes no mention of Ohm's Law. 'The law of Ohm, which is applicable under all circumstances, fully expresses the great fact:—*the intensity of an electric current, when a battery is in action, is directly as the whole electromotive force in operation, and inversely as the sum of all the impediments to conduction*' (R. Hunt, *Elementary Physics, an Introduction to the Study of Natural Philosophy* (London: Reeve & Benham, 1851), p. 256).
5. *among the lakes*: in the Lake District. Frankland had purchased a cottage in Windermere; see letter 1073.
6. *memory of our trip to Havre*: in June 1848.

From Thomas Archer Hirst 6 June 1855 Letter 1094

Queenwood | June | Wednesday. Morn^{g}

My dear John.

I am seated in the 1st Class room and around me are the lads working like horses at their examinations. It is a great pleasure to me to sit and look at their earnest faces. Most of them have scarcely looked off their work for upwards of three hours for instead of ceasing work at 12 as they might have done they begged unanimously to be allowed to work on until one. I wonder if one could analyse their motives as they work there whether we should entirely agree with such examinations. There is not a doubt of it that a desire to outstrip his neighbour[1] actuates every one. When they find themselves lagging it is the sense of shame at the consequences which is the most effectual spur. But I do not think there is anything objectionable in this emulation, except in some instances I do not think it leads to bad feeling or selfishness. I think the desire to excel and the wish to do justice to their inherent powers, may be, and is often entertained without the notion of beating any other person obtruding itself, and if so the exercise and discipline of an Examination must be altogether beneficial. I wish I could describe the lads to you as they work. On my left is Wardley,[2] who has without exception the finest head in Queenwood. There he sits nervously absorbed in his work, calculating and thinking with all his might and biting his nails the while. This lad is to be a cotton lord and to leave us this half year. He will carry away my prize and is another of those instances where real talent is likely to lie dormant on account of an unfortunate, inappropriate occupation.[3] Next him sits Aspden Dickinson[4] who is also about to leave us and become a manufacturer. Here the occupation is a suitable one. Aspden is rather dull but extremely persevering. He possesses considerable decision of character which if it does not lapse into mere obstinacy will be of good service to him. I expect his brother John Charles[5] on Saturday next. Next to Aspden sits Walter Stephens,[6] one of the most conscientious little workers I ever saw. Unfortunately he has but little power, I say unfortunately for if he had power he would use it well. He leaves us, too, and goes into business for which he is well fitted. Before he goes he must have a gift from Mr Edmondson in consideration of his many exceedingly good qualities. He is a favourite with all. The others in the room you know but little of. I will therefore pass them, and say a word in reply to your last letter. I never doubted the success of your lectures nor the manner you would employ that success. It was exceedingly satisfactory and pleasant to me to hear of it. Certainly I will remain at Broughton as long as you wish and willingly but I give you notice I shall drive you away as soon as I can, not for my sake or

Anna's (for we would both like to remain at Broughton the whole vacation) but for yours and Debus'. You have no business to be pottering about your Westminster Review articles[7] you ought to be away to Normandie and not keep Debus who also ought to be off and is really wishful to go as soon as possible. For these reasons I shall drive you away as soon as I can whilst at the same time I give you full leave to stop as long as you think fit. My brother[8] will also be with us.

Thine Affectionately. T. A. H.

RI MS JT/1/H/484

---

1. *a desire to outstrip his neighbour*: Quakers generally discouraged inter-personal competition.
2. *Wardley*: Possibly destined for the firm of Wardley & Unwin, 'cotton waste spinners', of Haslington, Lancashire, which went bankrupt in 1867 (*Manchester Courier and Lancashire General Advertiser,* 23 July 1867, p. 7).
3. *an unfortunate, inappropriate occupation*: As a 'cotton lord' Wardley would be wasting the academic training he had received at Queenwood College.
4. *Aspden Dickinson*: Aspden P. Dickinson (1839–85), a pupil at Queenwood, became a 'machine maker (employing 250 hands)' in 1881 Census for Blackburn, Lancashire (Ancestry).
5. *his brother John Charles*: John Charles Dickinson (1835–67), described as 'Cotton Spinner & Manufacturer' in 1861 Census for Blackburn, Lancashire (Ancestry).
6. *Walter Stephens*: not identified.
7. *your Westminster Review articles*: see letter 1092, n. 9.
8. *My brother*: John Henry Hirst (1826–82), Thomas Archer Hirst's eldest brother.

To Thomas Archer Hirst 8 June 1855 Letter 1095

My dear Tom.

You are a good hospitable fellow. Be assured of it if I could I should throw the claims of the Westminster[1] overboard, but I engaged to furnish half a sheet and must keep my engagement. I shall want you to do something for the Phil Mag.[2] next week—But I hope to be able to start from Southampton on Friday next.[3] I think I must ask Anna to have the kindness to ask her brother John[4] to have a look out for a lodging for me. Your last letter[5] interested me greatly—I think it would be no harm to tell the lads in earnest words that each might do his duty, irrespective of outstripping his neighbours—I would hope however that in the ardour of the work all jealousy is forgotten. I shall be with you tomorrow in good time, as I purpose starting by the 10 o/c train.

Your affectionate | John

RI MS JT/1/T/1006

1. *the claims of the Westminster*: see letter 1092, n. 9.
2. *something for the Phil Mag*: possibly a translation.
3. *Friday next*: 15 June, when Tyndall sailed from Southampton to France, via Jersey.
4. *her brother John*: John Martin.
5. *Your last letter*: letter 1094.

From Charles Lyell 11 June 1855 1096

53 Harley S^t^ | London | June 11. 1855

My dear Sir,

Your letter[1] was rightly addressed to me; Thursday[2] was the evening, & as our party was only beginning at the hour (10 ¼) which you mention I can only account for our not having had the pleasure of seeing you by supposing that the 'muscular power' which you & the coachman expended was exerted on the knocker of my friend Dr Fitton's house 53, Upper Harley S^t^.[3]

I hope another time we shall be more fortunate. If you think the naming & numbering of our Streets are a disgrace to the Post Office & worthy of the English at Balaclava[4] I quite agree with you

ever sincerely yours | Cha. Lyell

RI MS JT/1/L/47

1. *Your letter*: letter missing.
2. *Thursday*: 7 June 1855.
3. *Dr Fitton's house 53, Upper Harley S^t^*: Harley Street runs north-south, with Harley Street, Upper (where the physician and geologist William Henry Fitton (1780–1861) lived) extending to the north of Weymouth Street. Lyell's residence was south of Weymouth Street. On 7 June Tyndall recorded in his Journal: 'Went to Lady Lyell's at 10 o'clock but found my journey a vain one. Knocked loud enough to shiver the timbers of the door but nobody came. Returned without forcing an entrance' (RI MS JT/2/13c/751).
4. *worthy of the English at Balaclava*: The Battle of Balaclava was fought on 25 October 1854, as part of the war in Crimea. It included the infamous Charge of the Light Brigade in which, owing to the misunderstanding of an order from Lord Raglan, the Russian artillery was able to decimate the cavalry of the Light Brigade. Out of a total of about 670 men, some 110 were killed and a further 160 wounded, together with many of their horses. Lyell is here comparing the incompetence of the Post Office in failing to give streets unique names to Raglan's blunder, which is often cited as one of the greatest blunders of British military history.

To Thomas Archer Hirst 17 June 1855 1097

St. Helier's Jersey Sunday | June 17. 1855

My dear Tom

I want you to go to Mr John Chapman[1] of King William Street, Strand, not far from Charing Cross, and ask him to send the proofs of my article written for the Westminster Review[2] to you—and I want you to read the said article and if any typical[3] errors should occur, or if there be any thing which presents a difficulty, to the printer to set it right for me. If you shew Mr Chapman what I have here written that will be a sufficient introduction. He will give you the manuscript and if any doubtful point should arise a comparison between it and the proof will enable you to set it right.

Debus and I reached St. Helier's[4] yesterday. The wind lulled before we left Southampton and I promised myself a deep sleep & forgetfulness of all sea sickness—as I write the word my miseries come back upon me, and my stomach rumbles, in memory of what I suffered—nobody escaped—it was awfully rough. The wind during the day had lashed the water into motion and it took some time to tranquilize it. Once when I was very miserable Debus cast an eye upon me. Well John, Wie gehts? Schlecht genug, und Dir? Oh Gut![5]—some short time afterwards I heard him ejaculate 'Steward a basin'! and what followed I must leave to your imagination. At Guernsey we came into calm water, and seeing others indulging in the like and feeling awfully ravenous we had a chop and cup of tea. I felt strong but alas between Guernsey & Jersey the billows heaved themselves once more. I saw Debus leaning over a real,[6] the picture of physical dejection. The storm was manifestly brewing & soon the explosion came—although on the verge of one myself I could not help laughing at Heinrich.[7]

We are now pretty well. <u>I was</u>, before the writing of this note brought back the remembrance of my wrongs. I hope you find all comfortable in Chadwell St.[8] Love to Anna.

John

RI MS JT/1/T/604

---

1. *Mr John Chapman*: John Chapman (1821–94) was the proprietor and publisher of the *Westminster Review.*
2. *my article written for the Westminster Review*: see letter 1092, n. 9.
3. *typical*: typographical.
4. *St. Helier's*: St Helier is one of the twelve parishes on the island of Jersey and also its principal town, which has a south-facing harbour.

5. *Wie gehts? Schlecht genug, und Dir? Oh Gut!*: How are you feeling? Bad enough, and you? Oh fine! (German).
6. *a real*: presumably the rail of the ship.
7. *Heinrich*: Debus.
8. *Chadwell St*: Tyndall's lodgings in London were at 9 Chadwell St., Pentonville.

From Henry Moseley 18 June 1855 1098

Private Olveston, Nr. Bristol | 18 June 1855

Dear Sir,

An examination is to be held in July or the beginning of August[1] under the Board of Ordnance for appointments to Cadetships in the Royal Artillery.[2] Will you allow me to ask you whether in the event of an application being made to you from Lord Panmure you would be willing to undertake the office of Examiner in Physics?

The remuneration of the Examiners will depend on the number of candidates he will have to examine but will not be less than £50. I apprehend that the duties of the Examiner in Physics will not be heavy. Will you oblige me by an answer if possible by return of post addressed as below, and believe me

Yours faithfully | Henry Moseley

RI MS JT/1/3/884
LT Typescript Only

1. *held in July or the beginning of August*: The examination was originally planned to start on 11 July but was postponed until 1 August 1855.
2. *An examination . . . Royal Artillery*: Entry to the Royal Military Academy at Woolwich had traditionally been by patronage. However, after much controversy Parliament accepted the need for an open competition, the first such examination being held in August 1855. 'The object of the examination was to select twenty candidates between the ages of 19 and 21, to be appointed to Provisional Commissions, and forty between the ages of 17 and 19, to be admitted to the Senior Practical Class at Woolwich.' (*Report on the Examination for Appointments in the Royal Artillery and Engineers, Held at King's College, London, on the First of August, 1855* (London: Harrison, 1855), p. 4.) All candidates had to sit the examination in mathematics and also be examined in at least two other subjects from a list of nine, one of which was 'Natural Sciences'.

From Hector Tyndale 18 June 1855 1099

Philadelphia 18 June 1855

My dear John

With your good welcome letter 10th March[1] before me and my own conscience within me, much disturbed, I begin. To say that in all your 'blame, censure & condemnation' you are entirely in the right and I altogether in the wrong, makes poor amends for my idle 'omission', as you tenderly phrase it. In the miserable scrawl[2] I sent you a few days before yours came to me I spoke of a long and lingering idle fit. It has not passed off yet, in truth 'tis more—it has passed into almost sheer indifference to all things. If I spoke too much upon the theme—ego—before, and if I say too much upon it now, I crave your forgiveness. But in vindication of my love for you I must lay the blame upon my present state of mental health, and, of course, in doing so I must use the pronoun 'though believe me, my dear John, it is with no desire to dwell upon myself. Indeed that thing won't bear much comment. My whole wish is simply that you may continue to give me such of your thoughts and kind remembrances as I may be worth.

I sincerely regret that you have been ill, but ere now I trust you have again regained your elasticity and health. You have no right to be any other than well; no man who feels with nature so much can be unwell. But to a poor miserable devil like me—I don't speak now of my body, neither of myself proper, but only of my semi-self, the telegraph operating part of me—who discords Nature and untunes the elements in the infernal alchymy of my brain; to such a fellow nothing should be well—nor is it. To think that God's blessed Nature, His sunshine, air, water, trees and flowers—bathing and floating forever through the beautiful blue of Heaven, should be resolved, or be attempted even, into nasty, stinking gases, or into sterile elementary matter. And yet something like to this with human nature is what I do, or have done, thinking it a pleasure! And now the devil of analysis won't down or leave me, but having exhausted my powers in that infernal way, turns me in upon myself, and all these vile things, originating like nothing so much as coffin worms, within oneself, offend the inner senses of my soul—But no matter I have no right to sicken you with my maunderings and I shall not further.

I have said already that I was sorry to hear of your ill health, and yet, to tell you the truth, I ought not to be sorry, nor, perhaps, am I. As you say your letter was owing to it, or rather that when compelled to give up science 'thoughts of friends enter'. Now a headache isn't very pleasant company to enter memory with—and I fear a succession of such reminders would at last place both cause and effect in the same list—as bores.

I received two papers from you since my return which I read with much pleasure. The paper from Dr Coxe with an article on 'Types of Mankind'[3] was one of them. I thank the Doctor and please convey my kindest respects and pleasant remembrances to him as to Mrs: & Miss Coxe[4]—if etiquette in England allows of this. The work referred to, 'Types of Mankind', is not one to do honour to this or to any other country; inasmuch as the motive of it is a low one—the Pharisee's motive of self-justification and the glorification of one race above all others. Such a book, so far as it has any influence, is mischievous. I fear greatly, my fine Englishman, that 'our race' is so much engaged in building roads and other things of the kind, that we are prone to think all the roads from God's central light terminate with us alone—center all in 'Anglo-Saxon-dom'! I am not wholly sure that the great highway of humanity, for some generations at least, is not impeded, obstructed by this practical road-building race of ours. A life time spent upon so great a work even as the Britannia bridge[5] is not, perhaps, the noblest life time one could find. I do not know that the ladder of Jacob[6] is not seen, and mounted too, in glorious visions by men black as night—far away from sight or hearing of the 'great Anglo-Saxon race'. I am not sure that because a few grains of dust upon the 'Anglo-Saxon' speculum have been noted and classified (!) into white dust and black dust and round dust and square dust, we see any clearer or further towards the Great Center Heaven our souls travail for. But after all I suppose it is all right by this theory—that each dust speck, when we know it, becomes transparent and gives additional power to the lens by becoming one itself, throwing two rays from Heaven instead of one, making what you call polarized light, I believe—which is not true light after all!

Oh how I wish I had been with Dr Francis & yourself (my kindest regards to the Doctor always) on your Welsh trip.[7] But no matter the trout you ate and the hills you climbed and the great driving clouds you saw do me good through you. When you next go out of that nasty workshop of yours (I don't speak comparing it with other laboratories, among which it may be a perfumed garden for aught I know) keep a journal and send it to me, and dwell as much as possible on 'that rare and radiant maiden'[8] will you? Ah John the evil in you must be de-evilised (I don't advance any theory upon the origin of the name of the early playmate of humanity) before you can settle down in a staid gentlemanhood, like ______.[9] How can you speak so lightly of the needs of your lonely, solitary hours. 'Charm one from the skies'[10] indeed! I think I see you at it! The nearest approach you will make to a cloud-lady will be one from the hills—cloudwards perhaps—healthy, happy, good and a woman. . . . These don't come from thunders, lightnings or rain storms. —And then too, your 'life is so full of satisfaction except when smitten with debility' (is 'that rare and radiant maiden' one of your debilities?) that you never have the revelation 'of the need of a wife'! How many wives do you think you can get upon such terms? I tell you what, my fine bachelor, some of these days you may 'charm one from the clouds' and if you do, take care of the thunder storms that's all.

I am glad your cousin Payne has reached the great apex of the XIX pyramid A.C.[11] I suppose any one who stands there can see more than 'forty centuries'[12] ever saw. Yes gold is a wonderful thing, so wonderful that it would be worth the Philosopher's while to examine into the relations existing between it and the human ('Anglo-Saxon') mind. When you can ascertain the connexions between Mind and Matter, between 'the child of all the ages' and mighty gold, you will have done more than you did (beautiful as your experiments were) than when you drew the latent music from metals and from the earths. —I am glad to hear that one, at least, of your cousins has common sense, for it is not very common hereaway.

I do not think anything unpleasant or unkindly can arise between Faraday and yourself[13] because of a difference in views, unless, indeed, one, or both of you are desirous to quarrel because one likes mustard which the other does not. As you say Faraday is too great a man for such things to weigh with him beyond the intellect, and he is great enough for his simple opinion to shake the results of yours or any other man's researches. I would not, certainly, yield my religious convictions to his,[14] but I would, as certainly, like to clear up the difference in our views; and failing that I would feel that it was a difference of nature, which like other apparent differences is to be reconciled only by the Future. By the way if there is a simple relation anywhere of his general views of magnetic force, and if you have any easily 'come-at-able' relation of your own views, I shall be very glad to receive them, don't let the postage stand in your way but send me any and all you think will interest me. I asked you before to send me anything new, or able, on scientific matters and I am glad of this chance to refer to it again, and keep an account of any expenses you may incur and let me have it from time to time. Have you heard anything more of the Edinburgh professorship?[15] It would, I think, be a capital thing suiting you in every respect, particularly as it might give you a wider field of publication.

May 'that huge young fellow[16] with the black barb whom you took me to see' live long & happily with his chosen wife under 'his own vine & fig tree' without ever having the slightest occasion to desire to dress himself with the leaves of the latter. What you say of the English newspaper press entirely agrees with my own notions, but I apply it as well to the same gentry hereaway—Slavery-ridden, Custom-obeying, populace-courting always. But so it has ever been. The greatest comfort (!) being that the noble ones are growing (however slowly) in stature & in numbers. There is no Sodom[17] now, out of all the Sodoms, without, at least, 'three men' to save it from damnation.

Our early spring was dry & cold, latterly wet & cool, but thank God, the crops look finely here & promise well. I don't sell corn but I, in common with poorer devils, eat it & it has been enormously dear, causing, with the 'hard times' an unprecedented amount of suffering here. Allow me to throw up my hat (the hat of my minds eye) at the slow coming success of the Allies, 'long may they wave'. American sympathy (newspaper sympathy I mean) has been with Russia heretofore because of two things, I your courting of Austria the d—est tyrant in the world[18] they say, & II you didn't take Sebastopol[19] & sympathy goes with power. Beat Russia & annihilate Austria & then we'll come down.

Write soon. affectionately yours

Hector Tyndale

I haven't been so 'waked up' this long time as in writing this to you and it is with sincere regret I find the paper too small. If there is anything unsaid by me, consider it said & credit me with it. Yours Hector

RI MS JT/1/T/47

1. *letter 10th March;* letter 1052.
2. *the miserable scrawl*: letter 1063.
3. *The paper from Dr Coxe with an article on 'Types of Mankind'*: This was presumably a review by Coxe of J. C. Nott and G. R. Gliddon, *Types of Mankind: or, Ethnological Researches, Based upon the Ancient Monuments, Paintings, Sculptures, and Crania of Races, and upon Their Natural, Geographical, Philological and Biblical History: Illustrated by Selections from the Inedited Papers of Samuel George Morton and by Additional Contributions from L. Agassiz, W. Usher, and H. S. Patterson* (London, Paris, Leipsic, & Philadelphia: Trübner & Co, 1854).
4. *Mrs: & Miss Coxe*: Mary Anne Coxe and her daughter, whose given name(s) have not been identified.
5. *the Britannia bridge*: The bridge spanning the Menai Straights connecting the Isle of Anglesey to mainland Wales. Of innovative tubular construction, it was designed and constructed by Robert Stephenson and was opened in 1850.
6. *the ladder of Jacob*: After fleeing from his brother Esau, Jacob dreamed of 'a ladder set up on the earth, and the top of it reached to heaven: and behold the angels of God were ascending and descending on it'; Genesis 29:12).
7. *your Welsh trip*: see letter 1052, n. 4.
8. *that rare and radiant maiden*: see letter 1052, n. 15.
9. *like* _____: Tyndale left a blank.
10. *Charm one from the skies*: The phrase 'A charm from the skies' occurs in 'Home, Sweet Home', a popular song of the 1820s.
11. *the great apex of the XIX pyramid A.C.*: Tyndale seems to imply that owing to the wealth

Thomas Payne had amassed in Melbourne—see letter 1052—he had achieved an impressive level of material success as recognized by his contemporaries in the nineteenth century (XIX A.C., where A.C. = Anno Christi = Anno Domini). In terms of wealth he could thus be compared to the Pharaohs who constructed the pyramids. Letter 1052 also indicates that Payne visited Palestine, so he probably saw the pyramids.

12. *forty centuries*: During his combat with the Egyptian cavalry in July 1798, Napoleon Bonaparte urged his men, 'Soldiers, you are about to fight the rulers of Egypt; reflect that from these monuments [the pyramids] you are contemplated by forty centuries' (L. de l'Ardeche, *History of Napoleon,* 2 vols (New York: Appleton & Co., 1843), vol. 1, p. 127). Sources differ considerably over the details of this story.
13. *anything unpleasant . . . Faraday and yourself*: see letter 1052, n. 8.
14. *religious convictions to his*: Faraday was a Sandemanian, a small religious sect that cleaved to the Bible.
15. *Edinburgh professorship*: see letter 1052, n. 12.
16. *huge young fellow*: Hirst. Galton reported that he was 6ft 2½ins (V. L. Hilts, 'A Guide to Francis Galton's English Men of Science', *Transactions of the American Philosophical Society,* 65:5 (1975), pp. 1–85, on p. 80).
17. *Sodom*: The cities of Sodom and Gomorrah were destroyed because of the sins committed by their inhabitants (Genesis 18:20–19:26). The nature of the people's sin has been discussed extensively and is often attributed to a sexual perversion, 'sodomy' in particular.
18. *the d—est tyrant in the world*: Russia had expected Austria to be its ally in Crimea, as Tsar Nicholas had helped Austria in putting down the Hungarian Revolt of 1848. However, Austria threatened to support Britain and France, leading Russia to try to appease the Austrians by withdrawing from the principalities on the Danube, which Austria then occupied. By remaining neutral during the Crimean War, Austria became isolated and the emperor Franz Joseph 1 (1830–1916) was criticized by the protagonists on both sides for failing to assist with the international situation.
19. *you didn't take Sebastopol*: Despite the extensive bombardment by Britain and its allies for about nine months, the siege of Sevastopol was still ongoing.

From Peter Riess 19 June 1855 1100

Berlin 19 Juni 1855.

Verehrtester Freund

Nehmen Sie meinen verbindlichsten Dank für die vortreffliche Übersetzung des Aufsatzes über Influenz, den ich im Juniheft der phil. magaz. gefunden habe; es ist überall der Sinn so gut getroffen, daß ich kaum eine Ausstellung zu machen hätte. Nur einmal und zwar durch meine Schuld, ist ein Versuch nicht ganz deutlich. Seite 410 wird nämlich gesagt, daß eine Schellackplatte zwischen einer Flamme und einer Kugel "to and fro" bewegt worden, ich habe aber die Platte nur einmal (von oben nach unten) bewegt, und führe dies für Sie an, wenn Sie den Versuch einmal wiederholen wollen. Daß Sie unter meinen Arbeiten grade diese zur Übersetzung gewählt haben, ist mir einen Augenblick aufgefallen, weil darin eine Ansicht des großen Physikers bekämpft wird, welcher Englands Stolz ist, und dem ich nur mit Widerstreben entgegengetreten bin. Doch haben Sie gewiß den Aufsatz eben darum gewählt, weil Sie wissen, daß einem Faraday ein mit Gründen unterstützter Widerspruch erfreulicher ist, als gedankenloses Nachbeten. Grade die falschen Ansichten über die Influenz, die aus Faradays Arbeiten herausgelesen worden sind, obgleich sie gar nicht darin stehn, haben große Verwirrung angestiftet, von der de la Rive, Melloni, Knochenhauer hinlängliche Beispiele geliefert haben. Noch im vorigen Jahre hatte ich eine recht verdrüßliche Discußion mit Humboldt, der eine Arbeit Mellonis gedruckt zu sehen wünschte, die des verdienten Physikers ganz unwürdig ist; ich habe den Druck bei uns verhindert, der aber leider in den Comptes rendus der Pariser Academie statt gefunden hat. —Eine interessante Neuigkeit in der Electricitätslehre ist Gaugains electrisches Ventil das Ihnen aus den Compt. rend. bekannt sein wird, ich habe davon Veranlaßung genommen, eine Untersuchung über den Durchgang elektrischer Ströme durch verdünnte Luft anzustellen, die ich der Academie mitgetheilt habe, und die (leider während meiner Abwesenheit von Berlin) in den Monatsberichten gedruckt werden wird. Sollten die Separatabzüge nicht vergeßen werden, so werde ich Ihnen ein Exemplar schicken, freilich ein schwacher Dank für die vielen werthvollen Abhandlungen, durch deren freundliche Zusendung Sie mich erfreut haben. Daß es Ihnen gut geht, hoffe ich in Kurzem von Dubois zu erfahren. Ihren berliner Freunden geht es gut, nur daß ich, eines langwierigen, aber wie der Arzt behauptet, nicht gefährlichen Halsübels wegen ins Bad muß, und den Umgang eines meiner beßten Freunds, Dirichlets, verliere. Er hat der Versuchung nicht widerstehen können den Stuhl einzunehmen auf dem Gauß geseßen hat. Sie werden, wie ich höre, nach dem Continente kommen, werden Sie nicht Berlin einmal wiedersehen wollen? Berlin ist nur 26 Stunden von Paris entfernt. Aber kommen Sie nicht vor Anfang August, wo ich aus dem Bade zurück zu sein hoffe. Mit freundlichstem Gruße

Ihr | P. Rieß

Berlin, 19 June 1855.

Most honoured friend

Accept my most profound thanks for the excellent translation of the essay on induction that I have found in the June issue of the Philosophical Magazine;[1] the meaning has been struck so well throughout that I would hardly have to make any criticism. Only in one place, and indeed through my fault, is an experiment not entirely clear. Namely on page 410 it is said that a shellac plate was moved 'to and fro' between a flame and a ball, but I only moved the plate once (from above to below); I point this out for you if you want to repeat the experiment one day. That you have chosen to translate precisely this of all my works struck me for a moment, because in it a view is challenged of the great physicist who is the pride of England,[2] and whom I have countered only with reluctance. But you have doubtless chosen the essay precisely for this reason, because you know that for a Faraday, a contradiction supported by arguments is more agreeable than mindless parroting. Precisely the incorrect views about induction which have been read into Faraday's works, although they are not in them at all, have caused great confusion, of which de la Rive, Melloni,[3] Knochenhauer[4] have provided sufficient examples. Just last year I had a quite vexing discussion with Humboldt,[5] who wished to see a work of Melloni's published that is quite unworthy of the deserving physicist; I prevented it being printed here, but unfortunately this occurred in the Comptes Rendus of the Paris Academy.[6] —An interesting novelty in electrical theory is Gaugain's electrical valve[7] which will be known to you from the Comptes Rendus, and I have taken occasion from it to carry out an investigation of the passage of electrical currents through rarefied air, which I have communicated to the Academy[8] and which (unfortunately during my absence from Berlin) will be printed in their monthly reports.[9] Should the offprints not be forgotten [by the printer], then I shall send you a copy, a poor thank-you admittedly for the many valuable papers through the kind sending of which you have delighted me. I hope to learn shortly from Dubois[10] that you are well. Your Berlin friends are well, except that I have to go to a spa because of a prolonged, though, as the doctor maintains, not dangerous, throat condition, and shall lose contact with one of my best friends, Dirichlet.[11] He has not been able to resist the temptation to accept the chair that Gauss had occupied.[12] You will, as I hear, be coming to the continent;[13] will you not want to see Berlin once again?[14] Berlin is only 26 hours away from Paris. But do not come before the beginning of August when I hope to be back from the spa. With kindest greetings

Your | P. Riess

RI MS JT/1/R/29

1. *the essay . . . Philosophical Magazine*: P. Riess, 'On the Action of non-Conducting Bodies in Electrical Induction', *Phil. Mag.*, 9 (1855), 401–13.
2. *of the great physicist who is the pride of England*: Faraday.
3. *Melloni*: The Italian physicist Macedonio Melloni (1798–1854), who had served as director of the Physics Conservatory in Naples, was best known for his researches on radiant heat.

4. *Knochenhauer*: Karl Wilhelm Knochenhauer (1805–75), the director of the Realschule at Meiningen, published some 50 scientific papers mainly on electrical subjects.
5. *Humboldt*: The eminent explorer and man of science Alexander von Humboldt (1769–1859), was widely known for his expedition to Latin America, his multi-volume *Kosmos* and his idea that the natural world constituted a unity.
6. *Comptes Rendus of the Paris Academy*: Founded by François Arago in 1835, *Comptes Rendus de l'Académie des Sciences* carried the proceedings of the Paris-based Académie des Sciences, which included many important contributions to science. Melloni's contested paper was 'Recherches sur l'induction Électrostatique', *Paris, Comptes Rendus,* 39 (1854), pp. 177–83; see *Faraday Correspondence,* 4:2955.
7. *Gaugain's electrical valve*: Paper by the electrical researcher Jean Mothée Gaugain (1818–80) entitled 'Note sur un Appareil Electrique qui Fait Fonction de Soupape', *Paris, Comptes Rendus,* 40 (1855), pp. 640–42.
8. *the Academy*: der Akademie der Wissenschaften zu Berlin (the Berlin Academy of Sciences), founded in 1700.
9. *printed in their monthly reports*: Riess's 'Ueber den Durchgang elektrischer Ströme durch verdünnte Luft' was first published in *Monatsberichte* (monthly reports) of the Akademie der Wissenschaften zu Berlin for June 1855, pp. 393–409.
10. *Dubois*: Du Bois-Reymond.
11. *Dirichlet*: The German mathematician Peter Gustav Lejeune Dirichlet (1805–59), who had been at the University of Berlin, moved to the University of Göttingen following Gauss's death.
12. *the chair that Gauss had occupied*: Carl Friedrich Gauss (b. 1777), the eminent German mathematician who had been a professor at the University of Göttingen, had died on 23 February 1855.
13. *You will… be coming to the continent*: On the day this letter was written Tyndall arrived in France on his tour, which included jury service at the Exposition Universelle in Paris.
14. *see Berlin once again*: Tyndall had previously visited Berlin from 21 April to 22 June 1851 and from 13 July to 22 August 1853.

To Michael Faraday — 1 July 1855 — 1101

Hotel Manchester | 1 Rue De Grammont | Paris, 1st, July.

My dear Mr Faraday

The sun shines straight into my bedroom this glorious Sunday morning, but the light he sends which beautifies the world and makes the heart of man joyful is accompanied by heat which dulls the brain and makes a man lazy. Two principles operate upon me at present, the old duality of the moral world—the love of ease and indulgence which would prompt me to stretch myself on my bed, and the sense of duty which urges me to write to you. For the present however the latter has triumphed, and I hope it will triumph to the end. The close of the day I saw you last saw me in Hampshire where I remained for a week to get some writing done, and on Friday the 15th at midnight we loosed from the Docks at Southampton and steered for Jersey. I say 'we' as I have a chemical companion[1] along with me. Our passage to Jersey was in the highest degree abominable, I never suffered so much in my life. The wind had been high during the day, but the evening was calm and pleasant, and I promised myself quite an agreeable passage. The sea however continued to swing furiously although the cause of its oscillation had ceased to act: we were tossed up and down and the consequences may be imagined by those who have had the felicity of experiencing the more exalted phases of sea sickness. All on board were ill—the strongest gave way and the stomachic perturbation was universal. We reached Jersey safely and were obliged to continue there until the following Tuesday as there was no boat previously to France. I say 'obliged' for though Jersey might be a pleasant place for 3 days residence still with the exception of one day the weather was so bad that we were confined to our hotel. On Tuesday the 19th we set sail for St Malo[2] and reached the place after a passage of four hours and a half. Here we gave our portmanteaus in charge to the Messageries[3] who had them transmitted to Paris there to await our arrival, and with sufficient baggage to do for a foot journey of a fortnight we commenced our campaign through France. At St Malo we saw the tomb of Chateaubriand.[4] His bones are bedded in a rock on an island which when the tide recedes can be reached by footpassengers from the main land. When we were there the sun was shining on his tomb and the billows breaking and moaning against the rugged rocks beneath it. The place I believe had been one of the scenes of the poet's boyhood and at his own request he was buried here. From St Malo we walked to Dinan[5] in Brittany—a place beautifully situated—The country round is richly wooded, while the contour of the land and the colouring of the rocks are beautiful. A magnificent viaduct has recently been thrown over a ravine near the town, which adds its picturesqueness to that of the scenery. Most of the stone we saw was a dilapidated granite, and it was manifestly stratified; at least its crystalline plates were so arranged as to lie flat as if they had been pressed into their position by the superincumbent weight. The rocks which presented this appearance were quite friable, and their aspect was perhaps calculated to throw some light upon the obscure question of slate cleavage. From Dinan we proceeded to a place called Dol. Thence to Avranches,[6] whence we paid a visit to the celebrated Mont St Michel[7]—a rock to attain which we had to traverse a league of flat strand which is covered at High Tides, and to wade here and there knee deep through water. On this wild rock a monastery was built ages ago, a most splendid piece of architecture which is now in a state of capital preservation.

Its life was an oscillation between the tempests of war and the calm of religion—It is now a prison, and the splendid rooms, the Knights chamber and the refectory of the monks are converted into weaving rooms, where the prisoners work. I never saw a richer country than that which stretches from Dinan to Dol and from Dol to Avranches: Here and there you have views of vast extent and great beauty. The country is well wooded and through the foliage the rich green of the crops and the enclosures reveals itself. Apple trees are strewn every where and you meet cider at all turns. The people appear to like it much, indeed in many of the poorer cabarets you can get nothing but cider. To me it is not much pleasanter than vinegar. The villages at a distance with their brown and aged roofs look very picturesque, but they are vile within—In nothing is the difference between English and Continental life so strikingly exhibited as in the villages. Throughout the region which I have mentioned dirt is universal—Comparing the glorious country with the dwellings of the people, the words of Heber's hymn were often in my mouth.

> 'Man alone is vile'[8]

This inattention to cleanliness is by no means the result of poverty. In dirty houses you often see solid pieces of furniture with old massive brass ornaments which are scrupulously polished while there are holes in the dirty earthen floor. The people do not appear to have that sense of household harmony which is everywhere present among the wives of England and the consequence is the presence of such incongruities as I have mentioned. As regard cleanliness comfort and good order the Hotel de Londres at Avranches is pleasant to think of. The way to the heart of a savage is said to be through his stomach and certainly the excellence of the fare in the hotel spoken of, combined with the moderation of the charge has given it a pleasant place in my memory. The scenery about Avranches is delightful. We always made a point of visiting the churches in each town. There is something soothing in the cool tranquil air of these places, and however one may dissent from their present uses, a solemn and religious feeling is natural in such places. I like them best empty. When the priest is present his theological mechanics ruins the earnestness and tranquility of the impressions otherwise attainable. You rarely find the churches without a stray worshipper—sometimes a dozen or more, almost all of them are women. Hence we might argue that they are either the chief saints or the chief sinners in France. Some of them you see with earnest countenances deeply engaged in their devotions, many, however mutter their prayers as mechanically as if they were repeating their multiplication table. They yawn and look unutterably vacant. At Avranches we had a solemn High Mass conducted for the soul of an inhabitant who had fallen before Sebastopol.[9] A coffin and all adjuncts were there, but the

Crimea held the body. This Romish faith[10] must have had a wonderful, and I think on the whole a beneficial effect upon a half savage world. It seems vigorous now in France, but it is only seeming—its days are numbered, but with the instinct of an old man on the brink of the grave it still clings tenaciously to life. Women are its chief supporters at the present day and I suppose the reason is that they have more feeling and less intellect than men—I think I ought to ask you not to read this heretical utterance to Mrs Faraday or Miss Barnard. But I must hasten. Having passed through Caen, Havre, Rouen we reached Paris yesterday—I have as yet seen nobody but purpose calling upon some people to day. And now I will ask you to excuse the infliction of this badly written letter. With kind remembrances to Mrs Faraday and Miss Barnard

Believe me most faithfully Yours | John Tyndall

I shall leave the hotel tomorrow. And if you should have any thing to say to me the address | 14 Rue du Cirque.[11] Would be safest | J.T.

RI MS JT/1/12/4029–33
Faraday Correspondence, 4:3000
Typed Transcript Only

---

1. *a chemical companion*: Heinrich Debus.
2. *St Malo*: An historical walled port city at the entrance to the River Rance in Brittany, facing the English Channel.
3. *Messageries*: A commercial company that undertakes the transportation or delivery of goods, messages, or people (*OED*).
4. *the tomb of Chateaubriand*: The Royalist writer and politician François-René Chateaubriand (1768–1848), a leading figure in the romantic movement in France, was buried on a small tidal island close to St Malo.
5. *Dinan*: A spectacularly located port and walled town on the River Rance in northern Brittany.
6. *Dol . . . Avranches*: Tyndall was travelling eastwards from Dinan roughly parallel to the coast.
7. *Mont St Michel*: An island close to the Normandy coastline housing an historic abbey and monastery.
8. *Man alone is vile*: The hymn writer Reginald Heber (1783–1826) who served as Bishop of Calcutta from 1823. His hymn 'From Greenland's Icy Mountains' includes the lines: 'Though every prospect pleases,/ And only man is vile' (*The Poetical Works of Richard Heber* (London: J. Murray, 1842), p. 90). The second line is sometimes rendered 'And man alone is vile'.
9. *had fallen before Sebastopol*: a local man who had served in the French army during the siege of Sevastopol in Crimea.

10. *Romish faith*: Roman Catholicism.
11. *14 Rue du Cirque*: The address of the offices used by British exhibitors and jury members.

To Thomas Archer Hirst 1 July 1855 1102

14 Rue du Cirque[1] Paris | 1s[t] July 1855.

My dear Tom.

I have just finished the writing of my journal and after it a long letter to Faraday.[2] I am tired and therefore will make this letter to you shorter than it otherwise would have been. Your enclosure[3] reached me yesterday on my arrival here: I almost regret that I did not give you a post on my route so that I might have received the letters earlier. But I have replied to them and if it is too late I am content. Better to meet them thus than have them lost altogether as they would run the chance of being in the little country posts. On the night I left you we loosed from Havre and had an abominable passage. The sea roaring furiously and all on board were ill. I never suffered so much in my life, once in the midst of my woe my eyes fell upon Debus. [‘] Well Heinrich how do you feel?’ I asked [‘]oh gut[’][4] was his reply—a few minutes afterwards however I heard him ejaculate [‘]Steward a basin![’] and what followed I leave to your imagination. We remained in Jersey till the following Tuesday. We had one fair day there, but one intensely wet one. From the moment however when we set foot on the Coast of France the weather has been magnificent and we have made use of it. Since my railway campaigns I do not remember finding myself in such vigorous health as during the last fortnight. At St. Malo we gave our portmanteaus to the Messageries[5] and took with us just as much luggage as would answer for a fortnights journey. We always walked when it was possible and thus enjoyed the magnificence of the country at our leisure. The word magnificence is not an exaggerated term, for the country through which we have passed is truly glorious. On the whole Debus & myself have got on most happily—he has been in the mumps[6] once or twice, and I would not have mentioned this were it not that some trivial bit of folly has put him in the mumps this morning. We have had some oscillation in the way of fare—sometimes we have lived and lodged sumptuously sometimes the lodging has been indifferent. We went from St Malo, where we saw the tomb of Chateaubriand on its lonely island with the waves rocking and moaning against the rocks beneath—to Dinan—Here the country is delightful. From Dinan we went to Dol. from Dol to Avanches where we lived for 2 days like princes and found no sting in the bill. From Avanches we visited the celebrated Mont St Michel. Thence to Vire, to Caen where we saw Catholic processions, High Mass and the tomb of William the Conqueror.[7]

Thence to Havre. Thence to Rouen where rest the bones of Richard Coeur de Lion.[8] Yesterday we reached Paris, failed in obtaining lodgings in the Hotel des Etrangers & in the hotel Violet[9] but at length found good accommodation in the Hotel Manchester Rue Grammont. I have as yet seen nobody. I will call upon John Martin today. Tomorrow I will try to take lodgings. I have said yes to both the questions asked by Mr Moseley[10] & the London Ins^{tn}.[11] Don't you think I was right? good bye love to Anna

Ever Yours John

RI MS JT/1/T/605

1. *14 Rue du Cirque*: see letter 1101, n. 11.
2. *long letter to Faraday*: letter 1101.
3. *Your enclosure*: enclosure missing.
4. *oh gut*: oh fine (German).
5. *Messageries*: for this term and the geographical locations that follow, see notes to letter 1101.
6. *in the mumps*: a fit of melancholy or ill humour (*OED*).
7. *William the Conqueror*: King William I (c. 1028–87) was the first Norman ruler of England and reigned from 1066 to 1087.
8. *Richard Coeur de Lion*: King Richard I of England (1157–99), who reigned from 1189 to 1199, was known as Richard Coeur de Lion or Richard the Lionhart because of his reputation as a fierce warrior.
9. *Hotel des Etrangers & in the hotel Violet*: The expensive Hotel des Etrangers, 3 Rue Vivienne; Hotel Violet was a commercial hotel in Passage Violet to the north of the boulevards.
10. *questions asked by Mr Moseley*: re accepting the examinership for the Board of Ordnance; see letter 1098.
11. *London Ins^{tn}*: Tyndall had been invited to give further lectures at the LI (Journal, 20 June 1855: RI MS JT/2/13c/777).

To Sarah Tyndall[1] [5] & 6 July 1855 1103

My dear mother

You will naturally wish to hear of my safe arrival in Paris after having completed the journey I proposed to myself through Normandy and Brittany. I reached Paris in excellent health on Saturday last.[2] I told you I believe that I was appointed a juror to the Exhibition here, my duties have already commenced and I expect to be kept here three weeks at least. I shall endeavour to return to London at the end of the present month; and shall be very glad if I can do so. For though Paris is a most beautiful place, and possesses more opportunities for amusement than any other city in the world, though its palaces are grand and its people pleasant my simple desire is to get back to my work. There is however a great advantage in being here for some time, for I have an opportunity of meeting with eminent men from all quarters of the world and whom I previously knew by their writings alone.

My health continues very good—far better indeed than it has been within the last two years—my trip through the country during which I often walked 30 miles a day did me great service. I am stopping in a hotel at present, and if you write to me as soon as you receive this I shall be sure to get your letter here. Address it to me

Hotel Manchester | Rue de Grammont No 1. | Paris.

I wish you would let me know at the same time how you are off for money.

Your affectionate son | John

July 6th, 1855. | Prepay your letter: the postage if it be under a quarter of an ounce is four pence.

RI MS JT/5/16/159
LT Typescript Only

---

1. *Sarah Tyndall*: Tyndall's mother, Sarah (née Macassey/Macasey/McAssey) (1793–1867).
2. *Saturday last*: 30 June 1855.

From Charles Hawker[1] 9 July 1855 1104

Office of Ordnance | 9 July 1855

Sir,

I am directed by Lord Panmure to offer you the examinership[2] (Professor Moseley consulted you about) of candidates for admission to the Royal Artillery and Royal Engineers[3] which is to be held on Wednesday the 1st of Aug. in the hall of King's College London.

I have the honor to be | Sir | your obedient Servant | Charles Hawker.

John Tyndall Esq | 14 Rue du Cirque, Paris.

RI MS JT/2/7/149
LT Typescript Only

---

1. *Charles Hawker*: Charles Frederick Hawker (1821–1902) was a senior clerk in the Board of Ordnance at the War Office.
2. *examinership*: Tyndall 'Replied and accepted the offer, Friday 13th July' (Journal: RI MS JT/2/13c/777).
3. *candidates . . . Royal Engineers*: see letter 1098, n. 2.

To Michael Faraday 10 July 1855 1105

14 Rue Du Cirque Paris.[1] | 10th, July 1855.

My dear Mr Faraday

I repeated your question[2] 'what remembrance is there of Arago?' [3] yesterday to a friend, and the reply was that there is no remembrance. He left so many opponents behind him in the Academy[4] that all unity of action is destroyed, and the sum collected was too small for what was proposed.[5] I was at the Academy yesterday, and conveyed your sentiments of affection to Biot. He enquired very kindly after you—He was kind enough to place me in a good position, from which as he remarked I might see 'les personages'.[6] At the end of a seat near me I saw a fine looking old man, and noticed that many of the members, on passing him squeezed him silently by the hand, while tears appeared in the eyes of some of them. The old man appeared to have been smitten by some calamity with which his friends sympathised—I learned afterwards that it was Thenard[7] that he had lost his son quite recently and had suffered a second heavy domestic affliction some time before. It was the first day of his appearance since his son died. Regnault was president and once he said that the 'parole' was with M. Le Prince de Canino–[8] 'Le Prince de Canino n'existe pas' responded the person addressed—'Monsieur Bonaparte s'il vous plait'[9] I do not know whether it is that the title is abolished or that Lucien's republican tendencies caused him to disavow it. At the conclusion of the meeting I felt great pleasure in making the acquaintance of Prof DelaRive who is here with his wife.[10] He will remain here for eight or nine days and then proceed to London. I had never seen him before but his kind frank countenance agreed with the image that I had previously formed of him. I find in Wartmann an exceedingly agreeable companion: he and I walked together for several hours yesterday—we were at Versailles[11] on Sunday with Wheatstone. From all I can see Wartmann is not only well instructed in science but possesses qualities of heart which are pleasant and refreshing to those who require something more than the mere culture of the brain. He gave me a curious picture of the strifes and heartburnings existing among the scientific men of Paris. I could not help comparing this life of ambition, this grasping at and feverish longing after the honours of the world, with the quiet of your existence, and I hope I derived profit and strength from the comparison. Arago appears to have been unfortunate in this respect, and from all I can gather regarding him he appears to have lacked that stability of soul, that reliance upon higher things than mere worldly renown, which I think ought to make the true philosopher. I read some time ago his life of Ampère[12] and I could not help thinking at the time that he was incapable of appreciating fully Ampère's character. The latter appeared to me to possess qualities in a high degree which in Arago were at least rudimentary. LeVerrier I am told moves through the Academy as through a vacuum—he has no connexion with anybody. Thus do we find this august body split up into antagonisms which are scarcely worthy of ordinary illeducated citizens. Thenard is one of those whom all parties respect—the universal sympathy evidenced yesterday was a proof of this. He has in fact pursued science for the love of science, and has not used it merely as a stepping stone to worldly advancement.

While I write Wheatstone, Edmond Becquerel[13] and other members of the section are in the next room and Wheatstone's voice rings in my ears: He talks as if he were inspired being apparently carried quite beyond his own control. He evidently rates pretty highly those honours and marks of recognition and I am inclined to think that he pays in some measure the inevitable penalty, and often has an uneasy mind. Foucault has set up his pendulum in the Exhibition:[14] I saw him this morning for the first time: he has an arrangement of electromagnets underneath the pendulum by which its power is sustained, and the time of oscillation indefinitely prolonged. The fortnight I spent in Normandy was of immense value to me since my railway campaignings[15] I have not felt a greater infusion of physical energy than during that tour. I still continue very strong and hope to be able to keep my mind tranquil and consequently my health good while I remain at Paris. I shall endeavour to get back at the end of the month as indeed I long to be at my work again.

Thank you very much for communicating my request to Anderson[16]—He sends me my letters regularly, and now with kindest remembrances and best wishes for the health and happiness of Mrs Faraday Miss Barnard and yourself

Believe me | most faithfully Yours | Tyndall

RI MS JT/1/12/4034–6
Faraday Correspondence, 4:3004
Typed Transcript Only

1. *14 Rue Du Cirque Paris*: see letter 1101, n. 11.
2. *your question*: presumably contained in a (missing) letter from Faraday. 'Heard from Faraday and wrote to him' (Journal, 10 July 1855: RI MS JT/2/13c/777).
3. *remembrance . . . of Arago*: François Arago had died on 2 October 1853.
4. *the Academy*: The Académie des Sciences, founded in 1666, was the foremost scientific society in France.
5. *the sum . . . proposed*: Soon after Arago's death a committee of eminent people, principally scientists, had been formed in order to erect a monument to his memory.
6. *les personages*: people of importance, of high rank (French).
7. *Thenard*: The renowned chemist and teacher Louis Jacques Thénard (1777–1857). His

wife, Jeanne Victoire, née Humblot-Conte, had died on 27 April 1855, and his thirty-four year old son Henri had also died earlier in 1855.

8. *M. Le Prince de Canino*: The biologist and ornithologist Charles Lucien Bonaparte, 2nd Prince of Canino and Musignano (1803–57), was a nephew of the Emperor Napoleon I.
9. *Le Prince de Canino n'existe pas . . . Monsieur Bonaparte s'il vous plait*: The Prince of Canino doesn't exist, Mr. Bonaparte if you please (French). The prince thus expressed his wish to be known simply as Mr. Bonaparte.
10. *his wife*: Auguste de la Rive's first wife had died in 1850. On 24 May 1855 he married Louise-Victoire-Marie, née Fatio.
11. *Versailles*: The Palace of Versailles was the highly ornate chateau with extensive formal gardens built originally for Louis XIV. As one contemporary noted, 'no excursion to Paris would be considered complete without having paid it a visit; for it truly is a structure in which all the sumptuous luxury of a period renowned for its magnificence has been profusely lavished' ([F. G. Bush and T. A. Day], *The Excursionist's Guide; or, Three Days in Paris* (London: T. Bosworth, [1850]), p. 19).
12. *his life of Ampère*: F. Arago, 'Eloge d'Ampère Prononcé lors de la Séance Publique de l'Académie des Sciences le 21 Août 1839', in J. A. Barral et al, *Oeuvres Complètes de François Arago,* 16 vols (Paris: Gide et J. Baudry, 1854), vol. 2, pp, 1–116.
13. *Edmond Becquerel*: see letter 1042, n. 6.
14. *Foucault has set up his pendulum in the Exhibition*: the pendulum, which demonstrated the earth's rotation, was mounted in the Palais de l'Industrie; see letter 1052, n. 3.
15. *my railway campaignings*: Tyndall was referring to his work as a railway surveyor, principally in Yorkshire, 1844–47.
16. *my request to Anderson*: A note to Charles Anderson, presumably contained in the missing letter, mentioned in n. 2, above.

To Thomas Henry Huxley [11?] July 1855 1106

14 Rue de Cirque[1] Paris | Wednesday morning before 6 o'c

My dear Huxley

Until yesterday[2] I was not aware that you had been chosen for the Fullerian at the Royal Institution.[3] I have been wandering through Normandy without a home or local habitation to which a letter could be addressed, and this I suppose is the reason why I did not hear of your success earlier. I shake you by the hand my dear Fellow[4] with a satisfaction I would venture to say almost equal to your own on receiving the intelligence. The fates once seemed to point to our connexion in a distant land:[5] we are now colleagues at home, and I can claim you as my scientific brother. May the gods continue to drop fatness upon you, and may your next great step[6] be productive of all the felicity which your warmest friends or your own rebellious heart can desire.

Believe me, dear Huxley, | Ever yours sincerely | John Tyndall
Send the enclosed[7] to Francis for me.

IC HP 1.19

---

1. *14 Rue de Cirque*: see letter 1101, n. 11.
2. *Until yesterday*: It is possible that Tyndall received news of Huxley's appointment on 3 July 1855. However, as several London newspapers carried that information in their issues of 4 or 5 July, its more likely that he heard about it on 10 July.
3. *the Fullerian at the Royal Institution*: At their meeting on 2 July, the managers of the RI had elected Huxley the Fullerian Professor of Physiology. Huxley was the seventh incumbent of this professorship, which had been endowed in 1833 by the philanthropist John Fuller.
4. *Fellow*: Tyndall appears to have capitalized the word 'fellow' to make Huxley a fellow Fellow of the Royal Soceity.
5. *our connexion in a distant land*: Like Tyndall, Huxley had previously applied for a post—in his case the chair of natural history in 1851—at the University of Toronto.
6. *your next great step*: After an eight-year courtship, Huxley married Henrietta Ann Heathorn at All Saints' Church, St John's Wood, on 21 July 1855.
7. *the enclosed*: enclosure missing.

From Henry Moseley 12 July 1855 1107

The Deanery Bristol | 12 July

My dear Sir,

You will before this time have received an official letter from Mr Monsell[1] with reference to the examination to be held at King's Coll London on the 1st of August at 10 o'clock of Candidates for appointments in the Artillery and Engineers.[2]

The object of my present communication is to invite you to a conference of the Examiners[3] to be held at King's College[4] on Monday the 23[d] instant at 2 o'clock. I am also to ask you to be good enough to prepare by that time a series of questions such as you may think best adapted to the purposes of such an examination. Bearing in mind that the knowledge of the Candidates in experimental science is probably of a very elementary kind, acquired under unfavourable circumstances. The number of Candidates who have (up to this time) applied to be examined in this subject is 38.[5] Arrangements will be made to put the questions to press on the 23rd inst., and I am to ask that you would be good enough to bring them with you then. You will have an opportunity then of consulting with your colleagues as to other details of the examination. It is probable however that one day will be set apart for the examination[6] in your subject and you will probably deem it expedient to set two papers one in the morning and the other in the afternoon. Will you allow me to suggest that you should bring with you as long a list of questions as possible, because we may be able to place in your hands information then which will guide you in the selection

Yours my dear Sir truly | Henry Moseley.

I shall be very glad to hear from you addressed as above

You will bear in mind that the reading of the Candidates has received no special direction to this examination.[7]

RI MS JT/1/3/886
LT Typescript Only[8]

---

1. *an official letter from Mr Monsell*: letter missing; William Monsell (1812–94) served as clerk in the Board of Ordnance, 1852–57.
2. *the examination . . . Engineers*: see letter 1098, n. 2.
3. *conference of the Examiners*: Tyndall attended that meeting on 23 July (Journal: RI MS JT/2/13c/779).
4. *King's College*: King's College, which stands on the Strand in London, was founded in 1829 as an Anglican institution, in opposition to University College (f. 1827) which admitted Dissenters, Jews and other non-Anglicans.
5. *The number of Candidates . . . is 38*: The final number examined in 'Natural Sciences' was 41.
6. *one day will be set apart for the examination*: The 'Natural Sciences' papers were sat on Thursday, 9 August 1855.
7. *no special direction to this examination*: as this was the first occasion when the examination was held.
8. Moseley sent this letter to Paris and a similar letter (1108) was sent to the RI in London.

From Henry Moseley 12 July 1855 1108

Deanery Bristol | 12 July

My dear Sir,

I have written to you in Paris[1] referring to an official letter addressed to you there from the Ordnance and asking you to prepare before the 23 inst. a series of questions in Experimental Science for the Exam. which is to take place on the 1st of August; also inviting you to a Conference of the Examiners on the 23rd inst. at King's Coll at 2 o'clock, bringing the questions with you which are to be put to press on that day. I write this duplicate addressing it to London[2] in order that it may find you if you have left Paris.

Yours my dear Sir | faithfully | Henry Moseley.

RI MS JT/1/3/889
LT Typescript Only

1. *I have written to you in Paris*: see letter 1107.
2. *addressing it to London*: sent to the RI.

To Thomas Archer Hirst 13 July 1855 1109

Hotel Manchester | Rue de Grammont N° 1. | 13th July 1855

My dear Tom

I have written to you[1] since you wrote to me, I now write to you again and thus make you twice my debtor: My time is filled in Paris. Nearly every day Wertheim,[2] Wartmann and myself peregrinate among the philosophical instruments,[3] and sometimes at the day's end the whole of us feel greatly fagged. Today is a holiday for some of our group, otherwise I should not have time to write to you—Paris is always gay, but gay beyond the average at present, and night after night the tide of life swings and murmurs on the Boulevards and in the Palais Royale.[4] I usually dine there: Wartmann and I are pretty constant companions, and I find in him a man with whom it is pleasant to converse: we ramble together and pay visits together. Wheatstone is usually with us in the evening—and thus we constitute a trio who manage to spend their spare time together in pleasant amity. Any English juror who wishes will be presented to the Emperor[5] by Lord Cowley[6] the English Ambassador. I should like to see the man, but this liking will scarcely tempt me to go through the form of a presentation. The Prince Napoleon[7] gives a soirée every Saturday evening, and thither the jurors repair in considerable numbers: I have an invitation for the Saturdays of July and intend to go there tomorrow. The Sunday which followed the arrival of Debus and myself we spent together at St Cloud. Last Sunday Wheatstone Wartmann and myself spent at Versailles and saw the grandes eaux.[8] I had an invitation to tea on Wednesday evening from Mr Denison[9] and Mr Labouchere[10] although I had never spoken to the latter and had only exchanged a word or two with the former in the jury room. Sir David Brewster, Sir Wm Hooker, Sir Geo. Clarke,[11] Wheatstone, Prof. Willis[12] and myself were the guests, and we spent a pleasant rational evening together. You know Brewster &

Wheatstone are at war to the knife regarding the stereoscope.[13] I contrived to get between them at tea and thus relieved them from an unpleasant proximity to each other. I am of course on good terms with both of them. The official offer of the examinership[14] of candidates for the Royal Artillery and Royal Engineers reached me yesterday, and I have written back accepting it[15]—The examination takes places in the Hall of King's College London on the 1st of August next so that I must be at home at the end of this month. I do not know whether the post is likely to be a permanent one: the word 'examinership' was used in the official letter. I have seen John Martin several times but as yet have had no opportunity of dining with him; for this he is more to blame than I am, for I made an appointment with him a few days ago and he was so far behind time that I accepted the proposal of Wheatstone who chanced to pass me at the time to dine with him. Last night I was at Duboscq's. He had Brewster, Amici,[16] Steinheil, Ettingshausen, Moigno,[17] Wheatstone, and others. He went through his usual experiments with the electric light. Tonight Wheatstone Wartmann and myself go to Le comte de Moncel[18] who has been lately experimenting with Ruhmkorff's coil and who is to shew us his experiments. I wish either that you had sent me those letters you left at Chadwell Street, or that you had left them at the Institution: I have written to Mrs Whitehouse[19] about them but as yet have heard nothing of them. In a note sent to Francis yesterday I again alluded to them and hope he will make some enquiry about them. I wish you would write to Francis on the subject and say to him at the same time that I should be obliged if he would send me some separate copies of my paper direct. On last Monday I was at the Institute[20] and old Biot kindly placed me in the centre whence as he said I could observe 'les personnages'.[21] Regnault was president—one incident occurred which I dont understand--The Prince of Canino the Emperor's Cousin[22] was called upon by his usual title—he replied 'Le Prince de Canino n'est pas ici. Monsieur Bonaparte s'il vous plait'. Thenard was there and was the object of general sympathy as he has recently lost his son. LeVerrier, Becquerel,[23] Despretz, Cauchy[24] and others were there—I noticed one man with a frank open countenance who spoke with many present: when the meeting was breaking up. I found him at my side and he requested Wheatstone introduce us to each other: It was DelaRive: I felt great pleasure indeed in making his acquaintance. Foucault I have also seen—he has his pendulum swinging in the Exhibition:[25] the ball is of iron and an electromagnet is placed in the position of equilibrium. This pulls the ball downwards and at the moment the ball passes the centre the magnetism ceases. In this way he is able to prolong the oscillations indefinitely. I had a cosy old bachelor's dinner once with Despretz, and a quiet ramble afterwards. My health still continues very good, but I should like much to escape for a fortnight to the country. The day is hot and my head is not in the letter writing phase—so you must be content with the dish of incongruous gossip which I here lay before you. I am beginning to yearn for my magnets, the year has been sadly broken into already but I must not complain for the time has not been lost. Remember me affectionately to Anna Good bye

As ever | John

RI MS JT/1/T/606

1. *I have written to you*: presumably letter 1102.
2. *Wertheim*: The Austrian chemist Theodor Wertheim (1820–64) held the chair of chemistry at the University of Pest, Hungary.
3. *peregrinate among the philosophical instruments*: Tyndall, Wertheim and Wartmann all served on the jury for class VIII ('Arts relating to the exact Sciences, and to Instruction').
4. *Palais Royale*: Built for Cardinal Richelieu, the Palais Royale was currently the home of Prince Jérôme Napoleon, the Emperor's uncle; see n. 7, below.
5. *the Emperor*: Louis-Napoléon Bonaparte (1808–73); from 1852 he was Napoleon III, the first president of the Second Republic. The Exposition Universelle was mounted by the Emperor in response to the Great Exhibition held in London in 1851.
6. *Lord Cowley*: Henry Wellesley, 1st Earl Cowley (1804–84), was a diplomat who served as the British Ambassador in Paris from 1852 to 1867.
7. *Prince Napoleon*: Joseph-Charles-Paul Bonaparte (know as Prince Jérôme Napoleon, 1822–91), served as president of the Imperial Commission of the Exposition Universelle. He was a cousin of Napoleon III.
8. *Versailles and saw the grandes eaux*: see letter 1105, n. 11. The Grandes Eaux (at Versailles), 'or principal waterworks, may be considered as an illustration in a fanciful way of some mythological or legendary history or event—the sculptor and the professor of hydraulics having united in the production of the piece. The sculptures are executed in marble and lead; they are mostly of colossal proportions, and their number is wonderful' ('A Day at Versailles', *Leisure Hour,* 4 (16 August 1855), pp. 516–20, on p. 519).
9. *Mr Denison*: The liberal politician Evelyn Denison (1800–73), who later became speaker of the House of Commons, served as one of the jurors for Class III (Agriculture) at the Exposition Universelle in Paris.
10. *Mr Labouchere*: The liberal politician Henry Labouchere had served as president of the Board of Trade from 1838 to 1841 under Viscount Melbourne and from 1847 to 1852 under Lord John Russell. He had been greatly involved in mounting the Great Exhibition of 1851.
11. *Sir Geo. Clarke*: Sir George Clerk of Penicuick (1787–1867), was a Tory politician who had served as vice-president of the Board of Trade and Master of the Mint in the second Peel administration.
12. *Prof. Willis*: Robert Willis (1800–75) was professor of natural philosophy at Cambridge

University and professor of mechanics at the Government School of Mines. He was both an exhibitor (of a 'Mechanical apparatus for instruction') and a juror for Class VII ('Machinery for Woven Fabrics') at the 1855 Exposition Universelle.

13. *Brewster & Wheatstone . . . stereoscope*: Brewster and Wheatstone were engaged in a bitter priority dispute over the invention of the stereoscope. See also Tyndall's two articles 'On Binocular Vision and the Stereoscope', *Journal of the Photographic Society*, 3 (1856), pp. 96–102 and 116–21.
14. *official offer of the examinership*: letter 1104.
15. *I have written back accepting it*: see letter 1104, n. 2.
16. *Amici*: The Italian astronomer, microscopist and botanist Giovanni Battista Amici (1786–1863), who served as astronomer to the Grand Duke of Tuscany and was the director of the observatory at the Royal Museum in Florence.
17. *Moigno*: see letter 1027, n. 10.
18. *Le comte de Moncel*: Théodose du Moncel (1821–84) contributed to both archaeology and the science and application of electricity. He was an exhibitor and prize-winner at the Exposition Universelle.
19. *I have written to Mrs Whitehouse*: letter missing.
20. *the Institute*: The Institut de France was established in 1795 to encompass several learned societies including the Académie des Sciences, founded in 1666.
21. *les personnages*: see letter 1105, n. 6.
22. *The Prince of Canino the Emperor's Cousin*: see letter 1105, n. 8. The prince was a nephew of Napoleon I.
23. *Becquerel*: Antoine César Becquerel (1788–1878), who was a member of the Institut de France and had made significant contributions to the fields of electricity and electrochemistry.
24. *Cauchy*: The royalist and eminent mathematician Augustin-Louis Cauchy (1789–1857) who, after the political convulsions in France, was appointed professor of mathematical astronomy at the faculty of sciences at the Sorbonne in 1852.
25. *his pendulum swinging in the Exhibition*: see letter 1105, n. 14.

From Henry Moseley 14 July 1855 1110

Deanery Bristol | 14

My dear Sir,

Your letter addressed to the Ordnance[1] has been forwarded to me. A letter of mine[2] which you will have already received will have answered several of your questions. There are only 38 Candidates who have offered to be examined in the Natural Sciences (an unfortunately comprehensive term which I hope to be able to interpret with more precision by the 23rd).[3]

The questions are to be divided into sections of two or three questions in each section; and the Candidates limited to answering one question only in a section.

Thus the questions may be made to cover a large surface whilst the number of answers may be limited within a manageable compass. If you came prepared with 8 sections of questions for the morning paper and as many for the evening; two of these sections in each paper being on the elements of chemistry; it will abundantly suffice.

Yours faithfully | Henry Moseley

The reading of the Candidates has received no special direction to this examination.[4]

RI MS JT/3/891
LT Typescript Only

---

1. *Your letter . . . Ordnance*: letter missing; presumably the reply to the letter to William Monsell of the Board of Ordnance referred to in letter 1107, n. 1.
2. *A letter of mine*: letter 1107.
3. *by the 23rd*: the date of the Examiners' meeting at King's College; see letter 1107.
4. *The reading . . . this examination*: The candidates had not been advised to study any particular textbooks.

From Henry Moseley 16 July 1855 1111

Deanery Bristol | 16th. July 1855

Sir,

I beg to communicate to you the following arrangements proposed to be observed in the examinations of Candidates for appointments in the Royal Artillery and Engineers.

1st. The questions to be divided into sections,[1] each containing three questions, and no Candidate to answer more than one question in a section.

2nd. The questions in each section to be arranged as far as may be possible, in the order of difficulty, and to have reference to the same general division of the subject.

3rd. A paper of eight sections to be set for the morning and the same for the afternoon of the day assigned to the examination in your subject.

4th. Your estimate of the merits of each paper to be recorded upon it by one of the following words

Excellent, Good, Fair, Moderate, Imperfect, Nil, or by the initials of those words.

5th. A number of marks (hereafter to be determined upon) to be assigned to each of the degrees of merit specified above, and the sum of the numbers of marks assigned to each Candidate's answer to fix his place on the class list.

I have the honor to be | Your faithful Sevt. | Henry Moseley.

RI MS JT/1/3/890
LT Typescript Only

---

1. *The questions to be divided into sections*: For Tyndall's questions on the 'Natural Sciences' see *Report on the Examination for Appointments in the Royal Artillery and Engineers, Held at King's College, London, on the First of August, 1855* (London: Harrison, 1855), pp. 93–104. For the examination held on the morning of 9 August, Tyndall set 9 questions in 3 sections on heat; 6 questions in 1 section on the mechanical properties of air; 3 questions in 1 section on magnetism; and 6 questions in 2 sections on chemistry. For the afternoon examination he set 9 questions in 3 sections on light; 9 questions in 3 sections on voltaic electricity; 6 questions in 2 sections on chemistry; 3 questions in 1 section on sound; 4 questions in 1 section on meteorology; 4 questions in 1 section on astronomy; 4 questions in 1 section on geology; 4 questions in 1 section on mineralogy and crystallography; 3 questions in 1 section on metallurgy; 3 questions in 1 section on botany. Thus 13 different subjects (counting the two chemistry exams as one subject) were examined by 73 questions, divided into 23 sections.

To Thomas Archer Hirst 22 July 1855 1112

London Sunday night

My dear Tom

You may be in London for ought I know though you are hidden from me. Mrs Whitehouse tells me that you were here on Friday evening—had I known of your probable presence we might have contrived to meet each other—I left Paris on Saturday morning and reached London last night. I have come over thus suddenly to attend a conference to be held at King's College on Monday regarding these examinations.[1] I wrote to you from Paris but you probably had not received my letter when you came here on Friday—I find London altogether a quiet place after leaving the noise of Paris. My time there was very pleasantly spent. Wartmann and myself rambled together almost every evening and spent our time very agreeably. With one day's exception due to cold from wet feet my health was very good—The stock I laid in in Normandy lasted me throughout. I have as yet seen nobody here. Of John Martin I saw very little in Paris[2]—I called upon him several times soon after my arrival there but somehow or other the link between us broke—my time did not suit him or his did not suit me and so each of us pursued his own track. I managed to round off my work so as to leave Paris with a clear conscience. I may go back again but I hope I shall not be required. One week's ramble in the country would be of great value to me now. It would shake the rumble of cabs and omnibuses from my brain and dispel all thoughts of the comparative merits of the electric telegraph of England and Prussia.[3] Had I desired high life and loved to rub myself against great people I could not have been better situated than at Paris. I like to look at such people and to observe their manners and habits as a matter of instruction, but these once learned I recoil upon my old self and would not give up my own life for that of the best of them. You guessed aright regarding Hector's letter.[4] I will send it to you in a day or two—It is a remarkable document and reveals a state of mind which I should like to see altered but which it is very difficult to attack—The man's knowledge in fact swamps his faith and renders his life dry and unhappy. Give my love to Anna—She would have been in the third Heaven looking at the shawls, muslins & jewels of the exhibition.[5]

Your affectionate John

RI MS JT/1/T/915

1. *a conference . . . examinations*: see letters 1107 and 1108.
2. *Of John Martin . . . Paris*: see letter 1027, n. 8.
3. *the comparative merits of the electric telegraph of England and Prussia*: one of tasks facing the jurors at the Exposition Universelle.
4. *Hector's letter*: letter 1099, which was mentioned in missing letters between Tyndall and Hirst.
5. *shawls, muslins & jewels*: The Exposition Universelle included classes devoted to 'Goldsmiths' Work, Jewellery, and Bronzes' (Class XVII), 'Cotton Printing and Dyeing' (Class XIX), 'Woollen and Worsted Manufactures' (Class XX) and 'Silk' (Class XXI).

To Thomas Archer Hirst 23 July 1855 1113

July 23. 1855

My dear Tom

I wrote to you yesterday[1] (Sunday) but left my letter at home and lest I should not get back again before post time I scratch a line to tell you of my arrival in London—I started on Saturday morning[2] & reached London the same night. Today I am to attend a conference of the examiners[3] and it was this that brought me home so suddenly. I will take a run down to you soon as I want to get the rumble and turmoil of Paris clean out of my head. I suppose you will be soon opening[4]—Mrs Whitehouse informed me that you were in London on Friday last—You may be in it now for ought I know but my surmise is that you came up on some school business and returned immediately. Give my love to Anna—She would be in third heaven I am sure amid the shawls and jewels of the great exhibition.[5]

Your affectionate | John | Monday.

RI MS JT/1/T/607

1. *I wrote to you yesterday*: letter 1112.
2. *I started on Saturday morning*: from Paris on 21 July 1855.
3. *a conference of the examiners*: see letter 1107.
4. *soon opening*: The new intake of students to Queenwood College occurred in early August.
5. *shawls and jewels of the great exhibition*: letter 1112, n. 5.

To Thomas Archer Hirst 25 July 1855 1114

My dear Tom

I was just about to post the enclosed to Wright[1] when the thought occurred to me that you might possibly not like it. I therefore send it through your hands—I know very well that the privacy of a man's home may be invaded by his nearest friend, and besides this I know it would be too bad to think of eating up your substance for the length of time I purpose stopping in the country—If you would be content to make me pay my way I should be content to live with you, but I dont care whether I do or not for when I am within 5 minutes walk of you I have all that I desire. I was serious when I spoke to Edmondson today and I mean to turn his line of science to account but I should sleep more quietly & write more quietly at Broughton & hence my desire to fix myself there. On the whole I think it will be best for me to go to Wrights or if he has not room to pick out a little cottage for myself—I must leave these lodgings they are much too far from the institution so I purpose packing up my things in a few days, placing them in the Royal Institution & carrying my instruments with me to Broughton. I think you will see that my scheme is a reasonable one, and if so I will thank you to hand the note to Wright.

Affectionately Yours | John

RI MS JT/1/T/608

1. *the enclosed to Wright*: The enclosure to Wright (which is missing) may relate to his wedding on 28 July 1855, and/or to where Tyndall could stay during his proposed extended

visit to Broughton. On 29 July he went on a day-trip by excursion train to Queenwood and from 18 August to 5 September he stayed in the Lodge at Queenwood (RI MS JT/2/13c/779 and 783–86).

To Thomas Archer Hirst 25 July 1855 1115

My dear Tom.

It may gratify you to learn that your paper to the R.S. has been read[1] and I this day saw the proof of it printed[2]—it has been printed entire.

John | Wednesday,

RI MS JT/1/T/609

1. *your paper to the R.S. has been read*: read at the RS meeting on 21 June 1855, during Tyndall's absence from London; see letter 1058, n. 3.
2. *the proof of it printed*: by the firm of Taylor and Francis. However, its publication in *Roy. Soc. Proc.* would not take place for some months.

To George Gabriel Stokes 25 July 1855 1116

My dear Sir

The object of Mr Joule's investigation[1] is to enquire into the correctness of a law announced some years ago by Lenz and Jacobi[2] and expressed by Mr Joule in the following words:—'When two bars of iron of different diameters, but equal to one another in length, and surrounded with coils of wire of the same length, carry equal streams of electricity, the magnetism developed in the bars is proportional to their respective diameters.'[3] Referring to Poggendorff's Annalen vol. 61 page 262, I find the law as stated by its discoverers to be as follows:—'In solid cylinders of iron of equal lengths and upwards of ⅓ of an inch in diameter, the magnetism developed by currents of equal strengths, sent through spirals possessing the same number of coils, is proportional to the diameters of the cylinders.'[4] Instead of taking the same number of coils Mr Joule takes equal lengths of wire and winds them close round bars of different diameters: the number of coils must be therefore unequal—the same length of wire will go less often round a thick bar than round a thin one. I do not see therefore how experiments made with such an arrangement can furnish any evidence as to either the truth or fallacy of the Law of Lenz & Jacobi.

Mr Joule also observes upon the first page that he is not aware of any experimental researches having been made, since the publication of the law, which are calculated either to confirm or to confute it. Many such experiments are recorded in the Continental Journals. Among those who have made such researches may be mentioned Müller,[5] Hankel,[6] von Feilitsch[7] and Dub.[8] The most recent result with which I am acquainted is; that the magnetism developed in iron cylinders by equal currents, passing through the same number of coils, is proportional to the square roots of the diameters of the cylinders. (See Pogg. Ann. vol. 90. page 248 &c.)[9]

In page 3 Mr Joule compares the effect of a magnetized bar upon a magnetic needle at the respective distances of 6 and 12 inches; and states that the action at the former distance is 2.98 times the action at the latter.[10] In one of the experiments cited however the effect at 6 inches is more than 5 times the effect at 12 inches. The ratio 2.98 has nevertheless been used in all Mr Joule's subsequent experiments for the purpose of reducing the actions at 12 inches to the actions at 6. I cannot help thinking that such reductions must be very uncertain.

Mr Joule also states that the magnetism of the smaller bars 'is nearly proportional to the squares of their diameters'. One of these bars is ⅛ of an inch in diameter and another ¼ of an inch—hence the magnetism of the latter ought to be nearly 4 times that of the former. In one of the experiments cited however the magnetism of the latter is less than twice that of the former. Remarks of a similar nature might be made on the table at page 4.

I would beg to add that these observations of mine ought to be received with caution, for I confess that I have no clear comprehension of the purport of Mr Joule's experimental arrangements. This may possibly be traced to a desire on his part to express himself with brevity, or possibly to my ignorance of the subject in question.

I am my dear Sir | yours very faithfully | John Tyndall
25th July 1855

RS RR/3/155

1. *Mr Joule's investigation*: Faraday also criticised Joule's paper (*Faraday Correspondence,* 4:3002). The paper was resubmitted some months later and was again refereed by Tyndall (letter 1181) and Faraday (*Faraday Correspondence,* 4:3067). Despite these criticisms Joule's paper was published as 'Introductory Research on the Induction of Magnetism by Electric Currents', *Phil. Trans.,* 146 (1856), pp. 287–95. Page numbers below refer to the published version, which bears the date 20 June 1855 (p. 294), indicating that the paper was not substantially revised in the light of the referees' reports. A postscript (dated 21 December 1855) was, however, added containing further experimental results (p. 295).
2. *Lenz and Jacobi*: The Russian physicist Heinrich Friedrich Emil Lenz (1804–65) served as professor of physics and physical geography at the University of St Petersburg. Moritz von Jacobi (1801–74) was a German physicist and engineer who also

worked in St Petersburg. The law as published by Lenz in 'Ueber die Bestimmung der Richtung der durch Elektrodynamische Vertheilung erregten Galvanische Ströme', *Poggend. Annal.,* 31 (1834), pp. 483–94, on p. 485: 'If a wire moves in the neighbourhood of a galvanic current or a magnet, the galvanic current is induced in it, the direction of which is such that it would have caused a motion of the wire that would have been exactly the opposite of the one given it, provided it was movable only along the line of motion' (C. Jungnickel and R. McCormmach, *Intellectual Mastery of Nature: Theoretical Physics from Ohm to Einstein,* 2 vols [Chicago: University of Chicago Press, 1986], vol. 1, p. 137).

3. *When two bars . . . diameters*: Joule, 'Introductory Research', p. 287.
4. *In solid cylinders . . . the cylinders*: M. Jacobi and E. Lenz, 'Ueber die Gesetze der Elektromagnete', *Poggend. Annal.,* 61 (1844), pp. 254–80, on p. 262.
5. *Müller*: The German physicist Johann Heinrich Jakob Müller (1809–75), who taught at the University of Freiberg and published extensively on optics, photography and magnetism, including 'Ueber die Magnetisirung von Eisenstäben durch den Galvanischen Strom', *Poggend. Annal.*, 79 (1850), pp. 337–44.
6. *Hankel*: Wilhelm Gottlieb Hankel (1814–99) was professor of physics at the University of Leipzig who contributed a number of papers on electrical subjects, including thermo-electricity and atmospheric electricity.
7. *von Feilitsch*: Tyndall probably had in mind F. C. O. von Feilitzsch, 'Ueber den Magnetismus elektrischer Spiralen von verschiedenem Durchmesser', *Poggend. Annal.,* 79 (1850), pp. 564–68.
8. *Dub*: Julius Dub (1817–73), who pursued electrical research in Berlin.
9. *See Pogg. Ann. vol. 90. page 248 &c*: J. Dub, 'Gestze der Anziehung der Elektromagnete', *Poggend. Annal.*, 90 (1853), pp. 248–61 and 436–56.
10. *2.98 times the action at the latter*: Joule, 'Introductory Research', p. 289.

To Thomas Archer Hirst 27 July 1855 1117

My dear Tom.

The principal point is the securing of a place to work.[1] that is already safe and the rest is a trifle—dont bother yourself any more about the matter. I will settle all when I come down,[2]

affectionately | John

Friday. | Give Wright the letter.[3]

RI MS JT/1/T/610

1. *securing of a place to work*: see letter 1114.
2. *when I come down*: Tyndall arrived at Broughton on Sunday, 29 July, for a one-day visit.
3. *the letter*: see letter 1114, n. 1.

To Hector Tyndale 27 July 1855 1118

London July 27. 1855

My dear Hector.

When your letter[1] reached England I was in the midst of Normandy with a knapsack on my back seeking the health and vigour which had ebbed from me during the preceding session. Debus was my companion. We went first to Jersey, thence to St. Malo, thence to Dinan in Brittany, thence to Avranche, through Dol, whence we crossed the sands on a visit to Mont Saint Michel—thence to Mortain, thence to Vire thence to Caen, from Caen by water to Havre, thence to Rouen, thence to Paris.[2] The weather was glorious, and the country through which we passed worthy of being lighted up by such a sun. Normandy presents the general aspect of a garden, and since the days of my railway campaigning I have not enjoyed such an amount of physical energy as I experienced during the fortnight I spent there. It served to sustain me for three weeks amid the heat and confusion of Paris whither I was called to act the part of Juror to the great exhibition.[3] During my stay there I could mark a daily and rapid advance towards the completion of the exhibition and it is now thought to rival, if not to surpass that of London in 1851.[4] I should have remained there longer, and may be compelled to return, but I received an official letter offering me the examinership in Physics of candidates for the Royal Artillery & Royal Engineers[5] which offer I did not think it wise to decline; and as the examination was to commence on the first of August and a conference was to be held in relation to the subject on the 23rd of this month, I was compelled to start so as to be present at the latter. This accounts for my presence here and it also accounts for the fact of my not having written to you earlier, for your letter is of a character to prompt me to seize the pen and reply to it at once.

It is a remarkable letter Hector, and would require a week's discipline in prayer and fasting on my part to reply to it aright. That 'chronic indifference' is a thing I do not like—I remember some years ago suffering from something of the kind and being willing almost to accept the lowest superstition—any thing in fact that could stir up my emotion and rouse within my heart the affections of human nature, in exchange for it. I once saw in a picture a monk with his forehead in the mud prostrate before an image of the virgin[6] or something of the kind, and I wished to be like him. There is a species of knowledge sometimes possessed by the mind, which, while it completely annihilates the bugbears of superstition, also strikes at the very essence of faith and hope and love, and may perhaps do more mischief to the human soul than superstition itself. I will not say that that analytical spirit to which you refer is bad in itself, but it requires sturdy checks to prevent it from ending in painful vacuity. You have said so little about it that I am not quite sure whether I understand you aright, nor indeed am I certain of my own ability to understand you, for the understanding of such things implies an experience of them. All I feel is that there are sources of courage strength and consolation open to the human soul which are beyond my powers of analysis, they reveal themselves to me with an evidence equal to that of any physical fact, and I return to them sometimes with a certain willful obstinacy in opposition to that analysis which would fritter them away. In one thing I believe neither you nor I can hope for satisfaction, and that is in a life of idleness. The possession of force implies the existence of an object on which to exercise it, and if we cannot find proper scope for our powers in the world without, they will turn and feed like cankers upon the world within. Why man there is a wonderful significance in that little remark that you have attached to your letter. 'I have not been "waked up" this long time as in writing this letter'. You must make that waking up perennial and then all will be right. I could see as your letter progressed the gradual dissipation of the clouds which beset your brain at the commencement, until at length you gave me utterances worthy of yourself. I remember many years ago an experience of this kind which induced me to write a rhyme in which the following lines occur:—

> 'Labour is worship—so some sage has said
> And surely it preserves from many an evil
> For though it may not lift to heaven the head
> It keeps the heart from wandering to the devil!'[7]

In fact I had been idle, and cloudy and discontented,[8] and my heart stood a perilous chance of wandering to the devil, when something occurred that called me to <u>act</u>—It was the beneficial effects of this action that I endeavoured to extol in the doggrel from which the above is a quotation. You know how Carlyle defines that unpleasant feeling which you now experience—'It is force that has not yet found its way'[9]—You may ask where is the way? That is a problem which you yourself must solve, but the only plan that I have ever found effectual in finding the right sphere of action is to do the duty which lies nearest at hand, for this very act clarifies the vision and enables a man to see more clearly what his duty really is.

But you will wish an end to this impertinence of mine—and now with regard to other matters:—I have already sent you some scientific papers which you do not appear to have received:[10] I have also given directions to have a memoir[11] sent to you which has just been published and which formed the Bakerian Lecture before the Royal Society for the present year—I think the paper was posted yesterday. With regard to Faraday, there is no fear of any permanent difference between him and me—each succeeding day knits us more closely together. I love the man, and when I think of his great achievements, of his childlike simplicity, of his humility, of the absence of all assumption of impatience, of all trace of egotism which is a perpetual offence in the character of men of narrower natures, and a perpetual source of discomfort to themselves; when I think of all this Faraday's character is to me incomparable, and my scientific rebellion[12] is swamped by the affections which such a character engenders. But there is no reason why I should use the word rebellion—for there is a remarkable unity of thought between Faraday and me. He said himself to me not long ago 'Tyndall I believe there is no difference between us, I believe when we understand each other that we shall perfectly agree'.[13] I believe so to, and even should we differ in opinion I feel satisfied that such a difference will never modify that friendship which constitutional sympathy has established between us.

I commenced this letter when I arose this morning, and at the end of the last paragraph the clock struck 7¼: I breakfast at 8 o/c and the three quarters between 7¼ & 8 I have devoted to a plunge into a reservoir of cold water. I have paid my 6 pence & returned shivering though still in a glow. I do this for the sake of my head which is not as strong as I should like to see it. You say that I have no right to be unwell; I wish to heaven you could abolish my right to be unwell, I wish you could cleanse my cranium of that perilous stuff which sometimes weighs upon my brain—My great desire is to work, but I am perpetually baulked in this desire by headache and bodily debility,—Depend upon it the miseries of this life are pretty equally divided, and each of us has his own share of the load. I sometimes am forced to fetter myself and to sit still, not daring to think of any thing which requires serious thought. You might see me sometimes pale as a ghost, so pale indeed that you might imaging me upon the grave's rim, ready to drop into it for ever. But when I throw science overboard and return to that sweet nature whom I have always loved, and who has always been kind to me my force returns and with the return of force the return of thoughts which soon break me down again. Thus does my life oscillate, and the conviction is often forced upon me that I am far less strong than I was 5 years ago, and worse that I can hardly hope for a return of that vigour which was once mine and which no work seemed able to impair. When I said that the thoughts of friends entered when my head became bad, I meant to say that such entered as comforters, and hence are not likely to be regarded as bores. It is good for us to be smitten at times—It makes us feel as men—perhaps if I enjoyed unbounded health, and could pursue my work without check or hindrance, I might become less human—A man's sufferings are often wholesome lessons, sorrow, as Charles Mackay says, can give the eyes new lustre,[14] and through toil and trial and pain we work our way to the heaven of peace. All you say about the rare and radiant maiden[15] I entirely respond to. If I ever choose one I shall endeavour to choose a woman. I know it is true the choice was made—all my friends tell me so, still as ever I hang my fate upon the chapter of accidents and resign myself to the guidance of the gods. I have heard nothing of the Edinburgh professorship[16] for some time, and the reason is that the health of the present occupant of the chair continues sufficiently good to enable him to hold it. Payne has got married[17]—I received his card[18] in Paris, but I do not know where he is. I am not disposed to regard that problem regarding the influence of gold as highly as *[some]* seem to do. I think the world still possesses a few men of the Fabricius and Cincinnatus stamp[19], and their characters are much fairer to me as ideals towards which an approximation ought to be made than the characters of those who sweat for gold. I question my own ability indeed to solve the problem to which you refer; for I have never felt, and never hope to feel that might of gold which must be felt before we can render an account of it. Full well I know that gold cannot satisfy the aspirations of the human soul—full well I know that I have been sometimes a better, higher, and happier man with a slender purse and a bleak look out, than when my purse has been full and prospects fair. Dont mistake me—I like gold and will earn it—I shall have enough of it for my wants, but I will not seek it as an object of ambition, my work is of another kind & whatever gold I may possess must be made subservient to that work. Had I the splendid crown diamonds at present exhibited in Paris, by Jove I would soon have the best of them between the poles of a magnet, gold cannot purchase peace of mind, nor the warmth of a friendly heart and both of these are in my estimation infinitely beyond what it can purchase. And now I wont bore you any further. Goodbye my dear Hector & believe me

ever yours affectionately John.

RI MS JT/1/T/1443

---

1. *your letter*: letter 1099.
2. *We went . . . to Paris*: for the places mentioned see letter 1101.
3. *the great exhibition*: see letter 1064, n. 3.
4. *that of London in 1851*: the Great Exhibition of the Works of Industry of All Nations held in Hyde Park, London, from 1 May to 11 October1851.
5. *examinership . . . Royal Engineers*: see letter 1104.
6. *picture . . . the virgin*: not identified.
7. *Labour . . . the devil!*: These lines are from a poem of Tyndall's, probably dating from the

1840s, which has no formal title but starts: 'With cloudy head and discontented breast' (RI MS JT/8/2/1/33–35). Some of the lines also appear in another poem, that begins 'Don't you remember love' (RI MS JT/8/2/1/31–32).

8. *cloudy and discontented*: See previous note.
9. *It is force that has not yet found its way*: ' . . . and yet all misery is faculty misdirected, strength that has not found its way' (T. Carlyle, *Past and Present* (London: Chapman & Hall, 1843), p. 357).
10. *some scientific papers . . . received*: not identified.
11. *a memoir*: see letter 1029, n.1.
12. *my scientific rebellion*: Tyndall had differed from Faraday over the explanation of diamagnetism; see letters 1052, n. 8 and 1099, n. 13.
13. *Tyndall . . . we shall perfectly agree*: This quotation does not appear in Tyndall's Journal or in other letters or in any of his publications. However, on 9 March 1855 Tyndall wrote in his Journal: Faraday 'believes there is no difference between him and me on matters of fact and he should be sorry to see a difference arising through misconception' (RI MS JT/2/13c/733).
14. *sorrow . . . eyes new lustre*: 'Sorrow shall give thine eyes new lustre' (C. Mackay, 'Melodies and Mysteries' in Mackay, *Poems* (London: W. S. Orr & Co, 1848), p. 90). The Scottish poet, writer and journalist Charles Mackay (1812–89), who was known to Tyndall, had regularly contributed to the *Illustrated London News* since 1848. Mackay's son had attended Queenwood College when Tyndall taught there.
15. *the rare and radiant maiden*: see letters 1052, n. 15 and 1063, n. 3.
16. *the Edinburgh professorship*: see letter 1052, n. 12.
17. *Payne has got married*: Thomas Payne married Rosalie Hemphill on 3 July 1855 in the parish church at Aughadoey, Londonderry (*Morning Chronicle,* 11 July 1855, p. 8).
18. *his card*: card missing.
19. *men of the Fabricius and Cincinnatus stamp*: Gaius Fabricius Luscinus (fl. 3rd century BC) was a Roman general and statesman who, having refused to be bribed, has often been cited as an exemplar of incorruptibility. Lucius Quinctius Cincinnatus (519–430 BC) was a Roman farmer, aristocrat and statesman who exemplified civic virtue; having been called away from his farm to defend Rome against its enemies, at the end of the campaign he immediately returned home and refused all honours.

From Thomas Romney Robinson[1] 28 July 1855 1119

July 28. 55 | Observatory | Armagh

My dear Sir

I have read with much pleasure and improvement the Bak^{n} Lecture[2] which you were so kind [to send me.] Both in the sagacity of its reasoning and the elegance of its experiments it is well worthy of your fame. In one respect it will be very useful, that of forcing on the mind full conviction of the futility of the existing hypotheses[3] which attempt to explain magnetic phænomena. On that point I feel quite adrift.

In one point only I think Paramagnetic and Diamagnetic forces seem to agree, in being lessened by an increase of temperature. Has any comparison been made of fluid and frozen mercury in the diamagnetic relation?

I hope to meet you at Glasgow next month.[4]

Yours ever | T. R. Robinson

J. Tyndall Esq M D[5]

RI MS JT/1/R/31

1. *Robinson*: Thomas Romney Robinson (1793–1882), an Irish astronomer and physicist, was the director of the Armagh Observatory and president of the Royal Irish Academy.
2. *the Bak^{n} Lecture*: the Bakerian Lecture; see letter 1029, n. 1.
3. *the existing hypotheses*: In his Bakerian Lecture (pp. 14–15) Tyndall had addressed the hypotheses that Poisson and Ampère developed to explain the phenomenon of diamagnetism.
4. *at Glasgow next month*: at the forthcoming meeting of the BAAS.
5. *M D*: Robinson must have mistakenly thought that Tyndall was a doctor of medicine, whereas he had been awarded his PhD at the University of Marburg in 1850.

To James Glaisher[1] [July?] 1855 1120

Saturday

My dear Glaisher

Have you a spare copy of your beautiful snow-crystals?[2] or could you lend me a copy for a few days?

Yours very truly | John Tyndall

James Glaisher Esq.

Glaisher(C)/1/4, Fitzwilliam Museum, Cambridge

1. *James Glaisher*: James Glaisher (1809–1903) was the Superintendent of the Department of Meteorology and Magnetism at the Royal Observatory, Greenwich. Active in several scientific societies he pursued research on astronomy and meteorology and determined the variation of the dew point with humidity.
2. *copy . . . snow-crystals*: Probably refers to Glaisher, 'Snow Crystals in 1855', *Quarterly Journal of Microscopical Science,* 3 (1855), pp. 179–85. This paper, which was illustrated with impressive drawings by Cecilia Louisa Glaisher (1828–92), was published in the journal's

July issue. Glaisher also delivered a paper 'On the Severe Weather at the Beginning of the Year 1855: and on Snow and Snow-Crystals' at the AGM of the British Meteorological Society on 22 May 1855. It was published later that year (or possibly even early 1856) in *Report of the Council of the British Meteorological Society, read at the Fifth Annual General Meeting, May 22, 1855* (London: n.p., n.d), pp. 16–30. Tyndall referred to Cecilia's illustrations of snow particles in his journal on 20 June 1855 (RI MS JT/2/13c/761).

From Henry Moseley 1 August 1855 1121

King's Coll. 1 Aug.

My dear Sir,

I regret that I shall not be able to attend here on the day fixed for your examination[1] to explain the arrangements we have thus adopted in detail. Dr. Graves[2] has however kindly undertaken to do so for me

Yours truly | H. Moseley

RI MS JT/1/3/892
LT Typescript Only

1. *the day fixed for your examination*: The examinations in 'Natural Sciences' were held on the morning and afternoon of Thursday, 9 August 1855.
2. *Dr. Graves*: Charles Graves (1812–99) was professor of mathematics in the University of Dublin and one of the three examiners for the papers in mathematics.

From Henry Moseley 2 August 1855 1122

King's College | 2 August

My dear Sir,

The wide designation (Natural Science)[1] of the subjects in which you examine is a source of some little embarrassment to us. Some very few Candidates have I believe applied to be examined in Geology and a very few in Botany and other subjects. I think you have the names of those Candidates. It has occurred to me that it would stop some cavilling and would not perhaps be very troublesome to you to get from some friends three or four questions[2] in each of these subjects and to get 10 or 20 copies of them printed or lithographed; so that they might be put (in addition to the other questions) before the Candidates of whom I speak. The answers (if any are given) we might get looked over by competent persons; and any expense which might be incurred in printing the papers or otherwise the Ordnance will pay.

I have made this suggestion with great deference and with a feeling that I have no right to make you responsible for any other portion of the examination than that in your own subjects; but with a feeling also that you will appreciate the difficulty of the circumstances under which this examination is conducted and that you will therefore accept my apologies for the trouble I am giving you.

Yours truly | H. Moseley.

RI MS JT/1/3/894
LT Typescript Only

1. *The wide designation (Natural Science)*: On 3 August Tyndall recorded in his Journal: 'A letter from Mr Moseley asking me to prepare a few questions on all the subjects mentioned by the candidates and which have been all mashed together under the head of "natural science". I should say there is nearly a dozen such subjects including Entomology, Zoology, Meteorology, geology, botany, &c &c. The natural sciences are in a sad mess in England. People appear to have no clear idea of what the region of physics is. They mix it up with chemistry, and in the present instance every thing else is mixed up with them—and for all these sciences thus taken en masse we have the same figure as for a single one of the classical languages! The absurdity makes a man a partisan' (RI MS JT/2/13c/780).
2. *some friends three or four questions*: The examination included sections on both geology and botany (see letter 1111, n. 1), but it is not clear whether the questions on these topics were set by Tyndall.

From Carlo Matteucci 3 August 1855 1123

Pisa 3[d] August 1855

My dear Mr Tyndall

I am much obliged by your having thought of me with regard to your Bakerian Lecture[1] and I have written immediately to Masson[2] begging him to forward it without delay by post, for I believe that the expense of postage from France to Italy for printed papers is not so very great. I hope that you have received the Nuovo Cimento[3] & that you will continue to do so pretty regularly. It would be of great importance to us to receive the Phil. Mag. which you are good enough to send us en courant. I have still another request to make with regard to the exchange of our publications. Through the medium of Prof. Grove I received regularly for two or three months the Proceedings of the R. Society & those of the R. Institution—these for the last three or four months have been discontinued and I should esteem it as a particular favour if you would have the goodness to ask the Secretary[4] to continue to send me these papers as before by post—These publications are very interesting to us for the compilation of our journal[5] and you will easily conceive also that I am desirous to see Mr du Bois Reymond's lectures.[6] As a means of expediting any more voluminous publication extrapost[7] I send you the address of my Brother in law Mr R. Young[8] 3 Furnival's Inn. London.—who might occasionally have an opportunity of forwarding a small parcel to Paris or directly to Italy.

Believe me my dear Sir | Sincerely yours | C Matteucci

RI MS JT/1/M/57

1. *your Bakerian Lecture*: see letter 1029, n. 1.
2. *Masson*: The Paris publisher Victor Masson specialised in scientific and medical books.
3. *the* Nuovo Cimento: Carlo Matteucci and Raffaele Piria founded the scientific journal *Il Nuovo Cimento* in 1855. At their meeting on 2 April 1855, the managers of the RI had agreed to purchase *Il Nuovo Cimento* for its library.
4. *the Secretary*: John Barlow was the secretary of the RI.
5. *our journal*: *Il Nuovo Cimento.*
6. *Mr du Bois Reymond's lectures*: see letter 1066, n. 4. The two men were bitter rivals as du Bois-Reymond had severely criticised Matteucci's contributions to the field of electrophysiology.
7. *extrapost*: Extrapost 'implies that a person has a private carriage with post horses [harnessed] to it' (H. Bertold, *A Comparative Dictionary, of the German and English Languages* (London: printed for W. Joy), p. 54). Those using such carriages, which were surplus to the regular postal service, were required to obtain a permit.
8. *Mr R. Young*: Robert Young, an engraver, whose sister, Robinia Elizabeth (1811–97) had married Matteucci in 1846.

From Élie Wartmann 5 August 1855 1124

Paris, le 5 août 1855

Mon cher Tyndall,

J'ai bien reçu votre lettre accompagnant celle du Colonel Sabine, et je me suis empressé d'en communiquer les conclusions à Sir David Brewster et aux autres membres du Jury. L'opinion unanime, d'accord avec la vôtre, a été de proposer au groupe d'adresser des remerciements tout spéciaux au Comité directeur de l'Observatoire de Kew, mais de ne pas proposer de médaille pour cet établissement.

Je suis aussi passé chez le portier de l'hôtel Manchester: il n'y avait pas de lettres pour vous.

Nos travaux de jurés touchent à leur fin. L'horlogerie est terminée, les balances aussi: les instruments d'optique astronomique sont en majeure partie examinés, ainsi que les microscopes. C'est Mercredi prochain que s'assemble le groupe qui doit prendre connaissance de nos propositions. Nous regrettons très vivement l'absence d'un grand nombre de jurés de la Classe VIII. Sir David nous a quittés définitivement, l'alderman Carter n'est pas revenu, ni le Docteur Steinheil, et le prof Dove n'est pas arrivé. Est-ce que vous ne viendrez pas combler cette lacune ? Wertheim ne paraît plus : des occupations l'absorbent, en sorte que toute la besogne se partage entre MM Mathieu, Seguier, Froment, Brunner, Barbezat et moi. Quatre Français, deux Suisses, c'est trop peu.

Vous avez sans doute achevé vos examens, et j'espère que tout s'est passé à votre satisfaction. Avant de reprendre vos instruments chéris, venez donc faire quelques tours de Champs Elysées.

Mes compliments affectueux à MM Faraday, Sabine, Grove et Gassiot, et pour vous, cher Docteur, l'assurance de mon affectueux dévouement.

E. Wartmann

P.S.: Excusez le retard de ma réponse. J'étais tellement occupé que je n'ai pas pu y songer plus tôt.

Paris, 5th August 1855.

My dear Tyndall,

I have received your letter[1] sent together with one from Colonel Sabine, and hastened to communicate the conclusions to Sir David Brewster and other members of the Jury.[2] The unanimous view, which accords with yours, was to suggest that the group would communicate some special acknowledgments of the committee to the director of the observatory at Kew,[3] but not to award any medals to that establishment.

I have also dropped by the doorman at the Hotel Manchester: there was no letter for you.

Our work as a jury is coming to an end. The clockwork has been completed, and so have the scales as well: the astronomical optical instruments have been examined for the most part, as well as the microscopes. The group that must be informed of our proposals will gather next Wednesday. We thoroughly deplore the absence of a large number of jury members from Class VIII. Sir David has left us once and for all, Alderman Carter[4] did not return, nor did Doctor Steinheil, and Prof. Dove has not arrived. Would you not consider joining in to fill this gap? Wertheim no longer comes out [to the Exposition]: he is absorbed in some activities, as a result the entire chore is shared between Messrs Mathieu, Seguier, Froment, Brunner, Barbezat[5] and me. Four Frenchmen and two Swiss are insufficient!

You have probably finished your examinations[6] and I hope everything went to your satisfaction. Before picking up your beloved instruments again, come explore the [Exposition Universelle on the] Champs Elysées.[7]

My affectionate compliments to Messrs Faraday, Sabine, Grove and Gassiot, and for you, dear Doctor, be assured of my affectionate regards.

E. Wartmann

P.S: Excuse my delay in replying. I was so busy that I could not think about it earlier.

RI MS JT/1/W/6

1. *your letter*: letter missing.
2. *the Jury*: The jury for Class VIII ('Arts Relating to the Exact Sciences, and to Instruction') at the Exposition Universelle, Paris. Its members comprised Jean-Baptiste Philibert Vaillant (1790–1872), David Brewster, Claude-Louis Mathieu (1783–1875, French politician, astronomer, member of the Académie des Sciences and advocate of the metric system), Armand Séguier (1803–76, member of the Académie des Sciences, Paris), Paul-Gustave Froment (1815–65, French instrument maker), Jean Brunner (1804–62, Swiss instrument maker working in Paris), Theodor Wertheim (see letter 1109, n. 2), John Carter (1804–78, chronometer maker and civic leader), Heinrich Wilhelm Dove, Elie Wartmann, Edouard Barbezat (dates unknown, clockmaker of Chaux-de-Fonds, Switzerland), Carl August von Steinheil and John Tyndall.
3. *the director of the observatory at Kew*: John Welsh; see letter 1038, n. 1.
4. *Alderman Carter*: see n. 2, above.
5. *Wertheim . . . Mathieu, Seguier, Froment, Brunner, Barbezat*: see n. 2, above.
6. *your examinations*: Tyndall's role as examiner for entrants to the Royal Artillery and Engineers; see letter 1098, n. 2.
7. *Champs Elysées*: An impressive tree-lined boulevard in Paris linking the Place de la Concord and the Arc de Triomphe.

From Élie Wartmann — 8 August 1855 — 1125

Paris le 8 Août 1855

Mon cher Tyndall,

En réponse à votre aimable lettre d'hier, je me hâte de vous dire que dans le groupe tenu aujourd'hui, on a décidé définitivement des récompenses à accorder aux instruments de précision tels que les balances, aux instruments de physique et de chimie, à la chronométrie, etc. Il ne reste à statuer que sur les instruments d'optique et les microscopes, et sur les collections relatives à l'enseignement.

MM Dove et Magnus sont arrivés, mais Mr Steinheil est toujours absent.

Il va sans dire que votre présence à Paris sera avantageuse pour les travaux du jury: toutefois je ne la considère plus comme indispensable.

J'ai remis les exemplaires de votre mémoire destinés à MM Ruhmkorff, Leverrier et Billet: les deux derniers demeurent au Bureau Suisse jusqu'à ce que j'aie découvert les adresses des personnes à qui vous voulez les faire parvenir.

Je ne vous dis pas adieu, mais au revoir, car même si les travaux du jury sont terminés avant que vous ayez fini vos examens, vous aurez besoin de quelque repos et vous viendrez le chercher aux Champs Elysées.

Votre bien affectionné et dévoué, | Elie Wartmann

Prof Tyndall

Paris 8th August 1855

My dear Tyndall,

In response to yesterday's kind letter,[1] I hasten to tell you that in today's group meeting, we have definitely decided on the awards to bestow for precision instruments such as scales, for instruments [used in] physics and chemistry, for chronometry, etc. All that remains is to make a decision regarding the optical instruments and the microscopes, and for the collections related to teaching.

Messrs Dove[2] and Magnus have arrived but Mr Steinheil is still absent. It goes without saying that your presence in Paris would be beneficial to the Jury's activities: however I no longer consider it essential.

I have submitted the copies of your memoir[3] to Messrs Ruhmkorff,[4] Leverrier and Billet:[5] the last two [copies] remain in the Swiss office until I find out the addresses of the people you want me to send them to.

I am not bidding you farewell, but au revoir, for even though the Jury's work will be completed before you finish your examinations,[6] you will need some rest and you will come to seek it in the Champs Elysées.

Yours, with my affection and devotion, | Elie Wartmann

Prof Tyndall

RI MS JT/1/W/7

1. *yesterday's kind letter*: letter missing; presumably a reply to letter 1124.
2. *Dove*: see letter 1124, n. 2.
3. *your memoir*: probably Tyndall's Bakerian Lecture (see letter 1029, n. 2).
4. *Ruhmkorff*: The German instrument maker Heinrich Daniel Ruhmkorff (1803–77), best known for devising the Ruhmkorff coil, which produces electrical pulses of high voltage.
5. *Billet*: Félix Billet (1808–82) was professor of physics at the Faculté des Sciences at Dijon (1843–73) and he worked principally in the field of optics.
6. *your examinations*: see letter 1124, n. 6.

To unidentified 9 August 1855 1126

Royal Institution | Albemarle St | London | 9th Aug. 1855

Gentlemen.[1]

In response to your request[2] I beg to state that I at present entertain the desire and intention to be present at the Glasgow meeting of the British association.

Your obedient Servant | John Tyndall

Wellcome MS 7777/01

1. *Genttlemen*: the members of the BAAS who were planning the forthcoming meeting.
2. *your request*: letter missing; presumably from the 'Gentlemen' (n. 1) enquiring whether Tyndall would be attending its annual meeting to be held in Glasgow, 12–19 September 1855, and thus whether he would be present at the meeting of council and able to serve as a secretary to Section A ('Mathematical and Physical Science').

From Henry Moseley 14 August 1855 1127

Deanery Bristol | 14 Aug.

My dear Sir,

Will you allow me to hope that I may have the returns of your examination[1] in the course of this week?

Yours faithfully | Henry Moseley.

RI MS JT/1/3/895
LT Typescript Only

1. *your examination*: The 'Natural Sciences' papers for the competitive examination for appointments in the Royal Artillery and Royal Engineers had been held five days earlier, on 9 August.

From George Biddell Airy 15 August 1855 1128

Royal Observatory Greenwich | 1855 August 15

Dear Sir

I have to thank you for some papers of great value,[1] which I found here on my return from Scotland.

In my science I find it necessary to make a little dash of speculation sometimes. Allow me to send you a paper[2] of which I have just received copies and which may bear in some degree that character

I am, my dear Sir, | Faithfully yours | G B Airy

D[r] Tyndall

RGO MS 6/376.725

---

1. *some papers of great value*: presumably some of Tyndall's publications, including his Bakerian Lecture (see letter 1029, n. 2). In his Journal for 16 August, Tyndall wrote: 'He [Airy] says on his return from Scotland he found some papers of mine, of great value, waiting for him' (RI MS JT/2/13c/782).
2. *send you a paper*: On 16 August Tyndall recorded in his Journal that he had 'Received a note from Airy accompanying his last Memoir' (RI MS JT/2/13c/782). The memoir has not been identified.

To George Biddell Airy 16 August 1855 1129

Royal Institution of Great Britain | Aug 16th 1855.

Dear Sir

Your paper[1] reached me yesterday and your note[2] this morning, for both of these I offer you my sincere thanks.

Your speculation[3] is a beautiful one and seems to me a fair, though daring, deduction from the hypothesis of internal fusion—indeed the facts alluded to must react upon this hypothesis and strengthen it.

Believe me your kind appreciation of my labours excites more than an idle gratification within me—it augments my force and cheers me on.

I have recently succeeded in establishing the converse of some of the experiments alluded to in the paper I sent you,[4] making the magnets moveable and the magnetized bar of bismuth fixed—the opposite actions of the two ends of the bar upon the suspended magnets is perfectly manifest.[5] Both Faraday and DelaRive have seen the experiment[6]—they seemed much pleased and expressed themselves perfectly convinced of the polar state of the diamagnetic bar.

I took the liberty of sending you a short notice[7] this morning, previous to the reception of your note.

Believe me dear Sir | most faithfully yours | John Tyndall

RGO MS 6/376.726–7

---

1. *Your paper*: see letter 1128, n. 2.

2. *your note*: letter 1128.
3. *Your speculation*: not identified.
4. *the paper I sent you*: Tyndall's Bakerian Lecture; see letter 1028, n. 1.
5. *making the magnets moveable . . . perfectly manifest*: Tyndall published these experiments in his 'Further Researches on the Polarity of the Diamagnetic Force', *Phil. Trans.*, 146 (1856), pp. 237–59.
6. *Faraday and DelaRive have seen the experiment*: On 14 July Tyndall recorded in his Journal: 'Met DelaRive and Faraday in the morning and made experiments on diamagnetic polarity for them. They were greatly pleased, and Faraday expressed his perfect satisfaction with the experiments' (RI MS JT/2/13c/781).
7. *a short notice*: probably J. Tyndall, 'The Polymagnet', *Phil. Mag.*, 9 (1855), pp. 425–40.

From Henry Moseley — 16 August 1855 — 1130

Deanery Bristol | 16 Aug.

My dear Sir,

I beg to acknowledge the receipt of your returns of the examination.[1] Will you have the kindness to send me an account of the expenses you have incurred in respect to the additional questions.

I heartily concur in your opinion that Physical Science is most unjustly neglected in the education of this country. In fact this examination affords remarkable evidence of the disproportionate place assigned to classical literature.[2] All the Examiners lament the neglect of their respective subjects except the Classical examiners.[3]

Yours my dear Sir | truly | H. Moseley.

RI MS JT/1/3/896
LT Typescript Only

---

1. *your returns of the examination*: Tyndall's marks for the papers on 'Natural Sciences' examination sat on 9 August as part of the Examination for Appointments in the Royal Artillery and Engineers.
2. *classical literature*: Of the 171 candidates, 152 sat the Latin papers and 102 were examined in Greek. All candidates sat the Pure Mathematics exams. 137 took English Composition and 137 French. By contrast 41 took Natural Sciences and 22 Moral Sciences.
3. *the Classical examiners*: The examiners in Latin and Greek were Robert William Browne, George Butler and Goldwin Smith. Two of them stated that 'they were not prepared to find among them [the candidates] so many good classical scholars, and one of them, whose experience gives great weight to his opinion, considers the candidates selected to be generally young men of distinguished ability' (*Report on the Examination for Appointments*

in the Royal Artillery and Engineers, Held at King's College, London, on the First of August, 1855 (London: Harrison, 1855), p. 18).

From Jules Duboscq 22 August 1855 1131

21, Rue de L'Odéon | au fond de la Cour[1]
Paris le 22 Août

Monsieur

J'ai l'honneur de repondre a votre lettre du 16 courant et je vous donne ci contre le prix des appareils necessaires pour un cours complet d'optique. Cependant, je n'ai pas mis dans cette liste les appareils que j'ai déjà eu l'honneur de vous fournir et qui sont indispensables.

Dans l'attente de vos orders, je vous prie Monsieur de recevoir mes salutations les plus empressées et de me croire votre très dévoué serviteur | J. Duboscq

Liste des appareils pour répeter toutes les expériences d'optique.

√ 2 lentilles de 33 centimetres de foyer concave et convexe—60"

1 diaphragme à ouverture rectiligne—20"

1 id à fleche pour la refraction—10"

√ 3 miroirs plan, concave et convexe

√ 1 prisme à angle droit reflexion totale—45"

1 prisme de flint très pur pour les raies—60"

1 prisme *[a]* flacon pour avoir un spectre avec divers liquides de dispersion differente—45"

1 prisme achromatique a 3 verres—50"

1 prisme a angle variable pour l'achromatisme de la reflexion totale—200"1 lentille et ses diaphragmes pour les aberrations de spéricité et de réfrangibilité—90"

verre violet et dessin pour l'experience de M. Stokes—20"

1 Cuve a diagonale pour la refraction—36"

Cuve pour la decomposition de l'eau et pour la recomposition métallique—20"

× 1 appareil a deux corps pour la projection des photographies, objet d'Histoire naturelle

globules du sang. &c &c—250"

appareil pour les anneaux coloré de Newton—40"

appareil pour les anneaux a liquides—72"

id id les lames epaisses—50"

appareil pour projeter la double refraction, couleurs complementaires de Arago, les

cristaux a axes, les lames de chaux, les verres trempés comprimées &c—250"
presse pour comprimer—20"
presse pour courber—20"
pince a chauffer—18"
× Disque de newton transparent—30"
√ Lentille cylindrique avec prisme pour recomposer la lumiere—60"
Appareil pour les hyperboles mobiles—40"
2 parallelepipedes fresnel—40"
Collection de cristaux *[a axes]* verre trempés chaux sulfatée &c *[aussi]* 20"
Photographes objets microscopiques et autre, la pièce 5

21, Rue de L'Odéon | at the bottom of the court[1]
Paris, August 22nd

Sir,

It is my pleasure to respond to your letter of the 16th of this month,[2] and I am providing you herewith the price of the apparatus necessary for a complete optics course. However, I have not included in this list the apparatus that I have had the honour of supplying you, and which are indispensable.

While waiting for your orders, I am asking you, Sir, to accept my most considerate greetings. Sincerely, your much devoted servant | J. Duboscq

List of the apparatus used in performing all the optical experiments.

√ 2 lenses concave and convex each of focal length 33 centimetres—60"[3]

1 diaphragm with a rectilinear aperture—20"

1 ditto with an arrow [indicating] refraction—10"

√ 3 plane mirrors, concave and convex

√ 1 right-angle prism for total reflection—45"

1 very pure flint prism for the [dispersion of] rays—60"

1 prism with flask to obtain a spectrum with various liquids of different dispersion—45"

1 achromatic prism with 3 glasses[4]—50"

1 prism with a variable angle for the achromatism and the total reflection—200"

1 lens and its diaphragms for [measuring] spherical aberration and aberration by refraction—90"

Violet glass and a drawing for Mr. Stokes' experiment[5]—20"

1 diagonal vessel for refraction—36"

Vessel for the decomposition of water[6] and its metallic recomposition—20"

1 device with two [parts?] for the projection of photographs, natural history objects, blood cells, &c &c—250"

Apparatus for Newton's coloured rings[7]—40"

Apparatus for producing rings in liquid—72"

ditto for thick plates[8]—50"

× Apparatus for projecting double refraction, Arago's complementary colours, crystals with axes, livers of lime, compressed tempered glass etc.—250"

a press for compression[9]—20"

a press for bending—20"

pliers for heating—18"

× transparent Newton's disc[10]—30"

√ Cylindrical lens with prism for the recombination of light—60"

Device for [displaying] movable hyperbolic fringes[11]--40"

2 Fresnel rhombs--40"

A collection of crystals [with axes], tempered glass, sulfate of lime &c [as well] 20"

Microscopic object photographs and others, [priced] per item—5"

RI MS JT/1/D/180

---

1. *21, Rue de L'Odéon | au fond de la Cour*: The address forms part of a printed heading which also includes the wording 'Exposition de Londres | Council Medal | J. Duboscq' (Duboscq had been awarded a Council Medal for the optical instruments he displayed at the Great Exhibition in London in 1851) | 'Élève & Succr de Soleil' (Duboscq was the student and successor of the eminent optician and mathematical instrument maker Jean Baptiste François Soleil, who was also his father-in-law) | 'Les lettres adressées à la maison | J. Duboscq, doivent être affranchise' (Letters written to the house of J. Duboscq must be postmarked).
2. *your letter of the 16th of this month*: letter missing.
3. *60"*: The numbers at end of each line indicates the price, in francs. In the RI Managers' Minutes (vol. XI, f. 117) for 5 November 1855, it was 'Resolved, That a sum not exceeding £40 [about 1,000 francs] be placed at Dr. Tyndall's disposal, to purchase apparatus from M. Duboscq at Paris.'
4. *achromatic prism with 3 glasses*: Probably a compound achromatic prism composed of 3 pieces of glass.
5. *Mr. Stokes' experiment*: In his paper 'On the Change of Refrangibility of Light' (*Phil. Trans.*, 142 (1852), pp. 463–562), Stokes had shown that, while rays of some colours passed through a flask containing a solution of sulphate of quinine unhindered, others had been changed in colour, while the fluid became fluorescent.
6. *Vessel for the decomposition of water*: a voltameter.
7. *Newton's coloured rings*: Book 2 of Newton's *Opticks* (1704 and later editions) contains numerous experiments with these rings.
8. *thick plates*: Newton discussed these in book 2, part 4, of his *Opticks*.

9. *a press for compression*: for compressing, say, samples of glass in order to alter their optical characteristics.
10. *Newton's disc*: A wheel divided into seven sectors, each painted one of the seven colours of the spectrum. When rotated the disc appears white; showing that white is a mixture of these colours.
11. *Device . . . hyperbolic fringes*: For displaying the polarisation of light by crystals.

To Michael Faraday 27 August 1855 1132

Queenwood near Stockbridge | 27th Aug. 1855

My dear Mr. Faraday,

The thought of you wandered across my mind this morning before I rose, and I said in my heart I will commence the day by writing to him. I thought of writing to you after that terrible thunderstorm[1] which visited this neighbourhood, but I heard that it stretched over London also, consequently I infer over Sydenham and that you yourself were a witness of its grandeur. I dont think I have ever seen lightning so incessant and vivid. The evening was beautiful. I watched the grouping of the clouds in the west after the sun had sunk; it was magnificent, and eastward the moon, almost full, hung her shield at intervals in the blue space between the masses of cloud—all was calm towards 8 o'clock and summer lightning, as I thought it, played fitfully on the south west. As the evening advanced however the lightning drew nearer and rumbling thunder became audible, and towards 10 o'clock the spectacle was grand. I wandered with a friend[2] to a summit in the neighbourhood and watched the flashes along the horizon, these were reflected from the clouds above and an intense brilliancy shivered at intervals over the heavens and the earth. The yellow light of the gas in the windows of the house seemed utterly extinguished by the superior glare of the lightning. The flashes and peals drew nearer and as a matter of precaution my companion and myself withdrew from the height and sought shelter in the house; here we stood in a porch watching the fitful but almost incessant glare, a wild gust of wind rushed in upon us carrying heavy rain along with it and compelled us to close the door. Louder rolled the cannonade and one crash broke above us which almost seemed intended to shiver the building to atoms. About 120 boys[3] were in bed at the time; I thought of the little fellows and went upstairs to comfort them if necessary. The younger ones were greatly terrified and some of them were crying bitterly; I talked cheerily to them and when the vivid gleams broke in upon them exclaimed 'beautiful!' By degrees they also began to consider the flashes as things of beauty and not of terror, but some of them were greatly surprised to find that when they shut their eyes and covered their heads with the bed-clothes they still saw the flashes. I waited with them until I could assure them that all danger was past, until there was no cloud near to give us even a taste of the returning stroke;[4] but long after this the discharge continued along the eastern horizon, and like quivering red hot bars the flashes sped up and down. Francis who was down here on Sunday[5] told me he noticed circular flashes, and was very confident of the reality of the fact: I myself noticed several times the clouds surrounded by a fiery rim intensely luminous, this I mentioned to Francis but he would not admit it as being the explanation of what he saw. You drew attention to the same source of error many years ago.[6]

The people here have contrived to make me very comfortable in a little lodge[7] which stands at the entrance to the college lawn and which serves on Sundays as a meeting house for the few quaker youths in the establishment.[8] I have planted my apparatus. The work goes on slowly, for it is terribly difficult The greatest care is required to render the facts safe and when they are safe one does not know what to make of them. My strength oscillates as usual: when I give up work for a day and swing my limbs over the breezy downs I return with an amount of force which seems inexhaustible, but two days thought produces a wonderful diminution of my energy. I write at present upon the lees[9] of the vigour which I acquired on Sunday last in the New Forest.[10] Francis, the chemist here,[11] and I myself started early and wandered along bowery roads and through fine parks, until we found ourselves at noon knee deep in the fern of the forest. Visited the stone which stands upon the spot where Rufus was shot by Walter Tyrrell.[12] The people though they generally bear the character of barbarians were extremely civil and intelligent: there was an admirable precision in their directions when they showed us our way. The same quality was exhibited by the children. It was really remarkable to observe the exactitude with which they described the twists turnings and landmarks along the forest tracks. The day was glorious; the sun smote the foliage and seemed to drip as liquid gold into the shaddows beneath, the breezes made music amid the branches and as the rust cleared gradually away from the inner man our hearts responded to the sound. I had left home a decrepit wretch but as I advanced I found my vigour rush upon me. Passed through sunny little hamlets and by brown thatched cottages around which roses and runners and laurels clustered till finally we reached Lyndhurst;[13] it was 2½ o'clock, and having breakfasted early we put up at an inn and fortified ourselves with the wholesome country fare. Thence to the nearest station,[14] thence to Southampton, then to Dunbridge[15] from which we had a walk of four miles. The full harvest moon sailed above us through the cloudless sky, frosting the beech leaves with silver, and near her was the evening star. The air was clear and bracing and we finally crossed our respective thresholds leaving a day behind us to which memory can scarcely furnish a parallel.

As Francis sat beside me at dinner on Saturday in the same room with the boys; a question buzzed through the little host whether that was Prof. Faraday! Francis said to me at the time that the prospect of such a number of fine little fellows would be a pleasure to you, and the thought then and there occurred to me that supposing you wished to change your scene for a day you might possibly turn your thoughts and steps hitherward. One sunny day upon the chalk would do you good, one calm night spent away from the turmoil of city life. I can promise you a good bedroom with linen sweet and white. The place is also remarkable as furnishing evidence how nature deals with those who do violence to her mental laws, that here as in physics she is inexorable. This place was the Harmony Hall of the socialists;[16] but the memory of them is clean gone from the place. There is a gas-house here where the hydrocarbon principle is applied and that wonderful Boghead cannel[17] is distilled for the illumination of the establishment, there is a new and singular washing machine—there are the boys and the playgrounds which would be pleasant to your eyes: there is a most remarkable avenue of yew trees adjacent gloomy and glorious at the same time—there is another of elm trees which might have suggested the pillared aisles of our Cathedrals, and there will be gladness in all our hearts and a welcome in every eye should you set your foot within the establishment.

I had a letter some days ago from Frankland[18] in which he urges a promise I made to him some time ago to spend three days with him in the lake district: this I intend to do, and we then go together to Glasgow to the meeting of the Association.[19] I shall probably leave this place about the 6th. and spend a day in London, so as to gather a few things together that may possibly interest the physical section. I have had a letter from Duboscq[20] containing a list of apparatus for a course of lectures on optics concerning which I shall ask your counsel at a future day. Now I think I have sufficiently afflicted you with this long and badly written scrawl. The bell has rung—it is 8 o'clock and the boys are moving to breakfast. Desiring my best wishes to Mrs. Faraday and Miss Barnard I say for the present goodbye.

affectionately yours | John Tyndall

Mr. Best[21] one of our members wrote to me when in town asking me to open certain proceedings of the Hants & Wilts educational union by a lecture—this I declined, and to day I have declined another invitation from the same gentleman and one also from Leeds. But lest he should think me unkind living so near Southampton I went over to the meeting on Wednesday last—purposing simply to shew them the respect of my personal attendance they however induced me to say something—and indeed the proceedings were well conducted so that when called upon there seemed no reasonable objection to my saying a few words. I hear an account of the meeting has got into the papers[22] so if you should see it what I have said will account for my appearance there.

RI MS JT/1/12/4037–41
Faraday Correspondence, 4:3020
Typed Transcript Only

---

1. *that terrible thunderstorm*: Tyndall also recounted the storm in his Journal entry for 23 August (RI MS JT/2/13c/784).
2. *a friend*: Heinrich Debus.
3. *About 120 boys*: pupils at Queenwood College.
4. returning stroke: Following the downward electrical discharge from the charged cloud to the earth, the return stroke travels upwards from the earth along roughly the same path.
5. *Sunday*: 26 August.
6. *source of error many years ago*: In 1838 Faraday challenged the view that 'Electrical discharges in the atmosphere [are] in the form of balls of fire.' Although balls of fire may appear in the atmosphere, Faraday doubted whether 'they have anything to do with the discharge of ordinary electricity, or are at all related to lightning or atmospheric electricity' (*ERE,* vol. 2, p. 523).
7. *a little lodge*: 'My work room [at Queenwood College] will be at the lodge and I have a capital bedroom in the house' (Tyndall's Journal, 14 August 1855: RI MS JT/2/13c/781).
8. *for the few quaker youths in the establishment*: Although Queenwood College had been founded by Quakers in 1847 and the headmaster, George Edmondson, was a Quaker, the vast majority of pupils were not from Quaker families.
9. *lees*: loss.
10. *New Forest*: Created c. 1079 by King William I for the royal hunt, the New Forest covers over 200 square miles, principally of woodland.
11. *the chemist here*: Heinrich Debus.
12. *Rufus was shot by Walter Tyrrell*: King William II (c. 1056–1100), generally known as William Rufus, was accompanied on a hunting party in the New Forest by Walter Tirel (b. 1065) who is said to have killed the King with one of his own arrows. The 'Rufus Stone', which allegedly marks the spot, stands near Minstead in Hampshire.
13. *Lyndhurst*: A village in the New Forest, 15 miles south of Broughton.
14. *to the nearest station*: Lyndhurst Road Station.
15. *Dunbridge*: A small village in the Test Valley, 4 miles south of Broughton.
16. *Harmony Hall of the socialists*: In 1839 a socialist community led by Robert Owen had obtained the property at Broughton, which they named Harmony Hall. However, this project did not prove viable and by 1845 most of its residents had departed.

17. *Boghead cannel*: Cannel is a bituminous coal, which burns with a very bright flame. Because of its richness in volatile matter, it is often used in the manufacture of coal oils and gas (*OED*). One of the regions possessing this type of coal is the Boghead area of Bathgate, West Lothian, Scotland. See also J. Tyndall, 'Comparative View of the Cleavage of Crystals and Slate Rocks', *Phil. Mag.*, 12 (1856), pp. 35–48, on pp. 46–47.
18. *a letter . . . Frankland*: letter missing.
19. *the Association*: the BAAS.
20. *a letter from Duboscq*: letter 1131.
21. *Mr. Best*: Samuel Best (1802–73), a member of the RI, was the rector of Abbots-Anne, Andover, and rural dean of Andover.
22. *an account . . . the papers*: The meeting of the Hants and Wilts Educational Society, held in Southampton on August 23, was extensively reported in the *Hampshire Chronicle* (25 August 1855, p. 3): 'Professor TYNDALL offered some judicious remarks on the beneficial influence of talent being brought to bear on the public services of the country.'

From Michael Faraday 3 September 1855 1133

ROYAL INSTITUTION OF GREAT BRITAIN. | 3 Sept. 1855

My dear Tyndall,

Esteem my feelings as in the exact inverse proportion of my note, which is no fit answer to your very kind and pleasant letter.[1] But I have been a good deal lowered by a persistent attack of diarrhoea, and which, though I think it has been stopped, has left its marks in giddiness and feebleness of head. I once had very feeble hopes of being at Glasgow[2] with you all, but they are all gone, and I find once more what I am good for:—nothing. I thought if I waited longer before I answered you, I might lose you at Queenwood, and only write you a more stupid note than the present. Your picture of Queenwood excites many past remembrances, but I have no hopes of converting them into futures. We have all been poorly here—Wife, Jane[3] and self—but we are all mending. Jane is away from us. My wife sends her kindest remembrances.

Ever yours, | M. Faraday

Prof. Tyndall | &c. &c. &c.

RI MS JT/1/12/4138
Faraday Correspondence, 4:3022
Typed Transcript Only

---

1. *your very kind and pleasant letter*: letter 1132.
2. *being at Glasgow*: at the forthcoming meeting of the BAAS.
3. *Jane*: Faraday's niece Jane Barnard.

To Michael Faraday 5 September 1855 1134

Dear Mr Faraday

Your note[1] reached me this morning an hour before I left Queenwood. I was glad and sorry at once on reading it. Most sincerely do I hope that you will soon be quite strong again. It would have been a delight to me had you resolved to go to Glasgow,[2] now I am not quite sure that I act right in going and your presence there would have been a kind of quieter to my conscience. I know I could employ my time far better at home, but then I should be deemed unkind. I had just got hold of my work—just established that affinity between me and it which is always a work of some pain to me when I am obliged to quit it, and to enfeeble that affinity by other employment. However as I grow older I shall see more clearly how I ought to act, how far conform with the requirements of others and how far follow out my own notion of things—good bye for the present—say good bye to Mrs Faraday for me

Most faithfully Yours | John Tyndall

RI MS JT/1/12/4042
Faraday Correspondence, 4:3023
Typed Transcript Only

1. *Your note*: letter 1133.
2. *resolved to go to Glasgow*: for the forthcoming meeting of the BAAS.

From John Pringle Nichol[1] 13 September 1855 1135

Observatory | Thursday

My dear Dr Tyndall,

The enclosed[2] ought certainly to have been in Kelland's report of Section A,[3] as read by the Duke yesterday.[4] Smyth[5] is with me, & quite concurs in opinion. I have sent it to Kelland & Phillips.[6] Will you—if you think as I do, kindly do every thing in your power to have the omission rectified in the printed reports—

Yours ever faithfully | J. P. Nichol
Dr Tyndall

RI MS JT/1/N/13

1. *Nichol*: John Pringle Nichol (1804–59) was a Scottish astronomer who helped popularize the subject and served as professor of astronomy at the University of Glasgow.
2. *The enclosed*: enclosure missing.
3. *Kelland's report of Section A*: The Cambridge-educated Philip Kelland (1808–79) was professor of mathematics at Edinburgh University. At the 1855 Glasgow meeting of the BAAS he served as president of Section A, which covered 'Mathematical and Physical Science'. It would appear that some material that Kelland should have supplied to the Duke of Argyll, in order to be incorporated into his presidential address, had not been sent.
4. *read by the Duke yesterday*: George John Douglas Campbell, 8th Duke of Argyll (1823–1900), who was a prominent liberal politician, served as the incoming president of the BAAS and delivered his presidential address on Wednesday, 12 September (*Brit. Assoc. Rep. 1855*, pp. lxxiii–lxxxvi).
5. *Smyth*: Charles Piazzi Smyth (1819–1900) served as Astronomer Royal for Scotland (1846–88) and as professor of practical astronomy at the University of Edinburgh. He contributed three papers to the 1855 BAAS meeting.
6. *Phillips*: The geologist John Phillips (1800–74), who was the deputy reader in geology at the University of Oxford, served as assistant general secretary to the Glasgow BAAS meeting.

To Sarah Tyndall 16 September 1855 1136

My dear mother

It is a long time since you heard from me—I write to you now from Glasgow, so you see I am a great wanderer, and besides this have has so much occupying my thoughts recently as to prevent me from writing to you. I have been here for some time but purpose returning to the south of England in the course of the present week. I have been attending a meeting of the British Association here and being one of the officers of the association[1] a great deal of work has fallen upon my shoulders—I shall be glad indeed when it is all over and I can sit down in peace. If you would write to me on Thursday next directing your letter[2] to me at

Queenwood | near Stockbridge | Hants.

I should get it at the end of the week when I return to Queenwood. I have taken up my quarters there for the autumn as the air is purer and the place healthier than London.

I am living here at present with my old friend[3] in whose company I first went to Germany. We had made an excursion over Cumberland together and came on together to Glasgow. We have lodged in the same house, and will return to England together in a few days,

Goodbye for the present | and Believe me | Your affectionate son | John

Sunday evening

RI MS JT/5/16/159
LT Typescript Only

---

1. *one of the officers of the association*: Tyndall sat on the council of the BAAS and served as secretary to Section A ('Mathematical and Physical Science').
2. *your letter*: letter (if written) missing.
3. *my old friend*: Edward Frankland.

To Thomas Archer Hirst 16 September 1855 1137

My dear Tom

Here I am alone: it is Sunday evening and the rays of the sinking sun are shining upon the gray slates and red tiles of Glasgow. I have had hard work since I came here—The old Secretary[1] who has transacted the chief business of the Physical Section[2] for a great many years is ill, and has staid at home. This has thrown an immense amount of work on my shoulders and quite neutralised the benefit I derived from my excursion amid the lakes. My own paper[3] came off yesterday—it has bothered me not a little but I am beginning to be able to look calmly at the matter. I had but little sleep—I believe none—the night before and my head was not strong—nevertheless when my time came I found my brain clear and my courage by no means low. I have been told indeed that my communication was unequalled. Indeed I got through it entirely to my own satisfaction—Whewell,[4] Sir David Brewster, Sir William Hamilton[5] & numbers of that stamp were present and they all cheered me at the conclusion. Brewster took me by the hand and said 'Oh what an admirable lecture you have given us' introducing me at the same time to Sir William Hamilton—Thomson was beside me as I spoke taking notes—I said that I was glad to see him there and hoped that we should finally come to an understanding with each other[.] He arose and replied. Whewell apparently desirous of throwing the weight of his great authority in favour of Thomson and thus bearing me down rose and spoke—Had he not done so all would have passed off well. But I with my sensitive head rose and replied to both him and Thomson speaking I think with considerable power & effect upon the audience—I went just a little too far and for this I feel dissatisfied with myself. I said something to the effect that Thomson had acted a safe part in suffering himself to be guided by Faraday, one of Thomson's friends[6] upon the platform grumbled; Whewell glowered upon me like an enraged lion—indeed Thomson's friends en masse seemed to be offended with me. The great body of the audience was however upon my side and this perhaps made matters worse. when I was aware of the discontent I stopped and said that if I had uttered any thing unbefitting the calmness of scientific discussion, I begged to apologize for it. The Chairman[7] said that every body present felt that there was no need of an apology: the audience applauded loudly. Still the impression which the whole affair has left upon me is by no means a pleasant one. Thomson indeed invited me to dine with him and meet Liebig[8] afterwards: but it will take some time to banish the remembrance of the thing from the minds of all concerned.

I shall get away towards home on Wednesday or Thursday. It is just possible that Frankland and myself—we live together—will spend Wednesday at Loch Lomond.[9] I am anxious to return & should perhaps have been happier had I not come at all. But this feeling is of brief duration in my case. The molecules have been stirred but they are already subsiding and in a short time I shall enjoy the serenity of a philosopher. and now I have said as much as you & Anna will care about hearing—Goodbye

Yours affectionately | John

Greet Debus for me and tell him all you remember of what I have told you.

RI MS JT/1/T/611

---

1. *The old Secretary*: John Stevelly (1795–1868), professor of natural philosophy at Queen's College, Belfast.
2. *the Physical Section*: The BAAS meeting was divided into seven sections and one sub-section. Section A covered 'Mathematical and Physical Science'.
3. *My own paper*: A summary of Tyndall's paper 'Experimental Demonstration of the Polarity of Diamagnetic Bodies' was subsequently published in *Brit. Assoc. Rep. 1855*, pp. 22–23. Tyndall also provided a fuller account of the proceedings in his Journal entry for 18 September 1855 (RI MS JT/2/13c/784–85).
4. *Whewell*: William Whewell (1794–1866) was master of Trinity College, Cambridge, having previously held the chairs of mineralogy and of moral philosophy. A man of wide learning he had contributed to the study of both mineralogy and tides and had written books on mechanics, architecture and foundation texts on the history and philosophy of science.
5. *Sir William Hamilton*: The mathematician William Rowan Hamilton (1805–65) was professor of astronomy at the University of Dublin and the Astronomer Royal of Ireland.
6. *one of Thomson's friends*: not identified.
7. *The Chairman*: Philip Kelland; see letter 1135, n. 3.
8. *dine with . . . Liebig*: The noted chemist and educator Justus von Liebig (1803–73) had transferred from the University of Giessen to the University of Munich in 1852. He opened the field of organic chemistry and made substantial contributions to agricultural chemistry, plant and animal physiology and the chemistry of foodstuffs, as well as introducing the laboratory-based teaching method. Liebig was the principal guest at the

farewell dinner that had been held in Glasgow for the Scottish chemist Thomas Graham, who had recently left the chair of chemistry at University College London.

9. *Loch Lomond*: the large freshwater loch (lake) some 20 miles north-east of Glasgow. It straddles the boundary between the Highlands and Lowlands of Scotland.

To Michael Faraday 24 September 1855 1138

Monday, 25th Sep, 1855,[1]

My dear Mr Faraday

I have a bad pen and no knife fit to mend it, so I fear it will give your eyes some trouble to make out my blurred sentences. It is very still here at present—The sun set in golden clouds half an hour ago—The hive of life which this roof usually covers has been out nutting all day and is not yet returned, so I am left alone to make use of the tranquil evening hour in writing to you. In passing through London <u>en route</u> to Glasgow I was glad to learn that your health was restored, on coming through on Saturday I was equally glad to hear the good tidings confirmed.[2] I thought of running down to Sydenham[3] to see you and Mrs Faraday, but a second thought suggested that I might only disturb you—indeed the idea sometimes occurs to me that I give you a great deal too much trouble to read letters some of which contain almost nothing that could be expected to interest you, nevertheless some blind instinct causes me to write, and to hope that I do not pester you. Previous to going to Glasgow Frankland, myself and a third friend[4] made four days journey through the lake district of Cumberland. Frankland himself has taken a small house in Windermere, has a boat on the lake and spends his time fishing and getting his thin cheeks tanned with the hue of health. We started one glorious morning, went to Coniston; thence through a beautiful valley to Langdale Pikes and put up at a farm house at the base of the mountain. The clouds gathered and rain fell during the night the wet weather prolonged itself into the following day, but taking advantage of a lighter hour we set out once more. We had to face a steep and rugged pass, and to clamber with great toil over the rough boulders strewed along the bed of a mountain torrent. The rain fell heavily all the time and drenched me through. We reached the summit, turned into a wrong track which cost us upwards of ten miles additional labour. We had to ascend a second pass almost as steep and rugged as the first, the rain still falling—sometimes our pathway was converted into a brook through which we trudged ankle deep in water. My companions were furnished with mackintoshes, I had a large horserug over my shoulders which drank in the rain like a sponge and became enormously heavy chiefly along its dependent edges which flapped wearily against my drenched limbs. Notwithstanding this I was the freshest of the party when the day was concluded. Having reached Wastdale[5] we got shelter in a farm house—there were no public houses—and I was happy enough to secure a pair of trousers belonging to a corpulant mountaineer into which I crept while my own were drying. Next day was a repetition of the same. Crossing Blackstair[6] to Buttermere the rain descended and drenched us once more. The exertion however kept a warm moisture on the body and although the toil was excessive I enjoyed it—it left no rust on the muscles. From Buttermere, which is a glorious place, we passed by Honister Crag to Borrowdale—saw Lodore,[7] crossed Derwent Water in a boat and put up at Keswick.[8] Thence to Penrith,[9] thence to Glasgow where I was soon plunged amid the duties of secretary. From this I was not able to extricate myself until the last journal was prepared. Stevelly[10] was ill and did not make his appearance. This rendered my duties very laborious. The meeting was very successful as regards numbers, but the papers brought forward were not remarkable. Indeed I left Glasgow with the opinion which I entertained on entering it, that I should have contributed more to the advancement of science by staying at home and doing my work than by going there. The longer I live the more I learn to value private in preference to associated effort. It is to the solitary worker that science has to look for its advancement rather than to the brilliant gatherings of a society however pleased that society may be with its own performance. I described the experiments which I made before you and De la Rive.[11] Thomson and myself had a short discussion afterwards, the memory of which was unpleasant to me for a day or two, but which I have now almost forgotten. I have been told that the discussion was misrepresented in the newspapers,[12] but this I do not know as I have never seen any report of it whatever. I thought at the time that a degree of partisanship was shown almost approaching to unfairness; but it is perfectly possible that I was a partisan myself and simply saw my own image in others. The whole matter seems infinitely small regarded from a proper point of view, so I will say no more about it.

Brewster read a paper on what he called the triple spectrum.[13] You know he imagines the spectrum to be composed of three distinct spectra Red, yellow and blue his views have been criticised by Whewell[14] and others, and in the papers brought forward at Glasgow[15] he makes a terrible assault upon his antagonists. It was most cleverly written, full of that kind of talent which excites your admiration without influencing your convictions. On Wednesday morning I was weary enough and having completed my duties Frankland and myself proposed spending a day on Loch Lomond.[16] We went, and while there learned that Loch Katrine, the scene of Scotts poem of the Lady of the Lake[17] was within a few miles of us. Walked to Loch Katrine across the country. Crossed the lake in an elegant little screw steamer, and spent the night in the jaws of the Trosachs. I never saw a scene of deeper loveliness than that presented by the head of Loch Katrine. Next day we went to Sterling and thence to Glasgow whence I started on Friday at 9 o'clock and reached London at about 10 o'clock the same evening.[18]

I have got to work, but it will take a day or two to get thoroughly into it as I was before starting. I am trying whether the total magnetic intensity of bismuth is increased by compression. This I find to be the case, though I must still make some corroborative experiment. What stands in handbooks regarding the increase of specific gravity needs I think careful examination.[19] With the power of compression, and the balance at my command here I have not been able to prove that any sensible change of specific gravity takes place. I shall however apply more delicate tests by and by—and now farewell for the present with kind wishes to Mrs Faraday and Miss Barnard.

Believe me | most sincerely Yours | John Tyndall

RI MS JT/1/12/4043–6
Faraday Correspondence, 4:3026
Typed Transcript Only

---

1. *Monday, 25th Sep, 1855*: As 25 September was a Tuesday, this letter was probably written on the 24th, the transcriber (Louisa or her assistant) having inappropriately recorded the date on the envelope.
2. *glad to hear the good tidings confirmed*: possibly in a missing letter.
3. *Sydenham*: In 1855 the Faradays, who had private accommodation at the RI, also rented a house in the south London suburb of Sydenham.
4. *a third friend*: John Helm Nicholson (1824–1901), who was the librarian at Owen's College, Manchester and a former school friend of Frankland's.
5. *Wastdale*: The steep-sided Wasdale valley in the western part of the Lake District, including Wastwater, which is fed by the River Irt.
6. *Blackstair*: probably the Black Sail Pass from Wasdale to Ennerdale.
7. *Lodore*: the attractive Lodore Falls on Watendlath Beck, which flows into Derwentwater.
8. *to Buttermere . . . at Keswick*: Tyndall travelled through dramatic scenery; first north from Wasdale to Buttermere and then north-east to Keswick, a sizeable market town at the head of Derwentwater.
9. *Penrith*: A substantial market town on the west side of the Lake District with good railway connections.
10. *Stevelly*: see letter 1137, n. 1.
11. *the experiments which I made before you and De la Rive*: see letter 1129, n. 6.
12. *misrepresented in the newspapers*: Tyndall was probably referring to the newspaper report: 'Some interesting experiments were accompanied by a clear and lengthy explanation. A somewhat sharp discussion resulted as to Professor Faraday's opinion on the polarity of

bismuth, in which Professor W. Thomson, Dr. Whewell, master of Trinity College, Cambridge, and Dr. Tyndall engaged' (*Morning Chronicle,* 18 September 1855, p. 3).

13. *Brewster . . . the triple spectrum*: David Brewster, 'The Triple Spectrum', *Brit. Assoc. Rep. 1855,* pp. 7–9.
14. *his views have been criticised by Whewell*: in W. Whewell, *History of the Inductive Sciences, from the Earliest to the Present Times,* 3 vols (London: J. W. Parker, 1837), vol. 2, pp. 360–61. See also report in the *Athenæum,* 6 October 1855, pp. 1156–57. One participant noted: 'We had a paper from Brewster on the theory of three colours in the spectrum, in which he treated Whewell with philosophic pity [. . .]. Whewell was in the room, but went out, and avoided a quarrel' (Unknown to J. C. Maxwell, 24 September 1855, in *The Scientific Letters and Papers of James Clerk Maxwell,* ed. P. M. Harman, 3 vols (Cambridge: Cambridge University Press, 1990–2002), vol. 1, p. 72)
15. *the papers brought forward at Glasgow*: As well as his paper on 'The Triple Spectrum,' Brewster also read optical papers entitled 'On the Binocular Vision of Surfaces of Different Colours', *Brit. Assoc. Rep. 1855,* p. 9; 'On the Existence of Acari in Mica', ibid., p. 9; 'On the Absorption of Matter by the Surfaces of Bodies', ibid., p. 9; and 'On the Phænomena of Decomposed Glass', ibid., p. 10.
16. *Loch Lomond*: see letter 1137, n. 9.
17. *Scotts poem of the Lady of the Lake*: *The Lady of the Lake* was a popular poem by Walter Scott (1771–1832) first published in 1810. Set in the Trossachs—to the west of Stirling and bordered by Loch Katrine—he evoked a romanticized picture of Scotland's history.
18. *the same evening*: 21 September.
19. *What stands in handbooks . . . examination;* see letter 1054, n. 5.

From Wilhelm Weber 25 September 1855 1139

Göttingen | 1855 | Sept. 25

Hochgeehrter Herr

Ich danke Ihnen verbindlichst für Ihre gütige Mittheilung vom 3 September, und freue mich daraus ersehen zu haben, dass der von Hrn. Leyser in Leipzig angefertigte Apparat zur Nachweisung der diamagnetischen Polarität Ihren Erwartungen vollkommen entsprochen hat. Es ist mir diese Nachricht darum besonders angenehm, weil ich selbst vor der Absendung des Apparats nicht nach Leipzig gekommen und daher ausser Stande gewesen bin denselben vorher zu prüfen.

Es ist mir von hohem Interesse gewesen zu erfahren, dass die Herren Faraday und De la Rive dadurch Gelegenheit gefunden haben, sich von der Richtigkeit der Sache zu überzeugen.

Von besonderem Interesse was es mir auch zu erfahren, dass es Ihnen gelungen ist, die diamagnetische Polarität des schweren Glases, woran Faraday den Diamagnetismus entdeckt hat, damit nachzuweisen—der beste Beweis dass diese Versuche von dem metallischen Leitungsvermögen des Wismuths nicht wesentlich abhängig sind.

Ihre Abhandlung "On diamagnetic force" in den Philosophical Transactions Vol. 145 habe ich mit grossem Interesse gelesen. Bei der grossen Sorgfalt mit welcher Sie die Thatsache der diamagnetischen Polarität von der Erklärung oder Theorie derselben gesondert und zunächst nur die Thatsache selbst ausser Zweifel zu stellen gesucht haben, erlaube ich mir Ihre Aufmerksamkeit darauf zu lenken, dass der Satz, welchen Sie (Seite 39) als eine Conclusion aus meiner Theorie anführen,

> that the magnetism of the two iron particles in the line of magnetization is increased by their reciprocal action; but that, on the contrary, the diamagnetism of two bismuth particles lying in this direction is diminished by their reciprocal action

meine Theorie gar nicht nothwendig voraussetzt, sondern auch, ganz unabhängig von meiner Theorie, unmittelbar aus der Thatsache der diamagnetischen Polarität selbst folgt, wenn dieselbe wirklich so statt findet wie sie von Ihnen, übereinstimmend mit mir, ausgesprochen wird. Was Sie daher zur Widerlegung jener Conclusion angeführt haben, müsste folglich als ein Einwand gegen die Thatsache der diamagnetischen Polarität selbst betrachtet werden. —Die diamagnetische Wechselwirkung der in der Linie der Magnetisierung liegenden Wismuththeilchen ist der Wirkung der magnetischen Erregungskraft nothwendig entgegengesetzt, so dass letztere durch erstere geschwächt werden muss, weil jene (diamagnetische) entgegengesetzt ist der magnetischen Wechselwirkung der in der Linie der Magnetisierung liegenden Eisentheilchen, durch welche bekanntlich die Wirkung der magnetischen Erregungskraft verstärkt wird; folglich muss auch die im Wismuth durch magnetische Erregung hervorgebrachte Modification—worin dieselbe auch bestehen möge—geschwächt werden, weil die Erregungskraft geschwächt worden ist.—

(Ich glaube, dass auch dieser Einwand gegen diamagnetische Polarität sich beseitigen lassen wird: die von Ihnen beobachtete Erscheinung muss nämlich offenbar aus ganz andern, mit der Compression des Wismuthstaubes ebenfalls verbundenen Umständen, erklärt werden; denn die diamagnetische Wechselwirkung ist, wie ich gezeigt habe, viel zu schwach um eine Wirkung hervorzubringen, die an Grösse mit der der magnetischen Wechselwirkung im Eisen irgend vergleichbar wäre).

Erlauben Sie mir, bei dieser Gelegenheit noch einige Bemerkungen beizufügen, welche geeignet scheinen meine Theorie der diamagnetischen Polarität in ein richtiges Licht zu stellen.

Meine Theorie der diamagnetischen Polarität setzt voraus 1) dass die Thatsache der diamagnetischen Polarität nicht bezweifelt werde 2) dass in Beziehung auf die magnetischen Erscheinungen für sich betrachtet Poisson's Theorie von zwei magnetischen Fluiden und Ampère's Theorie von Molecularströmen als gleichberechtigt anerkannt werden. Wer jene Thatsache leugnet oder Ampère's Theorie schlechtweg verwirft, der kann auch, wie ich selbst zugebe, meine Theorie der diamagnetischen Polarität nicht gelten lassen.

Verwerfen Sie aber Ampère's Theorie von beharrlichen Molecularströmen nicht schlechtweg, sondern sind Sie geneigt auf den innern Zusammenhang und die wahre Bedeutung dieser Theorie näher einzugehen, so werden Sie leicht erkennen, dass es keineswegs eine willkührliche von mir gemachte Annahme (assumption) sei, dass im Wismuth Molecularströme erregt (excited) werden, wenn die erregende magnetische Kraft wächst oder abnimmt; sondern dass die Erregung solcher Molecularströme eine nothwendige Conclusion aus Ampère's Theorie ist, welche Ampère selbst nicht hat machen können, weil ihm die Gesetze der von Faraday entdeckten Volta-Induction noch unbekannt waren.

Überall wo Ampèrersche Molecularströme existiren müssen durch wachsende oder abnehmende magnetische Erregungskräfte Molecularströme erregt werden, die sich mit den vorhandenen bald summiren bald davon subtrahiren werden.

Endlich erlaube ich mir noch zu den Worten Ihrer Abhandlung (Seite 39):

> to carry out the assumption here made, M. Weber is obliged to suppose that the molecules of diamagnetic bodies are surrounded by channels, in which the induced molecular currents, once excited, continue to flow without resistance

folgende Bemerkungen beizufügen.

Die Annahme von Canälen welche die Molecule umgeben, und in denen die elektrischen Fluida ohne Widerstand sich bewegen, ist eine schon in Ampère's Theorie enthaltene Annahme, welche keineswegs von mir (zur Erklärung der diamagnetischen Polarität) hinzugefügt ist. Ein beharrlicher Molecularstrom ohne solchen Canal involviert einen offenbaren Widerspruch, nach dem Ohm'schen Gesetze.

Ich wundere mich übrigens nicht, dass Sie eine auf die Annahme solcher Canäle gebaute Theorie als „extremely artificial, that if I imagine the general conviction of its truth cannot be very strong“ bezeichnen; ich stimme Ihnen in gewisser Beziehung bei; nur wünsche ich Sie zu überzeugen, dass dieser Vorwurf ursprünglich Ampère's Theorie treffe und auf meine Theorie der diamagnetischen Polarität nur in so fern übertragen werde, als letztere auf erstere gebaut ist. (Vielleicht werden Sie Sich an der Specialität einer solchen Annahme weniger stossen, wenn Sie die dabei zum Grunde liegende einfache, und durch eine gewisse Analogie der Molecule mit den Weltkörpern im Weltenraume sich besonders empfehlende Fundamental-Vorstellung von denjenigen Zusätzen unterscheiden, welche Ampère machen musste, um mit den ihm zu Gebote stehenden mathematischen Methoden einen strengen Calcul durchzuführen. Er war dadurch genöthigt, die Betrachtung auf lineare Ströme zurückzuführen, welche canalförmige Schranken nothwendig fordern, wenn jede Möglichkeit seitlicher Ausbreitung von der Betrachtung ganz ausgeschlossen und eliminirt werden soll).

Um meine Theorie der diamagnetischen Polarität in ein richtiges Licht zu stellen, wünsche ich also Sie davon zu überzeugen, dass diese Theorie keineswegs auf neue Annahmen (Hypothesen) begründet ist; sondern dass sie blos auf diejenigen Conclusionen gebaut ist, welche man aus Ampere's Theorie ziehen kann, wenn man die Gesetze der von Faraday entdeckten Volta-Induction und die Gesetze elektrischer Ströme von Ohm damit gehörig verbindet. Ich behaupte nämlich dass, wenn Faraday den Diamagnetismus nicht entdeckt hätte, möglicherweise derselbe durch die Combination der Ampère'schen Theorie mit Faraday's Gesetzen der Volta-Induction und Ohm's Gesetzen des elektrischen Stroms, wie sie in meiner Abhandlung enthalten ist, hätte entdeckt werden können.

Was aber die gerügte Künstlichkeit der Ampèreschen Theorie betrifft, so hoffe ich, dass die Mathematik Methoden finden wird, um die erwähnte Beschränkung der Betrachtung auf lineäre Ströme zu beseitigen und den Anstoss der Canalförmigen Strombetten zu heben. Alle unsere Moleculartheorien sind noch sehr künstlich: ich für mein Theil nehme an der Künstlichkeit der Ampère'schen Theorie weniger Anstoss als an andern Künstlichkeiten unserer Moleculartheorien, deswegen, weil bei Ampère der Grund der Künstlichkeit klar und deutlich vor Augen gelegt ist, und folglich auch die Aussicht und der Weg eröffnet ist, wie man endlich zur Elimination derselben gelangen werde.

Herrn Faraday bitte ich meine herzlichste Verehrung auszusprechen.

Mit ausgezeichneter Hochachtung ganz | der Ihrige | Wilhelm Weber.

Göttingen, Sept. 25, 1855.

MY DEAR SIR,

ACCEPT my best thanks for your kind communication of the 3rd of September;[1] I am gratified to learn that the apparatus executed by M. Leyser in Leipzig for the demonstration of diamagnetic polarity[2] has so completely fulfilled your expectations. This intelligence is all the more agreeable to me, inasmuch as before the apparatus was sent away, it was not in my power to go to Leipzig and test the instrument myself.

It gave me great pleasure to learn that Mr. Faraday and M. De la Rive have had an opportunity of witnessing the experiments, and of convincing themselves as to the facts of the case.[3]

It was also of peculiar interest to me to learn that you had succeeded in establishing the polarity of the self-same heavy glass with which Faraday first discovered diamagnetism.[4] This is the best proof that these experiments do not depend upon the conductive power of bismuth for electricity.

I have read with great interest your memoir 'On the Diamagnetic Force'[5] contained in the Philosophical Transactions, vol. cxlv. It has been your care to separate the *fact* of diamagnetic polarity from the theory, and to place the former beyond the region of doubt. Allow me, with reference to this subject, to direct your attention to a passage of page 39 of your memoir, which you adduce as a conclusion from my theory; the passage runs as follows:—

> The magnetism of two iron particles in the line of magnetization is increased by their reciprocal action; but, on the contrary, the diamagnetism of two bismuth particles lying in this direction is diminished by their reciprocal action.

This proposition is by no means a necessary assumption of my theory, but is rather a direct consequence of diamagnetic polarity, if the facts be such as both you and I affirm them to be. What, therefore, you have adduced against the above conclusion must be regarded as an argument against diamagnetic polarity itself. The *diamagnetic reciprocal action* of the bismuth particles in the line of magnetization is necessarily opposed to the *action of the exciting magnetic force*. The latter must be enfeebled, because the diamagnetic is opposed to the *magnetic reciprocal action* of iron particles which lie in the line of magnetization, through which latter it is known the action of the exciting magnetic force is increased. Hence also the *modification* produced in bismuth by magnetic excitement, whatever it may be, must be weakened, because the force of excitation is diminished.

(I believe, however, that this argument against diamagnetic polarity may also be surmounted. The phænomenon which you have observed must be referred to other circumstances, also connected with the compression of the bismuth. For the diamagnetic reciprocal action is, as I have shown,[6] much too weak to produce an effect which could be compared in point of magnitude with the reciprocal action produced in the case of iron.)

I take this opportunity of adding a few remarks for the purpose of setting my theory of diamagnetic polarity in a more correct light.

My theory assumes:—1, that the fact of diamagnetic polarity is granted; 2, that in regard to magnetic phænomena, Poisson's theory of two magnetic fluids,[7] and Ampère's theory of molecular currents,[8] are equally admissible. Whoever denies the first fact, or rejects the theory of Ampère, cannot, I am ready to confess, accept my theory.

But supposing that you do not reject Ampère's theory of permanent molecular currents, but are disposed to enter upon the inner connexion and true significance of the theory, you will easily recognize that it is by no means *an arbitrary assumption of mine,* that in bismuth molecular currents are excited, when the exciting magnetic force is augmented or diminished; but that the excitation of such molecular currents is *a necessary conclusion from the theory of Ampère,* which conclusion Ampère himself could not make, because the laws of voltaic induction, discovered by Faraday,[9] were unknown to him. In all cases where molecular currents exist, by increase or diminution of the magnetic exciting force *molecular currents must be excited,* which either add their action to, or subtract it from, the action of those already present.

Finally, permit me to make a few remarks on the following words of your memoir:—

> To carry out the assumption here made, M. Weber is obliged to suppose that the molecules of diamagnetic bodies are surrounded by channels, in which the induced currents, once excited, continue to flow without resistance.

The assumption of channels which surround the molecules, and in which the electric fluids move without resistance, is an assumption contained in the theory of Ampère, and is by no means added by me for the purpose of explaining diamagnetic polarity. *A permanent molecular current without such a channel involves a manifest contradiction,* according to the law of Ohm.[10]

I may further observe, that I do not wonder that you regard a theory which is built upon the assumption of such channels, as 'so extremely artificial that you imagine the general conviction of its truth cannot be very strong'.[11] In a certain sense I quite agree with you, but I only wish to convince you that this objection applies really to the theory of Ampère, and only applies to mine in so far as it is built upon the former. (You may perhaps find less ground for objecting to the speciality of such an assumption, if you separate the simple fundamental conception, which recommends itself particularly by a certain analogy of the molecules to the heavenly bodies in space, from those additions which Ampère was forced to make, in order to apply the mathematical methods at his command, and to make the subject one of strict calculation. He needed to reduce the case to that of *linear* currents, which necessarily demand channel-shaped bounds, if every possibility of a lateral outspreading is to be avoided.)

To place my theory of diamagnetic polarity in a truer light, I am anxious also to convince you that this theory is by no means based upon new assumptions (hypotheses); but that it only rests upon such conclusions as may be drawn from the theory of Ampère, when the laws of voltaic induction discovered by Faraday, and the laws of electric currents by Ohm, are suitably connected with it. I affirm, that, even if Faraday had not discovered diamagnetism, by the combination of Ampère's theory with Faraday's laws of voltaic induction, and Ohm's laws of the electric current, as shown in my memoir, the said discovery might possibly have been made.

In respect, however, to the artificiality of the theory of Ampère, I hope that mathematical methods may be found whereby the limitation before mentioned to the case of linear currents may be set aside, and with it the objection against channel-form beds. All our molecular theories are still very artificial. I for my part find less to object to in this respect in the theory of Ampère than in other artificialities of our molecular theories; and for this reason, that in Ampère's case the nature of the artificiality is placed clearly in view, and hence also a way opened towards its removal.

To Mr. Faraday I beg of you to present my sincerest respect.

Believe me, dear Sir, | Most sincerely yours, | WILHELM WEBER.

Professor Tyndall.

RI MS JT/1/W/14
Phil. Mag., 10 (1855), pp. 407–9[12]

---

1. *communication of the 3rd of September*: letter missing. Letter contained 'a sketch of some experiments executed with the instrument devised for me by M. Weber himself' and also a copy of Tyndall's Bakerian Lecture (letter 1029, n. 1) (J. Tyndall, *Researches on Diamagnetism and Magne-crystallic Action, including the Question of Diamagnetic Polarity* (London: Longmans, Green & Co., 1870), p. 225).
2. *apparatus executed by M. Leyser . . . diamagnetic polarity*: This apparatus, which Weber designed, was used by Tyndall in his 'Further Researches on the Polarity of the Diamagnetic Force', *Phil. Trans.,* 146 (1856), pp. 237–59, and described on pp. 240–43. Georg Moritz Ludwig Leyser (1816–81), who developed such instruments as thermometers and electrostatic generators, was the instrument maker and 'Inspector des physikalischen Cabinets' at the University of Leipzig.
3. *Mr. Faraday and M. De la Rive . . . the case*: see letters 1129, n. 6 and 1138, n. 11.
4. *the polarity . . . diamagnetism*: For Faraday's experiments on the action of magnets on heavy glass see *ERE,* 3: 30–34.
5. *On the Diamagnetic Force*: Tyndall's Bakerian Lecture (letter 1029, n. 1).
6. *as I have shown*: W, Weber, 'Ueber den Zusammenhang der Lehre von dem Diamagnetismus mit der Lehre von dem Magnetismus und der Electricitat', *Poggend. Annal.,* 87 (1852), pp. 145–89.
7. *Poisson's theory of two magnetic fluids*: Following his work mathematising electricity, in 1826 Poisson published two papers in which he offered a mathematical analysis of magnetism. 'His most fundamental physical assumption was that magnetic effects are

produced by two magnetic fluids that act on each other through fundamental forces that vary as the inverse square of the distance between their particles' (J. R. Hofmann, *André-Marie Ampère* (Cambridge: Cambridge University Press: 1995), p. 329). On pp. 38–39 of his Bakerian Lecture (letter 1029, n. 1) Tyndall argued that Poisson's theory was unable to account for the phenomenon of diamagnetism.

8. *Ampère's theory of molecular currents*: Ampère 'supposed the molecules of a magnetic body to be surrounded by such currents, which, however, in the natural state of the body mutually neutralized each other, on account of their confused grouping. The act of magnetization he supposed to consist in setting these molecular currents parallel to each other, and starting from this principle he reduced all the phenomena of magnetism to the mutual action of electric currents' (Tyndall's Bakerian Lecture (see letter 1029, n. 1), p. 38).
9. *the laws of voltaic induction, discovered by Faraday*: Faraday, *ERE,* vol. 1, pp. 2–7.
10. *the law of Ohm*: see letter 1093, n. 4.
11. *so extremely artificial . . . very strong*: quotation from Tyndall's Bakerian Lecture (letter 1029, n. 1), p. 39.
12. Published with the title '*On the Theory of Diamagnetism. Letter from* Professor WEBER *to* Prof. TYNDALL.' On pp. 409–10 Tyndall added his comments on Weber's letter:

'The foregoing letter possesses more than a private interest, and I have therefore laid it before the readers of the Philosophical Magazine. On one point in it only would I ask permission to make a remark, and that is the proposition, that the diminution of the excitement of a row of bismuth particles in the line of magnetization by their reciprocal action is 'a direct consequence of diamagnetic polarity'. M. Weber (I believe) founds this proposition on the following considerations:—

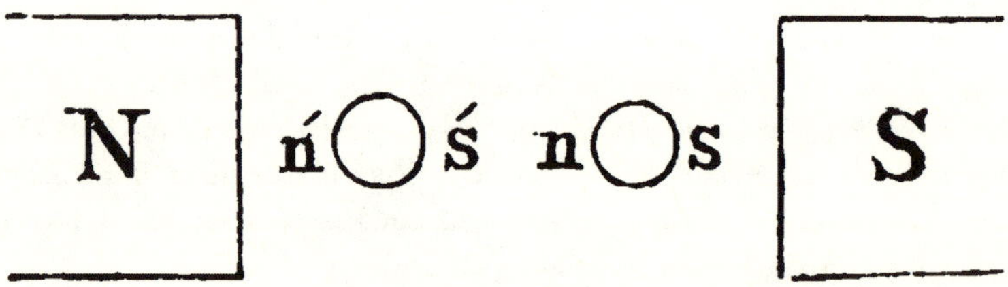

Let a series of bismuth particles lie in the axial line between the magnetic poles N and S: the polarity excited in these particles by the direct action of the poles will be that shown in the figure, being the reverse of that of iron particles under the same circumstances. But as the end *n* of the right-hand particle tends to excite a magnetism like its own in the end *n'* of the left-hand particle, and *vice versâ,* this action is opposed to that of the magnet, and hence the magnetism of such a row of particles is enfeebled by their reciprocal action.

Now it appears to me that there is more assumed in this ingenious argument than experiment at present can hear out. There are no experimental grounds for the assumption, that what we call the north pole of a bismuth particle exerts upon a second bismuth particle precisely the same action that the north pole of an iron particle would exert. Magnetized iron repels bismuth; but whatever the fact may be, the conclusion is scarcely warranted, that therefore magnetized bismuth will repel bismuth. Supposing it were asserted that magnetized iron attracts iron and repels bismuth, while magnetized bismuth attracts bismuth and repels iron, would there be anything essentially impossible, self-contradictory, or absurd involved in the assertion? I think not. And yet if even the possible correctness of such an assertion be granted, the proposition above referred to becomes untenable. It will be observed that it is against a conclusion rather than a fact that I contend. With regard to the fact, I should be sorry to express a positive opinion; for this is a subject on which I am at present seeking instruction, which may lead me either to M. Weber's view or the opposite. Be that as it may, the result cannot materially affect the respect I entertain for every opinion emanating from my distinguished correspondent on this and all other scientific subjects.'

From Hector Tyndale 27 September 1855 1140

Philadelphia | 27 September 1855

My dear John

Thank you for the letter[1] you dated to me just this day two months ago. Two months ago! Why in those two months we have travelled through and over 'all creation', as we yankees are said to say. We have gone with our egotisms[2] to the very source of all egoisms and pulled the beards of holy spirits sitting high in the eternal senate of Heaven. We have in those two months dived or climbed, whichever it may be, to the center of ourselves, and yet, strange to say, here we stand to day not a whit in advance of the position we had just two months ago! The telegraphic magnet speed is slow and wearisome in contrast with the rate at which we have rushed unwearied through the infinite space of self-consciousness. We have spoken, or dreamed that we have spoken (and is not that the same thing to <u>us</u>?) face to face with the Holiest, and yet here sit we writing by a light, the mere sunlight, a visible, physical thing, a decomposable compound more strange and more unknown to us than the great Universum[3] we have measured accurately within those two months. Here after two months of self converse; that is of universal egotistic converse with the highest and holiest we know, I sit down with face as blank and head as empty as before, to converse with you by brain talk and inky symbols! Why <u>are</u> we tied down by hands and pens and muddled brains when we desire and command that we may speak untramelled and unwearied by these slow contrivances? Oh were we not thus tied what worlds of flame and light and beauty we would send through eager space; what planetary bubbles, refracting the light of source and destiny, flooding dark chaos with light and awakening dead void to the new rapture of the sense of Light. Ah John what grand fireworks we should have, you and I and more of us, in this grand Versailles of imperial egotism–

But prithee stop me John with some forceful fact, check my wheels with a brick from the great edifice of science you are building or I may go bump into the middle of your egotism and make 'a smash' when I intended only to shake your hands.

There John I let it go, the most tiresome long straw in the wind blowing hereaway—Is it not an adverse trade wind?

Thank you again my dear fellow for your kind sympathy with me in my 'chronic indifference'.[4] You have been bored more than enough, before and now herein, with my gloomy egotism, and I cannot as a Christian man call on you again 'for dear charity's sake'.

I wish—please notice how often I wish—I had been with you and Debus in your walk through Normandy.[5] Consciousness would then have been content with existence—and contentment is health is it not? What pleasure to be thus free and careless of hours and rules made by rail road managers and respectable others. How capital the cider, the beer or the vin ordinaire[6] and the bread. How they seemed to assimilate with ourselves and to be the very thing we needed unconsciously—ah beautiful idea—angel symbolized by un-consciousness. The Sun, not so bright perhaps, the sky, not so clear may be, as these now over my head, but bright and clear and joy making. Yellow fields and green meadows border all the roads and birds singing everywhere on the summer clad trees. Clear vision and springey muscles, quiet soul and happy heart, easy mind and glorious health all come in from such a walk. Hail to you road-walker, Nature seeker, long may you live in power of health and strength and in the beatitude of Existence—pur et simple.[7] These things come not to every man, certainly not to every man here, and it is, perhaps, as much to this non-come-at-a-bility as to anything else that these walks seem so pleasant to me—The Sun, you know, lingers longest on the farthest cloud. Please present my kindest respects to Mr Debus. I have met here Mr Genth[8]—chemist—who is acquainted with him, having, I believe, been a fellow student with him under Bunsen. To Dr: Francis and to Dr: Coxe, should you see or write to the latter, please make my remembrances & respects.

Thank you for the 'Bakerian lecture'[9] I have read it with much pleasure, and though it is, perhaps, somewhat too technical for me to grasp in all the experimental details I still can easily and clearly follow you in all your ratiocination. 'Ye hev a way wid ye yer Honor'[10] of saying things very simply and clearly, that is very pleasant to a reader and when I come to study Physics I shall apply to you—Mind that.

You say 'I wish to Heaven you could abolish my right to be unwell.' Certainly I don't wish to destroy any right you may have—I being a very liberal believer in free will—but if you would piously walk more, take more breathings from the lips of the Nature you love so well; if in these pilgrimages, holy and religious, you would worship in companionship of a Saint or ordained priest of the Great Church, then the world of your Nature would become green, fresh and strong again (par parentheses the blind have led the halt before[11] so please excuse this) But I will on. 'You have no right to be sick' I have already said so; but it is of another kind of right I spoke. You have no such right taking Right as the opposite of Wrong, though you have it certainly, understood as an act of your own unrelated Will and as an eccentric privilege of that alone. Your avocation, and it is your part of ordained labour, is in accord with your desires, it is attractive to you. Nothing in it clashes or discords your harmony. You seek it as a pleasure while it serves you in acquittance of the labour required of you. Such I have no doubt is the intention of the law of labour, but this unrelated Will of Man has made labour seem a curse.

You say with much truth 'the only plan that I have ever found effectual in finding the right sphere of action is to do the duty nearest at hand.' But the difficulty is to find the 'nearest duty'. Or if that is found and done then the next nearest and the next until the search for duties becomes the vocation of the life, the labour of life, pleasant in itself, but empty of returns, perhaps, for the physical well being of those around us. Is civilization, in all it's demands for art and sensual surroundings, advantageous after all? But away with Metaphysics, defined by some one as 'whatever is not understand-able'. I will say only, and you may guess the rest if you will or deny the whole, he gains no honours nor reaps glory and he sleeps no satisfied sleep who lags rearward, whether from cowardice, laziness or inability. But he still preserves his right to grumble, to sickness and to discontent! At least these are not denied him and these at last become his pleasures, as was that of the overthrown rebel Titans[12] the belching of stinking fumes against the Heavens and the Great Victor Gods. This is very sad, but the rebellion goes on for ever. I have gained much satisfaction, inductively, from your lecture on 'the diamagnetic force'.[13] You have proved that Matter has resident force and is not wholly and imperatively influenced by external powers; e.g. paramagnetic and dia-magnetic actions arise from the force resident in the subjected matter, that the matter itself defines its own action and direction under the neutral or indifferent powers of the magnet <*word(s) missing*> fact, my dear John, goes a great way towards reconciling me with myself! <*word(s) missing*> as a 'laissez faire'![14]

and opposite as I may be to myself, I still may <*word(s) missing*> reconcile it as a Natural, and therefore, I hope, beneficial opposition. Oh for a downright, steadfast opposition, or for a faithful, hearty, concurrent Spirit. Either would be preferable, perhaps, to a mixt spirit, neither 'black' nor 'white' but only a mongrel 'grey'?

I am glad you have seen the great Exposition of Paris,[15] it must be glorious to see such an aggregation of concrete ideas. The London Exhibition of 1851[16] was grand and I doubt whether the French one is more grand, though I can readily believe there is more taste and higher art in the paintings & sculpture.

Bloody news from the Crimea.[17] Great havoc of Russian self love. Damaged Souls went to the Styx[18] border from thousands of butchered bodies. Even strong Will was broken when Sebastopol fell.[19] I am glad of it though looking at such ghastly sights is not gladdening generally. Would that Liberty and what Jesus and all great souls have fought for may be brought nearer to men from this bloody war. Let us wait & see.

You and I may differ much on forms of government perhaps, but I can conceive of no difference in the Great End desired.[20] It is not, it cannot be that this gory war is waged alone for Britain, or for France or even for poor, wretched Turkey. There must be, by God ordained, something behind and above the flaunting rags of nationality, and a nobler enthusiasm than is expressed in the hoarse cries of selfish laudation. Why are not armies clothed in the deepest mourning, as undertakers, who for the health of living souls bury out of sight and sense the dead bodies of lies and ancient tyrannies; whose business it is to ingrave old man of the mountain tradition[21] and all humbug, fiddling Neros.[22] Why do we not shout as gleefully to the successful body burier as to conquering armies? Let us shout in God's name but not in Moseses![23] Let us recognise the Victor whose body closes the yawning chasm. But let us not offend the Gods by brutal shouts coming from the bull ring and fitting enough for that. I fear however that the Spaniards do but act out the Matadore (or whatever his bull fighting name may be) who lives in the nature of all humanity. I who wrestle with the Gods and Angels, shall I gloat and glory over running blood whether shed by bulls or men? 'I, the foremost child of all the ages',[24] 'I to mate with squalid savage'[25] sang Tennyson then, how sings he now?—Enough, drag away the dead bulls and men, by gaily apparelled asses, to the sound of livelier music than those miserable groans of mine can make.

Mr Leland of this city has made, what is said to be, a very good translation of the 'Travel pictures'—of 'the witty & wicked Heine'.[26] It is published here in numbers and thinking you might like to see it I have directed the publishers here to send you the numbers as they are published, their house in London is Trübner & Co No 12 Paternoster Row. They will send it to you[27] as directed, no doubt, if they do not then please send for it, if you think it worth while to read. Heine is certainly witty, but it is wine wit mostly I apprehend. Pray excuse me for sending it, for I do not mean thereby any opinion upon your taste, for after all he is quite readable & funny. Do write soon to me and often as you can find time and I if I do not always respond as promptly as might be, attribute it to my moods & tenses.

And now my dear John I wind up the broken ends of my talk with you by sending the unknown remembrances of all my family, our family here to you, and by recommending your heart to that rare and radiant maiden,[28] your body to the roads and hill tops, your conceptive soul to the great views of Earth and Heaven to be seen only from those hill tops, and your kind memories to me. Good bye and take of your health here as you hope to go 'to a better place hereafter'.

affectionately yours | Hector Tyndale

RI MS JT/1/T/48

1. *the letter*: letter 1118.
2. *our egotisms*: possibly referring to human—even American—self-confidence in its own abilities.
3. *the great Universum*: the whole creation.
4. *chronic indifference*: see letter 1118.
5. *you and Debus in your walk through Normandy*: see letter 1118.
6. *vin ordinaire*: cheap table wine (French).
7. *pur et simple*: pure and simple (French).
8. *Mr Genth*: Frederick Augustus Genth (1820–93), who studied at Giessen and Marburg, emigrated to Philadelphia in 1848, where he founded a laboratory and established himself as an analytical chemist and mineralogist.
9. *Bakerian lecture*: see letter 1029, n. 1.
10. *Ye hev a way wid ye yer Honor*: You have a way with you your Honor; Hector humorously praises Tyndall's linguistic abilities.
11. *the blind have led the halt before*: cf. 'And if the blind lead the blind, both shall fall into the ditch' (Matthew 15:14).
12. *overthrown rebel Titans*: The Titans were the children of Earth and Heaven, the youngest being Chronos (Time). The power of Chronos was challenged by his son Zeus who, after prolonged battle, defeated his father and the other Titans; once defeated the rebellious Titans were imprisoned beneath the earth where they gave vent to their frustration.
13. *lecture on 'the diamagnetic force'*: Probably Tyndall's 'On the Diamagnetic Force', *Brit. Assoc. Rep. 1854*, pp. 14–17, or his Bakerian Lecture (n. 9, above).
14. *laissez faire*: A phrase expressive of the principle that government should not interfere with the action of individuals, especially in industrial affairs and in trade (*OED*).
15. *great Exposition of Paris*: the Exposition Universelle held in Paris from 15 May to 15 November 1855.

16. *The London Exhibition of 1851*: the Great Exhibition of the Works of Industry of All Nations held in Hyde Park, London, from 1 May to 11 October 1851.
17. *Crimea*: The war that had been raging since October 1853 between Russia and an alliance of Britain, France and the Ottoman Empire in the Crimean Peninsula on the northern side of the Black Sea.
18. *Styx*: In Greek mythology the Styx was the river dividing the earth from the underworld (Hades). Charon, the ferryman, transported the souls of the recently deceased across this river.
19. *when Sebastopol fell*: After nearly eleven months, the siege of Sevastopol ended on 9 September 1855 with the withdrawal of its Russian defenders. More than 100,000 soldiers died during the siege.
20. *the Great End desired*: possibly an end to the suffering of warfare.
21. *old man of the mountain tradition*: possibly the peddlers of revelations.
22. *fiddling Neros*: The Emperor Nero (37–68) is said to have 'fiddled while Rome burned'; a reference to the fire that destroyed Rome in the year 64, which his opponents claimed was started by Nero.
23. *Moseses*: Tyndale contrasts God with Moses, who was merely a leader of men.
24. *I, the foremost child of all the ages*: 'I the heir of all the ages, in the foremost files of time' (Alfred Lord Tennyson, 'Locksley Hall',1. 178).
25. *I to mate with squalid savage*: 'Mated with a squalid savage' (ibid.,1. 177).
26. *Mr Leland . . . Heine*: *Reisebilder* by the German poet Heinrich Heine (1797–1856) was first published in Germany in 1826. The English translation by Charles Godfrey Leland (1824–1903), with the title *Pictures of Travel,* was published in Philadelphia by John Weik in 1855 and soon afterwards by Trübner in London. In the translator's preface Heine is described as 'this "witty, wise, and wicked" writer' (p. 5). Leland's source for this phrase has not been traced.
27. *They will send it to you*: Tyndall subsequently received a copy and noted in his Journal, 27 November 1855: 'Read Heine's Reise Bilder which Hector has sent me' (RI MS JT/2/13c/806).
28. *rare and radiant maiden*: see letter 1052, n. 15.

From Edward Hughes 28 September 1855 1141

Greenwich Hospital | 28th Sept. 1855.

Dear Sir,

I regret exceedingly that you should have any cause of complaint about the manner in which your valuable contributions to my "Reading Lessons"[1] have been printed. The corrections[2] which you pointed out to me I instantly had made, as you will see by the accompanying copy,[3] but unfortunately my printer had struck off so many copies of the first impression that I could not afford to cancel them. The remaining copies of the impression however contain a list of 'Errata' at the end of each book, and are moreover differently bound.

You will find that the second edition of the First Book is not only revised but enlarged.

So anxious am I that there should be no cause of complaint, that I have not yet sent either of the books anywhere for review, and if your friend[4] will kindly wait a short time he can obtain the volume in the form in which I now send it to you, and in which I am about to send it to the Periodicals and Papers.

Should you discover any errors in your articles in the Second Book[5] I will thank you to point them out to me, and I will have them corrected immediately.

I am dear Sir | Yours faithfully | E. Hughes.

Dr. John Tyndall F.R.S.

RI MS JT/1/2/648
LT Typescript Only

---

1. *Reading Lessons*: E. Hughes (ed), *Reading Lessons. First Book* (London: Longman, Brown, Green & Longmans, 1855). Tyndall contributed three articles—'On Force and Motion', 'Collision of Bodies' and 'On Terrestrial Gravity'—under the heading 'Natural Philosophy', pp. 296–338.
2. *The corrections*: list missing.
3. *the accompanying copy*: not located.
4. *your friend*: probably Hirst.
5. *your articles in the Second Book*: E. Hughes (ed), *Reading Lessons, Advanced Series. Second Book* (London: Longman, Brown, Green & Longmans, 1855). Tyndall contributed six lessons—'Mechanical Properties of Liquids', 'Pressure and Displacement', 'Specific Gravity', 'Mechanical Properties of Air', 'Change of Atmospheric Pressure—the Barometer' and 'The Relation between the Volume and the Pressure of Elastic Fluids'—under the heading 'Natural Philosophy', pp. 289–322.

From Heinrich Buff[1] 29 September 1855 1142

Hochgeehrter Herr!

Ihre neue Abhandlung über Diamagnetismus, welche Sie die Güte hatten, an mich abzu[senden] ist bis jetzt nicht in meine Hände gelangt und ich muß deßhalb befürchten daß sie unterwegs verloren worden ist. Ich bewundere die Thatkraft womit der Sie trotz der Aufregungen des Lebens in der großen Hauptstadt von England, so viele wissenschaftliche Untersuchungen rasch hintereinander auszuführen vermögen. Von großem Interesse war es mir zu hören, [daß] Sie zu Ansichten geleitet worden sind, welche mit denen W. Weber's über Diamagnetismus übereinstimmen und daß mir es scheint, durch Ihre experimentelle Darstellung auch Faraday für die Weber'sche Theorie gewonnen worden ist. Diese scharfsinnige Theorie erleichtert ungemein die Erklärung der Erscheinungen. Zu meiner individuellen Befriedigung vermisse ich nur noch bei derselben oder doch bei ihrer letzten Grundlage, gleich wie bei den Ampère'schen Molekularströmen den ungezwungenen Ausdruck natürlicher Verhältnisse. —Es scheint Ihnen aufgefallen zu sein, daß Ihre, an sich betrachtet, den strengsten Anforderungen der Logik sowohl entsprechenden Untersuchungen über das Trevelyan-Instrument, [bei uns] nicht in gleichem Maße wie fast alle Ihre früheren wissenschaftlichen Arbeiten die Aufmerksamkeit angeregt haben. Der Grund liegt ohne Zweifel nur an der über das Faktum selbst in Deutschland verbreiteten Ansicht. Die von Faraday gegebne Erklärung ist nämlich bei uns längst in die Lehrbücher der Physik übergegangen, und war durch die spätere von Forbes aufgestellte Theorie niemals erschüttert worden. Nach dem Urtheile meines Freundes Zamminer, von welchem die kurze Berichterstattung in dem Jahresberichte für 1854 herrührt, bedurfte es daher keineswegs eines so großen Aufwandes an Kraft um das, was, so wie er nach den in Deutschland gemachten Erfahrungen zu glauben berechtigt war, doch niemand für richtig gehalten hatte, zu widerlegen.

Übrigens ist, ganz abgesehen von der Streitfrage, die verschiedenseitige Auffassung des Trevelyan-Versuches ganz sicher von allgemeinem Interesse gewesen; auch ist es wichtig daß die Anstellung des Versuches mit nichtmetallischen Körpern (Unterlagen) zuerst von Ihnen ausgeführt wurde. Bei der Vergleichung der Seebeck'schen Untersuchungen mit den Ihrigen [waren] diese nur aus dem vorher angegebenen Grund leider unbeachtet geblieben. Sollte es in Ihrem Interesse liegen von dieser, meiner Bemerkung öffentlich Gebrauch zu machen, so würde dieß genügende Veranlassung sein, damit auch im nächsten Jahresberichte darauf Rücksicht genommen werde.

Bei Ihren so wohl begründeten Ansichten über die Ursache der Trevelyan-Erscheinung hat es mich überrascht, daß Sie bei einer anderen Wärme-Erscheinung, ich meine den Leidenfrost'schen Versuch, die Theorie einer Wärmeabstoßung auf Abstand gleichwohl noch festzuhalten scheinen. So muß ich wenigstens schließen aus einer gelegentlichen Bemerkung in einer Abhandlung des Herrn A. Church. Dieser erwähnt einen Versuch, den er in Ihren Vorlesungen gesehen habe, durch welchen Sie zeigen daß das Wasser in Tropfenform den Übergang des electrischen Stromes zur glühenden Platte unterbreche, und sagt daß Sie durch diesen Versuch bewiesen hätten, daß die glühende Wand eines Metallgefäßes von dem darin befindlichen, nicht [anhängenden] Wasser durch einen Zwischenraum getrennt sei.

Den erwähnten Versuch habe ich ebenfalls oft und unter verschiedenen Abänderungen wiederholt, vermag demselben aber keineswegs eine so große Beweiskraft beizumessen. Ich finde nur daß der Übergang der Electricität [durch] Wasser in Tropfenform ungemein erschwert wird, und zwar nur aus dem Grund, weil die Puncte wechselseitiger Berührung zwischen Wasser und Metall sich auf eine, verhältnißmäßig sehr kleine Anzahl reduciren. Der Strom kann dadurch allerdings scheinbar unterbrochen werden, ob es geschieht oder nicht hängt aber nach meinen Erfahrungen, für eine gegebene Flüßigkeit, ganz von der Empfindlichkeit des Galvanometers, von der Stärke der [electromotivischen] Kraft und endlich von der Größe des Tropfens ab. Das spheroidal gestaltete Wasser verhält sich gegen die heiße Metallfläche ähnlich, wie Quecksilber gegen eingetauchte Metalldrähte, die dann nicht benetzt werden. In einer schon vor mehreren Jahren bekannt gewordenen Abhandlung (Annalen der Chemie und Pharmacie B. 77 S. 1) habe ich verschiedene, übrigens sehr leicht zu [nehmende] Verfahrungsweisen mitgetheilt, die Fortpflanzung des electrischen Stromes durch Wasser in Tropfenform sichtbar zu machen. Das <*word illeg*> bietet aber, wie mir später erst eingefallen ist, das zuerst von Mackrell beobachtete electrische Erglühen dieser Metalldrähte, wenn dieselben als Electroden einer sehr starken electrischen <*word illeg*> in heißes Wasser eingetaucht werden. Bei dem negativen Pole könnte man einwenden, daß er während der Glüherscheinung von Sprühlicht umhüllt ist, dieß ist jedoch bei dem positiven Draht nicht nothwendig der Fall, und dann zeigte im Zustand des Aufglühens das Leidenfrost'sche Phänomen in ganzer Vollständigkeit; gleichwohl hat der Strom und sogar die electrolytische Wasserzersetzung Fortdauer, denn wenn die positive Electrode kein Sprühlicht <*word illeg*>, ist das an derselben entwickelte Gas reiner Sauerstoff. Die Fortdauer des Stromes in diesem und in ähnlichen Fällen ist vielleicht kein ganz sicherer Beweis für das Stattfinden einer unmittelbaren Berührung, weil das zwischen dem Wasser und der glühenden Wand befindliche Gas, der Dampf [jetzt] durch die starke Erhitzung einen geringen Grad der Leitfähigkeit annimmt. <*word illeg*> geht es mir nur darum darzuthun, daß das Ausbleiben des Stromes in anderen Fällen nicht als Beweis für das Vorhandensein eines Zwischenraumes zwischen Metall und Wasser benutzt werden darf.

In der That bietet der späroidale Zustand des Wassers, in seinen wesentlichen Eigenthümlichkeiten näher betrachtet, Beweisgründe genug, welche gegen die Isolierung und Abstoßung als nothwendige Bedingungen der Tropfenform sprechen. Erlauben Sie mir Ihnen diejenigen, welche ich für die wichtigsten halte, hier kurz zusammenzustellen.

In einem mit <*word illeg*> [anhängenden] Ruß (Kohlenschwarz) überkleideten silbernen Löffel behauptet das Wasser die Tropfenform schon bei gewöhnlicher Temperatur, genau so wie bei starker Erhitzung des Löffels.

Der auf einer heißen Flüßigkeit rollende Tropfen, z.B. ein auf Wasser gebildeter Äther-Tropfen biegt die Oberfläche der Flüßigkeit ein, ähnlich wie Quecksilber durch den Druck eines Glasstabs eingebogen wird.

Große Tropfen nehmen eine oscillirende Bewegung an, die oft mit merkwürdiger Regelmäßigkeit stattfindend, die Bildung sternförmiger Figuren bewirkt, deren Entstehung ohne eine theilweise Berührung des Wassers mit seiner Unterlage ganz unmöglich sein würde.

Man erhält die größten Tropfen nicht auf solchen Unterlagen, welche das beste Ausstrahlungsvermögen, sondern welche die beste Leitfähigkeit für Wärme [besitzen], Silberschaalen sind daher <*word illeg*> dazu geeignet.

Die Erscheinung gelingt ausschließlich nur mit verdampfbaren Flüßigkeiten. Sie erfordert eine um so stärker erhitzte Unterlage, je höher der Siedpunct der Flüßigkeit liegt. Der Versuch gelingt daher bei Anwendung schlecht leitender Platten, z.B. wie Porzellan oder Glasgefäßen leichter mit Aether als mit Wasser. Die niedrigste Temperatur bei welcher eine Unterlage von der heiß [aufgegossenen] Flüßigkeit nicht benetzt wird, muß stets höher sein als der Siedpunct dieser Flüßigkeit. Z.B. in einer Silberschaale behauptet bei 75° (des hunderttheiligen Thermometers) Wasser, bei 137° Alcohol und erst bei 144° reines Wasser, die sphäroidale Gestalt.

Mit diesen Bemerkungen will ich nicht die Ansicht aussprechen, daß der Leidenfrost'sche Tropfen nicht auch von der erhitzten Unterlage getrennt sein könne. Ich glaube vielmehr, daß eine solche Trennung sehr oft eintritt und daß selbst größere Tropfen durch die Gewalt der entwickelten Dämpfe von der Fläche der Schaale losgerissen werden können. Nur behaupte ich, daß eine solche Isolirung nicht die nothwendige Bedingung des Eintritts der Erscheinung ist. Dieß beruht auf meiner Meinung [auf] einer durch die Hitze bewirkten Veränderung im relativen Verhältnisse der Anziehungen, welche die Gefäßwand gegen die flüssigen Theile und diese gegen einander ausüben.

[Die] Wand eines Gefäßes wird von einer eingegoßenen Flüßigkeit bekanntlich dann benetzt, wenn die Anziehung zwischen dem Stoffe des Gefäßes und dem der Flüßigkeit größer ist, als die Anziehung der flüssigen Theile untereinander. Im umgekehrten Falle behauptet die Flüßigkeit die Tropfenform <*word illeg*>. Die wechselseitige Anziehung der Körpertheile vermindert sich ohne Zweifel bei erhöhter Temperatur. Erhitzt man eine Silberschaale mit dem darin befindlichen Wasser, so bleibt die Wand der Schaale benetzt, auch während die Flüßigkeit siedet. In der That ist aber auch [unter diesen Umständen] die Temperatur des Silbers nur wenig höher als die des Wassers. Die wechselseitige Anziehung der Silber und Wasser Theile hat sich daher ziemlich gleichmäßig mit derjenigen der Wassertheile untereinander vermindert, das relative Verhältniß der Anziehungen blieb fast ungeändert.

Wird die [Silbermasse] über den Siedpunct des Wassers hinaus erhitzt, so muß ihre Adhäsionskraft zu dem Wasser stärker abnehmen als die Cohäsionskraft der auf ihrem Siedpuncte <*2 words illeg*> flüssigen Theile. Es wird also früher oder später ein Temperaturpunct eintreten, bei welchem die Cohäsion die Adhäsion überwiegt. Die Benetzung hört dann auf und es bildet sich die Späroidalgestalt, ganz im Einklange mit den gewöhnlichen Capillargesetzen.

Vielleicht stimmen Sie geehrter Herr mit dieser Betrachtungsweise der Sache überein und dann werden Sie, so hoffe ich, diese, wie ich jetzt sehe, etwas weitläuftig gewordene Auseinandersetzung gütigstens gefälligen. Mit hochachtungsvoller Ergebenheit habe ich die Ehre zu zeichnen,

H. Buff | Giessen am 29sten September 1855.

Esteemed Sir!

Your new paper on diamagnetism,[2] which you had the kindness to *[post]* to me, has not come into my hands thus far, and I must therefore fear that it has been lost on the way. I admire the energy with which you are able to carry out so many scientific investigations quickly one after the other despite the excitement of life in the great capital of England. It was of great interest to me to hear *[that]* you have been led to views which agree with those of W. Weber on diamagnetism,[3] and that it appears to me that through your accounts of experiments Faraday has likewise been won over to Weber's theory. This astute theory facilitates the explanation of the phenomena immensely. For my own satisfaction, the only thing I find lacking with the theory, or at least with its latest grounding, is the unforced expression of natural conditions, just as is the case with Ampère's molecular currents.[4] —You appear to have noticed that your investigations into the Trevelyan instrument,[5] which, considered in themselves, comply so well with the strictest requirements of logic, have not inspired attention *[here]* to the same degree as almost all your earlier scientific works. The reason for this lies without doubt only in the widespread view in Germany about the fact itself. You see, the explanation given by Faraday[6] has long passed into the physics textbooks here, and had never been shaken by the theory later put forward by Forbes.[7] In the judgment of my friend Zamminer,[8] from whom the short item in the annual report for 1854[9] stems, it therefore did not in any way require such a great effort to disprove what—as he was justified to believe after his experiences in Germany—nobody at all had thought to be correct.

Incidentally, quite apart from the dispute, the diverse perception of the Trevelyan experiment has quite certainly been of general interest; it is also important that conducting the experiment with non-metallic bodies (underlays) was first carried out by you. In comparing Seebeck's investigations[10] with yours, the latter [had] unfortunately remained unnoticed only because of the reason stated previously. Should it be in your interest to make use publicly of this my present comment, then this would be sufficient occasion for consideration of it to be taken in the next annual report as well.

WHEN I consider how well grounded your views are with regard to the origin of the phænomenon described by Trevelyan,[11] I am surprised to find, with reference to another of the phænomena of heat, namely the experiment of Leidenfrost,[12] that the theory of a repulsive action at a distance by heat seems to find favour in your eyes. At least I must conclude so much from a casual remark in a memoir upon the subject by Mr. A. Church.[13] Mr. Church mentions an experiment, which he witnessed in your lectures,[14] by which you show that water in the spheroidal condition interrupts the passage of an electric current to the red-hot metal underneath,[15] and observes that you regard this experiment as a proof that the surface of the vessel is separated by an interval from the drop within it.

The same experiment has been often made and observed under various conditions by myself, but I am not able to ascribe to it the same force of proof. I only find that the passage of electricity through water in the spheroidal state is rendered extremely difficult, and solely because the points of mutual contact between water and metal are reduced to a comparatively small number. The current can thereby certainly to all appearance be interrupted; whether this occurs or not, depends, according to my experience, for a given liquid, on the delicacy of the galvanometer, on the strength of the electromotive force, and finally on the size of the drop itself. The spheroidal water behaves towards the hot surface as mercury towards a wire dipped into it, which is not wetted by the liquid metal. In a memoir published several years ago (*Annalen der Chemie und Pharmacie,* vol. lxxvii. p. 1),[16] I have communicated different modes of experiment, which are, moreover, very easily devised, for the purpose of rendering visible the propagation of an electric current through water in the spheroidal condition. The most evident, however, as I learned afterwards, is the electric glow first observed by Mackrell[17] on thin metallic wires, when they are used as electrodes of a very powerful electric circuit and immersed in hot water. In the case of the negative pole, it might be objected that during the continuance of the red heat it is surrounded by a brush (*sprühlicht*); this, however, is not necessarily the case with the positive wire, and when it is absent, the wire exhibits at a red heat the phænomenon of Leidenfrost perfectly. The current, and even the electrolytic decomposition, also continue; for when the positive electrode disperses no brush, the gas developed upon it is always pure oxygen. The continuance of the current in this and similar cases is perhaps not a certain proof of the existence of direct contact; for the layer of gas, or vapour between the water and the red-hot surface, may at this high temperature assume a small degree of conductibility. I am, however, only desirous of showing, that the extinction of the current in other cases cannot be regarded as a proof of the existence of an interval between the water and the surface by which it is sustained.

Indeed, the spheroidal condition of water exhibits in its most essential phænomena sufficient evidence against the idea that insulation and repulsion are necessary conditions of the spheroidal state. Permit me to lay before you briefly those proofs which I regard as the most important.

In a silver spoon coated with soot, water asserts the spheroidal form at ordinary temperatures, exactly as when the spoon is raised to a high temperature.

The drop which rolls upon the hot surface of a liquid, for example a drop of æther upon water, bends the surface of the liquid, as a mercury surface is bent by the pressure of a rod of glass.

Large drops assume an oscillatory motion, which often continue with remarkable regularity, and cause the formation of star-shaped figures: without a partial contact of the water with the surface beneath, the production of these figures would be an impossibility.

The greatest drops are not obtained upon such surfaces as radiate heat best, but on those which possess the highest capacity of conduction; surfaces of silver are therefore preferable to all others for these experiments.

The phænomenon is produced with volatile liquids only; the higher the boiling-point of the liquid, the more strongly heated must be the surface on which it rests, in order to hold it in the spheroidal state. The experiment therefore succeeds better with æther than with water when bad conductors are used as vessels, such, for example, as porcelain and glass. The lowest temperature at which the heated surface is not wetted by the liquid poured upon it, must always be higher than the boiling-point of the liquid. For example, in a silver basin at 75°C. æther, at 137°C. alcohol, and at 144°C. water first assume the spheroidal form.

In making these observations, I will not assert that Leidenfrost's drop cannot be separated from the heated surface which supports it. I believe, on the contrary, that even drops of considerable magnitude may be torn from the surface by the force of the developed vapour. I only affirm that such insulation is not the necessary condition for the production of the phænomenon. This is caused, in my opinion, by the alteration produced by heat in the relative attractions exerted by the surface of the vessel upon the liquid, and by the liquid particles upon each other.

The surface of a vessel will, it is known, be wet by a liquid poured into it when the attraction of the matter of the vessel for the liquid is greater than that of the particles of the liquid for each other. In the reverse case the liquid asserts the spheroidal form. The mutual attraction of the particles of the body is without doubt diminished at a high temperature. If a silver basin be heated with the water within it, the surface continues wet even when the liquid boils. In fact, in this case the temperature of the silver basin is only a little higher than that of the water. The mutual attraction of the silver and the particles of water has therefore been diminished almost in the same proportion as the attraction of the water particles for each other; the relation of the attractions remains constant. If the silver basin be heated beyond the temperature of boiling water, its adhesion for water must diminish in a greater degree than the cohesion of the particles of the liquid which still remains at its boiling-point. Hence sooner or later a temperature will be attained, at which the cohesion overpowers the adhesion. The wetting of the surface then ceases, and the spheroidal condition sets in, quite in harmony with the ordinary laws of capillarity.

Perhaps, my dear Sir, you will be inclined to agree with this manner of viewing the subject; if so, I entertain the hope that you will excuse this somewhat elaborate statement of the question. With great esteem, I have the pleasure to subscribe myself,

Yours most sincerely, | H. BUFF. | Giessen, Sept. 29, 1855.

RI MS JT/1/E/106
Phil. Mag., 10 (1855), pp. 350–3[18]

---

1. *Heinrich Buff*: A student of Liebig's, Buff (1805–78) was professor of physics at the University of Giessen. He published extensively on both physics and chemistry, many of his research papers being on electrical topics.
2. *Your new paper on diamagnetism*: Tyndall's Bakerian Lecture (see letter 1029, n. 2).
3. *views . . . of W. Weber on diamagnetism*: W. Weber, 'Ueber den Zusammenhang der Lehre vom Diamagnetismus mit der Lehre von dem Magnetismus und der Elektricität', *Poggend. Annal.*, 87 (1852), pp. 145–89. Tyndall discussed Weber's theory in his Bakerian Lecture, pp. 38–40.
4. *Ampère's molecular currents*: see letter 1139, n. 8.
5. *your investigations into the Trevelyan instrument*: Arthur Trevelyan (1802–78) published 'The Method of Using the Apparatus for Exhibiting Vibrations Caused by Heat, 1829', which was printed in Edinburgh in 1850 together with two other essays by Trevelyan. On 27 January 1854, Tyndall referred to Trevelyan's instrument in his Friday Evening Discourse at the RI ('On the Vibration and Tones Produced by the Contact of Bodies Having Different Temperatures', *Roy. Inst. Proc.,* 1 (1851–54), pp. 356–59) and also in the paper he delivered at the RS ('On the Vibrations and Tones Produced by the Contact of Bodies Having Different Temperatures', *Phil. Trans.,* 144 (1854), pp. 1–10). See also n. 11, below, and letter 1149, n. 8.
6. *explanation given by Faraday*: In a lecture at the RI on 1 April 1831, Faraday explained the sounds by 'the tapping of the hot mass against the cold one underneath it, the taps being in many cases sufficiently quick to produce a high musical note': see Tyndall's discussion of the work of Faraday and Forbes (next note) in his RI lecture 'On the Vibration'.
7. *the theory later put forward by* Forbes: J. D. Forbes, 'Experimental Researches Regarding Certain Vibrations which Take Place between Metallic Masses Having Different Temperatures', *Transactions of the Royal Society of Edinburgh,* 12 (1834), pp. 429–61.
8. *my friend Zamminer*: Friedrich Zamminer (1817–58) was professor of mathematical physics at the University of Giessen.

9. *item in the annual report for 1854*: *Jahresbericht über Fortschritte der [. . .] Chemie, Physik, Mineralogie und Geologie für 1854* (Giessen: J. Ricker'sche, 1855), p. 113. Buff and Zamminer were among the editors of this report, which cited Tyndall's 'On the Vibration'.
10. *Seebeck's investigations*: E.g. A. Seebeck, 'Ueber die Erregung von Tönen mittelst der Wärme', *Poggend. Annal.*, 51 (1840), pp. 1–43; 'Beobachtungen über einige Bedingungen der Entstehung von Tönen', *Poggend. Annal.*, 53 (1841), pp. 417–35.
11. *phænomenon described by Trevelyan*: Trevelyan had contributed three papers on the vibrations produced by heated substances: 'Notice Concerning Certain Vibrations of Heat', *Edinburgh Journal of Science*, 6 (1832), pp. 141–43; 'Notice Regarding some Experiments on the Vibration of Heated Metals', *Transactions of the Royal Society of Edinburgh*, 12 (1834), pp. 137–46; 'Further Notice of the Vibration of Heated Metals', *Phil. Mag.*, 6 (1835), pp. 85–87.
12. *the experiment of Leidenfrost*: The work of Leidenfrost, discussed by Buff, was first published in Leidenfrost's *De Aquae Communis Nonnullis Qualitatibus Tractatus* (Duisburg: Ovenius, 1756), part of which has been translated by C. Wares as 'On the Fixation of Water in Diverse Fire', *International Journal of Heat and Mass Transfer*, 9 (1966), pp. 1153–66. 'Leidenfrost studied the boiling of small liquid masses on a hot surface. If the surface is hot enough, the masses are supported on a film of vapor generated by evaporation from the lower surface of the mass' (p. 1153).
13. *a memoir . . . by Mr. A. Church;* A. Church, 'On the Spheroidal State of Bodies', *Phil. Mag.*, 7 (1854), pp. 275–28; see also volume 4, letter 0902.
14. *an experiment, which he witnessed in your lectures*: Church referred to Tyndall's experiment proving 'the existence of a space between water in the spheroidal state and the containing vessel [. . .] by the complete interruption that space offered to the passage of a galvanic current' ('On the Spheroidal State', p. 275).
15. Footnote in published text: 'M. Poggendorff was, I believe, the first to perform this experiment. |—J.T.'
16. *a memoir . . .* : H. Buff, 'Ueber die Theorie des Leidenfrost'schen Versuchs und die Versuche von Boutigny', *Annalen der Chemie und Pharmacie*, 77 (1851), pp. 1–17.
17. *the electric glow first observed by Mackrell*: George Mackrell (b. c. 1796; listed as 'merchant' in 1841 census (Ancestry)), 'An Account of Experiments with a Constant Voltaic Battery, in which Metals were Ignited in an Acidulated Solution', *Proceedings of the London Electrical Society*, 1 (1841–43), pp. 6–9.
18. Beginning with the third paragraph ('WHEN I consider how well . . . '), the above is Tyndall's translation, as published in the November 1855 issue of *Phil. Mag.* with the title 'On the Experiment of *Leidenfrost.* Extracted from a Letter from Professor BUFF to Professor TYNDALL.' On pp. 353–54 Tyndall appended the following comments on Buff's letter:

   'With reference to the subject of Prof. Buff's interesting letter, I may be permitted to say that I have never published anything on the experiment of Leidenfrost, though I have taken some pains to instruct myself as to the cause of this singular phænomenon. At an early period of railway travelling, Mr. George Stephenson conceived the idea of making his locomotives rest on steam springs. The steam within the boiler formed an elastic cushion for the support of the boiler. My belief is, that the spring suggested by the great engineer is precisely that which supports the drop in the experiment of Leidenfrost; the liquid, I believe, rests upon a cushion of its own vapour. Some of my reasons for agreeing with those who hold this belief are as follows:—1. The incessant motion of the drop necessitates a continuous expenditure of force, and this force is only attainable by the production of vapour underneath the drop, which, as is well known, speedily diminishes in bulk in consequence of the evaporation. 2. In a deep concave vessel, the vapour escapes laterally from the liquid as it encounters the least resistance in this direction; but if the liquid be suffered to spread itself over a comparatively flat surface, or better still, if the surface be rendered a little convex towards the centre of the drop, the vapour finds the resistance in a lateral direction greater than that in a vertical direction, and the steam actually breaks through the thin liquid layer above it and escapes at the centre. The existence of the vapour beneath the drop is thus easily reduced to demonstration. 3. If an interval really exists between the drop and the surface, it occurred to me that if this interval could be made manifest to the eye it would tend to settle the question. With this view I made the following experiment:—A silver basin was inverted, and the polished convex surface slightly dinted so that a drop might rest upon it. A fine platinum wire was stretched vertically behind the basin and united with the poles of a small voltaic battery; the wire was in this way rendered vividly luminous. A drop of deeply black ink was placed upon the surface of the basin; bringing the eye on a level with the bottom of the drop, and looking towards the platinum wire, the latter was seen through the space between the drop and the bottom of the basin*. ([Footnote:] * A little spirit mixed with the ink renders the experiment more easy.) In this way the under surface of a drop, half an inch or more in diameter, may be examined, and it will be found that the liquid is at no point in contact with the metal. While, therefore, I agree with M. Buff in rejecting the explanation which attributes the phænomenon to a propping-up of the drop by the calorific rays, I think the facts justify the opinion that an interval exists between the liquid and the metal. This interval, as proved by Poggendorff, is sufficient to interrupt an ordinary voltaic current; but being filled with vapour at a high temperature, it is quite conceivable, as surmised by M. Buff, that a small portion of a strong current may pass through it. M. Buff's observation, that the phænomenon is not exhibited except by those liquids which are capable of furnishing the necessary vapour, is also confirmatory of the above view: and the fact that good conductors, which quickly supply the heat necessary for evaporation, show the phænomenon better than good radiators which do not furnish this supply, appears to be equally corroborative of the opinion which I have ventured to espouse.'

From Andrew Ramsay[1] 30 September 1855 1143

Edinburgh 30 Sept 1855

My dear Tyndall

Your note[2] reached me at Haddington[3] where I was extra busy by day & night.

I have never been on the spot you describe,[4] but I know something of the rocks on the banks of Lochlomond.[5] If not perfect Gneiss the rock you saw belongs to the Gneissic family,[6] viz Mica Quartz & felspar interlaminated in such twisted forms as those.

But there are many varieties of Gneiss and sometimes some one mineral is absent, scarce, imperfectly developed, or replaced by another.

The country you were in belongs to the great Gneiss & Mica schist district of Scotland. Gneiss & Mica schist often pass into each other. Unless I saw a specimen I could not positively name your rock.

By the middle or end of this week I expect to be in Wiltshire and Hampshire, one or both. If you are still there and want a ramble write to | 28 Jermyn St.[7]

I send this by Albemarle St[8] lest you should have left Queenwood

Ever sincy | Andrew Ramsay

I spent two evgs with Samuel Brown at Haddington[9]—certainly the most wonderful converser I ever met. His description of Forbes[10] was unrivalled & unrivalable.

RI MS JT/1/R/3

---

1. *Andrew Ramsay*: Andrew Crombie Ramsay (1814–91) was a Scottish geologist who served as director of the Geological Survey of the United Kingdom and held the chair of

geology at the Royal School of Mines in London. They had first met at a Red Lions dinner in January 1853 (RI MS JT/2/13b/596).

2. *Your note*: note missing.
3. *Haddington*: The historic Royal Burgh of Haddington is 20 miles east of Edinburgh.
4. *the spot you describe*: A location—probably near Loch Lomond—visited during Tyndall's recent visit to Scotland.
5. *Lochlomond*: see letter 1137, n. 9.
6. *Gneissic family*: Gneiss is a metamorphic rock (i.e., rock produced by chemical and physical forces operating at depth, such as pressure, heat, or volatile content) composed, like granite, of quartz, feldspar or orthoclase, and mica, but distinguished from it by its foliated or laminated structure (*OED*).
7. *28 Jermyn St*: As director of the Geological Survey Ramsay was located at the Museum of Practical Geology in Jermyn Street.
8. *Albemarle St*: the RI.
9. *Samuel Brown at Haddington*: The maverick Scottish chemist Samuel Brown (1817–56) was born at Haddington, 20 miles east of Edinburgh. 'His [Brown's] conversational powers, in particular, were more than remarkable' (Anon., Review of 'Experiments on Chemical Isomerism [etc.]', *North British Review,* 26 (1856–57), pp. 376–406, on p. 400).
10. *His description of Forbes*: Brown's reflections on the death of Edward Forbes were published in ibid., pp. 380–82.

From Michael Faraday 6 October 1855 1144

Sydenham | 6th Octr 1855.

My dear Tyndall

I was put into a very mixed mood by your last letter;[1] glad to hear from you, that you were out of the turmoil, had enjoyed the beauties of the Lakes, and was happily at home again: but sorry for some annoyances which I saw you had met with at Glasgow.[2] These great meetings, of which I think very well altogether, advance Science chiefly by bringing scientific men together and making them to know and be friends with each other; and I am sorry when that is not the effect in every part of their course. I know nothing except from what you tell me, for I have not yet looked at the reports of the proceedings:[3] but let me as an old man, who ought by this time to have profited by experience, say, that when I was younger I found I often misinterpreted the intentions of people, and found they did not mean what at the time I supposed they meant and further, that as a general rule it was better to be a little dull of apprehension when phrases seemed to imply pique and quick in perception when on the contrary they seemed to convey kindly feeling. —The real truth never fails ultimately to appear, and the opposing parties are, if wrong, sooner convinced when replied to forbearingly than when overwhelmed. All I mean to say is that it is better to be blind to the results of partizanship, and quick to see good will. One has more happiness in oneself, in endeavouring to follow the things that make for peace. You can hardly imagine how often I have been heated in private when opposed, as I have thought, unjustly and superciliously and yet have striven and succeeded I hope in keeping down reply of the like kind: and I have I know never lost by it. I would not say all this to you if I did not esteem you as a true philosopher and friend.

I have not been altogether idle but I am of necessity very slow now. I have not read the journals, because, when able I have been at work and writing a paper.[4] The latter goes on slowly, but I think will be a useful contribution of facts; and may help to advance the logic of magnetism a little though not much. The secret of Magnetic action is like a Sebastopol[5] at least in this point that we have to attack it in every palpable direction and make our approaches closer and closer on all the sides by which we can force access. My working is mainly with magnecrystals and the effects of heat on them.

You say it is still 'here' but do not say where the 'here' is; and my memory is so treacherous that now I have written this letter I shall not be able to send it until I go to London. But whenever you get it believe me to be as ever

Yours Very Truly | M. Faraday

Profr Tyndall | &c &c &c

My wifes best remembrances.

Institution of Engineering and Technology, SC MS 2/1/6
Faraday Correspondence, 4:3027

---

1. *your last letter*: letter 1138.
2. *turmoil . . . at Glasgow*: As Tyndall recounted in letter 1138, he found the recent BAAS meeting in Glasgow very demanding, the papers rather disappointing and he had been troubled by his altercation with Thomson.
3. *reports of the proceedings*: that appeared in the daily and weekly press.
4. *writing a paper*: Faraday, 'Experimental Researches in Electricity.—Thirtieth Series. Constancy of Differential Magnecrystallic Force in Different Media. Action of Heat on Magnecrystals. Effect of Heat upon the Absolute Magnetic Force of Bodies', *Phil Trans.,* 146 (1856), pp. 159–80.
5. *like a Sebastopol*: The siege of Sevastopol during the Crimean War lasted almost a year . The British, French and Ottoman forces made numerous attacks before the city was eventually taken on 6 September 1855.

From Charles Anderson[1] 9 October 1855 1145

Royal Institution Oct. 9th. 1855.

Sir,

On Saturday evening I received 5 iron bottles from Mr Newman,[2] and yesterday I made a cylinder of bismuth in a glass tube 3 inches long. I then cut it into 2 pieces of 1½ inches each, I put 1 of the pieces into one of the iron bottles filled it with water. Screwed it up and put it into a freezing mixture in a very few minutes there was a pretty loud explosion.

When I looked the bottle was broke into 5 pieces, I allowed the bismuth to remain in the freezing mixture till all the ice was dissolved. I then took it out dried it and with the other piece I send of to Dr Frankland the same time I post this

I remain[3] | Your humble servant | Charles Anderson

RI MS JT/1/12/4207
LT Typescript Only

1. *From Charles Anderson*: It would appear that this note from Anderson was sent to Tyndall via Faraday, possibly as an enclosure.
2. *Mr Newman*: John Newman (c. 1783–1860), optician and instrument maker of 122 Regent Street who frequently supplied apparatus for the RI.
3. *I remain*: followed by '(Sir)' in typescript, but probably added by LT.

From Edward Hughes 13 October 1855 1146

Greenwich Hospital | 13th Oct. 1855.

My dear Sir,

I shall feel greatly obliged to you if you will let me know whether you have done anything towards the article for the 3rd. Book[1] viz 'Undulatory Motion and Sound' I am under the impression that after I wrote to you from Edinburgh[2] you commenced the 'Article' and I trust by this time it is completed, as the sale of the first two Vols. encourages me to bring out the Third Vol. immediately—I have all the Articles for this Vol. in type except yours and Dr. Frankland's.

I enclose a re-set-up of the 'Broad Sheet' from which you will see that there are no less than 25 different contributors to the first two Vols.[3] Your wishes expressed in your last note[4] shall be attended to—hoping to hear from you shortly

Believe me | Yours faithfully | E. Hughes.

Dr. Tyndall, F.R.S.

RI MS JT/1/2/649
LT Typescript Only

1. *the 3rd. Book*: Tyndall contributed a chapter on natural philosophy covering 'Undulatory Motion' and 'Sound' to E. Hughes (ed), *Reading Lessons, Advanced Series. Third Book* (London: Longman, Brown, Green & Longmans, 1856), pp. 238–66.
2. *I wrote to you from Edinburgh*: letter missing.
3. *'Broad Sheet'... first two Vols*: missing. However, the list of contributors to all five volumes of *Reading Lessons* bound into some copies contains nearly fifty names.
4. *your last note*: note missing.

From Edward Hughes 17 October 1855 1147

Greenwich Hospital | 17th Oct. 1855.

My dear Sir,

Dr. Frankland is now at work on the Article for the Third Volume,[1] he has referred me to works for wood-cuts—If you will kindly do the same I can then let you have proofs of the Cuts with your letter-press—I have sent the 1st and 2nd Vols. to the 'Philosophical Mag.' for review[2] as you wished.

Yours faithfully | G. Hughes
Dr. Tyndall.

RI MS JT/1/2/650
LT Typescript Only

1. *Dr. Frankland... Article for the Third Volume*: Frankland contributed six lessons on chemistry to E. Hughes (ed), *Reading Lessons, Advanced Series. Third Book* (London: Longman, Brown, Green & Longmans, 1856), pp. 139–81.
2. *'Philosophical Mag.' for review*: Hughes's *Reading Lessons. First Book* (London: Longman, Brown, Green & Longmans, 1855) was reviewed in *Phil. Mag.,* 10 (1855), pp. 437–41. The anonymous review of Tyndall's chapter began: 'Professor Tyndall's [lessons], on Natural Philosophy, are in every respect admirable. The style is attractive, lucid, and vigorous' (p. 440).

From Henry Moseley 18 October 1855 1148

PrivateOlveston Vicarage | Nr. Bristol | 18 Oct.

My dear Sir,

With reference to the Examination for the R[oyal]. Artillery and Engineers to take place after Xmas,[1] I am desirous to consult you in respect to the appoin[tmen]t of a colleague to yourself in the department of the Natural

Sciences. The whole of the experimental sciences must I apprehend fall to your share if you are good enough to take the office again, but you will want a colleague as it regards the sciences of observation, moreover it will I think be considered necessary that each should so far cooperate with the other in the examination as to verify its results in regard to each Candidate. What I mean is, that whilst you set the questions only in the experimental sciences you should nevertheless look over those in the sciences of observation as well as your own; and that your colleague besides his own should look over your papers. I apprehend that the elementary character of the examination will make this arrangement practicable and it seems the only means of securing the guarantee which the public will I think require of two opinions on the merits of every Candidate. Your colleague ought to be a man of recognised position and eminence. The most eminent man indeed that we can get.[2] It will be rather desirable than otherwise that he should come from a distance. And as we have no examiner from Scotland, He might with advantage perhaps come from thence; or from Ireland. I have heard much of Prof. McCoy of Belfast.[3] Would he be a good man?

Yours faithfully | Henry Moseley.

RI MS JT/1/3/898
LT Typescript Only

1. *the Examination . . . Xmas*: The second competitive examination for entry to the Royal Artillery and Royal Engineers was held at Burlington House, London, from 21 January to 1 February 1856. The examination on 30 January covered 'Experimental sciences—Chemistry, Heat, Electricity, including Magnetism'; on 31 January it was 'Natural sciences—Mineralogy and Geology' (*Southern Reporter and Cork Commercial Courier,* 10 January 1856, p. [1]).
2. *The most eminent man indeed that we can get*: After consulting with Francis and Arthur Henfrey, Tyndall recommended Huxley, who was subsequently appointed (Tyndall's Journal: RI MS JT/2/13c/796 and 808).
3. *Prof. McCoy of Belfast*: Frederick McCoy (1823–99) had served as professor of mineralogy and geology at Queen's College, Belfast. However, in 1854 he was appointed to the chair of natural sciences at the newly founded University of Melbourne, Australia.

From Edward Frankland 22 October 1855 1149

Owens College | Oct. 22/55

My dear Tyndall

I have made careful determinations of the Sp gravities[1] of the two specimens of Bismuth with the following results:

N0.1. Bismuth submitted to prepare Sp. Gr. 9.8501.

N0.2. Bismuth not submitted to prepare Sp. Gr. 9.8468.

I do not however think that you can draw any safe conclusion from this isolated experiment because I find that the errors of observation might amount to .0035 which is rather more than the difference between the two specific gravities.

Although I could easily weigh to the 1/10 milligram in air yet I found it impossible to estimate less than ½ milligram in water, hence the liability to the above error. If the pieces were nearly ten times as heavy the error might probably be reduced to 1/7 or 1/8 of the above amount.

Although I was 10 days in Bristol I found it impossible to get over to Queenwood. The more I think of that British Association[2] the more am I convinced of the deep injury it annually inflicts upon science; with the exception of the Red Lion Dinner[3] and Loch Katrine[4] which are the two bright gems of the occasion I look back upon the whole thing with unmitigated regret. When we go to the Tyrol next year[5] we must contrive to take in the Naturforscher-versammlung in Vienna[6] as I should much like to see how they manage those things in Germany. On my way back to Windermere I bought the Lady of the Lake[7] & have since read it through with enthusiasm.

I hope you will seriously think about taking up your quarters at Windermere next Summer. You will find on enquiry that you can travel, to & from London 3rd class by fast trains, this privilege is accorded to all stations North of Lancaster. I feel sure you would find the locality very congenial for thought, work, and recreation; I have facilities there for performing any chemical operation of moderate character, so that you need be at no loss in that respect, & we could seize a fine day now and then for an aquatic or mountain excursion, when I will give you full liberty to pour out your poetical spirit to the most boundless extent. Do think seriously of the matter.

You once promised me a 'rocker'[8] made on your improved plan; could you direct Anderson to send me one down during this week together with a few square inches of German Tinder[9] for igniting Iron Wire in Oxygen, I cannot get it here.

Very sincerely yours | E. Frankland.

RI MS JT/1/F/44

1. *Sp gravities*: specific gravities. In an earlier (missing) letter Tyndall had presumably asked Frankland to determine the specific gravities of two specimens of bismuth that he was using in his experiments on diamagnetism. The magnetic properties of specimens of bismuth under different degrees of compression—hence, different specific gravities—were of interest to Tyndall, Faraday and a number of their contemporaries: see Tyndall's Bakerian Lecture (letter 1029, n. 2) and letter 1054.

2. *that British Association*: the recent meeting in Glasgow of the BAAS.
3. *the Red Lion Dinner*: The Red Lion Club was one of the 'Fringe groups [at the BAAS] devoted to scientific comradeship and disciplinary specialization', in this case botany and zoology. It 'was started by some young naturalists who were unhappy with the expense and oligarchy of the [1839] Meeting at Birmingham. They dined daily at the Red Lion' (J. Morrell and A. Thackray, *Gentlemen of Science: Early Years of the British Association for the Advancement of Science* (Oxford: Oxford University Press, 1981), p. 128). Frankland and Tyndall attended the Red Lion dinner in Glasgow on 18 September 1855. 'Had a most pleasant and cordial meeting of the Red Lions. Monckton Miles was in the Chair, beside him was a fine old Swedish Count, he was rather amazed when he first heard the Lions roar. Gassiot, Ramsay, Lankester, Best, Col. James, George Nelson, David Forbes and others. The evening passed off in a very hearty genial manner' (Tyndall's Journal: RI MS JT/2/13c/791).
4. *Loch Katrine*: A freshwater loch (lake) in the Trossachs, 30 miles north of Glasgow.
5. *the Tyrol next year*: 'The Tyrol [. . .] is the most westerly of the German territories of the empire, and borders upon Bavaria and Switzerland. It is also the most lofty and rugged of all the alpine regions of Austria' (H. Murray et al, *The Encyclopædia of Geography,* 3 vols (Philadelphia: Lea & Blanchard, 1840), vol. 2, pp. 101). On 12 June 1856 Tyndall departed for the Continent, including a visit to the Tyrol.
6. *Naturforscher-versammlung in Vienna*: see letter 1087, n. 10. Tyndall attended the meeting, which ran from 16 to 22 September 1856.
7. *the Lady of the Lake*: see letter 1138, n. 17.
8. *rocker*: A rocker is a metal block or bar having one surface convex with a groove along its length, made in order to demonstrate the ability of heat to produce audible vibration (*OED*). Tyndall had first experimented on rockers while in Berlin in August 1853 (RI MS JT/2/13b/613). See letter 1142, n. 5, for Tyndall's lectures and publications on rockers.
9. *German Tinder*: Amadou or German tinder, is prepared from species of fungus *Polyporus* and *Boletus,* that grow on trees, employed as a match and a styptic (*OED*).

From Henry Clifton Sorby[1] 30 October 1855 1150

Broomfield | Sheffield | Oct. 30th /55

My dear Sir,

I trust to be able to send you a paper on the cleavage of limestones[2] before very long. I have not been at home above a few weeks, and have been otherwise engaged; but now will set to work and write it. I have drawn some illustrations for it on the wood blocks, and shall send them up to London to be cut to-day. I shall perhaps order the engraver to let you have them without first sending them here. If so, where should they be sent? Please tell me.

My paper will be almost entirely confined to the question 'is slaty cleavage the result of a mechanical crystalline cause?'

I should be much obliged if you will inform me what kind of charts those were that you communicated to the Brit. Ass. from Captain Fitzroy.[3] Are they in detached sheets, so that I could purchase them for the seas round Britain? Is there any catalogue of them, that I could get to see? You know perhaps that the branch of geology that I am studying requires me to know a good deal of the actual directions of the winds of the present period.

Trusting to have the pleasure of seeing you when I am next in town.

I remain | Yours very truly | Henry Clifton Sorby.

RI MS JT/1/4/1344
LT Typescript Only

---

1. *Henry Clifton Sorby*: The geologist Henry Clifton Sorby (1826–1908) made extensive use of the microscope and spectrascope to study the structure of rocks and crystals.
2. *a paper on the cleavage of limestones*: H. C. Sorby, 'On Slaty Cleavage, as Exhibited in the Devonian Limestone of Devonshire', *Phil. Mag.,* 11 (1856), pp. 20–37.
3. *charts . . . from Captain Fitzroy*: The naval commander and meteorologist Robert Fitzroy (1805–65), who had commanded *H.M.S. Beagle* during Charles Darwin's voyage, had presented 'Wind-charts of the Atlantic, Compiled from *Maury's* Pilot Charts' at the 1855 meeting of the BAAS (*Brit. Assoc. Rep. 1855,* pp. 39–40). This published report makes no mention of Tyndall's role.

From Thomas Archer Hirst 4 November 1855 1151

Broughton | Nov 4th / 55

My dear John.

I have been engaged all the week in my spare time in reading Faraday's 3rd vol.[1] I find it exceedingly tough work, though exceedingly instructive. In experimental descriptions Faraday is admirably lucid, but in Theoretical matters decidedly obscure. There is an absence of mathematical strictness in his explanations which makes them exceedingly unsatisfactory[.] It is impossible to appreciate these explanations without fathoming, in some manner, his notions of lines of force, and he talks of them so loosely, now giving them a kind of physical reality, and again a merely representative character that it is exceedingly difficult to avoid confusion. In order to understand the very simplest action of a diamagnetic particle in the magnetic field I tried to get a correct interpretation of that often quoted passage on page 33 'its tendency is to move from stronger to weaker places of magnetic force'.[2] The first question is what is meant by 'the strength of a place of magnetic force'[3] and how is it estimated; the second question is, what is the nature of the mutual action between the particle and the magnet which gives to the former this tendency, and what is the state of the particle which causes such mutual action. For this tendency is a mere effect uninteresting in comparison to the state of the particle which causes it. Yet strange to say Faraday often speaks as if this tendency were a final cause[4] of phenomena. Well now to answer the first question we must examine the definition of lines of force in their representative character merely. The clearest definition I find on pages 328 and 329.[5]

Lines are to be drawn from a magnet in directions indicated by a small magnetic needle moved so as always to be a tangent to its line of motion. These lines are supposed to proceed in a certain direction indicated by the direction in which a marked pole of the needle points. Thus each line represents the direction of the magnetic force at any point. It is also conceived to represent a constant intensity, so that the 'strength of any place of magnetic force' is to be estimated by the number of lines which pass through any space of constant area which we may suppose spherical in form. I conceive Faraday's notion to be something of this kind, though I cannot yet see how we can adapt his notion to mathematical expression.

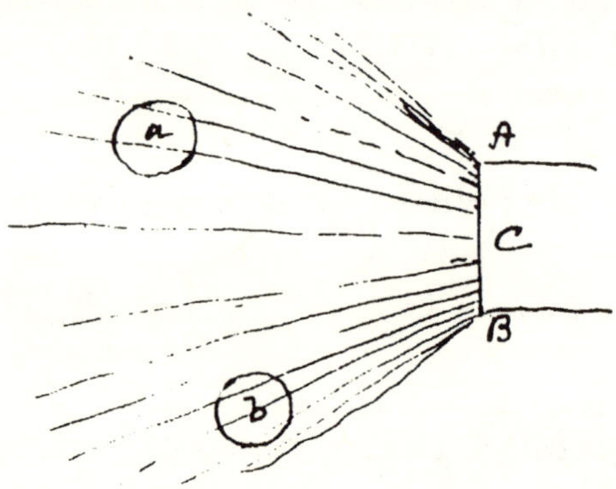

Consistently with Faradays notion, after having settled arbitrarily how many lines of force are to be drawn from a given magnet into the surrounding space, they must be distributed along the face of the magnet according to the intensity of the part from which they start, that is to say, more lines must be drawn from A and B than from C according as the magnetic intensity at A & B exceeds that at C. Now let a and b be two spheres at different parts of the field having equal radii then the comparative strength of these places is to be considered proportional to the number of lines which intersect the spheres. If instead of an extended magnet a magnetic point be taken this law will be found to coincide with the law of the inverse square of the distance.

According to Faraday's law of paramagnetic and diamagnetic action the sphere a[,] if paramagnetic and free to move, would change its position so that the greatest number of lines should pass through it and if diamagnetic so that the least possible number should pass through it. Understanding Faraday thus I turn now to the experimental verification of his law. I take his simplest experiment with two pointed poles and an elongated piece of bismuth suspended in the axial line midway between them. I imagine then, roughly, the nature of the lines of force between the poles (a very uncertain way of proceeding I know) which for simplicity I consider to be mere points one possessing N the other S polarity. Between such points the bismuth sets equatorially and the action appears to agree with the law that every particle of bismuth strives to move from stronger to weaker places of force.

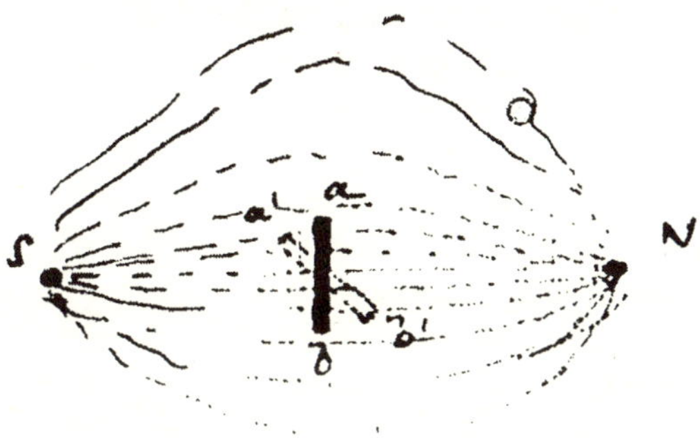

For it appears probable that a is a weaker place of force than a' and b than b'.

But with a single pole the explanation of the action, which is similar, is not so easy. For if a' b' is the original and ab the final position then although all particles between a' and c have moved from places of stronger to places of weaker force those between b' and c have moved from places of <u>weaker</u> to places of <u>stronger</u> force.

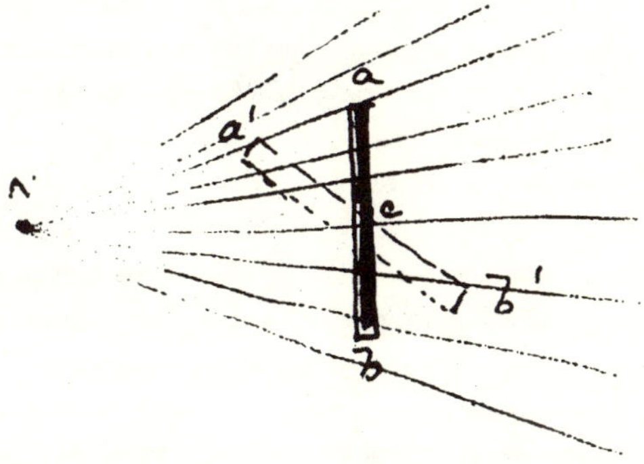

Faraday never mentions that the tendency of the particles between a' and c is stronger than the <u>anti</u> tendency of the particles between b' and c.

Indeed I am not sure that he would admit this for on page 344 he says 'The amount of magnetic is determinate for the same lines of force, whatever the distance of the point or plane, at which their power is exerted is from the magnet.'[6] On page 349 he says 'The relative amount of force, or lines of force in a given space is indicated by their number in that space.'[7] If so then a glance at the last two figures will shew that the amount of force in the space occupied by the bar in its final position is greater than the amount in any other position for more lines of force cross it. The whole bar then has moved in both cases from a place of weaker to a place of stronger magnetic force. Is this not curious? At any rate it shews that although his law may have led Faraday into no error it is an extremely obscure law not at all fitted to define accurately the action of diamagnetic bodies in all circumstances or rather it is a law very easily misunderstood and not very easily understood.

How are you suited with lodgings? How is your health? This week's close reading has had its usual effect upon my bowels[8] from which I am only just recovering. Anna joins me in love.

Yours affectionately | T A Hirst

I have written this letter for my own benefit rather than yours in order to force myself to close thoughts on the subject.

RI MS JT/1/H/219

1. *Faraday's 3rd vol*: The third volume of Faraday's *ERE* was published by Taylor and Francis in early April 1855 (*Publishers' Circular,* 18 (1855), p. 151). This volume contained Faraday's papers published in the *Phil. Trans.* between 1846 and 1852, together with his contributions to *Roy. Inst. Proc.* and *Phil. Mag.*
2. *its tendency is to move from stronger to weaker places of magnetic force*: The context is Faraday's examination of the behavior of samples of heavy glass in an electromagnetic field. The full paragraph reads: 'So, the indicating particle would move, either along the magnetic curves [produced by the electromagnet], or across them; and it would do this either in one direction or the other; the only constant point being that its tendency was to move from stronger to weaker places of magnetic force' (*ERE,* vol. 3, pp. 33).
3. *the strength of a place of magnetic force*: This is not a direct quotation from Faraday but a reformulation of 'stronger [. . .] places of magnetic force' and 'weaker places of magnetic force', in the passage quoted in n. 2.
4. *final cause*: The term 'final cause' (telos, Greek) is from Aristotelian philosophy and concerns the aim, purpose or endpoint of some action. The question whether natural phenomena possess final causes was controversial in the nineteenth century, especially after the publication of Darwin's theory of evolution in 1859. Hirst here detected a sense of purpose in Faraday's account of a particle's motion in a magnetic field.
5. *The clearest definition I find on pages 328 and 329*: The text of *ERE,* vol. 3, pp. 328–29

consists of the opening paragraphs of Faraday's Twenty-Eighth Series (published in *Phil. Trans.*, 142 (1852), pp. 25–56, on pp. 25–26), which includes: 'A line of magnetic force may be defined as that line which is described by a very small magnetic needle, when it is so moved in either direction correspondent to its length, that the needle is constantly a tangent to the line of motion; [. . .] or it is that line which coincides with the direction of the magnecrystallic axis of a crystal of bismuth, which is carried in either direction along it.'

6. *The amount . . . the magnet*: Faraday's text reads: 'The *amount* of magnetic force, as shown by its effect in evolving electric currents, is determinate for the same lines of force, whatever the distance of the point or plane, at which their power is exerted, is from the magnet' (*ERE*, vol. 3, pp. 344–45).
7. *The relative amount . . . that space*: In *ERE* Faraday referred to his earlier use of field diagrams to represent 'the idea of *lines of force*', in which 'the direction of the *line of force* is shown at once, and the relative amount of force, or of lines of force in a given space, indicated by their concentration or separation, *i.e.* by their number in that space' (vol. 3, p. 349).
8. *its usual effect upon my bowels*: 'One apparently trivial point brings me down terribly and that is constipation' (Tyndall's Journal, 1 November 1855: RI MS JT/2/13c/800).

To Thomas Archer Hirst 5 November 1855 1152

Monday

My dear Tom

Your letter[1] you say was intended for your own improvement, but there is a power in the world which overrules the designs of selfish men for the general benefit[.] I read your letter with great interest and pleasure. If you have sufficient time I have no doubt that you will make that review a truly serviceable and beautiful one. It is amusing to see how many write to Faraday asking him what the lines of force are. He bewilders even men of eminence, for the very fact of his making these lines of force the medium of his theoretic sight and his having done so much with them convinces the generality of people that they are the final cause of magnetic phenomena. Your remarks on this point I think very correct & very wise—The fact is that an experiment made with reference to the lines of force is virtually made with reference to the principle on which the lines of force themselves depend, and the deeper scientific mind looks through the lines of force after this principle. Faradays achievements are due to his immense earnestness and great love for his subject and this very mistiness which serves to obscure the verity of matters may have its compensations by rendering the subject attractive and thus wooing [?] a man to work at it with more fervour. I heard Biot once say that he could not understand

Faraday—& if you look for exact knowledge in his theories you will be disappointed—flashes of wonderful insight you meet here and there. But he has no exact knowledge himself, and in conversation with him he readily confesses this—In my next paper[2] I shall have to say something of these lines of force. Meanwhile it will be better for you to possess yourself of Faraday's notions as far as this is possible. We will talk the matter over afterwards.

My health thank the gods is excellent. I have kept my eye upon the bowels for some time and without betaking myself to physic have brought them into a good condition. Believe me Tom for the last three years I have not felt more full of hope and courage. I am in extremely comfortable lodgings[3] near the Institution & near Hyde Park. My address is 96 Mount Street Grosvenor Square.

I spent a pleasant day with Bevington yesterday—He desires to pay you a visit and I have volunteered to accompany him on the last Saturday in November[4]—I have ventured to promise him that you would be able to afford him a bed on Saturday & Sunday night—I will stow myself at Wrights or elsewhere. Goodbye Tom

& affectionately | John

RI MS JT/1/T/935

1. *Your letter*: letter 1151.
2. *my next paper*: J. Tyndall, 'Further Researches on the Polarity of the Diamagnetic Force', which was received by the RS on 27 November 1855, read at a meeting on 20 December and published in *Phil. Trans.,* 146 (1856), pp. 237–60, with a shortened version in *Roy. Soc. Proc.*, 7 (1854–55), pp. 555–58.
3. *lodgings*: On October 31, 1855, Tyndall moved into new lodgings at 96 Mount Street, about a third of a mile west of the Royal Institution (Journal: RI MS JT/2/13c/799).
4. *the last Saturday in November*: Tyndall went to Queenwood with Bevington on 1 December 1855 (Journal: RI MS JT/2/13c/807).

From Edward Frankland 13 November 1855 1153

Owens College | Manchester Nov. 13/55

My dear Tyndall

Many thanks for the rockers,[1] I used them in my lecture this morning & they performed very well, to the great delight of my students who were greatly amused by the attempts of the musical one to vary its tone.

You promised me a Bakerian lecture[2] at least half a dozen times but never yet fulfilled your promise. Send it by return of post like a good fellow along with that extract from your journal[3] which you promised my wife & for which she has been teazing me ever since.

You also promised me the 'Red Tape Worm',[4] between Sawry and Hawshead;[5] don't you recollect? A propos, we did not see the floating island on the Priest's Pot,[6] a friend of mine saw it a fortnight after we were there; it was pointed out to him by the farmer, who saw it blown across the lake last year & it is now partly stranded on the opposite side of the pot to the place where we thought it to be.

The Bristol affair[7] is still in abeyance and I feel my reluctance to make such a change increase rather than otherwise. I would not entertain the idea for a moment had I a desirable set of students, but to be eternally dragging mere children through the 5 groups[8] & no further is mere mechanical work, scarcely superior to breaking stones, and I feel it to be for me a waste of time. However this session I have got together a little knot of workers around me, who I hope will greatly relieve the monotony of my usual routine. viz. My two Assistants Dr. Guthrie[9] & Mr Wanklyn[10] (a former pupil of mine) Dr. Russell[11] & Dr. Muspratt[12] (not the great Sheridan[13] fortunately) but his youngest brother who has worked with Liebig[14] for several years. With Guthrie I am so far very well pleased he is an earnest & industrious follower of science, the very opposite in every respect to his predecessor.[15] He is working at Organo-metallic bodies. Russell is investigating the gases in the interior of different flames he finds that by using my machine he can get through a year's work in a couple of months & he will therefore work here during the Winter instead of returning to Heidelberg.

Wanklyn has formed Cadmium ethyl & is working at this & other organo-metallic compounds with great zeal & industry, & Muspratt comes to me next week to learn Assaying. I am also pitching Commerce to the devil as much as I can afford & working hard at my researches. I have applied to the Royal[16] for another grant this year, which, if I get it, will enable me to push on my experiments greatly.

My little wife has just increased the number of my olive branches by presenting me with a little girl last Saturday week.[17] Both mother & infant I am happy to say are as well as possible.

Believe me | ever sincerely yours | E. Frankland.

P.S. Our Red Lion friend Mr Jennings[18] sent me the copy of 'Groves of Blarney'.[19] He wrote from a London Hotel so I could not reply & thank him. If you ever see him pray present my best thanks & kind regards.

P.S. When would you like to have my letter on Bunsen's researches?[20] There is another matter connected with the Royal Society which I would earnestly lay upon your soul, as the Germans say. The other day Joule sent me a certificate to sign in favour of Mr James Young the Paraffin Oil man:[21] he is described as being eminent as a chemical manufacturer, the discover of several new manufacturing processes & as being attached to science & anxious to promote its progress. Now I know Young very well & also his so-called discoveries, of the latter, Paraffine & Paraffine Oil are the most celebrated & on them you already know my sentiments, but if you like I will send you the references to Reichenbach's memoirs[22] so as to enable you to judge for yourself. His other discovery is the Stannate of Soda process[23] which is nothing more than the repetition of a well known laboratory process on a large scale. It is in fact merely fusing the ore with caustic soda to render it soluble. On these two inventions rests all Young's so-called eminence as a chemical manufacturer; as to his being attached to science & anxious to promote its progress I say emphatically that he is entirely destitute of these qualities as they are understood in the Royal Society; his attachment & anxiety refer only to such portions of science as can be made available for putting money into his own pocket. Of course I returned his certificate to Joule[24] with as strong an expression of opinion as I have just made. Such admissions into the Royal Society will bring great discredit upon it especially in this district, where the true characters of such men are known.

With kindest regards & the compliments of the season in which my wife cordially joins

Ever yours sincerely | E. Frankland.

RI MS JT/1/F/45

---

1. *the rockers*: see letter 1149, n. 8.
2. *a Bakerian lecture*: a copy of Tyndall's paper cited in letter 1029, n. 1.
3. *that extract from your journal*: not identified.
4. *Red Tape Worm*: A humorous verse written by the recently-deceased naturalist Edward Forbes, but allegedly composed by a 'Government Clerk, who at an advanced age had awakened to the knowledge of the fallacy of the Superannuation Fund'. Parts of the poem were published in several contemporary periodicals, such as J. H. B., 'Biography of the Late Professor Edward Forbes', *Monthly Journal of Medicine*, 20 (1855), pp. 75–92, on pp. 87–88.
5. *between Sawry and Hawshead*: Sawrey and Hawkshead are villages in the Lake District that Tyndall and Frankland visited in early September before heading to the meeting of the BAAS in Glasgow.
6. *the floating island on the Priest's Pot*: Esthwaite Water, 'which is two miles long, and half a mile broad;—a quiet sheet of water, with two promontories stretching into it, which appear like islands, nearly dividing it into a chain of ponds. A round pond at the northern end of the lake, connected with it by a narrow creek, exhibits a strange phenæmenon. It has a floating island [. . .] actually bearing trees: and this island is carried by strong winds from the one side to the other. The name of the pond is Priest's Pot' (H. Martineau, *A Complete Guide to the Lakes* (Windermere: J. Garnett, 1855), p. 29).

7. *The Bristol affair*: Frankland had been undertaking consultancy work for Cox, Score & Co., a manufacturer of alkali and vitriol of Netham, Bristol, and presumably had been offered further employment.
8. *through the 5 groups*: 'In the systematic examination of the metallic oxides, the substances [. . .] are arranged in five groups, by the successive employment of HYDROCHLORIC ACID, SULPHURETTED HYDROGEN, AMMONIA, SULPHIDE OF AMMONIUM, and CARBONATE or AMMONIA' (J. F. Forbes, *First Steps in Practical Chemistry, for Agricultural Students, Farmers, and Bleachers* (London: Longmans, Brown, Green, Longmans and Roberts, 1858), p. 3.
9. *Dr. Guthrie*: Having studied at University College, London, and the University of Heidelberg, Frederick Guthrie (1833–86) received his PhD from the University of Marburg in 1855 and then worked as an assistant to Frankland at Owens College, Manchester.
10. *Mr Wanklyn*: James Alfred Wanklyn (1834–1906) had been apprenticed to a chemist in Manchester.
11. *Dr. Russell*: William James Russell (1830–1909) had studied chemistry at University College, London, and then worked as Frankland's assistant (1851–53), before moving to the University of Heidelberg where he obtained his PhD in 1855. He had then returned to Manchester.
12. *Dr. Muspratt*: The industrial chemist Edmund Knowles Muspratt (1833–1923), who had studied under Liebig and gained his PhD from the University of Munich before returning to the family alkali firm on Merseyside.
13. *great Sheridan*: A student of Liebig's at the University of Giessen, the Irish chemist James Sheridan Muspratt (1821–71), whose family owned a highly successful alkali manufacturing business on Merseyside, had founded the Liverpool College of Chemistry in 1841.
14. *Liebig*: see letter 1137, n. 8.
15. *his predecessor*: either C. J. Tufnell or B. Wilhelm Gerland.
16. *the Royal*: the RS, which received a £1000 p.a. grant from the government for the promotion of science. In 1855–56 Frankland received a grant of £100 for 'the continuation of his researches on Organo-Metallic Bodies' (*Return of the Application of the Sum of £1,000 Annually Voted to the Royal Society for Each Year since the Last Return; viz. the Year Ended 5 April 1855*, p. 1; House of Commons Sessional Papers, 1862 [192], XXX.651).
17. *My little wife . . . Saturday week*: Frankland's wife, Sophie, gave birth to a daughter, Sophie Jeannette, on November 3, 1855. Her middle name was taken from Jeannette du Bois-Reymond (see letter 1067, n. 4), Emil du Bois-Reymond's attractive young wife.
18. *Our Red Lion friend Mr Jennings*: Tyndall and Frankland met Thomas Jennings (dates unknown) a soda water and magnesia manufacturer from Cork at the Red Lion Club (see letter 1049, n. 3).
19. *Groves of Blarney*: an Irish melody by Richard Alfred Millikin or, possibly, the drama of that title by Anna Maria Hall.
20. *my letter on Bunsen's researches*: not located.
21. *Mr James Young the Paraffin Oil man*: The Scottish chemist James Young (1811–83) had

developed a method to produce paraffin oil by the distillation of coal. Joule had circulated a certificate in order to gain signatories in support of Young's admission to the RS.

22. *Reichenbach's memoirs*: The chemist and polymath Carl Reichenbach (1788–1869) discovered paraffin, which he discussed in several publications including 'Beitrage zur näheren Kenntniss der trockenen Destination organischer Körper', *Journal für Chemie und Physik,* 59 (1830), pp. 436–60; 'Ueber das Paraffin', ibid., 61 (1831), pp. 274–91.
23. *the Stannate of Soda process*: While working as a chemical engineer in Manchester, Young developed an improved method for producing sodium stannate (Na_2SnO_3), which was used in calico printing. At the 1851 Great Exhibition Young displayed 'Mineral oil. Paraffine. Stannates, with models of apparatus' (*Official Descriptive and Illustrated Catalogue,* 3 vols (London: Spicer Bros, 1851), vol. 1, p. 187).
24. *I returned his certificate to Joule*: It would appear that the certificate supporting Young's admission to the RS was not submitted in 1855. However, in 1873 Young was admitted, with Frankland's name at the top of his list of supporters.

To Thomas Archer Hirst 20 November 1855 1154

My dear Tom

I have read your article[1] as you wished though it dropped down upon me at a moment when I was very busy. Your health however concerns me more than your article. I should not, rash as I am, attempt to do what you did and attempt composition while perspiring in bed. I should make a sad mess of my composition under the circumstances. With regard to the visit I have no choice whatever of a time. Of course if I consulted myself I would not go at all so soon, it was Bevington's own proposal but there was nothing rigid in our agreement and I doubt not the first Saturday in December will be to him as pleasant as the last in November[2]—I will leave the matter between you and him, and shall be ready at any time you and he may agree upon.

And now for your article—I will tell you simply what I think—without waiting for qualifying expressions and you are to take all I say in good nature.

The first page or two dont strike me as being your best. The style is not I think the offspring of sober intellectual conviction, but makes rather an effort at being brilliant. The parallel with Newton I think is faulty. Newton grasped all that had been done before him on gravity & reduced it all to a simple law. Faraday has been the originator of new facts which some subsequent Newton must tie together. The phrase 'compared the effects of each' sounds bad to my ear. I think the style in these first pages and at some other places is more like that of a master to his disciples than that of a man telling to cultivated men what a book contains. Only a little change is required but a little will be of service. 'Of this kind is Faradays conception of Lines of Force'—you are not talking about 'kinds'—you mean I suppose to say that it is an illustration of what you have been stating. 'Committing ourselves to his or our own theoretic views' is a curious way of saying what you intend to say. What do you mean by 'fulfilling the sincere desires of Faraday himself?' I should rather say following the example of Faraday.

Why 'blackened at one side'—the blacking has nothing to do with the polarization—have you ever known it to be done?—The glass of your polariscope is, or ought to be, black glass. Any glass would polarise as well.—'Celebrated one'—I would say celebrated investigation. 'Knew the direction of the pencil' I would omit the word direction.—Aperture has not two ps I excuse Anna—it is rather a scientific word.—'Called the plane of incidence' 'Called the angle of polarization' You must manage to convey this knowledge without letting people know that you are teaching them. The phrase 'what is more' does for boys, not I think for review readers. 'A direct relationship between magnetism & light'—I doubt it. I would strike out 'direct'—You speak somewhere of a thing being important in two ways and still you have a 'thirdly'—You draw no clear line between the 'Rotation' and the discovery of diamagnetism. It ought I think simply to be stated that a piece of the same heavy glass suspended between the poles behaved so and so, and led to the discovery of this new force. Reich made no inference about polarity—he had no idea of the bearing of his own experiment. Weber applied it for him.—You are too explanatory about the attraction of a sphere of soft iron—don't forget—that your readers wont forgive you for assuming too much ignorance on their part They may be ignorant but to teach them requires a little tact and management. I know Mr Hopkins[3] well—a few short pithy lines, hitting the silly public which now demands a third edition of this stupid book[4] would be just the thing—You have dwelt too long on the book I think. Dont mention Baden Powell's lecture.[5] Your ought I think to speak of the book as if it were a disgrace to criticism to waste time upon it. 'To us this appears unfounded'—a very sober affirmation and one calculated to make a man say 'indeed!' after wading through such humbug. Both on Hopkins & Harrington[6,7] you have spent too much time. Had you not written that notice for the Phil. Mag. I dare say you would have said more about the Reading Lessons.[8] But you have I think said enough. Indeed now comes an unpleasant point viz.

The Review contains about 550 words to the page. Your notice is therefore far too long. Leave out or shorten immensely that long introduction to the paper on the rotation of the plane of polarisation. The experiment with the models in your allusion to me might just be mentioned, not described in full. I think the case described in the memoir I wrote in Berlin[9] of the crystal being caught between the forces of the magnet is a bit fit for a review. But you must not wander too far from your subject. You must keep close to 8 pages, and must calculate at the rate of 550 words a page. I know when you look over the notice once more, not while you are perspiring in bed—you will be able to make it shine like burnished silver. Lest you should be savage against me I send you a pretty sketch of Loch Lomond[10] as a peace offering. Indeed were I not so busy I would try and butter what I had to say a little. At present it is like MRS—[11] brown bread, coarse, but perhaps not unwholesome.

I have a german lamp[12] before me. I lighted it for the first time last night. It shed light upon my memory as well as upon my eyes and brings the living & the dead before me. Full as my brain was of magnetism I could almost have written a poem upon it. This I will do some day.

My health continues good. I won't read this over Tom though I know it is full of blunders. Just casting my eye on the outside page I saw the word too (the adverb) spelt two.[13] This reveals to me what a mess I am sending to you, but I know you are long suffering, and trust to the tenderness of those all-important flexible tubes—your bowels! I will see about the little scales when I have a little more breathing time.

Good bye | I kiss you & wish you well again | Your loving. John.

You say somewhere that I have 'corroborated' Weber's experiments. This I have not done but by a totally different course of experiment I have reached the same result. In my next paper[14] you will find the corroborations ample enough.

With regard to gases there is a great omission in the notice—you have alluded to Faraday's first experiments only. See note at page 219 and about the region 185.[15]

RI MS JT/1/T/612[16]

---

1. *your article*: A revised version of Hirst's article was published as part of 'Contemporary Literature—SCIENCE', *Westminster Review,* 65 (1856), pp. 254–61.

2. *not the first Saturday . . . November*: see letter 1152, n. 4.
3. *Mr Hopkins*: The Welsh geologist Evan Hopkins (1810–67), author of *On the Connexion of Geology with Terrestrial Magnetism* (London: R & J. E. Taylor, 1844).
4. *this stupid book*: 'Mr. Hopkins has inflicted upon us a new edition of his absurd book' ([Hirst], 'Contemporary Literature', p. 145). A second edition of Hopkins' book had been published in 1851 and a third (under review) in 1855.
5. *Baden Powell's lecture*: Possibly *Essays on the Spirt of the Inductive Philosophy, the Unity of Worlds, and the Philosophy of Creation* (London: 1855) by Baden Powell (1796–1860), the Savilian professor of geometry at the University of Oxford who wrote on both scientific and theological topics.
6. *Harrington*: Probably *New Theories of Light and Heat* (London: printed by W. Tyler, 1855) by a dentist named George Fellows Harrington (1812–95).
7. An asterisk at this point in the letter is linked to a footnote: 'let the hypothesis stand—it is rich!' This may refer to another Harrington, the sceptic in *Eclipse of Faith; or, a Visit to a Religious Sceptic* (London: Longman, Brown, Green and Longman, 1852) by Henry Rogers, a Congregationalist who held a chair at Spring Hill College (Birmingham). Although the phrases in Tyndall's footnote do not appear in Rogers' book, this Harrington entertains a number of theological hypotheses. John Henry Newman had challenged Rogers' speculations.
8. *the Reading Lessons*: Hirst briefly reviewed *Reading Lessons, First Book* and *Reading Lessons. Advanced Series. Second Book*—both edited by E. Hughes and published by Longman, Brown, Green & Longmans, 1855): 'We do not say that his [Hughes's] efforts have been crowned with complete success, but the list of writers whom he has engaged is sufficient guarantee for the general excellence of the lessons' ([Hirst], 'Contemporary Literature, p. 261).
9. *the memoir I wrote in Berlin*: J. Tyndall and H. Knoblauch, 'On the Deportment of Crystalline Bodies between the Poles of a Magnet', *Phil. Mag.*, 36 (1850), pp. 178–83 and 37 (1850), pp. 1–33.
10. *a pretty sketch of Loch Lomond*: sketch at head of this letter.
11. *Mrs*—: probably Mrs Wright, Tyndall's and Hirst's Halifax landlady.
12. *a german lamp*: not identified.
13. *the word too (the adverb) spelt two*: Tyndall subsequently altered 'You are two explanatory . . .' —in para. 4—to 'You are too explanatory . . . '
14. *my next paper*: see letter 1152, n. 2.
15. *See note at page 219 and about the region 185*: In the note appended to *ERE,* vol. 3, p. 219, Faraday discussed his and Edmond Becquerel's work on the magnetic nature of oxygen. Faraday offered experiments on the diamagnetic properties of gases 'about the region [of] 185'.
16. This letter was written on headed paper showing 'Loch Lomond—Ben Arthur or the Cobler'. Ben Arthur (or the Cobbler) is 884m in height and stands at the north-west end of Loch Lomond. Tyndall obtained this sheet of paper during his visit to Loch Lomond in late September 1855; see letter 1137, n. 9.

To William Thomson 20 November 1855 1155

20th Nov. 1855.

My dear Sir

If I can render you any assistance in preparing instruments for the lecture which you have so kindly consented to give us[1] it will give me great pleasure to do so.

We are well furnished with electromagnetic apparatus.

You can if you wish make Peltier's experiment[2] in a manner visible to all.

You can also produce fire by friction, and heat by the compression of air. We have contrivances for such experiments.

We have also a very effective manner of showing the heat produced by congelation.

We are well furnished in the line of Frictional Electricity.

You will however mark out your own course, and I can only say that I shall be happy to render you any assistance in my power.

I may take this opportunity of referring to a matter which struck me a few evenings ago, and on which a word from you will doubtless make me clear.

From your proof that the intensity of a magnetic field increases towards the centre of curvature[3] I should infer that if the lines of force were parallel straight lines the intensity at right angles to them would be constant. I have a steel horseshoe magnet here in which the lines of force run sensibly parallel from leg to leg almost from top to bottom, yet such a field is not one of constant intensity, for the force increases from the bend towards the poles.

When we examine such a field closely we even find that the lines of force are slightly curved, the centre of curvature being towards the bend, and not towards the poles.

According to this the intensity increases as we recede from the centre of curvature.

If you have not plenty of time at your disposal, don't trouble yourself thinking on this.

I have just finished a paper on polarity[4] which I purpose sending to the Royal Society in a few days. I am now entangled in compression experiments.[5]

Believe me | Very sincerely yours | J. Tyndall

RI MS JT/1/5/1544–5
Transcript Only

1. *the lecture . . . give us*: Thomson had been invited to deliver a Friday Evening Discourse at the RI on 29 February 1856. His lecture, entitled 'On the Origin and Transformations of Motive Power', was later published in *Roy. Inst. Proc.*, 2 (1854–58), pp. 199–204.
2. *Peltier's experiment*: The French physicist Jean Charles Athanase Peltier (1785–1845) performed an experiment in 1834 in which he demonstrated that cold is produced by an electric current flowing from antimony to bismuth. This experiment complemented the discovery thirteen years earlier by Thomas Johann Seebeck (see letter 1142) who found that a temperature difference between two different metals produced a voltage.
3. *your proof . . . centre of curvature*: W. Thomson, 'Elementary Demonstrations of Propositions in the Theory of Magnetic Force', *Phil. Mag.*, 9 (1855), pp. 241–48.
4. *a paper on polarity*: Tyndall's paper 'Further Researches on the Polarity of the Diamagnetic Force' was received by the RS on 27 November and read at its meeting on 20 December 1855. It was then published in *Phil. Trans.*, 146 (1856), pp. 237–59.
5. *compression experiments*: 'At work at the compression of oxalate of cobalt and oxalate of iron' (Tyndall's Journal, 19 November 1855: RI MS JT/2/13c/804). Some of his compression experiments were published in 'On the Relation of Diamagnetic Polarity to Magne-crystallic Action', *Phil. Mag.*, 11 (1856), pp. 125–37.

From William Thomson 22 November 1855 1156

2 College, Glasgow | Nov 22, 1855

My dear Sir

I am much obliged to you for your kind offer[1] to give me apparatus and assistance for experimental illustrations of the subject for my proposed lecture, of which I shall thankfully avail myself. I shall be as much pleased to <u>see</u> as to show fire made by friction, and all the other illustrations you mention will be valuable.

I am anxious to show the <u>stunning</u> effect of an electromagnet on a rotating mass of copper.[2] If I can get a copper spinning top made and conveniently arranged for spinning between the poles of a Ruhmkorff[3] or other powerful electromagnet, it will be curious to see it struck suddenly dead when the magnetism is excited; and I would like too if possible to get a mechanical arrangement to force it round at a rapid rate & to show the elevation of temperature it experiences. I propose to have a square head at the top of the spindle, and a key (like a clock or a large watch key) put on it & kept turning by a short stiff india rubber cord of which the lower end would be attached to the key and the upper end to the lower extremity of a vertical shaft to which a quick motion would be given by machinery. If after starting the motion by the mechanism it is desired to leave the top to itself, all that would be required would be to lower the stand on which it rests, when its own weight would clear it of the key. If I get this plan to succeed, I suppose I may rely on a quick motion apparatus. As soon as I have got it tried, I shall write to you as to the arrangements I would wish to make.

I do not know exactly how the difficulty you state[4] as to the application of my proposition to the lines of force between the branches of your steel magnet is to be explained. I *[would]* expect that if the lines of force are really curved with their convexity towards the open space

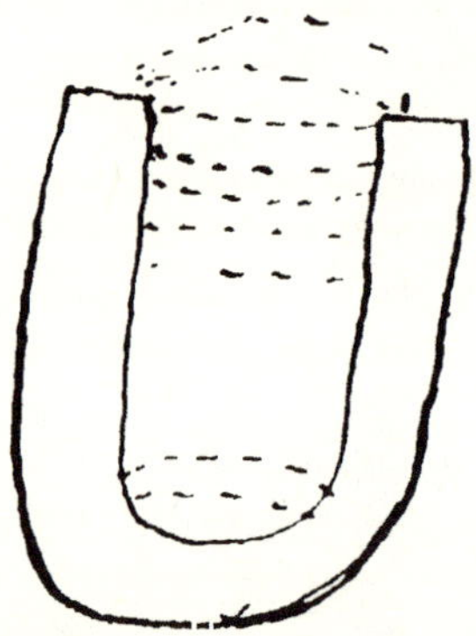

between the poles, the intensity of the force would in that part of the field, in which they are so bent, increase towards the middle of the bend; and this might be tested by taking a very small needle (say, ¼ of an inch of watch spring) hung by a single fibre, and making it oscillate in the different positions. But I would be much surprised if such curvature of the lines of force is found anywhere except near the bend, and I should expect curvature rather the other way in the space between the terminal halves of the branches. I hope, if not before, we shall be able to settle the difficulty when I see you in London.

I am glad you are going on with compression experiments, and I hope you will get satisfactory results. I have been testing the effect of longitudinal extension on electric conductivity, which in the case of iron wire I find decidedly a diminution although very minute. I mean to try lateral extension and compression.

Once more thanking you for your offered help

I remain, Dear Sir, | Yours very truly | William Thomson

RI MS JT/1/T/14

---

1. *your kind offer*: see letter 1155.
2. *show . . . mass of copper*: The experiments Thomson displayed during his lecture at the RI

are described in 'On the Origin and Transformations of Motive Power', *Roy. Inst. Proc.,* 2 (1854–58), pp. 199–204.

3. *a Ruhmkorff*: a Ruhmkorff coil for producing pulses of high voltage electricity.
4. *the difficulty you state*: see letter 1155.

From Heinrich Debus 17 and 25 November 1855 1157

Queenwood College | am 17 Nov. 1855.

Mein lieber Tyndall!

Mit vielem Vergnügen habe ich aus Deinem Briefe ersehen dass Du in den letzten Wochen in Geist und Körper vollkommen gesund gewesen bist. Nur diejenigen Menschen denen die Natur von Zeit zu Zeit den Gebrauch ihrer lebendigen Kräfte erschwert oder gar untersagt wissen den Werth eines klaren Kopfes und edlen Körpers im Zustand der vollkommenen Gesundheit recht zu schätzen. Wohl erinnere ich mich jener Zeit als ich Zahnwehe und Kopfwehe, Leibwehe und alle sonstigen Übel die des Menschen Geist und Körper beunruhigen nur von Hörensagen kannte, und wohl wünsche ich sehr oft die Kräfte meiner Jugend mit in das hohe Alter hinaufzunehmen.

In Beziehung auf Deine Arbeiten kann ich nur wiederholen was ich schon so oft gesagt und gedacht habe, dass Du mit ausserordentlicher Schnelligkeit die besten Untersuchungen ausführst. Man weiss nicht ob man den ungewöhnlichen Fleiss oder das Talent am meisten schätzen soll. Ich wünsche dass Dir diese herrliche Thätigkeit auch für die Zukunft zu Deiner eigenen Glückseligkeit und zum Besten der Wissenschaft erhalten bleiben mag. Zuweilen lege ich mir die Frage vor wie es kommt, dass ich selbst so wenig leiste ob mir gleich äusserlich alle Mittel gegeben zu sein scheinen. Es ist schwer selbst eine solche Frage zu beantworten; denn die Schwächen und Ir[r]thümer, die Mängel und Gebrechen unter denen man leidet hat man oft lieber als die guten Seiten des eigenen Selbst. Oder vielleicht nicht lieber sondern sie sind einem bequemer als das Regiment der Vernunft über unsere eigene Herrlichkeit[.]

am 25 Nov 1855

Ich wurde am letzten Sonntag am Schreiben dieses Briefes unterbrochen und so vollende ich heute gerne, so das was mir vor 8 Tagen nicht vergönnt war zu thun.

Auch ich habe mich seit Deiner Abwesenheit angelegentlich mit meinen wissenschaftlichen Arbeiten beschäftigt und bisher recht zufriedenstellende Resultate erhalten. Nur bedauere ich dass es nicht schnell genug geht sondern dass der mechanische Theil der Arbeit gerade wie Ballast in meinem Geiste liegt der Phantasie und dem denkenden Verstandt Zügel anlegt und beide gewaltsam zur Erde niederdrückt. So z.B. reducirt sich meine wöchentliche Arbeit jetzt auf 3 Verbrennungen d.h. organische Analysen von denen jede ohngefähr den besten Theil von einem ganzen Tag fortnimmt. Doch dies ist der einzige Weg auf dem uns die Natur einen Blick in ihr Schaffen und Treiben gestattet und darum keine Klage mehr darüber.

Als ein Jüngling arbeitete ich aus Liebe zur Sache oder vielleicht auch zur Befriedigung der Freude die ein junger Mensch immer empfindet wenn er Etwas seinen Geist bildendes seinen Blick erweiterndes sich aneignet. Noch erinnere ich mich des Vergnügens das ich in meinem 16tn Jahre beim lesen eines Satzes empfand der mir wie ein Lichtstrahl unbekannte Regionen eröffnete. In der Reinheit dieser Freude liegt ihre Grösse. Später arbeitete ich mehr aus Gewohnheit und jetzt als Mann fange ich erst wieder an mich mehr und mehr mit meinem Gegenstand zu identificiren und denselben neben der Bildung meiner eigenen Seele als den liebsten und unsterblichsten Theil meiner Selbst zu betrachten.

Atkinson ist seit 10 Tagen hier in Queenwood; und arbeitet im Laboratorium. Ausserdem übersetzt er für Francis. Derselbe hat die Assistentenstelle bei Brodie angenommen eine Stellung, die ihm, wie ich erwarte, sehr nützlich nicht nur in wissenschaftlicher sondern auch in anderer Beziehung sein wird. Brodie ist eine gute Acquisition für Oxford. So freut es mich dass Oxford einen guten Mann und ein guter Mann eine gute Stelle bekommt. Ausse[r] dem geht es in Queenwood seinen alten Gang. Von Mr Haas mit seiner Disciplin und dem Blowing up System, Mr Wright mit den Accounts, Mr Hearst mit der Photographie &c &c &c und wie die Herren sich alle heissen mögen kann ich leider Nichts berichten denn sie scheinen noch unter denselben Umständen, mit demselben Effect und in derselben Weise zu leben als früher.

Ich habe noch eine grosse Menge Oxalate of Iron mit herzlichen Grüssen, bin ich | Dein H. Debus

Queenwood College | 17 Nov. 1855

My dear Tyndall!

With much pleasure did I see from your letter[1] that in recent weeks you have been completely healthy in mind and body. Only those people for whom nature, from time to time, impedes or even inhibits the use of their vital powers, can properly appreciate the value of a clear head and a noble body in a state of complete health. I remember well that time when I knew toothache and headache, bodyache and all the other ills that disturb a person's mind and body from hearsay only, and well do I very often wish that I could take the powers of my youth with me into advanced old age.[2]

With respect to your works, I can only repeat what I have already said and thought so often, that you carry out the best investigations with extraordinary speed. One does not know whether one should treasure your unusual industriousness or your talent most of all. I wish that you may maintain this magnificent activity in the future also, for your own felicity and for the benefit of science. Occasionally I put the question to myself as to how it happens that I produce so little[3] myself, even though outwardly every means seems to be given to me. It is difficult to answer such a question for oneself, for one often prefers the weaknesses and errors, the flaws and afflictions under which one suffers to the good aspects of one's self. Or perhaps not prefer, but one finds them more comfortable than the rule of reason over one's own magnificence[.]

25 Nov 1855

Last Sunday I was interrupted while writing this letter and so I shall gladly complete today what I was not granted the privilege of doing 8 days ago.

Since you left[4] I, too, have been thoroughly preoccupied with my scientific works and have so far obtained some quite satisfactory results. I only regret that things are not going fast enough, but rather that the mechanical part of the work lies just like ballast in my mind, puts the reins on my imagination and conscious intellect, and pushes both forcefully down to the ground. Thus e.g. my weekly work is now reduced to 3 combustions, i.e. organic analyses, of which each takes more or less the best part of an entire day. But this is the only way in which Nature grants us a glimpse into her workings and doings, and so no more complaints about that.

As a young man I worked for the love of a thing or perhaps also to satisfy the joy that a young person always feels when he acquires something that shapes his mind and widens his view. I still remember the pleasure that I felt in my 16th year when reading a sentence that, like a ray of light, opened up unknown regions for me. In the purity of this joy lies its greatness. Later, I worked more out of habit, and now as a man, I am only now starting again to identify more and more with my subject and to regard it as the dearest and most immortal part of my self, apart from the formation of my own soul.

Atkinson[5] has been here in Queenwood for 10 days, and is working in the laboratory. He is also translating for Francis and has accepted the job of assistant to Brodie,[6] a position that, as I expect, will be very useful to him—not only in a scientific respect, but in others as well. Brodie is a good acquisition for Oxford, and thus I am glad that Oxford is getting a good man and that a good man is getting a good job. Aside from that, things are going their old way in Queenswood. Unfortunately, I cannot report anything about Mr Haas with his discipline and his Blowing up System,[7] Mr Wright with the accounts, Mr Hearst[8] with the photography etc etc etc, and whatever all the gentlemen's names might be, for they still seem to live under the same circumstances, with the same effect, and in the same way as earlier.

I still have a large quantity of oxalate of iron, with warm regards I am | Your | H. Debus

RI MS JT/1/D/19

1. *your letter*: letter missing.
2. *into advanced old age*: Debus was aged 31 when he wrote this letter.
3. *I produce so little*: The Royal Society's *Catalogue of Scientific Papers* lists eight publications by Debus between 1848 and 1853.
4. *Since you left*: Tyndall left Queenwood on 29 October 1855.
5. *Atkinson*: The chemist and physicist Edmund Atkinson (1831–1900) was a Lancastrian (like Frankland) who studied at Owens College and then in Marburg (where he met Hirst in 1853), Heidelberg, Göttingen and Paris, obtaining his PhD from the University of Göttingen, before becoming a teacher at Queenwood College and then working as assistant to Benjamin Collins Brodie. He devised an improved method of producing glycol acetate.
6. *the job of assistant to Brodie*: On 13 November the Convocation of the University of Oxford had elected the chemist Benjamin Collins Brodie (1817–80) to the university's Aldrichian chair of chemistry (*London Standard,* 14 November 1855, p. 2). Atkinson, who had worked as Brodie's assistant in London, accompanied him to Oxford.
7. *his Blowing up System*: possibly Haas's tendency to 'blow his top'—lose his temper—during lessons.
8. *Mr Hearst*: Thomas Archer Hirst.

To William Thomson 26 November 1855 1158

Royal Institution, Nov. 26, 1855

My DEAR SIR,

THE communication from Professor Weber which appears in the present Number of the Philosophical Magazine,[1] has reminded me, almost too late, of your own interesting letter on the same subject published in the April Number of this Journal.[2] A desire to finish all I have to say upon this question at present[3] induces me to make the following remarks, which, had it not been for the circumstance just alluded to, might have been indefinitely deferred.

With reference to the mutual action of a row of bismuth particles, you say that 'it is perfectly easy to demonstrate that it *must* be such as to impair the "diamagnetization" when the line of the row is parallel to the lines of the force'[4] (the 'must', you will remember, is put in italics by yourself). From this you infer, that in a uniform field of force a bar of bismuth would set its length along the lines of force. Further on it is stated that this action is 'demonstrated with as much certainty as the parallelogram of forces'; and you conclude your letter by observing that 'the assertions which I[5] have made are demonstrable in every case without special experiment, . . . and are absolutely incontrovertible, as well as incapable of verification, by experiment or observation on diamagnetics'.[6]

Most of what I have to say upon this subject condenses itself into one question.

Supposing a cylinder of bismuth to be placed within a helix, and surrounded by an electric current of sufficient intensity; can you say, *with certainty,* what the action of either end of that cylinder would be on an external fragment of bismuth presented to it?

If you can, I, for my part, shall rejoice to learn the process by which such certainty is attained: but if you cannot, it will, I think, be evident to you that the verb '*must*' is logically 'defective'.

We *know* that magnetized iron attracts iron: we *know* that magnetized iron repels bismuth: this, so far as I can see, is your only experimental ground for *assuming* that magnetized *bismuth* repels bismuth, and yet you affirm that an action deduced from this assumption 'is demonstrated with as much certainty as the parallelogram of forces'.[7] Do I not state the question fairly? I can, at all events, answer for my earnest wish to do so.

It is needless to remind one so well acquainted with the mental experience of the scientific inquirer, that the very letters which you attach to your sketch, page 291,[8] may tempt us to an act of abstraction—a forgetfulness of a possible physical difference between the *n* of iron and the *n* of bismuth[9]—which may lead us very wide of the truth. The very term 'pole' often pledges us to a theoretic conception without our being conscious of it. You are also well aware of the danger of shutting the door against experimental inquiry on an unpromising subject; and when you apparently do this in your concluding paragraph,[10] I simply accept it as a strong way of expressing your personal conviction, that the action referred to is too feeble to be rendered sensible by experiment.[11]

Believe me, dear Sir, | Most truly yours, | JOHN TYNDALL.

Phil. Mag., 10 (1855), pp. 422–3[12]

---

1. *THE communication . . . Philosophical Magazine*: reproduced here as letter 1039.
2. *your own interesting letter . . . Journal*: reproduced here as letter 1053.
3. *A desire . . . question at present*: This letter to Thomson was written the day before the RS received Tyndall's paper 'Further Researches on the Polarity of the Diamagnetic Force'. Thus, far from this letter being Tyndall's last word on diamagnetism, it was a prolegomena to his RS paper in which he criticised Thomson's views. See n. 11, below.
4. *it is perfectly . . . lines of the force*: see letter 1053.

5. *I*: after which Tyndall inserted '[yourself]', i.e. Thomson.
6. *demonstrated . . . observation on diamagnetics*: see letter 1053.
7. *is demonstrated . . . parallelogram of forces*: Thomson had written: 'Still, I repeat, this action, although demonstrated with as much certainty as the parallelogram of forces, is so excessively feeble as to be absolutely inappreciable' (letter 1053).
8. *page 291*: the first diagram in letter 1053.
9. *the* n *of iron and the* n *of bismuth*: the letter *n* refers to the north pole.
10. *your concluding paragraph*: In the final paragraph of letter 1053, Thomson claimed that the principles he had there enunciated 'are absolutely incontrovertible, as well as incapable of verification, by experiment or observation on diamagnetics'.
11. *the action . . . by experiment*: In his 'Further Researches on the Polarity of the Diamagnetic Force' Tyndall demonstrated, contra Thomson, that the action could 'be rendered sensible by experiment'. Although Tyndall did not mention Thomson in that paper he clearly included him among those 'investigators, [who] converted this belief [that diamagnetic polarity was not demonstrable] into a certainty' (*Phil. Trans.*, 146 (1856), pp. 237–60, on p. 238).
12. Published under the title 'On Reciprocal Electrical Induction'.

From César Despretz 4 March & 29 November 1855 1159

Mon cher M. Tyndal

J'ai eu l'honneur de vous écrire il y a quelques temps pour vous dire que j'accepte le [Phal] magazine aux conditions que Mr. Taylor vous a données. J'ai mis dans ma lettre un bon sur le correspondant de Mr. Duprat, pour le montant des livres que j'ai reçus en novembre et pour lesquels j'ai déposé la valeur le lendemain de la reception. Je pense bien que j'aurai dans le [Phal] magaz[ine] les originaux de beaucoup de travaux des physiciens anglais.

Je me propose d'aller à Londres vers la fin de mai. Je ne connais guères à Londres que vous et que MMr Faraday, Grove, Wheatstone et Gassiot. Bab[b]age. On me dit qu'il est très agréable d'être admis à l'Atheneum qui je crois, est près de Regents Park.

Je ne vois, parmi les membres de cette société que Mr Faraday et Mr Wheatstone que je connaisse. Voudriez-vous demander à Mr. Faraday, s'il pourrait facilement me faire admettre à l'[Athaneum] pendant mon sejour prochain a Londres. Si cela présente quelques difficultés, n'insistez pas. Je viens de recevoir une lettre de Mr Grove. Il s'absente me dit-il pour six semaines.

Quand je recevrai les livres, je vous repondrai immediatement comme j'ai eu l'honneur de vous le dire dans ma lettre.

Agréez et faites agréez mes compliments affectueux à Mr Faraday

Votre dévoué C. Despretz. M. de l'Institut Le 4 mars 1855 Rue cassette 20

J'ai laissé long temps ma lettre sans vous l'envoyer. Je vous prie de vouloir bien prier le libraire de m'adresser la collection le plus tot possible. Je vais en avoir besoin.

Quant à l'[Athaneum], je ne voudrais pas que Mr. Faraday dût pour cela se donner beaucoup de peine. Je ne serai pas à Londres plus de trois semaines et pendant ce temps j'irai visiter quelques grandes villes situées sur les lignes des chemins de fer.

Recevez l'assurance des sentiments distingués avec lesquels je suis | votre dévoué | C. Despretz M. de l'Institut | 29 $9^{b}r^{e}$ 1855

My dear Mr Tyndal,

It was my pleasure to write to you some time ago to tell you that I accept the Phil. Magazine under the conditions that Mr Taylor gave you. I have enclosed in my letter a [credit note] [addressed to] Mr Duprat's correspondent, for the value of the books that I received in November and of which I noted down the price the day following their receipt. In the Phil. Magazine, I believe I will have many works by the English physicists in their original language.

I intend to travel to London around the end of May.[1] I hardly know anyone in London apart from yourself and Messrs Faraday, Grove, Wheatstone and Gassiot, [and] Bab[b]age. I am told that it is very pleasant to be received in the Athenaeum[2] which I believe is close to Regents Park.[3]

Apart from Mr Faraday and Mr Wheatstone, I cannot think of anyone that I am acquainted with amongst the members of this society.[4] Could you ask Mr Faraday if he could have me admitted to the Athenaeum during my forthcoming stay in London? If this causes some difficulties, do not insist. I have just received a letter from Mr Grove;[5] he writes he will be absent for six weeks.

I will respond to you immediately after I receive the books, as I have had the honour of telling you in my letter.

Please accept my affectionate regards and send them to Mr Faraday.

Faithfully yours| C. Despretz M. of the Institute | 4 March 1855 | 20 Rue Cassette

I have kept this letter for a while before sending it to you. I beg you to please urge the bookseller to send me the collection as soon as possible. I will need it.

As for the Athenaeum, I would not really want Faraday to bother about it. I will not be in London longer than three weeks, and, during this time, I will visit some of the large cities connected by railway.

Please be assured of my highest regards with which I am | your devoted | C. Despretz M. of the Institute | November 29th 1855[6]

RI MS JT/1/D/122[7]

---

1. *London around the end of May*: According to entries in Tyndall's Journal, Despretz visited London from 10 May to after 22 May 1856 (RI MS JT/2/13c/841–44); see letter 1219.
2. *the Athenaeum*: Founded in 1824 the Athenaeum Club was a private gentleman's club that attracted not only men of inherited status but also those eminent in science, literature and art. Faraday, who served briefly as its first secretary, retained his membership until 1852. Thus he was no longer a member at the time of this letter.
3. *close to Regents Park*: The Athenaeum Club is nearly three miles from Regent's Park. The closest park is St James's Park.
4. *members of this society*: members of the Athenaeum Club.
5. *a letter from Mr Grove*: letter missing. However, on 23 November Tyndall recorded in his Journal: 'Dined at Mr Barlow's [. . .] A conversation with [William] Grove afterwards about Despretz and other matters' (RI MS JT/2/13c/805).
6. *November 29th 1855*: Despretz's French text gives the date of the second part of this letter as '29 9^{b}r^{e} 1855'. As the Latin for the number nine is 'novem', the month has been read as 'novem^{b}r^{e}', not the ninth month (September).
7. As this letter was not sent to Tyndall until the second part—dated 29 November 1855—was written, it has been inserted into the sequence of letters at this point.

From César Despretz 30 November 1855 1160

Mon cher Mr Tyndall,

Dans la lettre que je vous adressais hier, je vous disais que j'aurais bientôt besoin de la collection du P^{al}. Mag. Une discussion que j'ai eue hier dans une commission me montre que j'en ai besoin dans ce moment.

Si donc le libraire est disposé à m'envoyer la collection, je le prie de me l'envoyer sans tarder.

Je suppose ici que vous ayez reçu la lettre que je vous ai adressée il y a un mois environ.

Si vous connaissez un bon ouvrage de méteorologie, fait récemment par un homme de talent, je vous prie de l'y joindre pourvu que le prix ne soit pas au-dessus de 40 francs.

Mille excuses, agréez d'avance tous mes remercîments.

Votre dévoué C Despretz

My dear Mr Tyndall,

In the letter I sent you yesterday,[1] I was telling you that I would soon have need for the collection of the Phil. Mag. It appears from a discussion that I had yesterday in a commission,[2] I realized that I need it right away.

Thus, if the bookseller is willing to send me the collection, I urge him to send it to me without delay.

I assume here[3] that you have received the letter that I sent you about a month ago.[4]

If you know of a good publication on meteorology, recently completed by a talented man, I ask you to include it as well [in my order], provided that the price does not exceed 40 francs.

A thousand apologies; please accept my gratitude in advance.

Yours truly, C Despretz

RI MS JT/1/D/126

1. *the letter I sent you yesterday*: letter 1159.
2. *a commission*: Despretz served on various committees at the Paris Académie des Sciences, including committees that refereed papers .
3. *here*: in this letter.
4. *the letter . . . a month ago*: letter missing.

From Carlo Matteucci[1] 3 December 1855 1161

Pise. 3 déc,1855

Mon cher M[n] Tyndall

Vous savez que j'ai écrit dernièrement au Prf [Professeur] M. Faraday pour qu'il vous comunique en douteux [*adjective illeg*] que j'avai concue sur les conclusions tirées de vos dernieres experiences sur la polarité diamagnétique. La lettre que F[araday] m'a répondu me fait presque soupconner [?][2] que vous aviez interprété cette demarche d'une maniere inattendue pour moi et que vous n'en aviez pas été content. Mon intention n'etait pas celà : j'ai cru de faire dans cette circonstance ce que j'aurai fait très naturlment si j'avais été a Londres et j'avai causé de science avec vous—Moi comme tous vous aurai dit, étant dans mon cas[.]

Je travaille depuis longtemps sur ce sujet, je travaille de cette maniere je crains que votre maniere ne fut pas la meilleure—Réfléchissez, et si vous le croyez, repétez mes experiences ou les votres[.]–Voilà tout : j'espere de m'etre bien expliqué et que vous ne pouvez juger ma demarche que telle qu'elle est, tout à fait amicale. Par les même sentiment, je m'empresse maintenant de vous ecrire que j'ai répété avec tous les soins vos experiences, quant a la disposition generale, mais en les rendant du moins suivant ma manier [manière] de voir, indépendantes de toute critique. Ainsi—je n'ai employé que des spirales sans fer doux. Le corps diamagnetique était un cylindre de poudre de bismuth et de résine. Le fil de suspension était un fil d'argent et je mesurai les forces par la torsion. Bien. Ce resultat est conforme au votre et j'ose le dire, du moins dans ma tête, il est incontestable ce qui n'était pas avant.

Lorsque les extrémités a et b de deux spirales *[repoussent]*[3] des poles de même nom le cylindre de quelque maniere qu'on opera, en fermant un circuit avant l'autre ou tous les deux en meme temps, le cylindre le porte et le fixe [fixé] sur a ou vers a. Cela veut-il dire que la polarité diamagnétique existe?

C'est autre chose. Je ne fais que vous apporter ici le résultat principal auquel j'etais parvenu et qui m'avait fait douter de vos experiences, pour vous prouver ce qui suivant moi est contraire a l'idée de la polarité diamagnétique.

Le cylindre diamagnétique qui est dans la spirale n'oscille pas dans ce cas autour de son centre, mais il se deplace le long de l'axe de la spirale.

Dans ce cas, si l'on met une seconde spirale b (bout à bout) contre la première, on trouve que le cylindre diamagnétique, quelque soit la direction du courant en a, est toujours repousse par l'autre spirale b et exactement de la même quantité quelque fut la direction du courant de a ou de b et cela en changeant les forces relatives des courants.

Notez que si le cylindre est magnétique, mais d'une matiere aussi faiblement magnetique qu'on peut la comparer au diamagnétisme du bismuth, la polarité magnetique dans les mêmes circonstances est trop marquée et avec des différences enormes.

Le sujet est bien *[word illeg.]*. Dans quelques jours ou un mois je ferais paraitre mon travail et vous le jugerez.

En attendant j'ai voulu vous faire savoir que j'avai trouvé que vos expériences vont bien et comme j'avai ecrit a Faraday et a Grove que j'en doutai, je vous prie de lui dir de ma part que je suis parvenu à une complete verification des experiences.

Agreez mon respect et attachement | C. Matteucci

Pisa December 3, 1855.

My dear Mr Tyndall,

You know that lately I wrote to Professor Faraday[4] so that he could let you know about the *[adjective illeg.]* doubts I had conceived concerning the conclusions you drew from your last experiments on diamagnetic polarity.[5] I suspect from the letter that F[araday] sent me back[6] that you had interpreted my efforts in a manner I was not expecting and that it had upset you. That was not my intention: I believed that I was doing in this circumstance was what I would have done very naturally had I been in London and discussing science with you face to face. Like anyone else in my position, I would have told you.[7]

I have been working on this topic for a long time, I work in this manner. I fear that your method is not the best. Think about it, and if you believe it, repeat my experiments or yours. That's all: I hope I explained myself clearly[8] and that you will judge my initiative for what it is, absolutely friendly. It is with the same feelings that I rush to write to you today that I repeated your experiments faithfully with regard to the general disposition, but making them, in my opinion, independent of any criticism. Hence, I have only used helixes [*spirales*][9] without soft iron. The diamagnetic body was a cylinder of bismuth powder and resin. The suspended wire was a silver wire and I measured the forces by torsion.[10] Good. This result is consistent with yours, and I dare say, at least in my mind, it is indisputable, unlike previously.

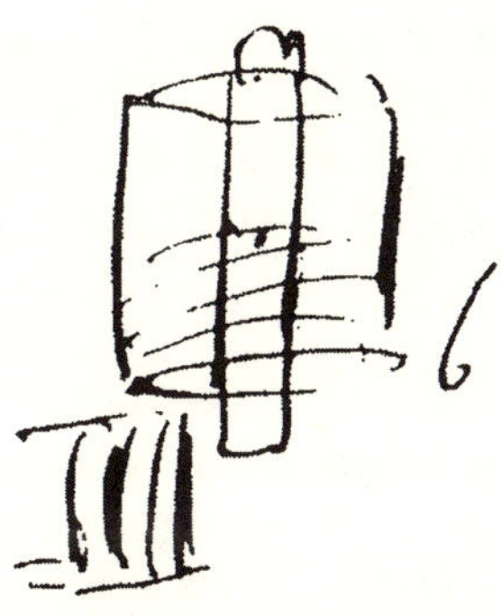

When the extremities a and b of two helices *[word illegible]* the same polarity by the same name, the cylinder, in whichever way it is operated, by closing a circuit before the other or one after the other, the cylinder is carrying it and fixing it on a or towards a.

Does that mean that diamagnetic polarity exists? That is something else [again]. I only intend to bring to you here the main result that I have achieved and which made me doubt your experiments, in order to prove to you what, according to me, is contrary to the idea of diamagnetic polarity.

The diamagnetic cylinder which is in the helix does not oscillate in this case around its centre but moves along the axis of the helix. In this case, if a second helix b is placed against the first one (end to end) we find that the diamagnetic cylinder, whatever the direction of the current in a, is always repelled by the other helix b and by exactly the same amount whatever the direction of the current in a or b and that by changing the relative forces of the currents [?]. Note that if the cylinder is magnetic, but made from a material that is as indistinctly magnetic that we can compare it to the diamagnetism of the bismuth, the magnetic polarity in the same circumstances is too pronounced and with enormous dissimilarities.

The subject is quite *[word illeg. difficult?]*. In a few days or a month, I will publish my work[11] and you will be able to judge it.

In the meantime I'd like you to know that I've found your experiments are true and because I wrote to Faraday and to Grove that I had doubts, I beg you to tell him from me that I am sending a complete verification of the experiments

With respect and affection | C. Matteucci.

RI MS JT/1/M/58

1. Accents have not been restored and grammatical errors are ignored except where the sense demands.
2. *soupçonner*: writing unclear; could be 'supposer' or 'frissonner'.
3. *reposoussent*: The verb is unclear and may be 'repressentent'.

4. *I wrote to Professor Faraday*: letter missing.
5. *doubts I had . . . experiments on diamagnetic polarity*: Matteucci had expressed these orally in his lectures at the University of Pisa in June 1853. These lectures appeared in French as *Cours spécial sur l'induction, le magnétisme de rotation, le diamagnétisme: et sur les relations entre la force magnétique et les actions moléculaires* (Paris: Mallet-Bachelier, 1854), with the critique of Tyndall's experiments, pp. 253–57. Matteucci presented a copy of this to Faraday, who passed the book to Tyndall in time for him to add 'A Note on M. Matteucci's objections' to the Bakerian lecture as it went to press. J. Tyndall, 'On the Nature of the Force by which Bodies are Repelled from the Poles of a Magnet', *Phil. Trans.,* 145 (1855), pp. 1–51 at pp. 41–43. When Tyndall republished his papers on diamagnetism in 1870, he was rather cavalier concerning his dispute with Matteucci, writing: 'Matteucci undertook to repeat the experiments recorded in my [Bakerian] lecture but failed to obtain the results. He considered the memoir a tissue of error from beginning to end, and thought my character as a scientific man so gravely compromised that he wrote to ask Faraday for advice as how he ought to act under the circumstances. Faraday showed me Matteucci's letter, and the result of our conversation regarding it is stated by Faraday himself [*Faraday Correspondence,* 5: 3033]. Weeks, it may have been months, elapsed without my hearing anything further about the matter; when at length a terse, frank letter [letter 1161] reached me direct from Matteucci, the substance of which was this: "I have written to Faraday, to Grove, and to Wheatstone, stating that your experiments were wrong. I now wish to give you the opportunity of correcting them, and of saying to these gentlemen that I have repeated all your experiments and found them true to the letter".' J. Tyndall, *Researches on Diamagnetism* (London: Longmans, Green, 1870), pp. 182–83.
6. *the letter that Faraday sent me back*: *Faraday Correspondence,* 5: 3033.
7. *Moi comme tous . . . étant dans mon cas*: This might also be rendered as 'I did what everyone in my position would have done'.
8. *I explained myself clearly*: in the *Cours* (n. 5).
9. *I have only used helixes*: We have translated 'spiral' as 'helix' since this is the term used by Tyndall in the Bakerian lecture.
10. *I measured . . . by torsion*: by means of a torsion balance.
11. *I will publish my work*: no such publication has been traced.

To Thomas Archer Hirst 6 December 1855 1162

6th Decr 1855[1]

My dear Tom.

I am sorry to add to your troubles but I can do nothing else than send you the accompanying note from Chapman which reached me five minutes ago in company with your review.[2] While urging some alteration of the latter he 'admires its tone & ability'. Shorten Faraday if you can & lengthen the others if you can, and let the whole be 600 words shorter than at present. Leave out all you have said about me and mine. This will simplify your labour. May the gods grant you patience & good temper

Ever your | John

Kiss Anna for me and let the act calm your own spirits—Life is full of troubles—I have lost the last two days over a quibble.

J. Chapman | on my article in the | Westminster[3]

[enclosure]

8 King Wm St Strand | Dec 6 '55.

My dear Sir,

I shall be glad if you will reduce the accompanying paper[4] about 600 words, it is rather more than a page too long.—The part should be cut out of the notice of Faraday, which I am sorry to see so disproportionately long. It almost assumes the form of a separate article, and entirely prevents justice being done to the other books—especially to Arago and Grove. The notice of Faraday is moreover little likely to be understood by the majority of the readers of the Westminster; so much space being occupied the exposition should certainly have been of a more popular character. Having said this much, permit me to add that I admire the ability and tone of the paper throughout.

Faithfully yours | John Chapman

Professor Tyndall.

RI MS JT/1/T/944

---

1. *6th Decr 1855*: The date is unclear. The LT Typescript suggests 6 or 7 December.
2. *your review*: Hirst began the review at Tyndall's request on 4 November 1855 (Hirst Journal, RI MS JT/2/32c/1181). It embraced the 3rd volume of Faraday's E*RE*; Edward Sabine's translation of François Arago, *Meteorological Essays* (London: Longman, 1855); a reissue of Evan Hopkins, *On the Connexion of Geology with Terrestrial Magnetism* (London: Richard Taylor, 1851); George Fellows Harrington, *New Theory of Light* (London: William Tyler, 1855); Edward Hughes, *Reading Lessons. First Book* (London: Longman, 1855); Robert James Mann, *Lessons in General Knowledge* (London: Longman, 1855); Julius Stockhardt, *Familiar Exposition of the Chemistry of Agriculture* (London: Henry Bohn, 1855); William Grove, *Correlation of Forces*, 3rd ed. (London: Longman, 1855); and Alexander Stephen Wilson, *Unity of Matter* (London: Samuel Highley, 1855). The review appeared under the heading 'Science' in *Westminster Review*, 65 (1856), 254–71.
3. *J. Chapman on my article*: annotation added by Hirst.
4. *accompanying paper*: proof copy missing.

From Thomas Archer Hirst 9 December 1855 1163

Broughton. | Decr 9th / 55

My dear John,

According to Chapman's request[1] I have shortened the notice of Faraday's book by 600 words and sent it to him. Without relinquishing the original idea of the review I could not make the part on diamagnetism shorter than it now is, to do so would involve rewriting the whole review which I have by no means time to do. I have cut out, as you desired, the part about yourself. If I ever write another review for the Westminster I shall have a better notion of what is desired than I had when I began this. The coming week will of course be a busy one for me, examinations[2] commence tomorrow. Shall we see you next Saturday Evening?[3] I enjoyed yours and Bevington's late visit[4] very much. I feel to know Bevington much better than I did. He is indeed a genial kind-hearted soul. It was a great delight to me to listen to his scientific arguments with you and to notice the amount of real intelligence and thought they displayed in him. It was especially beautiful to notice the absence of all pretension in him and the graceful manner in which after long contesting he could cede a point. I noticed too a peculiarity in your arguments with Bevington which was instructive. I do not know whether you are aware that you have a habit when arguing with many people of trying their tempers. A man with moderate self-esteem and an irritable temper could never stand against you in argument not necessarily because his cause was weaker than yours or his ability less but chiefly because he would be in danger of losing his wits with his temper. I have noticed this in many instances (occasionally when I have been your opponent) it arises I think from the position you take of testing your adversaries knowledge round about the point in question. He is thus thrown off his own line of rails, his self-esteem is perhaps a little wounded by his knowledge being questioned, his temper is aroused and—he flounders. Well this was not the case with Bevington. There was little or no self-esteem to wound, little or no temper to rouse and he succeeded better than I have seen other people do in keeping his feet before you.

Anna is still suffering from the cold & cough. She desires her love to be sent with mine.

Your affectionate | T.A. Hirst.

RI MS JT/1/H/220

---

1. *Chapman's request*: see letter 1162.
2. *examinations*: end of term school examinations at Queenwood College.

3. *next Saturday Evening*: Tyndall did not go to Queenwood on Saturday 15 December.
4. *late visit*: According to Hirst's Journal, Tyndall and Bevington had visited Queenwood on Saturday 1 December and returned to London on Monday 3 December.

To Thomas Archer Hirst 11 December 1855 1164

11th. Dec. 1855.

My dear Tom

I have read your pretty note[1] to me with a great deal of interest for it held up a glass before me which revealed myself to myself in a manner in which I had not appeared to myself previously. With regard to Bevington he is a man of clear and subtle mind and who has thought deeply on many matters—He has once or twice through inadvertence on my part placed me in difficulty and compelled me to ground arms. He chooses in his own quiet way the points over which he has thought and the probability I find is that his thoughts lead him to a correct conclusion. He is as you say a sweet soul—one of the best and cleanest hearted men that I have ever met. But you really make me aware of peculiarity in my own case that I was not before acquainted with—though I have no doubt that your observations are perfectly correct. Perhaps this trying of the temper of others may be referred to some fault in my own—some result of ill health, or perhaps a result of impatience with some <u>philistine</u> with whom you have heard me argue. But you say you have felt it in your own case. This I have scarcely ever observed, never has mortal father I think less to complain of regarding his son than I have in this respect regarding you. I know I have tried Debus's temper and Haas's temper sometimes, but this was generally when they shot illogically wide of the mark which they ought to have kept in mind. In general I have no thought of trying the temper of the man with whom I argue and if I do so it must be through some peculiarity of which I have not yet become conscious. About going down[2] I can say nothing as yet. It is almost too soon after my last visit. Should my thoughts take shape about it during the week I will let you know. I am very sorry to learn that Anna has not yet got rid of her cold, but if companionship in misfortune enables her to bear it more patiently, she has me for a companion. I have lately been greatly afflicted by a cold. Bence Jones prophesies dark things concerning me if I do not mend my ways. So I have agreed to follow his advice for the future. While walking towards the Institution[3] yesterday I noticed a nice little boy trotting along by my side. At length he looked up in my face & said it was very cold. 'It is my boy' I replied, patting his little shoulder, 'I live in Bond Street'[4] he said 'and have been at school—now I am going home'. Thus we talked. Once I was going to make the little fellow a present, but the goodness of his dress and superiority of his general appearance dissuaded me. At a street corner he vanished as if by magic and I have not seen him since. Trifling as the thing is it pleased me.

Today I was walking up Mount Street,[5] when a one horse carriage dashed by me. The horse lashed out furious kicks as he passed me and ran straight for a lamp post. The carriage came smash against it and I heard a scream from within. The horse continued kicking & plunging and the driver appeared to have lost all power over him. In an instant the handle of the carriage door was in my hand and in the next a fine girl in my arms—a little boy was also in the carriage whom I rescued. The horse was finally secured and I suppose by this time bound down to his good behaviour. The lady absolutely never looked at me. She escaped with remarkable velocity into an adjacent shop.

I told you I think that Matteucci had been writing to Faraday about my experiments, and stating that they were absolutely incorrect. I had a letter[6] from him today and the following are the concluding words of it.

'En attendant je voulais vous faire savoir que j'avais trouvé que vos expériences sont bien, et comme j'avais écrit a Grove et à Faraday que j'en doutais, je vous prie de lui dire de ma part que je suis parveneu à une complète verification des experiences.' | avec respect et attachement | C. Matteucci.

So much for the amend honorable, he finds me quite correct.

I have just finished a paper[7] which I wish you had time to read over, but you are so busy at present that it would be monstrous to ask you. Now I think I have written you a pretty amusing letter—this as much for Anna as for you.

yours affectionately | John

[Postscript]

Will you just ascertain in a quiet way whether Wright has got Tennyson's poetry—and if not I should like to present them to him at Christmas. You would like him to of course. He is a worthy little fellow and I like to make him feel now & then that I have an esteem for him.

RI MS JT/1/T/613

---

1. *your pretty note*: See letter 1163.
2. *About going down*: a reference to his visiting Queenwood College at some future date.
3. *Institution*: the RI.
4. *Bond Street*: a street running parallel with Albemarle Street immediately behind the RI.
5. *Mount Street*: a street in Mayfair where Tyndall moved in October 1856.
6. *letter*: Tyndall quotes from the end of letter 1161; see this for a translation.
7. *finished a paper*: 'Further Researches on the Polarity of the Diamagnetic Force' was read to the RS on 20 December 1855 and published in *Phil. Trans.*, 146 (1856), pp. 237–259.

From William Baker[1] 14 December 1855 1165

Lead Works Sheffield | Dec 14th 1855

My dear Sir

I beg to hand you 4 castings of our refined lead which I hope is sufficiently pure to answer your purposes.[2] My last analysis gave in 100 parts

Copper 000.0028
Iron 000.0042
Silver 000.0022

I cast them by pouring into a heated mould—continuing to pour until the metal solidified at the top in order to ensure as much as possible uniform density. I enclose also one waster[3] which I have immersed in dilute nitric acid for 24 hrs—as it shows very prettily the crystalline structure of the metal & may afford you clearer insight of the nature of the objects you are experimenting upon.

If you think a nearer approach to absolute purity is necessary for your investigations I shall be happy at some time to furnish you with a bar or two, which I might prepare at my leisure.

Yours very truly | William Baker
John Tyndall Esq.

RI MS JT/1/B/9

1. *Baker*: William Baker (born in London in 1830; Ancestry) was a professional chemist and manager of the White Lead Mill on the River Sheaf. The works had opened in 1759 to manufacture pigments for paint and pottery glazes (1861 Census).
2. *answer your purposes*: This possibly had reference to the Pelletier–Seebeck effect which Gustav Magnus had investigated with different metals in 1852. In his 'Reports on the Progress of the Physical Sciences' in January 1852, Tyndall had abstracted Magnus's paper 'On Thermo-Electric Currents', *Phil. Mag.*, 3 (1852), pp. 81–90. In a table summarising Magnus's results on p. 87, Tyndall noted that the results with lead were 'uncertain'. Perhaps this led him to seek purer samples?
3. *waster*: an imperfect specimen.

From James Bevington 15 December 1855 1166

Neckinger Mills[1] Bermondsey | London, 15th Dec. 1855

My dear Tyndall,

I have just put a cigar in my mouth that I may sit down at ease and have a chat with you.

I retain Hirst's note[2]—It is delightful to see two friends so familiar and yet so earnest that either can set to and canvass each other's weak points as though they were his own—I can't say that I have ever felt any thing in the tone of your conversation or argument to try the temper. this may be special with me because I feel that when we talk on scientific subjects I am but an explorer where you are familiar with all the tracks and landmarks, and I naturally defer to your opinion though I raise an objection out of a spirit of fun and for conversation—one better informed or more confident in his own opinion might possibly feel your mode of argumentation differently, but I cannot think it is a fault that has grown in you to any great excess and a little timely attention may easily set it right. Conversation is a delightful gift and tis well to aim at perfection in so charming an art. No one is so humble but that something may be learned from him and no one so intelligent but that he may be pleased with the mode in which a thought is communicated. It is a pity that Hirst does not give vent to his imaginings through that modest voice of his, but he seems preeminently the man of reflection and the writer and perhaps we should be jealous if he had more gifts in the same degree. However I am sure I don't half know him yet, a few amiable traits have come to me but I don't like him the less because there is so much to explore. To me he is like a country full of mountains and defiles[3] that would take a life to exhaust.

Can you spend Xmas day with us?[4] we shall dine as usual on turkey and plum pudding without any stranger to partake of the hospitable meal, save in the kitchen where I understand there will be posse of old retainers.

Mrs Bevington[5] bid me ask you to come but as I know you have many older friends and acquaintances and may have already appropriated the day I shall not calculate upon you till I hear.

Geoffrey[6] came home last night, overdone with cold and travelling. He kept his bed to-day but I hope will be himself again in a day or two.

I read aloud the essay[7] on the labours and writings of Faraday. It is a work of love well performed but I do not agree with the conclusion—The State in my view should confine itself to the preservation of its own and to the life of its own and to the life of its members—all the rest should be left to individuals and voluntary associations.

But I know my views are peculiar and perhaps not tenable as I would limit the guardianship of property by the State even more than people are now inclined. For instance I would cut away the patent laws and abolish much that has been done recently in protecting authors and copyright[8]—I think we should have a far better set of writers without copyright and turn to the right[9] about many a compiler to a more useful occupation. I don't find any trace of these laws in the writings of Moses and even in reference to the produce of labour he had some liberalities unknown to our laws[10]—In his country a stranger might pluck the ripe grape and eat his fill, so long as he refrained from pocketing.[11]

After this long yarn I will wish you a merry Xmas and say farewell my good fellow | J.B. Bevington.

RI MS JT/1/1/151
LT Typescript Only

---

1. *Neckinger Mills*: The river Neckinger, which rises in Southwark and drains into the Thames at St Saviour's Dock. is now one of London's 'lost rivers', having been enclosed during slum clearance in the 20th century. The Neckinger Mills, which were built originally as a paper factory, were transformed into tanneries by the Quaker family of the Bevingtons in 1805. Leather making continued there until the 1950s, when the firm moved to Leicester. (Geoffrey Bevington, *Bevington & Sons Ltd 1795–1995,* privately printed, Leicester: Bevington & Sons, 1991.)
2. *Hirst's note*: Letter 1163 which Tyndall had evidently sent to Bevington.
3. *defiles*: a narrow gorge between mountains.
4. *Xmas day with us*: Tyndall accepted this invitation. RI MS JT/2/13c/ 811.
5. *Mrs Bevington*: Mary Bevington (1804–73) was born in Ireland. No record of her maiden name or date of marriage has been found (Ancestry).
6. *Geoffrey*: Geoffrey Bevington (1839–72), the Bevingtons' youngest son who had been educated at Queenwood College.
7. *the essay*: This was probably Hirst's review for the *Westminster* Review. See letter 1162, n. 2.
8. *protecting authors and copyright*: The revised Copyright Act of 1842 had increased the protection of authors' writings from 28 years to 42 years, though in practice authors had copyright for the whole of their lives plus seven years. The Patent Law Amendment Act of 1852 had established a central Patent Office in London.
9. *and turn to the right*: sense unclear; possibly a mistranscription by LT.
10. *liberalities unknown to our laws*: possibly a reference to keeping the Sabbath, Exodus 20: 8–11.
11. *pluck the ripe grape . . . refrained from pocketing*: a possible reference to Leviticus 23:22.

From Jules Duboscq 15 December 1855 1167

Dec 15 55[1]

Monsieur

J'ai l'honneur de répondre à votre honorée par laquelle vous me demandez quand je pourrais vous livrer les appareil que vous m'avez commandé. J'ai aussi reçu une lettre de Mr Jones qui me donne la liste des appareils les plus pressé je vous ferais ce premier envoi dans le courant de Janvier, si vous aviez besoin pour vos leçon de quelques autres appareils soyez assez bon je vous prie pour me le faire savoir par un mot et je tâcherai de vous satisfaire.

Je vous enverrais avec les appareils un programme des cours de la Sorbonne ainsi qu'une petite Brochure. C'est un petit traité de manipulation de mes appareils qui a été fait par Mr Du Moncel un amateur.

Veuillez Monsieur recevoir je vous prie mes très humbles salutations et me croire votre très dévoué serviteur.

J. Duboscq

15 December 1855

Sir,

I have the honour to respond to your letter in which you ask me when I could deliver you the instruments that you ordered from me.[2] I also received a letter from Mr Jones,[3] which gives me the list of the most urgent devices. I will send you the first shipment during the month of January. If you should need some other instruments for your lectures, I ask you to be kind enough to let me know with a word, and I will try to satisfy your request.

Along with the instruments, I will send you a programme of courses at the Sorbonne, as well as a small brochure. It is a small treatise on handling my equipment, which was done by Mr Du Moncel,[4] an amateur.

Please receive, Sir, my humble greetings and believe me your much devoted servant.

J. Duboscq

RI MS JT/1/D/184

1. *Dec. 15* 55: The date is obscured and may be 15th or 16th of December.
2. *ordered from me*: letter not found, but see letter 1131.
3. *Mr Jones*: Henry Bence Jones acting in his capacity of secretary of RI.
4. *Du Moncel*: see letter 1109, n. 18.

To Peter Riess 15 December 1855 1168

15th Dec. 1855

My dear Prof. Riess.

I have just read your letter to Faraday[1] carefully over and cannot deny myself this pleasure of saying how much I have been gratified by the perusal of it. It is a plain manly statement, and believe me it is appreciated by Faraday as such. He read a portion of it for me this morning with evident gratification, and this evening he handed the entire letter over to me—It will appear in company with his letter to you in the next number of the Philosophical Magazine.[2] Believe me you judge Faraday right when you say that he prefers the utterance of an independent opinion to 'Gedankenlose Nachahmung'.[3] I do not agree with him in Magnetism, and he knows that right well. I do not believe in his lines of force as he believes in them; but he never thinks of swaying me, though a young man, of the weight of his authority. Your concluding remarks[4] were peculiarly pleasant to me, for they express the truth in a philosophic spirit. No doubt of it Faraday will work better with his self made tools than if he adopted the tools of another: but I have not the least doubt that every result which he has won by using these tools can be brought into accordance with theories very different from his own. It would be a work of extreme ease to go over all that he has written for some time regarding lines of force, conduction polarity &c. and to shew that every experimental result he has obtained follows in the most simple and evident manner from the fact that the forces with which he deals are polar forces. But as you say he will employ his own weapons best, and the noble use he has made of them ought to prevent us from wishing to see them changed. I had told him before he received your letter of the mistake of translation[5] to which you refer. The translation was not made by me, but on the whole it appears to be a very good one.

I have been working at diamagnetic polarity lately. Matteucci, Feilitsch, and others you know denied Weber's results, and proposed certain conditions for the rigorous demonstration of the principles. I have satisfied all their conditions by proving the polarity of diamagnetic conductors & insulation, solids & liquids. Heavy glass, sulphur, calcareous spar, statuary marble, nitre, phosphorous, wax and other insulators have all been proved to be polar.

In a series of letters to Faraday[6] & others Matteucci denied the results described in my last Bakerian Lecture[7]—He has made the amende honorable[8] by writing to me thus a few days ago.

'En attendant je voulais vous faire savoir que j'avais trouvé que vos experiences sont bien, et comme j'avais écrit à Faraday et à Grove que j'en doutais je vous prie de lui dire de ma part que je suis parvenu à une complete verification des experiences'![9]

Pray present my respects to Mrs. Riess[10] and to your family. It is long since I heard from my friend Prof. Magnus. Will you kindly say to him that the first part of his memoir on Hydraulics will appear in the next Phil. Mag.[11] Kindest greetings to Professor Rose[12] & Poggendorff,[13] and to DuBois when you see him.—and now goodbye.

| and believe me always | Yours very sincerely | John Tyndall

Staatsbibliothek zu Berlin (Stabi) Slg Darmstaedter 1855 Tyndall Be. 1–58, 48–9

---

1. *your letter to Faraday*: Faraday had written to Riess on 19 November 1855; Riess translated and answered it on 10 December 1856 as 'Über die Wirkung nicht-leitender Körper bei der elektrischen Induction', *Poggend. Annal.,* 97 (1856), pp. 415–41. Tyndall arranged for both letters to be published. See note 2.
2. *Philosophical Magazine*: M. Faraday, P. Riess, 'On the Action of Non-Conducting Bodies in Electric Induction', *Phil. Mag.,* 11 (1856), pp. 1–17; Faraday (pp. 1–10), Riess (pp. 10–17).
3. *Gedankenlose Nachahmung*: thoughtless imitation (German).
4. *Your concluding remarks*: After admitting he could not persuade Faraday about the insulation of conductors, Riess wrote: 'The great philosopher works best with the help of his own conceptions, his self–made tools, whose imperfections he avoids by dextrous applications. But these tools, so efficacious in his hand, are not only useless but very dangerous in the hands of others' (p. 17).
5. *the mistake of translation*: Riess's earlier article 'On the Action of Non-Conducting Bodies in Electric Induction' had been translated from *Poggend. Annalen,* 92 (1854), pp. 189–92 in *Phil. Mag.,* 9 (1855), pp. 401–13. It evidently did not meet with Riess's approval.
6. *letters to Faraday & others*: not in *Faraday Correspondence.*
7. *Bakerian Lecture*: see letter 1031, n. 2.
8. *amende honorable*: a public apology to restore the honour of one who was offended (French).
9. *des experiences*: 'In the meantime I wanted to let you know that I found your experiments to be well made, and since I have written to Faraday and to Grove that I doubted them, I beg you to tell them on my behalf that I came across a full verification of the experiments.' The quotation is from letter 1161.
10. *Mrs Riess*: Riess married Fanny Gütterbock in 1835. Their only daughter Rebecca married the physicist Georg Quincke (1834–1924) in 1863 (ADB).
11. *in the next Phil. Mag.*: G. Magnus, 'Hydraulic Researches', *Phil. Mag.,* 11 (1856), pp. 89–107.
12. *Professor Rose*: Heinrich Rose (1795–1864), Professor of Chemistry at the University of Berlin.
13. *Poggendorff*: Johann Christian Poggendorff (1796–1877), Professor of Physics at the University of Berlin and editor of the *Annalen der Physik* since 1824 (DSB).

From Lord Ashburton 20 December 1855 1169

The Grange | Dec. 20th

My dear Sir,

Would you give Lady Ashburton[1] and myself the pleasure of a visit from you at any time before the 14th of Jan. My station is Andover road.[2]

Your very truly | Ashburton.

RI MS JT/1/121
LT Typescript Only

1. *Lady Ashburton*: Harriet Mary Montagu (1805–57) married William Bingham Baring (Baron Ashburton from 1848) in 1823. She was an outstanding hostess and intimate friend of Carlyle's. ODNB.
2. *Andover road*: a railway junction near the Ashburton's country seat, The Grange.

To Thomas Archer Hirst 22 December 1855 1170

22nd. Dec. 1855.

My dear Tom

My mind was in a strait a few minutes ago whether I should open my journal and write there or write to you. Your personality however warmed my thoughts and decided the direction of my pen. I write although I have nothing to write about; Christmas is drawing near, and one ought to be genial cheerful and communicative whether one is or not. I think we are best off when some madness besets us, faith in some form or other, which enlivens our marrow and prevents us from falling into a leaden lethargy in which we resemble vegetables rather than men. There are few things more contemptible to me than this lethargy which I feel creeping on me sometimes and which converts me into a mere seeing fossil in God's creation. I have been reading this evening the silly lecture of Dr. Hare[1] of Philadelphia on Spirit rapping and Table turning, and when I closed the paper the thought presented itself to me, that even the folly of this man, prompting him as it does to intellectual and bodily activity, is more respectable than that putrid indifference into which men of deeper insight sometimes fall. I have been reading in the evenings lately the Life and Works of Goethe, written by Lewes of the Leader.[2] The books interest me deeply, especially where they give me the naked facts of Goethe's life, and thus enable me to interpret them for myself. The work is dedicated to Carlyle as a result of the intellectual guidance which he has given to the author. Appropos of Carlyle—We have often said that we should one day gravitate towards each other, and this is likely to be fulfilled very soon. I was surprised this morning by a note[3] from Lord Ashburton asking me to pay him & Lady Ashburton a visit at his seat in Hampshire. The invitation was kind and straightforward so I wrote back saying that I should join them on the 11th of January. I learned afterwards that Carlyle is to be their guest at the same time, and if so I shall feel a high pleasure in taking the old lion by the mane. Yesterday we had an animated dinner at the Philosophical Club:[4] a goodly company were present. Hooker, Graham,[5] Grove, Playfair,[6] Gassiot, Ansted,[7] Miller[8] of Cambridge whom I saw for the first time. Stokes, Hopkins,[9] Sabine, Wheatstone, Murchison,[10] Lord Wrottesley &c. &c:[11] a kind of gravitation seems established between myself and Wheatstone—at all events the fates or chances throw us very much together. At the meeting of the Royal Society afterwards I was called upon to give an account of my recent experiments which I did much to the apparent satisfaction of every body. The weather is intensely cold here and I suppose still more so in Hampshire. There is an east wind keen & cutting as a Saracen's sword[12] which searches a man to the marrow. I cannot complain of my health, the awful cold which smote me a fortnight ago is gone and I am able to work a good deal. To my inexpressible satisfaction I find that my bowels have returned to a normal condition, and this without medicine or any other aid save change of diet recommended to me by Bence Jones. I like Bence Jones I am happy to say more and more. He is a man like Goethe's Manager in the prologue to Faust, [13] thoroughly downright & practical, and although we cannot come into contact with each other at all points, still the man's practical solidity and withal real good nature have engendered in me a very pleasant feeling towards him. Doubtless each of us knows the other better than when we became first acquainted. I have promised to dine with Bevington on Christmas day: he wrote me a long letter,[14] referring to his visit to Broughton which I would send to you were it at hand. Faraday I am sorry to say is a little out of order, but it is an affection of the stomach which I trust will speedily pass away. His lectures are drawing near: he and I are still like father and son: this evening I had a religious discussion with him—yesterday I had a scientific one, and I think were you present you would not have detected much of that irritating quality which you have recently revealed to me as my possession. This quality I shall endeavour to put my heel upon, for it must be there or you would not notice it. I find my new lodgings[15] comfortable but still I think I will leave them before the end of the year: and for this reason. I agreed when I came to pay so much rent, and to be charged exactly as they paid for other things. I find however that whenever an opportunity is given for a little extortion that it is invariably made use of. To whatever they buy for me they always add a stiff percentage I dont like this—it makes my thoughts unpleasant. I have spoken about the matter once, tomorrow I think I shall feel disposed to give my landlord—a respectable tradesman who attends church punctually!—notice that I shall leave him. And now I have charged you with gossip, and I have only to crown it by wishing you a happy Christmas. Long may your soul enjoy its life: long may your heart beat warm, and forty years hence when I am in the dust and you stand looking after me across the bourne of life, may you forget the failings of your friend and think with the freshness of youth upon those qualities common to both which for a time welded you and him together.

as ever | John

RI MS JT/1/T/614

---

1. *Dr. Hare*: Robert Hare (1781–1858), Professor of Chemistry at the University of Philadelphia until his retirement in 1847. He was ridiculed for his *Experimental Investigation of the Spirit Manifestations* (New York: Partridge & Brittan, 1855) to which Tyndall referred.
2. *Lewes of the Leader*: G. H. Lewes, *The Life and Works of Goethe,* 2 vols. (London: D. Nutt, 1855). Lewes had founded and edited the weekly *Leader* since 1850.
3. *a note*: see letter 1169.
4. *Philosophical Club*: see letter 1085, n. 8.
5. *Graham*: Thomas Graham (1805–69), Scots chemist, Professor of Chemistry at University College London (1837–55), Master of the Mint (1855–69), and FRS (1836).
6. *Playfair*: Lyon Playfair (1818–98), Professor of Chemistry at the Royal School of Mines; Secretary for Science and Art in the Department of Science and Art (1853–58), and FRS (1848).
7. *Ansted*: David Thomas Ansted (1814–80), Professor of Geology at King's College, London, and FRS (1844).
8. *Miller*: William Hallowes Miller (1801–80), Professor of Mineralogy at Cambridge (1832–80), and FRS (1838). See letter 1076, n. 2.
9. *Hopkins*: William Hopkins (1793–1866), mathematician and geologist; private coach for Mathematical Tripos at Cambridge, and FRS (1837).
10. *Murchison*: Roderick Impey Murchison (1792–1871). In 1855 he was appointed Director General of the Geological Survey of Great Britain and Director of the Royal School of Mines and the Museum of Practical Geology in Jermyn Street; knighted (1846), and FRS (1826).
11. *Lord Wrottesley &c. &c.*: see Tyndall's Journal for 20 December 1855, RI MS JT/7/810.
12. *a Saracen's sword*: the type of sword used by the Saracens in the Holy Land when they fought the Crusaders. Used here metaphorically.
13. *Goethe's Manager in the prologue to Faust*: In the first Prologue to *Faust,* three characters (the Manager or Director, the Poet, and an Actor) discuss the purpose of theatre. The Manager's purpose is to make money. Bence Jones was one of the wealthiest physicians in London.
14. *a long letter*: see letter 1166.
15. *my new lodgings*: Tyndall moved from lodgings in Mount Street, Mayfair to ones in South Audley Street, Mayfair on 1 January 1856.

To Peter Riess 22 December 1855 1171

22nd Dec. 1855

My dear Sir

I have just one word to say with regard to your letter of the 19th.[1] I have no commercial connexion whatever with the Philosophical Magazine—My name stands on the cover at the request of the proprietors. When I meet an interesting paper I recommend its translation but I have no time to devote to translation myself. I recommended your paper[2] to be translated but beyond that I am not responsible. I sometimes do not even read the magazine for 2 or 3 months together so that whether things are rightly translated or not is altogether beyond my knowledge. All I can do now is enclose your errata to Dr Francis and to request him to attend to it.

I am acquainted with the paper of Quintus Icilius[3]—I am not aware that he has added anything to the results of Weber as he has operated with Bismuth only. It grieved me to hear of the loss sustained by my friend Knoblauch[4]—it is a terrible infliction and will be felt keenly by him.

To Du Bois offer my warm congratulations[5]—His letter to Prof Magnus[6] will I believe appear in the next number of the Philosophical Magazine—I said in my last note to you that Prof. Magnus's paper would appear in the next number. This Francis now informs me will be impossible. The paper however is translated and will appear in an early number.[7]

Yours very sincerely | John Tyndall

Staatsbibliothek zu Berlin (Stabi) Slg Darmstaedter 1855 Tyndall Be. 1–58, 50–1

1. *letter of the 19th*: letter missing.
2. *your paper*: P. Riess, 'On the Action of Non-Conducting Bodies in Electrical Induction', *Phil. Mag.*, 9 (1855), pp. 401–13.
3. *Quintus Icilius*: Gustav von Quintus Icilius (1824–85), 'Ueber die Polarität des diamagnetischen Wismuths', *Poggend. Annal.*, 96 (1855), pp. 81–109 [On the polarity of diamagnetic bismuth]. As a Privatdocent at the University of Göttingen, Quintus had assisted Weber in 1852. He eventually became Professor of Physics at the Royal Technical University in Hannover.
4. *Knoblauch*: Hermann Knoblauch (1820–95), German physicist who had studied with Gustav Magnus in Berlin. Professor of Physics at Marburg (1852–53), where he worked with Tyndall, and Professor of Physics at Halle (1853–95). Tyndall evidently knew that Knoblauch's son and wife had died before receiving letter 1180 below.

5. *my warm congratulations*: Du Bois-Reymond had been elected Associate Professor of Physiology at the University of Berlin.
6. *his letter to Prof Magnus*: 'On a Method of Exhibiting fine Galvanometric Experiments to a Large Audience', *Phil. Mag.*, 11 (1856), pp. 109–11.
7. *in an early number*: G. Magnus, 'Hydraulic Researches', *Phil. Mag.*, 11 (1856), pp. 89–107.

To Auguste Arthur De la Rive 23 December 1855 1172

Royal Institution, 23rd, Dec, 1855.

My dear Sir

From the interest you exhibited in the experiments on diamagnetic polarity during your visit to London[1] I think it will not displease you to learn the further progress of the enquiry, an account of which I had the pleasure of giving to the Royal Society a few days ago.[2] You know the disposition of the apparatus and you saw the experiments with bismuth and heavy glass. For the sake of clearing the subject of all doubt I thought it right to accept the conditions laid down by those who had opposed the notion, for the rigorous demonstration of diamagnetic polarity. One of these conditions, was that we must operate with insulated fragments of bismuth: Well I took a quantity of the powder of this metal, and exposed it to the air, it soon became so much tarnished as to be totally unable to conduct a voltaic current, but this powder enclosed in tubes gave me almost as great a deflection, and always in the same sense, as that of massive bismuth. Antimony is a better conductor than bismuth, but its diamagnetic power is feebler: the effect produced by this metal is less than that of bismuth, thus proving that it is upon its diamagnetic quality, and not upon its conductive quality, that the effect depends. Further, copper is perhaps fifty times a better conductor than bismuth, but its diamagnetic power is nearly nil: with this metal no sensible effect was obtained in my experiments; which, of course, could never be the case if the effect were due to induced currents. Passing on to positive insulators, I have added to the heavy glass; cylinders of sulphur, statuary marble, wax, calcareous spar, nitre and phosphorous; all these exhibit polarity equally distinct (though in a greatly diminished degree) to that of bismuth. A marble statue therefore, acted upon by the earth's magnetism, has its feet a south pole and its head a north pole, being the reverse of that of an iron statue under the same circumstances: and there cannot be a shadow of doubt that each human being as he walks the earth's surface is similarly polarized, for all the tissues of the his body are diamagnetic. I have extended the examination to liquids and find that solutions of muriate of cobalt, muriate of nickel and sulphate of iron enclosed in glass tubes and surrounded by an electric current, behave exactly as cores of soft iron would do, while distilled water, bisulphide of carbon and other diamagnetic liquids behave exactly as cylinders of bismuth. Were it necessary I should have no difficulty in increasing the number of bodies examined.

All the phenomena of magnecrystallic action can be deduced with the greatest ease à priori from the principle of polarity, and it is impossible to account for many of these phenomena in any other principle. It is surprising that this has not been more insisted on during the discussions which have taken place upon this subject. Some of the earlier experiments of Mr Faraday which he adduced to prove the notion he entertained at the time, that the magnecrystallic force was neither attractive nor repulsive are to my view perfectly demonstrative of the fact that the force must be both attractive and repulsive, or in other words must be polar. As this question has been already worked at by Professors Thomson and Plücker I do not think that I shall make it the subject of a special communication to the Royal Society, but I hope in an early number of the Philosophical Magazine to indicate in a brief and simple manner and from my own point of view the relation of the question of diamagnetic polarity to that of magnecrystallic action.

I have had a pleasing letter[3] from M. Matteucci lately. Soon after the receipt of the Bakerian Lecture[4] for the present year he occupied himself in repeating the experiments there described but found that he could not obtain my results. It was natural to suppose that I had made some grave mistake, and not wishing to expose my presumed errors publicly he wrote to Faraday and Grove upon the subject, asking their advice under circumstances which he considered embarrassing. M. Matteucci's letters[5] were placed in my hands by Faraday, but I had repeated my own experiments so often under such varied circumstances, and moreover in the presence of so many competent judges that I could not entertain a doubt of their accuracy. It was nevertheless pleasant to me to receive a few days ago from M. Matteucci himself a letter which concluded with the following words:—"En attendant j'ai envie de vous faire savoir que j'avais trouvé que vos expériences vont bien, et comme j'avais écrit à Faraday et à Grove que j'en doutais, je vous prie de leur dire de ma part, que je suis parvenu à une complete verification des experiences."[6] The indistinct words are equally indistinct in M. Matteucci's communication.

In the next number of the philosophical Magazine you will find an interesting discussion[7] carried on between Mr Faraday and Prof Riess of Berlin: it is very spirited but the spirit is also very good. Mr Faraday I am sorry to say is suffering a little from illness at present, but there is every prospect of its being very transitory. Next week he commences his course of juvenile lectures when I doubt not he will give the young people a great treat.

I saw with pleasure the 2nd volume of your great work on electricity[8] in the library of the Royal Institution yesterday, but as yet I have not had time to look through it.

My compression experiments have lately led me to a physico-geological question which at present possesses great interest, and an article on which by Mr Sorby will appear in the next number of the Phil Mag.[9] I mean the subject of slaty cleavage. It is a beautiful subject, and I hope ere long to be able to communicate to you some results connected with it. When you see Prof Wartmann would you kindly present my remembrance. I am also indebted to M. Soret[10] for some interesting memoirs; should you see him pray be kind enough to thank him for me.

Believe me dear Sir | most sincerely Yours | John Tyndall

RI/JT/1/335
LT Typescript Only

---

1. *during your visit to London*: De la Rive had been in London in May and August 1855. Tyndall had joined him in experiments on 14 August 1855. See Journal RI MS JT/2/13c/781.
2. *a few days ago*: J. Tyndall, 'Further researches on the polarity of the diamagnetic force', *Phil. Trans.*, 146 (1856), 237–259, read 20 December 1855.
3. *a pleasing letter*: see letter 1161.
4. *Bakerian Lecture*: J. Tyndall, 'On the Nature of the Force by which Bodies are Repelled from the Poles of a Magnet', *Phil. Trans.,* 145 (1855), pp. 1–51.
5. *Matteucci's letters*: This correspondence has not survived but is referred to in *Faraday Correspondence,* 5: 3033.
6. *verification des experiences*: The French quotation is the conclusion to letter 1161.
7. *discussion*: M. Faraday and P. Riess, 'On the Action of Non-Conducting Bodies in Electric Induction', *Phil. Mag.,* 11 (1856), pp. 1–17.
8. *your great work on electricity*: A. de la Rive, *Traité d'Électricité Théorique et Appliqué,* 3 vols. (Paris, 1853–58). The 2nd volume appeared in 1855. An English translation by the electrician Charles V. Walker was published 1853–58.
9. *Phil Mag*: H. C. Sorby, 'On Slaty Cleavage, as Exhibited in the Devonian Limestones of Devonshire', *Phil. Mag.,* 11 (1856), pp. 20–37.
10. *M. Soret*: Jacques-Louis Soret (1827–90), Swiss chemist, professor of medicine at the University of Geneva (*Gèneve, Mém. Soc. Phys.,* 31 (1890–91), pp. xxviii-xxxv). No letters

from Soret to Tyndall have been found for 1856; for later correspondence between 1866 and 1888, see the relevant volumes of *Tyndall Correspondence.*

From Edward Sabine 23 December 1855 1173

13 Ashley Place | S. W. | Sunday 23^{d} Dec.[1]

Dear Tyndall,

I put your note in my pocket and walked after Church to consult our President:[2] but he had gone to the Regent Park[3] for a walk.

I have no hesitation in expressing the opinion that you may set to work at once whilst your thoughts intent on the subjects. It is precisely a subject to which the Gov[ernment] Grant,[4] the funds of the Society itself, & the donation fund[5] are applicable: and I have no doubt whatsoever that your request for £100 from one or other of these sources will be acceded to.

I will however send this note (enclosing yours) thro' Dr. Sharpey,[6] who will give you his opinion.

Sincerely yours | Edward Sabine | John Tyndall Esq.

RI MS JT/1/S/544

1. *Sunday 23^{d} Dec.*: Dated by the fact that 23 December 1855 was a Sunday.
2. *our President*: John, Lord Wrottesley who lived at Blackheath to the south of Greenwich.
3. *Regent Park*: Regent's Park was one of the royal parks laid down by the Prince Regent (later King George IV) and the architect John Nash, from 1818 onwards. It was opened to the public in 1835.
4. *Gov[ernment] Grant*: An annual sum of £1000 administered by the RS from 1851 onwards on behalf of the government for the promotion of scientific research in the United Kingdom. For Tyndall's first grant from this source in 1852, see Sabine's advice in volume 3, letters 0613 and 0625.
5. *the donation fund*: an investment fund for the promotion of scientific research set up in 1828 through a legacy left to the RS by William Hyde Wollaston and to which Fellows donated. By the 1840s it produced an annual income of £120 per annum.
6. *Dr Sharpey*: William Sharpey (1802–80), professor of anatomy and physiology at University College London, and one of the joint secretaries of the RS between 1853 and 1872.

From Thomas Archer Hirst 24 December 1174

Broughton | Decr 24th/55

My dear John,

Thanks for your kind brotherly letter[1] which I read several times over. No failings of yours will ever overshadow my remembrance of you or ever affect the relation which always has and I am sure always will exist between us. I am almost sorry I ever mentioned your occasional manner of argument seeing you have thought so much about it. I do not think I should have done so had I alone noticed it, but within the last six months I have spoken to three or four people about you, good enough people in their way although incapable of seeing much further than mere failings, and I felt rather sorry to hear them express opinions or rather attribute qualities to you which were far below you. For my own part I am always too conscious that whatever you have to say is well worth hearing and considering to think much about the manner in which it is said. I know your heart and your motives far too well, to let any peculiarity of manner annoy me. But I might as well tell you that I have heard some of these good people[2] call you conceited, vain, pretentious and so forth. They do so it is true because they are unable to appreciate your character and are only able to fasten on a few superficial peculiarities of the same, and they are only worth listening to except for the purpose of attending to such little peculiarities of manner which if liable to misinterpretation had better be altered than not. But I have already bothered you too much about so trivial a subject. I should say there is not one of us who is not similarly talked about and with far more justice. The fact is I meet with so few who are not great admirers of you and I am so anxious that every body should be so that these few exceptions caused me to think more than usual about such scandal, and perhaps set me on the alert to see if there was any ground for their objections. As far as I am concerned I hope you will continue to be the same John Tyndall to me as you always were, I hope the welding will never weaken and that the conviction that you have in me as firm a friend as ever will cheer you as you grow older. For myself I am proud of your strength and of your successes. You may rest assured every new victory of yours is closely watched by me and rejoiced over as if it was my own. Go on John and God bless you.

The memoir[3] you have sent me is a valuable one and in my opinion well worth publication in the Phil. Mag. Is it similar to the one you have sent to the R. S.? After reading it over carefully I have only one or two suggestions to make. I think it would be best to be a little more explicit or complete on page 13. I would take more pains to shew that a disc

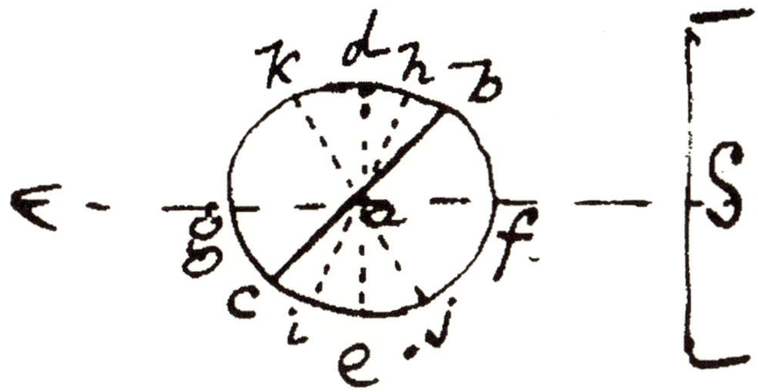

suspended from its centre *a* behaves as if it were a neutral body with a magnet or diamagnet placed along its axis bc. For as you have shewn this is a very important point. In your fig 2 you only consider two directions through the crystal. I think it would be better to mention all the intermediate directions.

If the magnetization along any line through a crystal is greater the less the angle it makes with the line of greatest magnetization. Then by comparing any two lines hi and jk equally inclined to the direction of the resultant of magnetic force it is clear that the magnetization along hi would be greater than along jk because the angle hab is less than the angle kab. The same being true of all lines in the quadrants daf and gae when compared with the corresponding lines in the quadrants fae and dag. The couples acting on the former would be greater than those acting on the latter and the crystal would turn as if a resultant couple were applied to the axis *bc*.

The explanation of Faraday's experiment with the sulphate of iron turning on a silken axis is very good indeed. With reference to your fig. 9 it would be best to state that the centre of the sphere ought to be placed nearly in the line of motion of the pole *n*. As far as the disposition shewn by your figure is concerned your explanation is very clear and precise, but it may possibly be thought to depend too much upon the figure by which the eye is apt to be misguided with respect to directions when actual distances are not given or plotted to a scale. With respect to the explanation referring to your fig. 10. It would be more complete if you were to mention the effect of the pole N on the other axes in the bar besides the extreme one *ns*. It is true by so doing your explanation would be more complicated, but it would be more satisfactory to your readers to know that you had considered the effect of the pole on all parts of the bar.

The memoir is an exceedingly interesting one and as usual very clearly written, it is very pleasing to see how all Faradays and Plückers results paradoxical as they appeared fall under the general & simple rule.

Yours affectionately | T. A. Hirst.

RI MS JT/1/H/221

1. *brotherly letter*: See letter 1170.
2. *some of these good people*: not identified.
3. *The memoir*: The memoir was published by Tyndall as 'On the Relation of Diamagnetic Polarity to Magnecrystallic Action', *Phil. Mag.*, 11 (1856), pp. 125–37.

From William Thomson 24 December 1855 1175

My dear Sir,

I have taken pains to make the enclosed letter[1] clear for the printers; and I trust you will excuse its form, as I have been under the necessity of finishing it in a steamer (at this moment shaking excessively) or else of letting it miss a post which I am anxious not to do, in order that it may if possible be in time for publication in the forthcoming Number of the Philosophical Magazine.

With the compliments of the season | I remain | Yours very truly | William Thomson

RI MS JT/1/T/32

1. *enclosed letter*: letter 1176

From William Thomson 24 December 1855 1176

Glasgow College, Dec. 24, 1855.

MY DEAR SIR,

I HAVE been prevented until to-day, by a pressure of business, from replying to the letter[1] you addressed to me in the Number of the Philosophical Magazine published at the beginning of this month.

You asked me the question, 'Supposing a cylinder of bismuth to be placed within a helix, and surrounded by an electric current of sufficient intensity; can you say, *with certainty,* what the action of either end of that cylinder would be on an external fragment of bismuth presented to it?'

In answer, I say that the fragment of bismuth will be *repelled* from either end of the bar provided the helix be infinitely long, or long enough to exercise no sensible direct magnetic action in the locality of the bismuth fragment. I can only say this with the same kind of confidence that I can say the different parts of the earth's atmosphere attract one another. The confidence amounts in my own mind to a feeling of *certainty.* In every case in which the forces experienced by a little magnetized steel needle held with its axis reverse along the lines of force, and a fragment of bismuth substituted for it in the same locality of a magnetic field, have been compared, they have been found to agree. In a vast variety of cases, a fragment of bismuth has been found to experience the opposite force to that experienced by a little ball of iron, that is, the same force as a little steel magnet held with its axis reverse to the lines of force; and in no case has a discrepance, from this law been observed. I feel therefore in my own mind a certain conviction, that even when the action is so feeble that no force can be discovered at all on the bismuth by experimental tests, such, in regard to sensibility, as have been hitherto applied, the bismuth is really acted on by the same force as that which a little reverse magnet, if only feeble enough, would experience when substituted in its place. Now there is no doubt of the nature of the force experienced by the steel magnet, or by a little ball of soft iron, in the locality in which you put the fragment of bismuth. One end of a magnetized needle will be attracted, and the other end repelled by the neighbouring end of the bismuth bar; and the attraction or the repulsion will preponderate according as the attracted or the repelled part is nearer. There is then certainly repulsion when the steel magnet is held in the reverse direction to that in which it would settle if balanced on its centre of gravity. In every case in which any magnetic force at all can be observed on a fragment of bismuth, it is such as the steel magnet thus held experiences. Therefore I say it is in this case repulsion. But it will be as much smaller in proportion to the force experienced by the steel magnet, as it would be if an iron wire were substituted for the bismuth core. Yet in this case the repulsion on the bismuth is very slight, barely sensible, or perhaps not at all sensible when the needle exhibits most energetic signs of the forces it experiences. You know yourself, by your own experiments, how very small is even the *directive* agency experienced by a steel magnet placed across the lines of force due to the bismuth core. You may judge how much less sensible would be the attraction or repulsion it would experience as a whole, if held along the lines of force; and then think if the corresponding force experienced by a fragment of bismuth substituted for it is likely to be verified by direct experiment or observation. I think you will admit that it is 'incapable of verification,' as well as 'incontrovertible' by any collation of the results of experiments hitherto made on diamagnetics. As to the concluding paragraph of my letter which you quote, you do me justice when you say you accept it as an expression of my 'personal conviction that the action referred to is too feeble to be rendered sensible by experiment.' I will not maintain its unqualified application to all that can possibly be done in future in the way of experimental research to test the mutual action of diagmagnetics under magnetic influence. On the contrary, I admit that no real physical agency can be rightly said to be 'incapable of verification by experiment or observation;' and I will ask you to limit that expression to experiments and observations hitherto made, and to substitute for the concluding paragraph of my letter the following statement, written for publication three days later, and published in the same Number of the Magazine as that to which you communicated my letter (Phil. Mag. April 1855, p. 247).[2] 'The mutual influence' between rows of balls or cubes of bismuth in a magnetic field, 'and its effects' in giving a tendency to a bar of the substance to assume a position along the lines of force, 'are so excessively minute, that they cannot possibly have been sensibly concerned in any phaenomena that have yet been observed; and it is probable that they may always remain insensible, even to experiments especially directed to test them.'[3]

I remain, my dear Sir, | Yours very truly, | WILLIAM THOMSON.

Dr. Tyndall.

Phil.Mag. 11 (1856), pp. 66–7
Printed heading: VII *On the Reciprocal Action of Diamagnetic Particles.* | *Letter from* Prof. THOMSON *to* Prof. TYNDALL | Glasgow College, Dec. 24, 1855.

---

1. *letter*: J. Tyndall, 'Letter to Prof. W. Thomson on Reciprocal Molecular Induction', *Phil. Mag.*, 10 (1855), pp. 422–23.
2. *my letter (Phil. Mag. April 1855, p. 247)*: W. Thomson, 'Elementary Demonstrations of Propositions in the Theory of Magnetic Force', *Phil. Mag.*, 9 (1855), pp. 241–48. See letter 1053.
3. *directed to test them*: Thomson was quoting from letter 1053.

From Henry Moseley 27 December [1855] 1177

32 Cumberland Terrace | Regents Park | 27 Dec.[1]

My dear Sir,

I will endeavour to alter the days appointed for your examination and will communicate with you further on the subject next week.

The scheme of our examination was drawn out by a Committee composed chiefly of officers of the Engineers and Artillery and have received the official sanction of the Ordnance I apprehend we have no alternative but to conform to it however much we may dissent.

Yours truly | Henry Moseley.

RI/MS/JT/3/900
LT Typescript Only

1. *27 Dec*: Unyeared like most of Moseley's letters. We have assigned it to 1855 on the basis that Tyndall was moving his lodgings and might have found the assigned examination date inconvenient.

To William Thomson 27 December 1855 1178

Royal Institution | 27th Dec. 1855.

My dear Sir,

The people at Red Lion Court[1] thoughtlessly forwarded your letter[2] to me without opening it, and thus lost the post which you saved. I took it back immediately and urged Francis strongly to publish it. This however he declares to be impossible this month. He may change his mind.[3]

I think the letter will pleasantly close the discussion, and if I have anything else to write about which I expect to have—I think the most satisfactory plan would be to write privately at first, & afterwards we could publish or not publish just as we thought necessary.[4] I have something to say with regard to the law of movement from stronger to weaker places of force & vice versa in the magnetic field; but at present I am too busy to take the matter up.

Wishing you many happy returns of the Season | Believe me | Most truly Yours | John Tyndall

RI MS JT/1/5/1549
LT Transcript Only

1. *Red Lion Court*: the premises of Taylor & Francis, publishers of *Phil. Mag.*
2. *your letter*: letter 1176.
3. *He may change his mind*: letter 1176 appeared in the January 1856 number of *Phil. Mag.* and not in December 1855.
4. *as we thought necessary*: Thomson does not appear to have responded to this suggestion.

To Thomas Archer Hirst 29 December 1855 1179

Friday night 29th Dec. 1855

My dear Tom

Your letter[1] reached me on Wednesday morning after my return from a hunt after lodgings,[2] and caused me to forget, during its perusal, the drizzle and the mud which had made my journey miserable. I mean to rest my head tonight, and while so doing to sing for my own amusement a lark-song in response to the noble organ tone of your letter. I am a miserable sinner Tom, I know; and deserved damnation in some form or other long ago. Those people who call me vain, conceited, and pretentious are quite right—and their tongues would refuse to utter an adjective base enough to depict me truly. I am proud, false, parsimonious, selfish, and I covet me neighbour's wife when she is handsome. There is no bit of rascality of which I am not capable; and now, as in all similar cases, when I have become a little exalted in life, my vices shew themselves in magnified proportions. The world will gradually become disgusted with me, and before long you will find me an object of general abhorrence. I thought of getting married this year, but the fact is my reputation stinks to such a degree that no girl will take me. I once thought there was some power for good within me, and indeed there is, but the love of evil is so predominant in my constitution as not only to neutralize the good to the last unit, but also to leave a clean balance of scoundrelhood which cannot fail to gladden the heart of Mephisopholes[3] when he makes it the subject of his infernal contemplation. It is not weakness merely that afflicts me; that feebleness of character which cannot save even good men's heads from becoming giddy upon sudden exaltation. It is a root of positive badness which sprouts into baneful luxuriance, and renders my every deed unwholesome. 'Conceited—vain, pretentious!' Little do the sweet lips know which uttered these simple adjectives the depth of the infernal within me. It is like calling Niagara[4] neat—John Haas courteous—or Helen of Troy[5] pretty—all which adjectives require to be multiplied by ∞ to give the correct mathematical expression of the fact. Now, there is a general confession for you—but not the confession of a smitten penitent, who from his scarred and humble marrow-bones asks God's pardon, but the confession of a reprobate who rejoices in his sin, and is far too much of a devil to entertain the slightest aspiration for the godlike and the good.

Why are these people who undertake to give me a character lenient to me? Partly, doubtless, because they have not seen enough of me of late years. I will venture to say that those who made use of the aforesaid adjectives have had only casual intercourse with me; for if they had waited to fathom my depths of depravity their sense of truth must have overcome even their tenderness, and prevented them from dealing in such descriptive sugar plums as 'conceited, vain, pretentious.' Nay I doubt not if they cast their glance back a little they will call to mind the partial exhibition of these fulsome qualities in days long gone by. I remember being once virtually charged by a Queenwood man[6] of warm imagination with a betrayal of confidence, with deliberately yielding of secrets confided to me in private correspondence. My reply, God pardon me, was that of Scipio,[7] who, when accused, tore the documents which would have cleared him; and this at a time when some pounds I had borrowed from you was the only plank between me and pecuniary perdition. Three days after I came to Queenwood from Germany[8] I had the pleasure of hearing my manner pronounced 'effusive' in a tone which caused me to smell like a dunghill in my own nostrils. I have been accused of crimes against the laws of honour, and religion, at the moment when my accuser was anxious to take the guilty one to his heart. I have been simply called impertinent and presumptuous, when, in reality, I deserved kicking for my impertinence and presumption. The fact is, Tom, you are not aware of half the folly that I have been guilty of in my day; nor do you know a tenth part of that forbearing love with which my delinquencies have been interpreted by my friends at Queenwood. Their hypotheses change concerning me as my position changes; but they are always marked with the same gentleness. At present the most reasonable and tender hypothesis is that I am puffed up—my sins being thus gently accounted for by persons who are far too noble to deny that were they in my position they would probably incur a slight danger in this direction.

I have been lately reading the life of Goethe.[9] That was the greatest scoundrel that ever lived, he beats me hollow Tom; though I, as aforesaid, am a rascal of no mean order. The following proportion holds good also:—

Wolfgang : John : : Weimar : Queenwood.[10] The leniency of the last term to me is only to be paralleled by the kindness of the third term towards Goethe. They only called him a <u>conceited upstart</u>—a <u>parvenu</u>[11]—a <u>fop</u>—and in after years, <u>imperious</u> and <u>unmannerly</u>. In these gentle terms did they express themselves regarding this most diabolical villain. Really Tom, as our friend Emerson says, there is more sweetness in the world[12] than we give it credit for; a truth which receives its most brilliant illustration, on different scales, at Queenwood and in Weimar.

And now Tom a sweet good night: it is near 12 o clock and I grow sleepy. I will weigh your hints upon my paper and adopt them when I can. The paper sent to the Royal Society[13] is of a totally different character, and it was the idea that the one I sent to you was not of sufficient importance to be sent to the R. S. which suggested to me the thought of sending it to the Phil. Mag.[14]—again good night Tom. Good night to my censors—sweet be their dreams and may the lap of fate be filled with Christmas puddings through long coming years for their especial jollification

As ever | John.

RI MS JT/1/T/615

1. *Your letter*: letter 1174.
2. *hunt after lodgings*: Tyndall moved from Mount Street, Mayfair, to South Audley Street, Mayfair on 1 January 1856.

3. *Mephistopheles*: the Devil or a demon. He was one of the principal characters in Goethe's *Faust*.
4. *Niagara*: The Niagara Falls in Canada, which Tyndall was to visit in 1872 and describe in a Friday Evening Discourse at the RI on 4 April 1873 (*Macmillan's Magazine,* 28 (1873), pp. 49–62).
5. *Helen of Troy*: the mythical daughter of Zeus and Leda who was supposedly the most beautiful woman in the world.
6. *by a Queenwood man*: not identified, but presumably a fellow teacher.
7. *Scipio*: Publius Cornelius Scipio Africanus (236–183 BC), Roman General who defeated Hannibal at the battle of Zama in the 2nd Punic War. In 187 BC his younger brother Lucius was accused of misappropriating tribute money paid to Rome by the defeated King of Syria. When Lucius (in his own defence) produced his account books in Senate, Scipio tore them to pieces and asked why bother about a missing 3000 Talents when 15,000 Talents had been received. Tyndall's source for this story is not known; he was unlikely to have read Livy's account of Scipio in Latin.
8. *to Queenwood from Germany*: Tyndall returned to teaching at Queenwood at the end of June 1851 but he made no record of the events mentioned in his Journal.
9. *Life of Goethe*: see letter 1170, n. 2.
10. *Wolfgang . . . Queenwood*: Tyndall is equating himself at Queenwood with Goethe living at Weimar.
11. *parvenu*: an upstart (French). In 1776 Goethe accepted the position of privy councellor to the young Karl August, Duke of Saxe-Weimar (1757–1828). There he faced a good deal of snobbery and criticism from the conservative court.
12. *there is more sweetness*: the source of this Emersonian paraphrase has not been identified.
13. *The paper sent to the Royal Society*: see letter 1129, n. 5.
14. *to the Phil. Mag.*: see letter 1174, n. 3.

From Hermann Knoblauch[1] 30 December 1856 1180

Halle. d. 30n Dec. 55

Mein theurer Freund.

Ich schreibe Ihnen heut in meiner Muttersprache. Denn was ich Ihnen heut zu schreiben habe, dafür habe ich in einer fremden Sprache, keine Worte. —Sie haben mich in Marburg zum letzten Male gesehen, als ich mit schwerem Herzen von einem mir theuer gewordenen Wirkungskreise schied; aber ich hatte ein reiches häusliches Glück, das mich in die neue Heimath begleitete. In dem damals geborenen Sohne, blühte uns eine herrliche Zukunft auf, für deren Eröffnung meine liebe Frau und ich dem Himmel, aus tiefster Seele, dankten.

Hier in Halle, aber, gerade um die Weihnachtszeit erkrankte unser armes

Kind und wir pflegten es Tag und Nacht, worauf es zu unsrer größten Freude genas, bis die Hitze es im Sommer von Neuem aufs Krankenlager warf, von dem die treueste Pflege, und die geschickteste ärztliche Hülfe, es nicht zu retten vermochten. Der Tod unseres damals einzigen Kindchens hat seitdem unser Leben mit einer nicht zu überwindenden Trauer durchgezogen, die durch ein 14 monatliches Uebel meiner lieben Frau, von der sie erst etwa vor einem Jahre hergestellt wurde, noch um so schmerzlicher wurde.—

Im August dieses Jahres nun war uns der Seegen zu Theil geworden, wieder einen Sohn zu besitzen. Der Gram um das Verlorene, hofften wir nun, werde sich in dem Anschauen des neugeschenkten Lebens vermindern und wir fingen wieder an, voll freudiger Zuversicht und mit dankerfülltem Herzen in die Zukunft zu blicken. Aber nicht lange währte unser gemeinsames Glück! 17 Tage nach der Geburt des Kindes wurde meine liebe Frau ganz plötzlich vom Schlage getroffen. Ich hatte sie eben noch, zwar angegriffen aber munter, verlassen mit den Worten, dass ich ungern an meine Experimente ginge, weil ich sie dann längere Zeit nicht sehen könne; sie hatte mir darauf selbst zugeredet an die Arbeit zu gehen und mir recht viel Glück dazu gewünscht indem sie noch hinzufügte sie dächte selbst an dem Tage noch recht fleissig zu sein, um ein Geburtstags-Geschenk für meinen Vater zu beenden.

Kaum aber hatte ich das Zimmer verlassen, als ich zurückgerufen wurde. Meine liebe Frau war mitten im frischen glücklichen Leben vom Schlage getroffen worden. Eine halbe Stunde darauf that sie den letzten Athemzug.

Sie können—da Sie mich seit Jahren näher kennengelernt haben—ermessen, wie dieser Verlust mich erschüttert, denn ich empfinde tief und vergesse nicht. Ich wünsche Ihnen von ganzem Herzen, daß der Himmel Ihrem Leben die harten Geschicke ersparen möge, die mir in der kurzen Zeit unserer Trennung beschieden gewesen sind. Und doch muß ich noch jetzt sagen, daß das kurze Glück, das ich trotz alledem genossen, selbst durch so namenloses Jammern wie er durch solche Verluste hervorgerufen wird, nicht zu theuer erkauft ist.—

Mir würde aller Lebensmuth entsinken, wenn ich nicht neben meinem Beruf noch einen Trost aus dem Anblick meines Kindes schöpfte, das mir jetzt alles ersetzen soll, was ich verloren habe und auf das ich nun meine ganze Zukunft baue. Möchte mir der Himmel dieses theure Vermächtniß meiner lieben Entschlafenen erhalten!

Es hat mir recht leid gethan, daß wir uns in diesem Sommer nicht, wie ich gehofft hatte, in Paris gesehen haben. Ich hatte mich, so oft ich nach der Ausstellung kam, in der englischen Abtheilung nach Ihnen erkundigt, Ihren Apparat gesehen und Ihr Portrait bei Duboscq mir zeigen laßen. Aber meine Zeit war wegen der Vorlesung, so knapp zugemeßen, daß ich fort mußte, als 3 Wochen abgelaufen waren. Es freute mich zu hören, daß mein Vater wenigstens noch das Vergnügen gehabt hat, Sie zu sehen. —Mein Aufenthalt in Paris ist fast durch Geschäftswege erschöpft worden; ich habe viel gelernt und für die Universitätssammlung bedeutende Erwerbungen gemacht. Da wir gleiche Zwecke hatten, so würde mir der Aufenthalt viel angenehmer gewesen sein, wenn wir zusammen dort gewesen wären.

Mit vielem Interesse habe ich die Abhandlungen, die Sie die Güte gehabt haben, mir durch Prof. Poggendorff zukommen zu laßen, gelesen. Die Trevelyan-Versuche haben durch ihre [Ihre] Untersuchungen eine überraschende Erweiterung bekommen. Die große Arbeit über Magnetismus und Diamagnetismus ist namentlich durch die erschöpfende Vielseitigkeit der Betrachtungen und die Mannigfaltigkeit der Versuche von durchgreifender Bedeutung. Ich habe mich in dieser Zeit viel mit Gedanken an Sie beschäftigt, auch wenn ich nicht eher an Sie geschrieben habe, was namentlich darin seinen Grund hat, daß jeder Brief in dem ich von meinem Unglück schreibe, die blutende Wunde immer von Neuem wieder aufreißt.

Ihr eben erhaltenes theilnehmendes Schreiben, in dem Sie mir, noch ehe ich selbst es Ihnen gemeldet, Ihre Theilnahme an meinem schweren Geschick aussprechen, mahnt mich, meine Zeilen nicht länger zurückzuhalten. Es hat mich innigst gerührt und ich danke Ihnen von ganzem Herzen dafür. Denn das Mitgefühl meiner Freunde thut mir sehr wohl.

Oftmals, und besonders in dieser Zeit, habe ich an die schöne Zeit zurückgedacht, welche wir in Marburg verlebt, auch an die Weihnachtsabende, die wir dort gemeinsam zugebracht haben. —Wie Sie von Prof. Frankland wißen werden, ist das gute Marburg in diesem Jahre vom Nervenfieber schwer heimgesucht. Frau Prof Fick; Syndicus Schotten; Frl. Weihrauch u. A. sind daran gestorben. Frl Mathilde Gerling, stud.[ent] Hüter, Frau Prof. Kürschner sind noch daran krank. So giebt es Leid überall!

Leben Sie recht wohl; haben Sie vielen Dank für Ihren freundlichen Brief und vergessen Sie nicht | Ihren | treuen, tief gebeugten Freund Herm. Knoblauch

Halle (6. Herkerstraße)

Ich bitte mich meinen Freunden allerseits in England bestens zu empfehlen.

Mein Vater, der in diesen Tagen hier war und gestern nach Berlin zurückgekehrt ist, lässt Sie bestens grüßen | K.

Halle, 30th Dec.55

I write to you today in my mother tongue: because that which I have to say to you I have in a strange language no utterance.[2] You saw me in Marburg last,[3] as with a heavy heart I separated myself from a sphere of action which had become innerly dear to me. But I had at the time a rich household happiness, which accompanied me to my new home. In the son then born to me we saw a blooming future, for the opening of which my dear wife and I thanked heaven from our deepest souls.

Here in Halle however just at Christmas our poor child[4] became ill; we tended it [him] day and night and to our great happiness saw it finally cured, until the heat of summer cast it anew upon the bed of sickness, from which the most tender care and the most skilful medicinal aid were unable again to rescue it [him] The death of our only child, introduced a sadness not to be surmounted into our lives; which was rendered more painful by an illness of 14 months of my dear wife from which she only first recovered about a year ago.

In August of the present year we were blessed with the possession of another son;[5] the sorrow for the lost one, we hoped now to see diminished in the contemplation of the life newly presented to us and we began again, full of cheerful confidence and with grateful hearts to look into the future. But our common joy did not endure long! 17 days after the birth of the child my beloved wife was struck suddenly by paralysis (plötzlich vom Schlage getroffen).[6] I had just left her, weak it is true, still cheerful and lively, with the words that I went reluctantly to my experiments for I could not hope to see her for some time. She counselled me in reply to go, wished me good success, and added that she hoped to be quite active on that day, preparing a birthday present for my father.[7]

Scarcely had I left the room when I was called back. My beloved wife, in the middle of her fresh and blooming life had been struck down. Half an hour afterwards she drew her last breath.

You who have known me closely for years may estimate how I was shaken by this loss: which I deeply feel and cannot forget. From my whole heart do I wish that heaven may spare your life the hard fate which has become mine since the brief time of our separation. And still I must say even now, that the brief happiness which despite of all I have enjoyed; even by this unspeakable woe which such a loss brings along with it, was not purchased too dearly.

All my life would sink were it not that besides my calling I have still a consolation in the presence of my child, that must now compensate me for all that I have lost, and on which I build my entire future. May heaven leave me this dear memento of the beloved sleeper who is gone.[8]

I was very sorry that we did not see each other in Paris this summer[9] as I had hoped. Every time I went to the exhibition, I asked after you in the English section; I saw your apparatus and had Duboscq's portrait[10] of you shown to me. But my time was so tight due to the lecture that I had to leave when 3 weeks had passed. I was pleased to hear that my father at least still had the pleasure of seeing you.[11] —My stay in Paris was exhausted almost entirely by business; I learned a lot and made some important acquisitions for the university collection. As we had similar objectives, the stay would have been much more pleasant for me if we had been there together.

I have read with much interest the papers that you have had the kindness to send me through Prof. Poggendorff.[12] The Trevelyan experiments[13] have got a surprising extension through your investigations. The large work on magnetism and diamagnetism is of far-reaching significance in particular through the exhaustive multiplicity of the considerations and the diversity of the experiments. I have been very much preoccupied with thoughts of you lately, even if I did not write to you sooner, which has its reason in particular in the fact that every letter in which I write about my misfortune always tears the bleeding wound open again anew.

Your sympathetic letter,[14] which I have just received, and in which you express your sympathy to me about my grievous loss, even before I mentioned it to you myself, admonishes me to hold back my lines no longer. The letter has moved me most deeply and I thank you with all my heart for it. The sympathy of my friends does me much good.

Often, and especially at this time, I have thought back to the nice time that we had in Marburg, also to the Christmas evenings that we spent together there. —As you will know from Prof. Frankland, good old Marburg was severely hit by typhus this year.[15] Prof. Fick's wife;[16] the advocate Schotten;[17] Fräulein Weihrauch[18] and others have died of it. Fräulein Mathilde Gerling,[19] the student Hüter,[20] Prof. Kürschner's wife[21] are still ill with it. Thus there is sorrow everywhere.

Farewell; thank you very much for your kind letter and do not forget | your | loyal, deeply stricken friend Herm. Knoblauch

Halle (6. Herkerstrasse)

I ask you to kindly greet all my friends in England.

My father, who was here the past days and returned to Berlin yesterday, sends you his best regards | K.

RI MS JT/1/K/21

1. *Hermann Knoblauch*: see letter 1171, note 4.
2. Tyndall translated the first part of Knoblauch's letter in his Journal for 2 January 1856 (RI MS JT/2/13c/812–13). The remaining part of the letter is our editorial translation.
3. *in Marburg last*: Tyndall moved from Marburg to Berlin on 21 April 1851; Knoblauch moved to the Chair of Physics in the University of Halle in the autumn of 1853.
4. *our poor child*: not identified.
5. *another son*: name not found, but he was killed during the Franco-Prussian war in 1870.
6. *Schlage*: *Schlage* strictly means a stroke.
7. *my father*: Karl Friedrich Knoblauch (1793–1859) who was a silk and ribbon manufacturer in Berlin.
8. Tyndall's translation ends here.
9. *in Paris this summer*: Knoblauch was referring to the 1855 Exposition Universelle in Paris where Tyndall had been a Juror.

10. *Duboscq's portrait*: Probably a photograph of Tyndall taken by Jules Dubosq. The photograph has not been traced.
11. *my father . . . pleasure of seeing you*: Tyndall failed to record this meeting in his Journal.
12. *Poggendorff*: see letter 1084, note 2. Tyndall had probably sent offprints of his Bakerian Lecture as well as the papers published in *Phil. Mag.* during 1855.
13. *Trevelyan experiments*: Tyndall had been introduced to 'Trevelyan's rocker' by Knoblauch in Berlin in August 1853 and experiments at the RI on the sounds produced by heated crystals had led to the paper read to the RS on 26 January 1854. See letter 1142, n. 11. For Trevelyan's rocker see, Ira M. Freeman, 'What is Trevelyan's rocker?' *Physics Teacher*, 12 (1974), p. 382.
14. *Your sympathetic letter*: letter missing.
15. *typhus this year*: Nervenfieber was the medieval German term for typhus, a bacterial disease that was endemic in Europe during the nineteenth century. The bacteria were spread by lice and fleas and caused high fever and delirium. The Marburg epidemic lasted from October to December 1855 and caused 110 deaths in the town.
16. *Prof. Fick's wife*: The wife of Adolf Fick (Frankland's brother-in-law) was born Emilie von Cölln (1829–1901).
17. *the advocate Schotten*: Karl Schlotten (1821–55).
18. *Fräulein Weihrauch*: Knoblauch's misspelling of Mathilde Wilhelmine Amoene Weyrausch (1829–7 November 1855), the daughter of Kaspar Weyrausch (1798–1860), the headmaster of the Knabenschule (primary school) in Marburg.
19. *Fräulein Mathilde Gerling*: probably an unmarried sister of Professor Gerling's.
20. *the student Hüter*: not identified.
21. *Prof. Kürschner's wife*: neither Professor Kürschner nor his wife has been identified.

1856

To George Gabriel Stokes 4 January 1856 1181

4th January–1856

My dear Sir

I have read Mr Joule's paper[1] carefully over—What my remarks upon the paper[2] in its former state were I do not accurately remember, but a comparison of those remarks with the paper in its present state condition would, I think, shew that what I formerly stated has not been rendered quite nugatory by the recent alterations. Were I permitted to express an opinion I would say that Mr Joule's eminence as an experimenter might be reasonably accepted as a sufficient guarantee that his communication is fit for the Philosophical Transactions; though I confess my own opinion is that the reputation of Mr Joule would not be increased by the publication of the paper. To my mind the reasoning appears loose, the conclusions do not seem to be in all cases warranted, and the mode of experiment seems doubtful. I say this with extreme reluctance, simply because it is my unpleasant duty to say it; and I hope the council will bear in mind, in forming their estimate of these remarks, that Mr Joule is a far older labourer in the domain to which his paper refers than I am.

I am dear Sir | very truly yours | John Tyndall
Prof. Stokes Sec. R. S.

RS MS RR/3/157

---

1. *Mr Joule's paper*: see letter 1116, n. 1.
2. *My remarks upon the paper*: letter 1116.

From James Bevington 14 January 1856 1182

Neckinger Mills, Bermondsey | London 14th. January 1856.

My dear Tyndall

I received your pleasant and familiar note[1] which like a lover's epistle has been carried about in my side pocket to charm away all the glummish and evil spirits that might attempt to invade my peace.

I consider it no slight acquisition to have one who can read favourably the yeast and ferment of the brain and judge that it proceeds from some sound material below—however crude and indifferently expressed.

I have received a bill of the Walnut Desk[2] which amounts to £3.

If you would like to have one less elegant and more accommodating for £2.15, you can have it changed without any trouble or inconvenience to me. Let me know and I will give the necessary directions.

Next Thursday week the Coventrys[3] and a few familiar acquaintances meet at my house;—from 7 to 9 we read Shakespear—Macbeth.[4] I shall most likely read the part of the ambitious sinner, you can take Macduff or Banquo,[5] if you can find your way to Wandsworth[6] for that part of the entertainment. At 9 we sup, and at the supper table we introduce a discussion.

'Is Sculpture or Painting the higher art?'

A Mr Mansell[7]—an Irishman takes the part of the Sculptors and the rest take the side they may fancy. This is done with a view to draw out some of the young folk so as to enable them to express any opinions they may have and to increase their confidence for general conversation. The subject is perhaps a little too deep to become as universal for converse as I could wish, but as it was fixed at our last meeting we must make the best of it.

Mrs Bevington[8] very much wishes that you could come to help us, but knowing your many engagements I cannot do more than say I shall be glad to see you.

You see it will do if you come at 9 o'clock, and we will keep a bed aired for you.

In the last number of the National Review there is an essay on the life and writings of Dr Young[9] which I am sure would interest you.

Let me hear from you my good fellow when the thoughts bend naturally towards the Neckinger.[10]

Remember me to Hirst and his wife when you write

and believe me | affectionately Yours | James B. Bevington | John Tyndall Esq. | Royal Institution.

P.S. | I am not yet familiar with your whereabouts at the West End[11]—or I should have addressed this to your sanctum.

RI MS JT/1/152
LT Transcript Only

---

1. *note*: note missing.
2. *bill of the Walnut Desk*: unclear; possibly Bevington had bought a desk for Tyndall's new lodgings.

3. *the Coventrys*: family not identified and not mentioned in Tyndall's Journal.
4. *Macbeth*: Shakespeare's Scottish play. Bevington held regular dinners at his house at which the guests would read and discuss plays and books or debate philosophical questions.
5. *Macduff or Banquo*: Banquo is the friend of Macbeth's whose descendants were prophesised to become Kings of Scotland by three witches. The ambitious Macbeth has Banquo murdered to prevent this but is then haunted by Banquo's ghost. Macduff, the Thane of Fife, is the eventual hero of the play when he slays Macbeth.
6. *Wandsworth*: The Bevingtons lived at Ivy House in Allfarthing Lane, Wandsworth. Neckinger Mills was Bevington's leather factory at Bermondsey.
7. *Mansell*: not identified and not mentioned in Tyndall's Journal.
8. *Mrs Bevington*: letter 1166, note 5.
9. *Young*: Thomas Young (1773–1829), professor of natural philosophy at the RI from 1801–3. See [Baden Powell], 'The Life and Writings of Dr Thomas Young', *National Review,* 2 (1856), pp. 69–97, an essay review of George Peacock, *Life of Thomas Young* (London: J. Murray, 1855).
10. *Neckinger*: Bevington was inviting Tyndall to reply when he thought of his leather factory again.
11. *whereabouts at the West End*: Bevington was unclear where Tyndall had moved to and wondered whether he should have addressed his letter to the RI.

To Thomas Archer Hirst 15 January 1856 1183

15th Jan. 1856

My dear Tom

I have I suppose about 10 or 15 pounds in the hands of the people at Red Lion Court who are my virtual Bankers.[1] On this I intended to live until I should make a haul from the London Institution[2] where I commenced lecturing yesterday this 'haul' will however not reach me for 6 weeks.

So much to account to you why I did not send you the money by return of post.[3] If I had it of course I would, and now for further shifts.

I have written off to Hughes[4] asking him to send me 30 pounds. If it comes, which I hope, I will send it to you immediately.

Meanwhile to facilitate matters I will try whether I cannot borrow 30 pounds from somebody. It is hard if ones credit cannot reach beyond this. I will write to you more at length tomorrow & send you a long ragged account of my visit to Lord Ashburton, and all I saw there—in fact I threw the thing while there into the form of a letter to you[5] which I will complete & send to you—hoping that you may have sight & patience to read it.

Your affectionate | John.

RI MS JT/1/T/865

1. *Red Lion Court who are my virtual Bankers*: Tyndall was referring to money held for him by Taylor & Francis for his translation work.
2. *The London Institution*: Tyndall began his series of second lectures on light at the LI in Finsbury Square on 14 January 1856. He continued to deliver these annual courses until 1860. Janet C. Cutler, *The London Institution 1805–1933* (PhD thesis, University of Leicester, 1976).
3. *By return of post*: Tyndall still owed Hirst money he had borrowed to complete his studies in Germany.
4. *Hughes*: see letter 1046, n. 3 and n. 5 for Tyndall's contribution to Edward Hughes (ed.), *Reading Lessons* (1855).
5. *a letter to you*: see letter 1187.

To Thomas Archer Hirst — 17 January 1856 — 1184

<u>17th Jan.</u>

Dear Tom.

I have got some money[1] and will send you 20 pounds by tomorrows post. I cant get cash to day. Wright tells me you are not at Broughton—well I suppose if I send it there all will be right[.]

as ever | John

RI MS JT/T/866

1. *I have got some money*: see letter 1183.

To Thomas Archer Hirst — 18 January 1856 — 1185

Royal Institution | Albermarlest | London | <u>18th</u> Jan. 1856.

My dear Tom,

Here you will find the halves[1] of two fives & one ten pound note—When you let me know that they have reached you I will send the corresponding halves.

affectionately | John <u>Tyndall</u>

Dr. Thos Hirst | Boughton near Stockbridge Hants

RI MS JT/1/T/616

1. *halves*: For security reasons Tyndall tore bank notes into halves which he sent by separate mailings. The practice had begun in the early days of coach travel when highway robbery

was a common danger. Travellers would cut banknotes or promissory notes in half, take one half on their journey and send the other half by a different coach. The halves were then joined together, cashed, and cancelled from further circulation.

To William Thomson 26 January 1856 1186

Royal Institution | 26th Jan 1857[1]

My dear Sir

It won't take more than a minute & therefore gives me no trouble to say that all you refer to in your letter[2] shall be at your command with the exception of the incandescence of the Thermoelectric pile.[3] Of this there is no chance & I fear you must give it up. Incandescence, combustion, and volatilization you can produce to any amount. If you like to produce some very beautiful effects by the induced current of the Leyden battery[4] it will also be in your power to do so.

Very truly Yours | John Tyndall

RI MS JT/1/5/1552
LT Transcript Only

---

1. *Jan 1857*: wrong year provided by Mrs. Tyndall.
2. *your letter:* letter missing. Thomson had been invited to deliver a Friday Evening Discourse at the RI on 29 February 1856 . See letter 1199.
3. *Thermoelectric pile*: a battery built from circuits of different metals. Usually, following the model built by Leopoldo Nobili (1784–1835) in 1835, it consisted of couples of soldered lengths of alternating bismuth and antimony arranged in series in blocks with the soldered sections exposed on one side. These could be heated to generate an electric current. The apparatus was considerably improved by Macedonio Melloni (1798–1854) after Nobili's death.
4. *Leyden battery*: The Leyden jar for storing static electricity had been developed in the mid-eighteenth century. By linking jars together to form a 'battery', powerful bursts of static electricity could be supplied for experimental purposes.

To Thomas Archer Hirst 12 January–7 February 1856 1187

[The Grange, Northington, Alresford, Hampshire] | 12th January, 1856[1]

My dear Tom,

To occupy myself before breakfast I will connect you with the running journal of events which cross me under my present circumstances. I left

London yesterday and travelled in the midst of a bunch of young sailors who were proceeding to Portsmouth, to Andover Road Station. There I saw Bennett & Miss Bennett[2] who are now visitors at Queenwood. From Andover Road, in the teeth of a very cutting wind I made my way in a gig to the seat of Lord Ashburton.[3] We soon diverged from the road and drove through miles of woodland, here and there the sound of the horses hoofs and the rattle of the wheels coming back with deep resonance from the trunks of the flanking trees, as if the ground underneath were hollow. At length I reached 'the Grange'—the name of the place. Met Lord Ashburton & two of his visitors and was very soon in a current of conversation with him. He was suffering from gout and about an hour after my arrival Lady Ashburton[4] came and turning a little wheeled instrument[5] on which he sat caused him to roll away to an adjacent room and keep himself quiet for a time. I was left with the two visitors—they conversed and Carlyle was their subject. I chimed in, and though I could agree with neither and neither with me still the subject established a point of contact between us. One of the gentlemen[6] I had met at my friend Pollocks previously; I knew I had seen him somewhere, but he was the first to recollect where; had tea at about 6. At 7 oC my acquaintance informed me that dinner would be ready at 7 ½.—tea before dinner—so I ascended to prepare myself for it. But previously I ought to have said that Lady Ashburton, in a corner of the room commenced reading in a low voice to a circle of persons near her a piece of Brownings;[7] by degrees her circle widened, and finally she had all in the room for listeners. At her left was Carlyle, leaning on the sofa. Carlyle's wife sat opposite to him. These were the only two in the room, with the exception of my older acquaintance—Spedding,[8] mentioned by Carlyle in his life of Sterling[9]—that I knew, even by sight. The piece was excellent, and called for laughter and comment from those around. Carlyle at intervals grunting out an assent—once he said 'that's just it, faith is the conviction in something not the discerning of its truth', or some words to that effect. He looked I thought rather kindly at me before we went to dinner—and I walked up intending to speak to him; some other person spoke to him at the moment and thus defeated me. Done with him, Carlyle turned to me & asked if I had come from Oxford to day—he thought I was a clergyman no doubt with my pale grave countenance. I said that I had come from London, and made a few straggling unsatisfactory remarks regarding my visit to Bonchurch.[10] In fact I was a little timid, which timidity did not wear away till the evening advanced. At dinner I was placed between two strangers, but contrived to get into a pleasant conversation with one of them, on Scott,[11] Wordsworth,[12] Longfellow[13] &c. He knew them all personally; but as yet I dont know him. After dinner Lady Ashburton came and said that Ashburton wanted to see me: I went to his Lordship's library where he was seated as before with feet supported, but as he told me in no pain. He is a man I should say about 55, tall, and thin with an animated, intelligent countenance, through which dawns a good deal of benevolence. We were in close conversation on educational matters for at least an hour. Lady Ashburton once came to say that he must be quiet, he begged for a quarter of an hour longer, which she granted. He is going to propose a resolution in the House of Lords[14] and wishes to be fortified with as much exact information as possible. He appeared to me to be a very noble fellow, full of interest for the improvement of the country, deploring our shortcomings and evidently ready to devote time, purse, and talent to their removal. At the end of the quarter of an hour Lady Ashburton came and announced its expiry—which he doubted. I bade him good night with the understanding that we should talk more on the subject tomorrow.

When I came into the room where the ladies & other guests were assembled, an old gentleman came and asked me had I been with Lord Ashburton. With him I had an hours conversation on the same subject. Meanwhile: a group collected; a meeting assembled, a table was placed, beside which sat a reporter and behind it a chairman: A thin fellow,[15] with a face full of intellect, got up and made a speech, his speech was in support of a resolution read by the chairman[16] for the abolition of molasses! It was in the true Exeter Hall style,[17] and full of wit and talent[.] Carlyle roared with laughter—the applause was loud and frequent, and thus with the most joyous good humour the thing ended. Music rang afterward: One feature strikes you every where and that is the thorough freedom of every body—each seems to do just as he likes: wanders where he pleases through the magnificent rooms; reads, talks, or admires the pictures just as it strikes his fancy. If this be the so called formality of aristocratic life, I must say that I have seen it a million times intensified in a party of trades people. The cold I suffered coming from Andover Road was too much for me; and I almost suspect that a flannel shirt which I put on in the morning was not well aired. I felt a pain in the right side of my chest which increased towards bed time. In bed it was very severe and kept me from sleeping all night. I almost feared that I should be compelled to remain in bed this morning—but here I am, up before any body else, and have written these 10 pages for your edification before breakfast. I will now quit my splendid bedroom, go down stairs and see what is to be had in the way of food.

Saturday morning 11 oC. Sun shining brightly and frost sparkling without.

Have just breakfasted. I had pictured when informed of Carlyle's presence here a general image of the manner in which I should make his acquaintance. How deceitful I have found all such pictures! Better a thousand times to picture nothing but to cope with circumstances as they come. Last night this creed impressed itself on me, this morning I acted upon it, and at the breakfast table found myself in the midst of an argument on homeopathy[18] with Carlyle. He imagined I doubt not that I was a believer in the system and heaped ridicule upon it. I pushed him here and there regarding the great effects of apparently trivial causes: Thus we got into a discussion which eventually engaged almost all the people at the table. What I have seen thus far of Carlyle makes me revere the old brute more than ever. Nobody seems to agree with him but he pushes his way through all, dealing out doom & praise with unhesitating faith. He took up a sheet of the Illustrated London News,[19] and opening a page full of the antics of ballet dancers and theatrical absurdities called Speddings attention to it. There Spedding he exclaimed—there is our latter day evangel! or something to that effect. He is pained he says to observe our blunders in this war,[20] and continues to be pained that fools still continue to think they can wage war and win battles by writing in the Times newspaper.

When I went downstairs this morning one of the first people I met was he who made the speech last night,[21] leading a sweet little girl, his daughter,[22] by the hand. The little thing in pronouncing the word No, said it thus N e a w! He caught her up in his arms—'You little wretch I'll break your head if you disgrace your family by such barbarous pronunciation'—He is full of humour, a clergyman to boot! a capital fellow manifestly. He rolled his maledictions upon the child with poetic fervour—and often with true poetic expression. Westmancott[23] the sculptor had just left, Miss Baring[24] asked Brookfield[25]—that is his name—if he was not a disappointed man—he had told her that if he had to begin life again he would not be a sculptor. Yes said Brookfield we are all disappointed. If I were to begin life I would take to one of those green caravans—those shows which perambulate the country, where I should have nothing to do but to be jolly playing a big drum.

All this will give you a clearer idea of the perfect absence of all formality & restraint than any mere assertion of the fact would do—and therefore I write it. Had a glorious walk after breakfast with Lord Ashburton's young physician.[26] We talked science the first half of our journey and Tennyson the last half—he is enthusiastic on Tennyson. In fact Tennyson himself was here two days ago reading his own Maud,[27] and delighting everybody: it required some persuasion to get him to come but once here he was delighted with his visit. He intends amplifying Maud in a certain direction. At lunch I sat beside Carlyle and there got into close conversation with the man. I forget at the moment what we talked about but it was cordial and human. Yes, now it occurs to me, it related to his visit to Berlin,[28] to his failure to obtain any great likeness of Frederick,[29] and to the fact that Rauch[30] in producing his statue had gone to no trouble to give us the image of the man, and had totally failed to do so. What we want in art is, a picture which should be true to the man's character, and no one can imagine Fritz's[31] character speaking through such a shell as Rauch has given him. After lunch another walk. I ought to say that during lunch I asked Carlyle quietly whether his vis à vis was the Spedding he alluded to in his Life of Sterling—Yes, yes, he exclaimed aloud that's Spedding! Everybody heard it. Mrs Carlyle looked towards me grave at first. I laughed and she burst out into laughter also.

She went to meet the guests I learned after breakfast, and asked was it not too bad of Carlyle to hit me so hard on the question of homeopathy. No matter I am getting closer & closer to Carlyle, and I shall reach him by and by. Sat for a time before dinner in a group of which Carlyle formed one. I tried to stir him by starting some theoretic views of the nature of life—mere physiological life; ascribing it to molecular force: on the incomprehensibility of life he was immoveable—it is a mystery which will for ever baffle us. Anecdotes passed to and fro. Afterwards our companion left and I found myself beside him alone: I asked him about Emerson. He warmed up and spoke solemnly, earnestly, and lovingly of his great brother. Lately he complains of his eyes[32] which makes his letters fewer than they use to be. It was most remarkable to see a man like Emerson amid such a bustling community—the highest political honours within his reach, but living in his own lofty sphere totally indifferent to such things: any utterance from such a soul must be remarkable. A man disdaining action, because there is no room for noble action, and proclaiming that he will sit still till the time for action comes; accepting our present indigence until a richer time appears. I loved the tones of Carlyles voice as he thus spoke of Emerson, and the solemn earnest sparkle of his eye. I spoke to him about Lewes's life of Goethe.[33] It was admirable as biographies go, but a true image of the man Goethe was not to be found there. Goethe's life as a writer he considered a tragic one. Such a man ought to have room to <u>act</u> in the world. I retorted that writing was really action—he replied it was a poor species of action. I asked supposing Goethe to have given himself to science wholly, as he did in part, would he have called his life a life of action. Not much higher than that of writing was his reply. Such a soul ought to have governed Germany—he ought to have been King of Germany. Thus we talked until the bell called us to dress for dinner. I know I am nearer to the man than I was this morning. I told you I would reach him and I am doing so, though in a crooked way. Luck and a little tact placed me beside Mrs Carlyle at dinner. I talked much to her and was greatly pleased with her. I told her of some of the criticisms of which her husband was the subject at which she laughed heartily; Carlyle is very different from Tennyson in this respect. Tennyson is extremely sensitive to criticism Carlyle is perfectly indifferent to it—though, remarked Mrs Carlyle I think I have noticed that it rather makes Carlyle dislike the author of the criticism—'I think I have noticed this' she said. I thought the admission very charming. Sometimes Carlyle receives anonymous letters [,] he reads the first line, and then dismisses them to the fire: it is wonderful the efforts that people make for his conversion. They don't abuse him, and are usually quite affectionate to him, warning him that he has now one foot in the grave. A short time ago, at Richmond,[34] he and Mrs Carlyle were waiting at the Railway station for a train. A young lady near asked her companion what book she had got—Oh she exclaimed, I have got something from that mad writer Carlyle. Her description of Kingsley[35] was exceedingly good. 'I often think that a low diet and the loss of blood would improve him greatly—he is such a tumultuous man: his health and spirits are so high and they make his life so rank and noisy that he quite confuses me'. She is sometimes amused by young ladies of 17 coming to see Carlyle, and saying to him 'Oh Mr Carlyle I have read all your books and understand them perfectly'. when at the time they have not lived long enough to understand anything. Carlyle is now at work at the life of Frederick the Great. She thinks he has written longer on the subject than is commensurate with his love and admiration for Fritz. He must love the man he paints wholly: he must have faith in his heroism otherwise he cannot write. Macaulay[36] takes pleasure in pulling men down, but Carlyle cannot do this—he cannot write under such inspiration. Macaulay has pulled down Marlborough.[37] Carlyle believes that he was a great man, but has not yet read Macaulay's attack upon him. Thus the dinner passed and a pleasant one it was: afterwards the time was spent in conversation till nearly 12 oC.

Sunday Jan. 13th. Wrote as usual before breakfast. Writing materials and all appliances which could administer to human comfort are to be found in my bedroom. Outside I have glimpses of flower beds and noble beeches, with the glimmer of a stretch of water encircled by its woods. At breakfast Hofmann sat between me and Carlyle, and we talked long of Goethe and of Faust. Carlyle thinks little of Faust in comparison with other utterances of Goethe. Beau Brummel[38] was a reality equal to Mephistopheles. A friend came to him one morning and asked him how he did. 'Suffering from cold' said he in his own imperturbable way. 'I have been sleeping with a damp woman all night!'. The tone of Carlyle's voice in repeating the anecdote was extremely rich: indeed both the expression and the mode of it could be put beside the best of Mephistopheles. A long walk before lunch, first alone, in company with Mr Cole[39] afterwards. At lunch again near Carlyle: did the other guests know how completely they were forgotten they would have deamed themselves slighted. He talked of the Chinese cousins[40] mentioned in his life of Stirling. A Chinese man is in search of a husband for his daughter: he is anxious to have a grandson that shall burn incense at his grave; and the writer takes occasion to describe the characters which the seeker falls in with. It was a book worth shiploads of such as are too common. A bad book was like a bad man, it ought to be avoided and no communication held with it. The library of George the third[41] was a model of a library. It is wonderful how far a small amount of intellect will go when we have conscience to lay it out well. George the third was smallminded but he was conscientious, availed himself of the best means, consulted the best authority, and in this way managed to get his books together. They are the most valuable part of the British Museum library. The principle which seems to guide our collector of libraries is not the intensive merits of books but the desire of possessing what nobody else has. This seems to be the motive in the British Museum. A selection ought to be made, the greater portion of the books ought to be put in the cellars down stairs away from the sight of those to whom they may do harm: here they might be read by those who take an interest in them, but up to the general light of heaven they should not come. They might be classified according to their vices into 'Damned N° 1. Damned N° 2. Damned N° 3 &c'. Sometimes during this conversation Carlyle's words rose to eloquence.

Branching off from his description of the Chinese man's search his eyes would gleam, his voice became more deep & solemn, while his fingers would squeeze & crush his napkin. At first I thought the latter act indicative of his intention to move away, but it was not—His fingers were the mere accompaniment of his thought, and he crushed his napkin repeatedly before the conversation ended. A walk afterwards with Mr Ellice,[42] and a talk on education; continued the conversation with Lord Ashburton when we returned. Moved into an adjacent room where the ladies and other guests were assembled: crystallised into little groups—I and Mrs Brookfield[43] formed one. She reminds me of some of those beings addressed by Tennyson in his earlier poems, and that you[44] used to admire so much. We talked for an hour: there was an odour of social cordiality, of warm good fellowship and happiness throughout the entire room. The servant at length came to say that the gig was waiting for me: I rose, & bade my beautiful companion goodbye. Lady Ashburton rose and bade me a cordial goodbye. Mr Ellice came forward and did the same. Hofmann was making his adieus at the same time. While thus uncoiling myself from the company I could see old Carlyle moving forward to throw himself in my way as I passed. As I reached him I had a cordial squeeze of the hand and an earnest goodbye. Lord Ashburton was the last and so I left the place with a perfume of kindness hanging round my thoughts and a pleasant memory of the pleasant hours I spent at 'the Grange'.

Your always affectionate | John | 7th Feb. 1856

RI MS JT/1/T/617

1. Although begun at The Grange (Lord Ashburton's country seat) on 12 January, Tyndall only completed the 32 pp letter on 7 February. Most of the letter is based on Tyndall's Journal for 11–13 January 1856 (RI MS JT/2/13c/817–20).
2. *Bennett and Miss Bennett*: William Bennett (1804–73), tea merchant, naturalist and 'latitudinarian Quaker' (as Tyndall referred to him). They had first met in July 1853 on a field trip organized by the Entomological Society. Tyndall was a frequent visitor to the Bennetts at their home in Betchworth, Reigate, in Surrey. The wife was Elizabeth Bennett (b. 1798) and the unmarried daughter Mary Bennett (b. 1836). See R. Desmond, *Dictionary of British and Irish Botanists and Horticulturalists* (London: BMNH, 1994). The Bennetts were presumably visiting Queenwood out of Quaker interest.
3. *Lord Ashburton*: see letter 1169.
4. *Lady Ashburton*: see letter 1169.
5. *a little wheeled instrument*: an invalid carriage or wheelchair.
6. *One of the gentlemen*: according to Tyndall's Journal entry for 11 January 1856, this was James Spedding (1808–81), the Francis Bacon scholar (RS MS JT/2/13c/815).
7. *Brownings*: Robert Browning (1812–89), famous English poet.
8. *Spedding*: see n. 6.
9. *Life of Stirling*: Thomas Carlyle, *Life of John Sterling* (London: Chapman & Hall, 1851). Both Sterling and Spedding had been members of the Cambridge Apostles while studying at Cambridge. While living at Hastings, in August 1838 Sterling formed a dining club in London so that he could meet his metropolitan friends once a month. The secretary of what became known as the Sterling Club was Spedding, as noted by Carlyle in his *Life of Sterling*, p. 159.
10. *Bonchurch*: the village on the Isle of Wight where Stirling was buried in the churchyard. Tyndall had visited the grave on 23 July 1852 inspired by reading Carlyle's memoir (RI MS JT/2/13c/ 577).
11. *Scott*: Walter Scott (1771–1832), Scottish poet and novelist.
12. *Wordsworth*: William Wordsworth (1770–1850), English poet.
13. *Longfellow*: Henry Wadsworth Longfellow (1807–82), American poet and Professor of Modern Languages and Literature at Harvard (1836–54).
14. *A resolution in the House of Lords*: We could find no record of this resolution in *Hansard*. Ashburton was a strong supporter of a national education system and admired the system of object lessons that the Revd Richard Dawes (1793–1867) had introduced at the village school at King's Somborne in Hampshire. See D. Layton, *Science for the People* (London: Allen & Unwin, 1973), pp. 49, 113, 119, 122–23.
15. *a thin fellow*: W. H. Brookfield. See n. 25.
16. *The chairman*: not identified.
17. *Exeter Hall style*: Exeter Hall had opened in the Strand in 1831 and had become well-known venue for religious and political meetings of a strident kind.
18. *Homeopathy*: a system of medical therapy developed by Samuel Hahnemann (1755–1843) in which like cured like, as opposed to allopathic medicine in which the pharmaceuticals

employed did not cause symptoms similar to the disease in question. The treatments were controversial because only infinitesimal quantities of the active ingredients were used. Tyndall was to recall this argument over homeopathy in 'Personal Recollections of Thomas Carlyle' published in the *Fortnightly Review* in January 1890 and reprinted in his *New Fragments* (London: Longmans, 1892), pp. 347–91.

19. *Illustrated London News*: A weekly illustrated magazine founded in 1842 by the engraver Henry Vizetelly, the salesman Henry Ingram, and the editor E. W. N. Bayley. One of the most successful of Victorian periodicals, it continued publication until 1989.
20. *This war*: The Crimean War (or Russian War) had been caused by Russia's occupation of Moldovia and Wallachia in 1853 in order to protect Orthodox Christians in the Ottoman (Turkish) Empire. The Turks, supported by the French, declared war on Russia in November 1853. Britain joined the French at the end of March 1854. Hostilities ceased in February 1856 and peace was agreed in Paris on 30 March 1856. The war became a byword for military incompetence and brought about reforms in the army and nursing. See letter 1096, n. 4.
21. *he who made the speech last night*: this was evidently Brookfield, n. 25 below.
22. *his daughter*: Magdalene Brookfield (1850–82).
23. *Westmacott*: Probably James Sherwood Westmacott (1823–1900), the son of the sculptor Henry Westmacott (1784–1861). J. S. Westmacott had exhibited 'Peri at the Gates of Paradise' at the Exposition Universelle in 1855.
24. *Miss Baring*: Either of Ashburton's two unmarried sisters: Luisa Baring (d. 1888) or Lydia Baring (d. 1868).
25. *Brookfield*: The Revd William Henry Brookfield (1809–74), Anglican preacher and education inspector.
26. *young physician*: not identified.
27. *Maud*: Alfred Tennyson, 'Maud', a poem first published in 1855. Subsequently Tennyson issued revised versions in 1856 and 1859.
28. *Berlin*: Carlyle had visited Berlin in 1852 to conduct research on the life of Frederick the Great.
29. *Frederick*: Frederick II (1712–86), King of Prussia from 1740 until his death. Carlyle's *History of Frederick II of Prussia, called Frederick the Great* appeared in six volumes (1858–65).
30. *Rauch*: Christian Daniel Rauch (1777–1857), German sculptor whose masterpiece was the equestrian statue of Frederick. Tyndall had been present at the unveiling of the statue in Berlin on 31 May 1851 (RI MS JT/2/13c/545).
31. *Fritz's*: Fritz was used as the popular German nickname for King Frederick II.
32. *lately he complains of his eyes*: Emerson suffered from rheumatism and declining eyesight in the 1850s.
33. *Lewes's Life of Goethe*: For Tyndall's reading of this biography, see letter 1170, n. 2.
34. *Richmond*: a town on the River Thames familiar to Tyndall from his visits to Richard Taylor.
35. *Kingsley*: Charles Kingsley (1819–75), historian and novelist.

36. *Macaulay*: Thomas Babington Macaulay (1800–59), English historian and Whig politician; author of 'An essay on Frederick the Great', *Edinburgh Review*, 75 (1842), 218–281, republished in his *Critical and Historical Essays from the Edinburgh Review*, 3 vols. (London: Longman, 1853), vol. 3, pp. 200–88.
37. *Marlborough*: John Churchill (1650–1722), military hero and Whig diplomat who was made 1st Duke of Marlborough by Queen Anne in 1689. In Macaulay's *History of England from the Accession of James II*, 5 vols. (London: Longman, 1848), Marlborough was treated as an arch-villain.
38. *Beau Brummel*: George Bryan Brummel (1778–1840), Regency dandy and friend of the Prince of Wales (later George IV). He had died a pauper in France.
39. *Mr Cole*: Henry Cole (1808–82), civil servant, one of chief organisers of the Great Exhibition of 1851, and a British Commissioner at the Exposition Universelle in 1855. Since 1853 he had been one of the two secretaries of the Department of Science and Art.
40. *Chinese cousin*: A reference to *Iu–Kiao–Li: or The Two Fair Cousins, a Chinese Novel*, 2 vols. (London: Hunt & Clarke, 1827) which was derived from a French translation of a Chinese tale made by the French sinologist Jean-Pierre Abel-Rémusat (1788–1832). The story concerns two characters who pretend to be poets by plagiarising the work of real poets. The story enchanted Sterling and Carlyle as well as Emerson and Goethe. See *Life of John Sterling* (n. 9), pp. 122–23.
41. *Library of George the 3rd*: King George III (1738–1820). His magnificent royal library, formerly in the British Museum, now forms a highly visible part of the British Library.
42. *Mr Ellice*: Either Edward Ellice the younger (1810–80), Liberal politician, free trader and advocate of Irish disestablishment; or his father Edward Ellice (1783–1863), politician and trader with Canada.
43. *Mrs Brookfield*: Jane Octavia Brookfield (née Elton,1821–96), literary hostess and novelist who had had an affair with Thackeray.
44. *you used to admire so much*: Tyndall may have been thinking of Hirst's admiration of Tennyson's *The Princess* which he read in Halifax in 1849.

From Henrietta Huxley[1] [7 February 1856?][2] 1188

| 14 Waverley Place | [London] Thursday

Dear Dr. Tyndall

You repent of your sins in so facetious a melancholy, that I can in no wise attribute to you—the depth of remorse you would have me believe—however we did <u>not</u> wait, for you on Sunday.[3] You expressed a doubt about coming and a doubt I have ever found to end in a negative. Consequently 'the mutton had' not 'more than a normal crisp'. Though I never knew before—that crisp, was its normal condition. As I particularly wish this explained will you come and dine with us on Sunday ¼ past 6. Hal wants to talk with you about the Examination papers[4]—and moreover you shall see my pretty sister Mrs George Huxley![5]

Do you know we are going to Germany this summer?[6] to ruralise in some quiet village which you shall recommend. As you know all about Düsseldorff[7] I have just read a long letter from my old schoolmistress,[8] who says she is looking forward to seeing me with 'großer Familiär'[9] it being an old promise of Hal's[10] that we should some day revisit my school.[11]

I shall not want the stinsels till the winter, so you can forget them a little longer, if it's more convenient to you—but bring 'Hiawatha' [12] with you. I cannot give you that copy but we will exchange. Hoping your 'moral marrowbones' don't ache with kneeling so long. I bid you rise pardoned.

Your sincere friend | Henrietta A. Huxley

Hal has been using my pens—which accounts for my unusual bad writing.

RI MS JT/1/9/2876
LT Transcript Only

---

1. *Henrietta Huxley*: Thomas Huxley married Henrietta Anne Heathorn (1825–1915), usually known as 'Nettie', at All Saints Church, Finchley Road, London, on Saturday 21 July 1855. Tyndall was not present and first met her at the Huxley's home in Waverley Place, St John's Wood, on 28 November 1855 (RI MS JT/2/13c/806).
2. The letter is undated. However Tyndall records in his Journal that he and Huxley discussed examinations on Sunday 10 February, making the invitation date of Thursday 7 Febuary plausible. The examinations for entry to the Royal Engineers had taken place on 30–31 January 1856, so both men were engaged in marking scripts.
3. *on Sunday*: If Thursday 7 February is the correct date for this letter, it would appear that Tyndall had failed to turn up for dinner the previous Sunday 3 February.
4. *the Examination papers*: Huxley had joined Tyndall as an examiner for Royal Artillery and Royal Engineers commissions in October 1855.
5. *Mrs George Huxley*: Huxley's brother George (1824–63), a barrister and businessman, had married Mary ('Polly') MacIntyre (b. 1827) in 1845. Tyndall first met her at the Huxleys on Tuesday 1 January 1856.
6. *this summer*: Tyndall joined the Huxleys in Switzerland in August 1856, both parties having travelled through Germany.
7. *Düsseldorff*: A German city lying at the junction of the river Düssel with the Rhine. Tyndall had visited it on his first Rhine journey to Marburg on 22 October 1848 (RI MS JT/2/13c/393).
8. *my old Schoolmistress*: not identified. Mrs Huxley had spent two years as a schoolgirl at Neuwied on the Rhine south of Bonn.
9. *Größer Familiär*: 'and your family' (German).
10. *Hal's*: Nettie Huxley's nickname for her husband and derived from Henry.

11. *my school*: the school at Neuwied.
12. *Hiawatha*: The newly-published popular narrative poem by Henry Wadsworth Longfellow, *The Song of Hiawatha* (London: T. Nelson, 1855).

From Lord Ashburton 10 February 1856 1189

Grange | [Alresford, Hampshire] | Feb. 10.

My dear Mr Tyndall

Your letter[1] has only just been forwarded to me with the accompanying report[2] which I now return with many thanks. Dr. Frankland will be a valuable witness.[3] You must give me more opportunity of making his acquaintance.

There is another evil besides the scarcity of lecturers, and that is the demand for them. What we want is masters not lecturers, class teaching, not professional enlightenment.

My heart is sick to see so much energy so much healthy craving for knowledge wasted in the purveyance and consumption of [crumbs] instead of food.

Yours very truly | Ashburton

I will send the report[4] shortly.

RI MS JT/1/121
LT Typescript Only

1. *Your letter*: letter missing. It was evidently written in response to Ashburton's request (letter 1187, n. 14) for information on science teaching.
2. *accompanying report*: it is unclear what this was, but possibly the RI series of lectures on science education, *Lectures on Education* (London: Parker, 1855), to which Tyndall had contributed an essay on physics education.
3. *Dr Frankland will be a valuable witness*: Tyndall had evidently promised to ask Frankland's opinion on the state of science teaching.
4. *the report*: it is unclear what this referred to.

From Thomas Archer Hirst 10 February 1856 1190

Broughton | Feby 10th 1856

My dear John.

Thanks for your long and very pleasant description[1] of your visit to 'the Grange' It was easy for me to realize the several pictures and conversations you described. I too have often pictured what I should say if I ever met Carlyle, but you are quite right in saying that such pictures are worthless. Heinrich

Heine[2] in describing his first interview with Goethe says 'I had for so many long winter nights thought over what lofty and profound things I would say to Goethe if I ever saw him. And when I saw him at last, I said to him, that the Saxon plums were very good! And Goethe smiled.'[3] I hope you and Carlyle may meet again and often if it be only for the sake of my reading your descriptions of the interviews. Anna and I are now reading similar descriptions by Miss Bremer in her 'Homes of the new World.'[4] One picture, of Emerson at home was very pleasant to me. Miss Bremer is a sensible observant woman, very amiable I should say, who amidst her intellectual habits retains her womanly trustfulness and feelings in due freshness. She has a high appreciation for Emersons noble and lofty nature, says frankly that she feels herself ennobled in his presence. She has too much of the womans nature in her to find consolation and satisfaction in his high self-reliant creed but she can admire it in him. In her interviews with him she evidently felt her own individuality endangered by his loftiness and made several not very happy but still very praiseworthy attempts to stand well in his eyes by an assertion of her own faith and her preference for it rather than his. I am myself not at all clear about Carlyles notion of action. If I understand you, he admires Emerson because he sits still and meditates seeing that he cannot act and retain his independence. I doubt whether that word cannot is in Emersons case correct. Men like Emerson can do great things when they feel so inclined. I rather think he feels his strength to be greater in meditation than action. And Carlyle himself what has he done and what is he doing? I question much whether any kind of action could have been more beneficial or durable than his, which as far as I know has consisted ever in writing. With Goethe I guess it was much the same.[5] Perhaps it would have been better for Germany could they have employed Goethe differently but under the circumstances Goethe did, doubtless, what was best.

But how are you? Hard worked I should say. I am glad to tell you we have almost 100 boys at Queenwood which is much better than we anticipated this half year. We are now in full work, and are already engaged in that more disagreeable task[6] we commenced when you were here. On the whole I think there is some improvement and with unceasing watchfulness I have now some hope that we shall keep comparatively pure in future.

Thine affectionately | T A Hirst

Anna send her love with mine. She is afraid you are forgetting her.

RI MS JT/1/H/222

1. *pleasant description*: letter 1187.
2. *Heinrich Heine*: See letter 1140, n. 25.

3. *I had for . . . And Goethe smiled*: Hirst probably quotes this anecdote from a review ('German Wit: Heinrich Heine') that appeared in the January issue of *Westminster Review,* 65 (1856), pp. 1–33 at p. 15.
4. *Miss Bremmer in her Homes of the New World*: Fredrika Bremer (1801–65) was a Swedish writer and advocate of women's rights. Her travels in America as an independent woman between 1849 and 1851 led to *Hemmen i den nya världen* (Stockholm: P. A. Norstedt, 1853) which was translated the same year by her English friend Mary Howitt as *Homes of the New World. Impressions of America,* 2 vols. (London: A. Hall, Virtue & Co., 1853).
5. *With Goethe . . . it was much the same*: Hirst was reflecting whether it would have been better for Germany if Goethe had stuck to writing rather than accepting a Court appointment at Weimar. See letter 1179 n. 11.
6. *more disagreeable task*: the outbreak of masturbation at Queenwood referred to at greater length in letter 1220.

From David Brewster 13 February 1856 1191

St. Leonard's College | St. Andrews | Feb. 13th. 1856.

My dear Mr Tyndall,

Having agreed to write a short Report for the Board of Trade,[1] on certain Optical Instruments in Class VIII, you would oblige me much if you could give me any information on the following points.

Why was no medal adjudged to Lord Ross[2] and Mr Lerebours[3] who exhibited their telescopes in models because they could not do it otherwise.

Can you give me the dimensions of Lerebours object glass[4] which received a great Medal of honour? Was it tested by the jury?

Was Lerebours telescope tested by the jury?

Why did Dr. A. Steinheil telescopes get no medal nor even honorable mention? They were superior and cheap Instruments.

Why did Merz of Munich[5] get no medal for his Achromatic object-glass, he being one of the first makers in Europe.

If you could give me any memoranda you may have on these Instruments, or refer me to any sources of information you would oblige me greatly.

I hope to send you soon, by a private hand, if possible, a packet of my papers.

I am, | My dear Mr Tyndall | Ever most truly yours | D. Brewster.

RI MS JT/1/1/156
LT Transcript Only

1. *Report for the Board of Trade*: D. Brewster, *Paris Universal Exhibition. Report on certain*

Optical and other Instruments (London: George E. Eyre & William Spottiswoode, 1857). (Copy among Brewster's papers in private possession.)

2. *Lord Ross*: Lord Rosse, William Parsons, Third Earl of Rosse (1800–67), astronomer noted for his 'Leviathan of Parsontown' at Birr Castle in Ireland completed in 1845. He was President of the RS 1848 to 1854.
3. *Mr Lerebours*: Nicolas Lerebours (1807–73) inherited the optical workshop that his father Noel Lerbours (1761–1840) had founded in 1789. The firm became the Maison Lerebours & Secretan in 1845 when the Swiss mathematician Marc Secretan became a partner. The firm was also involved in photography.
4. *Lerebours object glass*: According to *Rapports du Jury Mixte International,* 2 vols. (Paris: Imprimerie Impériale, 1856), vol. 1, p. 403, the objective was 38 cm in diameter.
5. *Merz of Munich*: Either Siegmund Merz or his brother Ludwig Merz, owners of the Munich optical firm that had been founded by Fraunhofer (d. 1826). Ludwig Merz also taught at the University of Munich.

From James Bevington 18 February 1856 1192

Neckinger Mills | Bermondsey | London. 18th Feb.1856.

My dear Tyndall

I have been commanded (and in so gentle a voice that to obey is a pleasure) to ask if you will join our evening party on Wednesday the 27th of February—Whether we shall read or what we shall read has not been fixed, but on this point we shall be glad of your advice. For discussion at the supper table we have to decide—'Whether Reading or Conversation is the most instructive.'

I shall advocate conversation and shall be glad to have an opponent worthy of my wit. First impressions are in favour of reading and unless some good reasons are shown to the contrary the majority will be against me. But I shall be content if only I can induce in any one to devote more attention to the cultivation of good conversation—In our moody England[1] I cannot help thinking it would be a great advantage if conversation were more general.

Geoffrey[2] and I attended at the Royal Institution and learned that there was something more in diamagnetism and paramagnetism than we could fully comprehend.[3]

A few days ago I wandered round that magnificent Palace at Westminster[4] and saw some of the notables of the land. There was that well abused and plucky Lord John Russell,[5] the Lord Palmerston,[6] and many worthies unknown to me—Some intelligent chaps and others whose lights shined internally—Coming away I observed one fellow[7] with rather long pendulums[8] to his gastronomic apparatus—His toga of fine black cloth hung gracefully on his shoulders. His physiognomy meditative but energetic. Doubtless he was brooding about philosophers in Germany, France and Italy, and how he would leave an unperishable name when the mighty palace was dust again. I stopped for an instant to catch his eye, but no sublunaries were at that time worthy of his thoughts. On he strode, with his big thoughts and was soon lost in the crowd of mortals. Methought one of these days I will catch him in his den. If I stop him now I shall deprive the world of some new idea and moreover I shall be late for the train.

If you could send three tickets for any lecture of Owen—or Lyell or any philosopher on a popular subject I should bring Mary & Geof.[9] in a fly and should be very much obliged.

Remember me to Hirst and believe me | affectionately Yours | J.B. Bevington.

RI MS JT/1/1/153
LT Transcript Only

---

1. *in our moody England*: Probably a reference to the growing dissatisfaction with and criticism of the conduct of the Crimean war. Although Sevastopol had fallen in September 1855 there remained the prospect that, after a hard winter, Russia might invade from the west. In fact, Russia sued for peace in March 1856.
2. *Geoffrey*: James Bevington's son, Geoffrey Bevington (1838–72).
3. *we could fully comprehend*: Father and son had attended Tyndall's lecture 'The Disposition of Force in Paramagnetic and Diamagnetic Bodies' on 1 February 1856. This was printed in *Roy. Inst. Proc.*, 2 (1854–58), pp. 159–64.
4. *Palace at Westminster*: the Houses of Parliament.
5. *Lord John Russell*: John Russell (1792–1878), English statesman, leader of the parliamentary Whig party and frequently opposed to Palmerston.
6. *Lord Palmerston*: Henry John Temple (1784–1865), third Viscount Palmerston, English statesman and leader of the Tory party; Prime Minister from 1856 to 1859.
7. *one fellow*: not identified, but probably a uniformed Palace official.
8. *pendulums*: moustaches.
9. *Mary & Geof.*: Bevington's wife Mary and son Geoffrey.

From George Wynne 18 February 1856 1193

Board of Trade | 18 Feby 1856

My dear Tyndall

Eddy,[1] since Scott Russell's difficulties,[2] is thinking of trying for the practical class at Woolwich,[3] I do not think he has much chance but as working for it will be all gain to him I am willing that he should try his luck, it will be necessary for him to read hard and I have thought that you might perhaps be able to direct me to a good mathematical master not merely to a good mathematician but to one who is likewise able to impart his knowledge.

Mrs Wynne from headaches and bad sleeping is but poorly I hope your sanitary condition[4] is better than when we last met, but I should fear,—

Ever most truly yrs | Geo Wynne

RI MS JT/1/W/107

1. *Eddy*: Edward Wynne, who was 18.
2. *Scott Russell's difficulties*: Edward Wynne had been apprenticed to his maternal uncle, the naval architect John Scott Russell (1808–82). (Russell had married Mrs Anne Wynne's sister Harriet Osborne in 1836.) Russell had contracted to build the iron hull of Brunel's *Great Eastern* at his Millwall dockyard in May 1853. He was subsequently placed in financial difficulties by a fire in September 1853, the death of his backer, the railway engineer Daniel Gooch in 1854, and the fact that he had under-estimated his costs. He was declared bankrupt in February 1856 and presumably had to release Wynne from his apprenticeship. (George S. Emmerson, *John Scott Russell. A Great Victorian Engineer and Naval Architect* (London: John Murray, 1977), pp. 105–8.)
3. *Woolwich*: the Royal Engineering College at Woolwich. It is unclear whether Wynne knew that Tyndall was involved in the entry examinations.
4. *sanitary condition*: (euphemism). As Tyndall's former employer, Wynne was undoubtedly aware of Tyndall's proneness to bouts of constipation.

To Thomas Archer Hirst 19 February 1856 1194

19th Feb. 1856

My dear Tom

I have been just to see Col. Wynne and this letter is the offspring of our interview. You have heard I daresay of the difficulties of Scott Russel[1] to whom Col. Wynne's second eldest son was articled.[2] Well owing to the confusion of his uncle's affairs[3] the youth is about to make an effort for the practical Class at Woolwich before which however he has to pass an examination [.] Col. Wynne wrote to me asking whether I could recommend him to any good mathematical teacher:—but I will send you his note which will explain the whole matter—I naturally thought of you & Queenwood and from what I told him regarding you & it he is very anxious if it be possible to put his son in your hands—He could attend your classes at the College and if you could give him a little private time afterwards it would just be the thing. I suppose he could hardly have a private room in the College, and possibly arrangements might be made for you to take him to yourself—for a few months. I know the youth thoroughly and have always admired him. He is full of the soundest good sense truthful, straightforward and altogether of an admirable disposition. Perhaps Anna would have no objection to the proximity of a countryman of hers who possesses such a character. Would you then write to me & say whether such an arrangement as I have here indicated could not be made, and give me an idea of the terms say for 3 months or half a year. Of course on this point you would have to consult Edmondson. I think it likely that he would also pursue French & drawing.

Ever yours John | write to me soon

RI MS JT/1/T/618

1. *Scott Russel*: letter 1193, n. 2.
2. *articled*: Tyndall substituted 'articled' for 'once apprenticed'.
3. his uncle's affairs: John Scott Russell's bankruptcy.

From James Bevington 20 February 1856 1195

Neckinger Mills. Bermondsey .| London 20th Feby 1856

My dear John

I shall consider you booked for Wednesday the 27th. A bed will be at your service, but if you should find other engagements so press that you will not be able to enliven us with your Johnsonlike presence,[1] then drop a line, we shall know it is not the Vagabondish spirit that we wish to encourage in you that keeps you away but some prude that calls you away in the name of duty or necessity.

The two tickets[2] will be as much appreciated as more.

Yesterday, at early noon I was fortunate to be attacked with tooth ache. My wife had provided me with Lewes Life of Goethe [3] & I made it an excuse of sitting over the parlour fire & enjoying this literary treat, notwithstanding my predilections for conversation.

How do you like your present lodgings?[4] Perfection is not be met with, but with your moderate requirements, I think you might to shift & change again & again till you find the suitable family.

Hill[5] says that Hirst owes him a letter, when you write you must tell him that this worthy disciple of his is pining for be[6] of his masters encouragements.

Yours as ever | James Bevington | To | J Tyndall

RI MS JT/1/B/95

1. *Johnsonlike presence*: Bevington was comparing Tyndall to Samuel Johnson (1709–84) the lexicographer whose speech and wisdom was made famous by his biographer James Boswell (1740–95).
2. *The two tickets*: tickets for a lecture at the RI.
3. *Lewes Life of Goethe*: see letter 1170, n. 2.
4. *your present lodgings*: Tyndall had moved to South Audley Street, Mayfair on 1 January 1856.
5. *Hill says*: Henry Hill; see letter 1035, n. 6.
6. *be*: probably a slip of the pen for 'lack'.

To Thomas Archer Hirst 20 February 1856 1196

20th Feb. 1856

My dear Tom.

I saw young Wynne[1] last night he had been down during the day to see his aunt—the wife of Scott Russell.[2] She deprecates his leaving his uncle. So he laid his reasons before her, and desired them to be laid before his uncle. If his uncle can solve his difficulty[3] satisfactorily he will remain with him—if not he will go to Queenwood—The thing will be very soon decided—He will go down, take his own lodging and arrange all matters himself—He will require no looking after as he is as sensible as his father, and has long been accustomed to be his own master. So for the present the matter I suppose must rest. I was five minutes too early with my letter yesterday[4]—otherwise I might have intercepted it.

I send you a note from Bevington.[5] There is an allusion to you in it.

Believe me | ever yours | John | Saturday

RI MS JT/1/T/619

1. *young Wynne*: Edward Wynne; see letters 1193 and 1194.
2. *the wife of Scott Russell*: Harriette Russell, née Osborne. See letter 1193, n. 2.
3. *his difficulty*: the insolvency mentioned in letter 1193, n. 2.
4. *my letter yesterday*: see letter 1194.
5. *a note from Bevington*: Given the speed and efficiency of the postal service, this was probably letter 1195.

To Thomas Archer Hirst 20 February 1856 1197

My dear Tom.

It is my lecture day[1] and therefore I have not time to do more than say that

I am sorry you have been smitten by that cold, though it is a malady suffered by numbers at present; and to say how pleased I am with the arrangement you have been good enough to make regarding Eddy Wynne. I sent your note to his mother[2] I am sure she will be greatly satisfied by the arrangement. All that I can give you in the way of those examination papers you shall have.

Kiss Anna for me. This sentence is a proof that I am better today than I was yesterday: for then I had little thoughts of Kissing.

Your affectionate | John

There is a paper in Pogg. by Reich[3] where he either confirms me or Matteucci. I have not yet seen the paper.

RI MS JT/1/T/970

1. *lecture day*: Tyndall was giving a series of weekly RI lectures on light.
2. *your note to his mother*: letter missing. Tyndall took Wynne to Queenwood on Saturday 1 March 1856 and Wynne lodged with the Hirsts in Vine Cottage. He attended Hirst's mathematics and physics classes, as well as those of Debus and Haas. In addition, Hirst gave him six private lessons a week from 7.00 to 8.00 am. Hirst's Journal (RI MS JT/2/32c/1188).
3. *a paper in Pogg. by Reich*: Ferdinand Reich (1789–1882), German chemist and physicist best known for his discovery of the element indium in 1868. He was Professor of Physics at the Freiberg Bergakademie. At Matteucci's request, he repeated Tyndall's experiments with bismuth using a torsion balance he had made to measure the density of the earth. The paper, 'Ueber die Diamagnetische Wirkung', appeared in *Poggend. Annalen,* 97 (1856), 283–96, and was translated by Tyndall as 'On Diamagnetic Action', *Phil. Mag.,* 11 (1856), pp. 249–56. In fact, Reich concluded that Tyndall had been correct. See Tyndall's Journal for 9 March 1856 (RI MS JT/2/13c/828).

To Thomas Archer Hirst 22 February 1856 1198

Feb 1856

My dear Tom

Your note[1] has just reached me. Thank you for it sincerely. I will write to you by tomorrow's post regarding my friend Eddy Wynne. He will go to Gale's[2] if you cannot with perfect convenience take him in. I have him to dine with me today and we will come to a decision. I think he will be with you on Monday—he is a boy who when he sees a thing is to be done will do it promptly. It would be easy to please him in every respect. When he went into his uncle's establishment[3] he seized his file and worked as a workman. That is the spirit of the lad in all he does. In fact he is one just fit to enact Abernethy's doctrine of necessary[4]—that is to live on 6 pence a day and earn it himself. He is I should say between 16 & 17 years of age.[5]

Your note is full of sweetness to me—I wish I had a little time to write to you on that subject. I had a most interesting conversation with my friend Huxley on the point a few days ago—he enlightened me much and I have something to communicate to you on the matter. Good bye my dear Tom till tomorrow

ever your affectionate | John | Friday

RI MS JT/1/T/868

---

1. *Your note*: letter missing but probably written by Hirst on 21 February.
2. *Gale's*: not firmly identified from Census data, but possibly Sarah Gale, the widowed innkeeper of the Tally Ho in Broughton. Hirst had experienced disciplinary problems with a senior boy named Gale in 1854 and this was possibly her son, George Gale (b. 1836). See letter 1035, n. 5.
3. *his uncle's establishment*: John Scott Russell's shipbuilding yard.
4. *Abernethy's doctrine of necessary*: The Scottish surgeon John Abernethy (1764–1831) is reputed to have told a patient suffering from gout that the cure was 'live on sixpence a day—and earn it'. The saying was later used by George Bernard Shaw in his 1906 drama, *The Doctor's Dilemma* (London: Constable, 1911).
5. *17 years of age*: Edward Wynne was, in fact, eighteen.

To William Thomson 23 February 1856[1] 1199

Royal Institution | Saturday 23rd Feb. [1856]

My dear Sir.

Up to the present moment I have been unable to write a word in reply to your last note.[2] On Thursday I have a lecture and shall be very busy; but on Friday I am quite at your service. We shall have a long day to prepare, and from the nature of the experiments I infer that we shall be able to make all arrangements comfortably & efficiently. We have a very beautiful means of showing the heat of congelation[3] which I will have prepared on Thursday, so that it shall be cool and ready for action on Friday. The subject as far as I can see is an extremely promising one, but remember you have only an hour at your disposal and if you made all the experiments on your list it would be at the rate of an experiment every four minutes. Would this afford you sufficient time for explanation? If out of the fifteen you were to choose ten, these ten would afford so many resting places for the general mass of the audience, and would at the same time allow you a tolerable freedom of exposition. With regard to the experiment on the effect of induced currents I devised the following means of estimating the comparative conductivity of bismuth and copper some time ago. A cube of each metal was taken, and suspended from a twisted string, above the cube and attached to it by a copper wire was a small pyramid with the base horizontal and its four sides were formed by 4 little triangular pieces of looking glass. The mirrors rotated with the cube, and the latter of course was placed between the poles of an electro-magnet. A beam of light was thrown upon the mirror and as it turned round—slowly at first—the images reflected from its sides followed each other in a circle of about 30 feet in diameter. As the motion quickened the patches of light blended themselves into a continuous line. On closing the circuit with copper the effect was astonishing—your own term best describes it the cube was 'struck dead.' The idea of the top is a beautiful one if you can realise it—if not we can leave the experiment in the above form.

Very Sincerely Yours | John Tyndall

RI MS JT/1/5/1550–51
LT Transcript Only

---

1. The letter was printed, and dated as 1856, in S. P. Thompson, *Life of William Thomson Baron Kelvin of Largs*, 2 vols. (Macmillan: London, 1910), vol. 1, pp. 313–14.
2. *your last note*: probably letter 1186. Thomson had been asked to deliver a Friday Evening Discourse at the RI on 29 February 1856 and Tyndall had agreed to help Thomson set up the experiments (*Faraday Correspondence*, 5: 3705 and 3079). See W. Thomson, 'On the Origin and Transformation of Motive Power', *Roy. Inst. Proc.*, 2 (1854–58), pp. 199–204.
3. *heat of congelation*: the heat abstracted when a fluid changes to the solid state as when water forms ice. Tyndall's demonstration was probably a version of the one he used later in the 1862 series of lectures on heat. This used a supersaturated solution of sodium sulphate kept quite still overnight. By simply dropping a single crystal of the salt into the solution, the audience could observe the solution slowly solidify. When a thermo-electric pile was held against the flask, the temperature rose rapidly from zero as solidification took place. J. Tyndall, *Heat Considered as a Mode of Motion being a Course of Twelve* Lectures delivered at the Royal Institution of Great Britain in the Session of 1862 (London: Longman, Green, 1863), pp. 153–54.

From Heinrich Debus — 24 February 1856 — 1200

Queenwood College, Stockbridge, Hants. | Feb 24, 1856|

Mein Lieber John!

Erst heute schreibe ich Dir weil die Zeit der wir armen Menschenkinder ja unterworfen sind mir ein Bischen knapp zugemessen war.

Ja so sind die Menschen, das was sie nicht verstehen können, das jenige welches in ihren kleinen Köpfen keinen Platz finden kann verschreien und verrufen sie und sogar das Höchste und Heiligste das wir je besessen haben wird in den Koth getreten und verachtet. Diese kleinen Philosophen oder richtiger Nichtphilosophen meinen dass ihre Weisheit unfehlbar und daher allmächtig sei. Mit Hochmuth und Selbstsucht wird demgemäss Alles abgehandelt und in eine Form gegossen die nur zu sehr alle Merkmale ihres Ursprungs an sich trägt. Du wirst bereits gefunden haben dass ich von G. H. Lewes und den Bekenntnissen einer schönen Seele spreche. Es ist das Schicksal der menschlichen Vernunft sich Fragen vorzulegen die sie nicht beantworten kann. Und doch kann sie auch diese Frage nicht zurückweisen weil eine eigne Wissbegierde sich immer wieder geltend macht und sie in ihrer Ruhe verfolgt. Diese höchsten Fragen sind Gott, Freiheit und Unsterblichkeit. Wo aber das Wissen aufhört fängt der Glaube an. Und je lebhafter die Begierde, je grösser der Wunsch etwas über jene hohen Gegenstände zu erlangen um so stärker ist in der Regel der Glaube. Wenn nun das Gemüth rein und edel dann setzt es sich zu dem Höchsten in ein schönes Verhältniss, es malt sich auf der einen Seite das Herrlichste das der Geist empfangen in möglichster Vollkommenheit aus und sucht das eigne unvollkommene Selbst danach heranzubilden und ins Gleiche zu setzen. Und wenn man nun auch sagt dass dies Alles nur ein Gebilde des eigenen Geistes sei, so thut dies Nichts insoweit als die Wirkung in Betracht kommt, denn ein solcher Glaube mag wohl für Wahrheit stehen und stehen bleiben. Denn die Liebe einer reinen frommen Seele ist eine bessere Liebe, ihre Freundschaft eine treuere Freundschaft, ihre Neigung lebhafter und inniger weil dies Alles durch den Glauben von einer höhern Quelle fliesst mit dem Vollkommensten verwandt und darum ehrwürdig heilig erscheint. Wenn es aber unter den Menschen einen giebt auf den jene Bekenntnisse einen wohlthätigen Einfluss gehabt haben, der dadurch glücklicher, ruhiger, heiterer und besser geworden ist wird denn Herr Lewes dieselben noch verspotten? Ich bin ein solcher Mensch. Mit der grössten Freude habe ich dieselben gelesen und eine grosse herrliche Wirkung davon empfunden. Denn auch ich habe die Erfahrung gemacht dass nur in dem man sich zu Gott wahrhaft erhebt nur indem man sich ihm ganz hingiebt, wahrhafte Selbstveredlung wahrhafte Besserung und damit höhere Befriedigung erreicht werden kann. Ist dies Misticismus bin ich ein Mystiker? Bin ich es wohl so mag man mir diesen Namen geben ich trage ihn mit Freuden. O wollten mir die höhern Kräfte einen schärfern Verstandt, eine lebhaftere Einbildungskraft und eine kräftigere Sprache geben, es sollte mir ein Spass sein diese Weltweisen in ihre richtigen Stellen zu versetzen. Auch ich verehre reine Philosophie, ja von allen Theilen des menschlichen Wissens ist sie mir das Liebste. Doch meine Philosophie liegt auf der Erde und wo sie sich in den Himmel erhebt und anfängt in das Leere und Unbegreifliche zu kommen wo sie Nichts wissen kann da lasse ich Ehrfurcht und Glaube an die Stelle des Schlusses und Beweises treten und achte und liebe jeden Menschen der mit der Herrlichkeit der Erde vor ihm an das hohe Wesen glaubt das Alles durchdringt und Allem zu Grunde liegt. Glaube ist das Verhältniss, die Formel unseres Selbst zum höchsten Wesen und Religion die Erkenntnis unserer Pflicht als göttlicher Gebote. Darum Ehrfurcht vor denen die Glauben im Herzen und Religion in ihren Thun haben. Goethe wollte keine mystische Person sondern ein reines frommes Streben ein Erheben zum Höchsten und die Wirkung dieser Handlung auf das Leben und die Beschaffenheit unseres Geistes schildern. Ferner wollte er den Einfluss eines guten Beispiels zeigen. Wie weit ihm dies gelungen ist mag natürlich eine andere Frage sein. Ich selbst glaube dass die Erzählung ein Bischen zu sehr in die Länge gezogen viele unbedeutende Begebenheiten aufgenommen die besser auszulassen wären. Wenigstens scheint es mir dass nachdem man einmal zum Ende des Zwecks gekommen ist eine kleine Abkürzung der Erzählung das Interesse erhöhen würde. Die Person welche dem Autor als Original diente war eine Fräulein von Klettenberg die gegen die Mitte des vorigen Jahrhunderts in Frankfurt lebte durch Frömmigkeit und einen edlen Lebenswandel ausgezeichnet und eine Freundin Goethes und seiner Familie war. Diese vortreffliche Dame hat grossen Einfluss auf Goethe gehabt und manches Gute zu seiner Bildung beigetragen.

Ich setze Dir einige Citate aus dem Kosmos hierher die dir vielleicht nützlich sein können.

Deutsche Ausgabe Bd I, Seite 110–162

Humbolt handelt hier von Cometen, Fixsternen, Planeten, Zodiacallicht, Sternschnuppen &c. &c. Ich glaube nicht dass Du viel brauchbares finden wirst. Seite 142 &c&c wo er von Zodiacallicht spricht scheint mir für Dich am interessantesten zu sein.

Ueber Nordlicht—Seite 198 &c.&c.

Ueber Polarlicht Seite 206—7 &c.&c.

Ueber Blitz—Seite 363.

Childrey über Zodiacallicht Seite 409 und die folgenden Seiten.

Gern würde ich Alles dies abgeschrieben haben. Es ist aber zu viel und Du kannst es auch besser im Original lesen und denn besser wählen wie es Deine Zwecke verlangen. In der Royal Institution giebt es gewiss einen Kosmos. Oder Du kannst das Buch sonst wo mit Leichtigkeit in London bekommen. Doch glaube ich nicht dass Du viel darin für Deinen jetzigen Bedarf finden wirst.

Nun lebe wohl lieber John | Mit herzlichen Grüssen | Dein H. Debus

Queenwood College, Stockbridge, Hants. | Feb. 24, 1856|

My dear John!

I am writing to you[1] only today because time, to which we poor human beings are of course subject, has been somewhat meagerly apportioned to me.

Yes, so it is with people—they decry and bring into discredit that which they cannot understand, whatever cannot find a place in their small heads, and even the highest and holiest that we have ever possessed is trodden into the muck and despised. These little philosophers, or more correctly non-philosophers, opine that their wisdom is infallible and therefore all-powerful. Accordingly, everything is treated with haughtiness and egotism and poured into a mould that only too much bears all the characteristics of its origin about it. You will already have realized that I speak of G. H. Lewes[2] and the Confessions of a beautiful soul.[3] It is the fate of human reason to pose questions to itself that it cannot answer. And yet it cannot reject this question either, because an inherent thirst for knowledge asserts itself again and again and pursues it in its tranquility. These highest questions are God, freedom and immortality. But where knowledge ends, faith begins. And the livelier the thirst, the greater the wish to acquire something about these lofty subjects, all the stronger, as a rule, is faith. If indeed the mind is pure and noble, then it places itself in a beautiful relationship to the highest; on the one hand, it imagines the most magnificent that the spirit can receive in the greatest possible perfection, and seeks to educate its own imperfect self in its image and to emulate it. And even if one now says that all this is just a creation of one's own mind, this does nothing in so far as the effect is considered, for such faith may well stand for the truth and remain doing so. For the love of a pure pious soul is a better love, its friendship a truer friendship, its affection more lively and more intimate because all this flows through faith from a higher source, akin to the most perfect, and therefore appears venerable and holy. But if there is one among men on whom those confessions have had a charitable influence, who, through them, has become happier, calmer, more serene and better, will Herr Lewes still mock them? I am such a person. I have read such things with the greatest pleasure and felt a great and magnificent impression from them. For I have also had the experience that only in so far as one truly raises oneself up to God, only in so far as one gives oneself entirely to Him, can true self-ennoblement, true improvement and therefore higher satisfaction be achieved. Is this mysticism; am I a mystic? If indeed I am, then one may give me this name, I shall bear it with joy. O, would the higher powers grant me a sharper intellect, a more lively imagination and a more powerful speech, it should be a pleasure for me to put these sages in their rightful place. I too venerate pure philosophy; indeed, of all parts of human knowledge, it is the dearest one to me. But my philosophy is an earthly one, and where it rises towards heaven and begins to enter the void and the incomprehensible where it can know nothing, there I let reverence and faith take the place of deduction and proof, and respect and love every person who believes, with the glory of the earth before him, in the high essence that pervades everything and lies at the foundation of everything. Faith is the relationship, the formula of our self to the highest essence, and religion the realization of our duty as a divine commandment. Therefore, reverence to those who have faith in their hearts and religion in their actions. Goethe did not want to portray a mystical person, but a pure and pious striving, an exaltation to the highest, and the effect of this action on life and the constitution of our spirit. Furthermore, he wanted to show the influence of a good example. To what extent he succeeded in this may, of course, be another question. I myself believe that the story was drawn out a little too much: too many insignificant events were included that were better left out. At least it seems to me that once one has come to the end of one's purpose, some shortening of the story would increase interest. The person who served as the original for the author was a Fräulein von Klettenberg[4] who lived in Frankfurt towards the middle of the last century, was distinguished by piety and a noble way of life, and was a friend of Goethe's and his family. This excellent lady had a great influence on Goethe and contributed much that was good to his education.

I shall set out for you here some quotations from Kosmos[5] that can perhaps be useful to you.

German edition Vol. 1, pages 110–162

Humbolt[6] deals here with comets, fixed stars, planets, zodiacal light, shooting stars &c. &c. I do not believe that you will find much that is useful. Page 142 &c&c where he speaks of zodiacal light seems to me to be the most interesting for you.

On the northern lights—page 198 &c.&c.

On polar light page—206–7 &c.&c.

On lightning—page 363.

Childrey[7] on zodiacal light page 409 and the following pages.

I would gladly have copied all of this. But it is too much and you can read it even better in the original and then better choose how it meets your needs. There is certainly a Kosmos in the Royal Institution.[8] Or you can easily get the book somewhere else in London. However, I do not think that you will find much in it for your current needs.

Now farewell dear John | With warm regards | Your H. Debus

RI MS JT/1/7/2412–14
LT Transcript Only

---

1. *you*: Dir (German) is informal address.

2. *G. H. Lewes*: George Henry Lewes (1817–78), English critic, philosopher and miscellaneous writer best known to Tyndall, Hirst and Debus as the author of *Life and Works of Goethe* (London: Dent, 1855).
3. *Confessions of a beautiful soul*: Goethe's account of the life of Susanna von Klettenberg in Book 6 ('Bekenntnisse einer schönen Seele') of his *Wilhelm Meisters Lehrjahre* (1795–96), an outstanding example of German Bildungsroman.
4. *Fräulein von Klettenberg*: Susanna Katherina von Klettenberg (1723–74), Moravian abbess and mystic who befriended Goethe at the age of nineteen and made a deep impression upon him. Her only writings were a few hymns which she published anonymously.
5. *Kosmos*: Alexander von Humboldt, *Der Kosmos: Entwurf einer physischen Weltbeschreibung,* 5 vols. (Stuttgart: Cotta, 1845–62). Two rival multi-volume English translations of the first volume appeared in 1846 and 1849: *Cosmos: Sketch of a Physical Description of the Universe,* trans. Mrs Elizabeth Sabine ('Under superintendence of Lieut. Col. E. Sabine'), 4 vols. (London: Longman, 1846–58); and *Cosmos: A Sketch of a Physical Description of the Universe,* trans. Mrs E. C. Otté, 5 vols. (London: Henry Bohn, 1849–58).
6. *Humbolt*: Alexander von Humboldt (1769–1859), German traveler and polymath whose popular account of natural philosophy, *Cosmos,* became a best seller. Tyndall had met him in Berlin on 26 May 1851 (RI MS JT/2/13c/544).
7. *Childrey*: Joshua Childrey (1623–70), English clergyman and antiquarian who first described zodiacal light (the white glow in the night sky caused by cosmic dust) in 1661.
8. *in the Royal Institution*: this copy of *Kosmos* is still in the RI Library.

From Olivia M. Jones[1] 7 May 1856 1201

31 Brook St | March 7

My dear Dr. Tyndal
How are you
Will you come and dine with Papa to morrow at half past six
Thank you for your letter[2] good bye my dear Dr Tyndal.
I remain your ever affectionate | L M Jones.
'Invit^n to dinner | little Lily Bence Jones'[3]

RI MS JT/1/J/131

1. *Olivia M. Jones*: nicknamed Lily (1845–1919), the 10-year old daughter of Dr Henry Bence Jones. She had two sisters, Millicent (1843–1932) and Mary (b. 1847), and two brothers, Henry (1844–1912) and Ralph (1845–66). Information from the 1851 Census.
2. *your letter*: letter missing.
3. *Invit^n . . . little Lily Bence Jones.* This was Tyndall's annotation. The letter was also transcribed into Tyndall's Journal (RI MS JT/2/13c/ 829).

From George Biddell Airy 8 March 1856 1202

Royal Observatory, Greenwich, 1856 March 8

Dear Sir

I have to present to you my thanks for two papers[1] which reached me this morning. I ought long ago to have acknowledged a copy of your quarto Royal Society paper,[2] and intended to do so, but I suppose that interruptions put it out of my thoughts.

My concerns with magnetism, especially in Gauss's[3] processes for absolute measure,[4] and in the investigation of the angular disturbances of compasses, have all been of accurate character. They have all implied the application of the ordinary rules of mechanics by means of the severest numerical calculations. And the exactness of the representation of the phenomena by the calculation, in various ways, has produced on me a distinctness of conviction that magnetism (I should say paramagnetism) under all circumstances acts on calculable mechanical principles, such as can hardly be felt without going through those steps.

It has been a matter of no small grief to me to find that, till a comparatively late time, a totally different theory, a theory of extreme vagueness,[5] has been advocated by the highest authority.

And now it is a matter of great satisfaction to me to find that, in all your investigations and explanations applying to diamagnetism, you have reduced all to a mechanical and calculable form.

I conceive that the position of the science is now very greatly altered, to an extent far beyond that to which your investigations ostensibly apply.

I am, dear Sir, | Very faithfully yours | G B Airy

Professor Tyndall

RI MS JT/1/A/28
Handwritten Copy: Airy Papers, MS RG0.6/694

---

1. *two papers*: Offprints of Tyndall's recently-published papers, one of which was clearly 'On the Relation of Diamagnetic Polarity to Magnecrystallic action', *Phil. Mag.*, [4] 11 (February 1856), pp. 125–37.
2. *quarto Royal Society paper*: J. Tyndall, 'Further Researches on the Polarity of the Diamagnetic Force', *Phil. Trans.*, 146 (1856), pp. 237–59.
3. *Gauss's*: Carl Friedrich Gauss (1777–1855), German mathematician, astronomer and physicist.
4. *absolute measure*: In 1832 Gauss established absolute units of distance (millimetre), mass (milligram) and time (second) in *Intensitas vis Magneticae Terrestris ad Mensuram*

Absolutam Revocata (Göttingae; Dieterichianis: Göttingae, 1833), one of the last scientific works to be published in Latin.

5. *a theory of extreme vagueness . . . advocated by the highest authority*: Airy was referring to Faraday and his field theory. It was the simple mechanical couples and calculability of Tyndall's work on magnetism that Airy (and later Herschel) liked as opposed to what even Tyndall referred to as Faraday's 'misty' concept of a 'field'. See Tyndall's earlier comment on Faraday's lines of force in letter 1152.

To Thomas Archer Hirst 9 March 1856 1203

9th. March, 1856

My dear Tom

I have had a note[1] from Col. Wynne in which he expresses his gratification on the arrangement that you have made regarding Eddy.[2] He also enclosed me a note of the latter[3] to him at the same time. Speaking of the money part of the matter he said it would be at the rate of £120 a year: no doubt he continued this will appear a large sum to you but when you remember the number of private lessons which I receive you will I am sure consider it extremely moderate. This in substance is what he said. I thought the figure large myself, but doubtless it is accounted for in the manner he indicates, and you have fairly looked at the question before deciding. At all events he could not be in better hands.

We had Thomson in town last week. He gave the Bakerian lecture at the Royal Society,[4] and gave us a Friday evening.[5] I heard a portion of the former but had to come away before it was finished.[6] I am told it was very unhappy. He spent his time it is said in describing experiments which failed; and one person present (Grove) remarked that if everybody did so the greatest blockhead would have the longest paper. The impression regarding our Friday evening is I believe that it was not successful. People say he had considerable help from Faraday, Anderson & myself and yet got on but poorly. A friend of his and mine[7] told me last night that he thought the lecture did him no good. I was amused one morning with a conversation in which I took part with Thomson. Lord Wrottesley, Faraday, Thomson, Cayley[8] and myself had been to breakfast at Col. Sabine's[9] and on returning Cayley asked Thomson something regarding the influences of wire drawing on the density of the substance. He at once replied, and gave his reasons for his reply.[10] They seemed quite convincing to him, but neither myself nor Cayley could see the matter with equal clearness. Coming up James's street[11] I suggested an hypothesis the admission of whose possibility would completely destroy all that he had said.[12] He paused: admitted the validity of my remark and exclaimed just as promptly as before—'The whole theory of elasticity is wrong; Stokes has been urging a thought similar to yours with regard to light:[13] no doubt the entire theory of elasticity must be modified to take in your notion [']–Well: I could not but think cheaply of a theory which could be so easily shattered, nor could I rate the convictions as high that were liable to such sudden changes though appearing so positive as long as they were uncontroverted.

Yesterday a remarkable note from the Astronomer Royal[14] reached me: I send it to you, but you will I hope send it back to me. I had sent him a copy of my last paper[15] in the Phil. Mag. and this drew the note from him. I also send you another document[16] still more pleasant than Airy's which you will also send back to me or keep until I come down.

And now with regard to my coming down: I wish you would engage that place at Mrs Gale's[17] for me. Your house[18] is full enough at present and I shall be near enough to you at Gale's cottage. I expect to be coming down early next week, that is 8 days or so from the present time.[19]

I think I have nothing more to say. Goodbye. Love to Anna

Ever yours affectionately | John | Sunday 9th March 1856

RI MS JT/1/T/876

---

1. *a note*: letter missing.
2. *regarding Eddy*: see letters 1193, 1194, 1196, and 1197, n. 2.
3. *a note of the latter*: that is, Eddy Wynne; the letter is missing.
4. *Bakerian lecture at the Royal Society*: W. Thomson, 'On the Electrodynamic Qualities of Metals', *Phil. Trans.,* 146 (1856), pp. 649–751, delivered on 28 February 1856.
5. *Friday evening*: see letter 1199, n. 1.
6. *it was finished*: According to Tyndall's Journal, he left early in order to attend a dinner engagement with the London ophthalmologist William White Cooper (1816–86) (RI MS JT/2/13c/ 827). One of Cooper's patients was the subject of Tyndall's RI lecture 'On a Peculiar Case of Colour Blindness', *Phil. Mag.,* 11 (1856), pp. 329–33.
7. *a friend of his and mine*: Tyndall's Journal offers no clue to this person's identity.
8. *Cayley*: see letter 1088, n. 2.
9. *breakfast at Col. Sabine's*: This was on 29 February before Faraday and Tyndall returned with Thomson to the RI to help him prepare experiments for his Friday Evening Discourse (RI MS JT/2/13c/ 827).
10. *his reasons for his reply*: Thomson clearly recognised that Cayley's question involved the theory of elasticity, the mathematical treatment of the deformation of solids and liquids under stress.
11. *James's Street*: a road adjacent to Covent Garden Market.
12. *all that he said*: Tyndall had explored the effects of stress and deformation on solid bodies and found it affected their magnetic properties. He explained this in terms of their molecular constitution.

13. *Stokes has . . . with regard to light*: According to the wave theory, light consists of transverse vibrations in the ether. Augustin Fresnel had also argued that this ether was a solid through which the earth moved unhindered. However, this model could not account for stellar aberration. Having earlier pursued research on hydrodynamics, Stokes criticised Fresnel's theory and during the period 1845–48 he advocated an elastic solid ether that, in the neighbourhood of the earth, was dragged along by the earth; a theory that could explain aberration. See G. G. Stokes, 'On the Constitution of the Luminiferous Aether, Viewed wuth Reference to the Phenomenon of the Aberration of Light', *Phil. Mag.,* 29 (1846), pp. 6–10.
14. *Astronomer Royal*: see Airy's letter 1202.
15. *my last paper*: letter 1202, n. 1.
16. *another document*: letter 1201 from Olivia Jones.
17. *Mrs Gale's*: see letter 1198, n. 2.
18. *Your house*: The Hirsts lived at Vine Cottage in Broughton and Mrs Gale in an adjacent cottage.
19. *from the present time*: Tyndall joined the Hirsts at Vine Cottage on Tuesday 18 March (RI MS JT/2/13c/830).

To George Airy — 10 March 1856 — 1204

Royal Institution | 10th March 1856

Dear Sir

The letter[1] with which you have favoured me has been a source of great satisfaction and encouragement to me.

Believe me I have not accepted the convictions of which my papers are the utterance without due caution. I have sometimes wavered and said to myself—There may be something in the matter which you do not see—look what great things have been accomplished in another way—You may be shutting out the light by adopting your present course.

But the more I have thought on the matter the more profoundly have I ever felt the truth of the view which you have expressed in your note.

On the other hand, as you very well know, a hypothesis may be valuable in two ways. It may, by its truth, help us to systematic arrangement or lead us to discovery. But it may also, by the attractiveness it possesses for the person who holds it, prove a source of energy and be a spur to action.[2] In this way while vague, or ever erroneous, it may be, and I doubt not has been, productive of glorious results.

Believe me dear Sir | most faithfully yours | John Tyndall
GB Airy Esqr

RGO MS 6/694

1. *The letter*: letter 1202.
2. *a spur to action*: Tyndall was referring to Faraday's field theory. See letter 1202.

From Mary Anne Coxe 23 March 1856 1205

Edinbro. | Sunday, March 23rd.

My dear Dr Tyndall

Sunday though it be, I must just write you one line, to thank you for the last reading you sent me,[1] and to tell you honestly that it was far and away beyond my capacity!

I write this to deprecate your wrath, before my husband faces you, which he hopes very soon to do.

He, expects to be in London on Thursday first[2] at Fenton's Hotel[3]—he and his colleague Mr Monteith[4]—they go up to meet their English Brethren,[5] before their report is printed—he will be excessively busy for they have many matters to discuss—but I am very sure he will find, or make time to see you nevertheless—I would have accompanied him, but my eldest daughter Mrs Ed. Whately[6] is now with us, and will remain with me till his return, at the end of as few days as may be (not less I fear than 14). When this said report is through the press, we hope to be in London together, on our way to the Continent, but as that will not likely be before June, I greatly fear that we shall find you flown—though who knows but we may stumble on a meeting across the Channel. You and he will talk over your mutual plans. I do envy him the sight of you, and one or two other friends—little as I like London in a general way—

> 'I loathe the Squares and Streets, And the faces that one meets
> Hearts that have no love for me.'[7]

Now good bye dear Dr Tyndall, and believe me always | very much Yours | Mary Anne Coxe.

RI MS JT/1/1/275
LT Typescript Only

1. *the last reading you sent me*: letter missing; presumably Tyndall had recommended that she should read a book or article.
2. *first*: this refers to Dr Coxe's first location in London and not to 1 April, which was a Tuesday (not Thursday).
3. *Fenton's Hotel*: a hotel at 63 St James's Street, London. It was a popular, respectable hotel, but not considered suitable for the gentry or aristocracy.
4. *Mr Monteith*: not identified.
5. *their English Brethren*: The Scottish Lunacy Commission which presented its *Report by*

Her Majesty's Commissioners Appointed to Inquire into the State of Lunatic Asylums in Scotland (HMSO; Edinburgh) to both Houses of Parliament in 1857.

6. *Mrs Ed. Whately*: The Coxe's daughter, née Teckla Coxe (b. 1830) had married the clergyman Edward W. Whately (b. 1824). He was the minister of St Giles Church, Norwich.
7. *I loathe . . . Hearts that have no love for me*: Mrs Coxe was quoting from Tennyson's *Maud, or, The Madness* (London: Moxon, 1855), Part 2, IV, verse 13. The poem probably interested Mrs Coxe because of her husband's involvement in psychiatry.

From Henry Moseley 29 March [1856] 1206

Olveston | 29 Mar.[1]

My dear Sir,

You will see from the enclosed[2] that we cannot have Prof. Christison.[3]

My only doubt about Prof. Henfrey[4] arises from our having already one King's Coll. Professor in our list of Examiners and must perhaps have a second. Prof Henfrey would make a third. As we must be above the suspicion of partiality for any particular Institution, I fear this would not do? Perhaps he does not know Zoology. I think however that would easily be learned in London.

Is the Mr Babington[5] of whom you speak of St. John's Coll Cambridge?

Yours truly | Henry Moseley.

RI MS JT/3/887
LT Typescript Only

1. *29 Mar*: The letter is unyeared. It has been tentatively dated as 1856 on the basis that it appears to be Moseley's reply to a missing letter in which Tyndall suggested a number of possible examiners for questions on botany.
2. *the enclosed*: enclosure missing.
3. *Prof. Christison*: Robert Christison (1797–1882), toxicologist and Professor of Medicine at the University of Edinburgh since 1822. Why Tyndall should have suggested him as an examiner for admissions to the Royal Military Academy is unclear.
4. *Prof. Henfrey*: Arthur Henfrey (1819–59; FRS 1852), Professor of Botany at King's College London since 1853.
5. *Mr Babington*: Charles Cardale Babington (1808–95; FRS 1851, botanist who taught at St John's College, Cambridge. The founder and secretary of the Ray Club in 1836.

To George Grey[1] 3 April 1856[2] 1207

Dr. Tyndall, F.R.S., a supplementary Juror at the Paris Exhibition,[3] writes thus strongly: In wishing success to the undertaking in which you are engaged, I express the feeling of every individual who would like to see England rescued from a position which has long been a reproach to her. The Committee of Industrial Instruction[4] is the outward and visible indication of a great public want, of a want which will assuredly find less pleasant means of utterance by and bye, if it be not met and satisfied in time . . . The results of such a movement are incalculable; among other things it would abolish the unnatural divorce at present existing between the cultivator of Science and the practical man; the former would learn that every practical realization of a thought furnishes a purchase for new speculative effort, while the latter would learn how society would be his 'practice,' had it never possessed a scientific base.

Parliamentary Papers, 1856 [2065] Third Report of the Commissioners for the Exhibition of 1851, to the Right Hon. Sir George Grey, Bart

---

1. *George Grey*: George Grey (1799–1882), English statesman and Home Secretary from 1855 to 1858.
2. *3 April 1856*: This was the official date of publication. We do not know when Tyndall actually wrote his report.
3. *Juror at the Paris Exhibition*: see letter 1070, n. 3.
4. *Committee of Industrial Instruction*: Better known as the Department of Science and Art that was created in 1853. The surplus of the Great Exhibition of 1851 was used by the Royal Commissioners of the Exhibition, and following the vision of Prince Albert, 'to increase the means of industrial education and extend the influence of science and art upon productive society'. Various Schools of Design had been set up and aided by government grants in 1837, These were amalgamated under the umbrella of a Department of Practical Art commanded by Henry Cole (1808–82) in 1852. A science section, with the chemist Lyon Playfair (1818–98) at its head, was soon added. The combined sections were placed under the Board of Trade in March 1853 as a Department of Science and Art with Playfair and Cole as joint secretaries. Its task was to 'promote instruction on technical questions relating to art and manufactures' (*Hansard*, 3rd series, 125, (4 April 1853), c. 522).

To Thomas Archer Hirst 6 April 1856 1208

Sunday 6th April

My dear Tom

Thank you for the Observer[1] though it does contain but little and that little I have not yet read. I wish I had met Col. Wynne down there[2]—I am sure I should have found many points of contact between you—these things are in some measure matters of habit. The points of contact of course must be nature's gift, but the discovery of them often lies within the domain of wit. Your judgement of the man however is a correct one.

My health still oscillates. I gave my first lecture on Thursday last,[3] and had some beautiful experiments. The audience still continues very large, very attentive and very kind; I met on Thursday Mr Mallet[4] whose name you have sometimes seen in the Phil. Mag. he described some beautiful experiments regarding the cleavage of ice and the manner of making them—I tried them yesterday and failed. But I dont know why I mention the fact. On Friday evening we had a lecture from a young fellow named Roscoe[5] who has been recently associated with Bunsen in an investigation on the chemical actions of Light. I can truly say that I carried all the discourse away with me & found no difficulty in understanding it, but the general impression was that it was a failure. Yesterday an artist[6] who is engaged on a painting of the theatre of the R.I. when the Princes[7] were there took my noddle[8] also to fit into it at some corner. Afterwards I called upon Claudet the photographer[9] to borrow some Daguerreotypes for my next lecture. I spent an interesting & constructive hour with him, during which however he managed to take me in half a dozen positions.[10] So that in one day I have had a regular descent of artists.

Scarcely has the discussion with Matteucci[11] passed away when I am attacked by Plücker in his private letters to his friends. Wheatstone—a great partisan of Plücker's—shewed me a letter[12] wherin he stated that I had in a seducing manner totally disfigured his ancient way of viewing the subject. and also insinuates that I had influenced Faraday's opinion of him. He has also written to Faraday[13] in the same strain. If he wished to retain Faraday's esteem he could not have done a more silly thing. I shall endeavor to avoid discussion with him, though he proposes to write an edition of his labours and to lay his claims before the Royal Society.[14] If one wished there is no man who could be more readily reduced to a scientific scarecrow than this same Mr Plücker. But it is worthless work to loose time in a strife of the kind and if he does not pressure very hard I will let him have his way.

With love to Anna Believe me Tom | affectionately yours | John

RI MS JT/1/T/889

---

1. *for the Observer*: *The Queenwood Observer* was an occasional newspaper written and printed by staff and students of Queenwood College. A few copies have survived at the Friends' Library in London. The issue Tyndall referred to evidently contained an essay by Hirst on 'Swinging'. See Hirst Journal, 13 April 1856 (RI MS JT/2/32c/1191).
2. *down there*: According to Hirst's Journal, Colonel Wynne had visited his son Eddy and Hirst at Queenwood on Friday 28 March (RI MS JT/2/32c/1191).
3. *I gave my first lecture on Thursday last*: the first of 11 lectures on light.
4. *Mr Mallet*: Robert Mallet (1810–81; FRS 1854) was an Irish railway engineer and geologist

who made pioneering studies of earthquakes. The meeting with Mallet was not recorded in Tyndall's Journal.

5. *a young fellow named Roscoe*: Henry Enfield Roscoe (1833–1915; FRS 1863), English chemist. After training at University College London he became a pupil of Robert Bunsen's at the University of Heidelberg where he worked on the chemical action of light (photochemistry). He lectured at the RI on 'The Measurement of the Chemical Action of Light' on 4 April 1856. See *Roy. Inst. Proc.,* 2 (1854–58), pp. 223–25.
6. *an Artist*: The artist was Alexander Blaikley (1816–1903), a Sandemanian and distant relative of Faraday's.
7. *the Princes*: Prince Albert had taken his two sons Albert Edward (1841–1910), Prince of Wales, and Alfred (1844–1900), the Duke of Saxa Coburg, to hear Faraday's Christmas lectures in December 1855. See Tyndall Journal, 27 December 1855 (RI MS JT/2/13c/811).
8. *my noddle*: my image. According to a key to this frequently-reproduced painting, which hangs outside the lecture theatre of the present-day RI, Tyndall is about half way up on the left hand side, three places to the right of two women. The clean-shaven image does not look like Tyndall at all!
9. *Claudet the photographer*: Antoine Claudet (1797–1867; FRS 1853), French photographer and inventor who kept a shop in Regent Street.
10. *half a dozen positions*: Claudet photographed Tyndall. It is not known whether any of these daguerreotypes survive uncatalogued at the RI.
11. *discussion with Matteucci*: Matteucci's former criticism of Tyndall's magnetic work. See letter 1161.
12. *shewed me a letter*: Wheatstone showed Tyndall the letter (which was in French) when he went to Hammersmith on Sunday 30 March 1856 (RI MS JT/2/13c/832).
13. *He has also written to Faraday*: Plücker had written to Faraday in English on 21 March 1856. *Faraday Correspondence,* 5:3109. Plücker accused Tyndall of 'fighting against a theory, which never was mine' and merely giving the very explanation of how uniaxial crystals behaved in the presence of magnets that he had given in 1849.
14. *lay his claims before the Royal Society*: He did so in J. Plücker, 'On the Magnetic Induction of Crystals', *Phil. Trans.,* 148 (1858), 543–47. Plücker had been elected a Foreign Member of the Royal Society in June 1855, mainly through the efforts of Wheatstone.

To Henry Provost Babbage[1] 7 April 1856 1209

7th April 1856

My dear Sir

I thank you for reviving my memory of 'observations'[2] in which I took great interest when I heard them delivered, and most heartily do I respond to the hope which you express at the conclusion.[3] The claims of such men ought to need no urging.

Most sincerely yours | John Tyndall

BL–37196–464

---

1. Henry Provost Babbage (1824-1918), the youngest son of Charles and Georgiana Babbage. After military service in India he returned to England and helped his father improve the designs of the difference and analytical engines.
2. *'observations'*: Tyndall had attended Henry Babbage's lecture on his father's calculating engine at the Institution of Civil Engineers on 4 December 1855. The lecture had just been published as Henry Provost Babbage, 'Scheutz' Difference Engine, and Babbage's Mechanical Notation', *ICEP,* 15 (1855–56), pp. 497–514.
3. *The hope which you express at the conclusion*: '[Babbage's machine] had, as it were, photographed the footsteps of time; and with power more enduring than electric fire, it had conferred fixity and permanence on the swiftest motion. . . . Whatever course might be chosen, [Babbage] left it with the fullest confidence in the hands of a Society whose object was essentially progressive' (n. 2, p. 514).

From William Snow Harris 14 April 1856 1210

Athenaeum | 14 April 1856.

Dear Dr. Tyndall,

I have carefully looked over the Paper I propose for the Phil. Mag.,[1] and have endeavoured to meet your views about it as to length. I find I can not manage to write the paper over again—for want of time but I have cut out at least 4 pages of it—so as to bring it under a more convenient space—the parts cut out are those to which you seemed to take exception—and it does not appear to me upon very due consideration, that it is now open to much controversy at all events. The enclosed[2] has just turned up among my papers will you favour me by accepting it: the record may be useful for reference—I have sent on the paper to the printer as requested by you—and I trust it may turn out useful to science and not be altogether unworthy of the Phil. Magazine—I have cut out the last figure relating to Doves Experiment[3] and will endeavour to make it a subject for separate consideration.

most faithfully yours | W. Snow Harris.

[Enclosure]

Since we met[4] I have examined carefully under competent authority M. Riess work and his Papers in Poggendorf[5]–and have satisfied myself that he does really estimate what he calls 'Density' of Electricity by the Electrometer—as given by De La Rive[6]—I find that De La Rive's statements are mere transcript of Riess's work—and Riess work is in great parts a mere repetition of his former Papers in Poggendorf. It is true he finds fault with the terms Tension &c. &c. but still he employs them and explains them under the term Density &c. . &c. . Riess in his papers I find does nothing but object to all I have done but does the same thing himself under another form. I am quite prepared with answers to all Riess has said.

Professor Tyndall | F.R.S. &c.

RI MS JT/1/2/469
LT Transcript Only

---

1. *the Paper I propose for the Phil. Mag.*: Even after cutting, Harris's prolix paper occupied 22 pages of the May issue of *Phil. Mag.,* W. Snow Harris, 'On a General Law of Electrical Discharge', *Phil. Mag.,* 11 (1856), pp. 339–60. The paper was an historical account of his own researches on static electricity since 1826 and a bitter attack on the work of Peter Riess (as suggested in the enclosure to Tyndall). Riess replied with a paper carrying the same title, *Phil. Mag.,* 11 (1856), pp. 524–27, in which he accused Harris of being out of date. Harris then replied with 'On Certain Phaenomena of Electrical Discharge', *Phil. Mag.,* 12 (1856), pp. 136–40 in which he accused Riess of being 'dictatorial and intolerant'. Riess answered back two months' later with 'On a Law of Electrical Heat', *Phil. Mag.,* 12 (1856], pp. 322–24, in which he accused Harris of personal affront. Harris had the last word at the end of the year with 'On Riess's Law of Electrical Heat', *Phil. Mag.,* 12 (1856), pp. 553–54. Tyndall had foreseen the consequences of publication when he wrote in his Journal, 'I have asked him [Harris] to condense it, saying that if published in its present form it would give rise to a discussion of which I do not see the end.' RI MS JT/2/13c/833.
2. *The enclosed*: the paragraph following the signature below.
3. *Dove's Experiment*: In 1842 Dove had investigated the effect on electrostatic induction when iron rods were placed in helical coils (H. W. Dove, 'Ueber die durch Annähherung von massive Eisen und eisernen inducierten elektrischen Ströme, *Poggend. Annal.,* 56 (1842), pp. 268–74), and his apparatus was discussed in A. de la Rive, *A Treatise on Electricity* (London: Longmans, 1853), p. 425. Snow Harris did not refer to the experiment in his later papers.
4. *Since we met*: According to Tyndall's Journal, Tyndall and Harris had met on 10 April 1856. 'I have asked him to condense it, saying that if published in its present form it would give rise to a discussion of which I do not see the end' (RI MS JT/2/13c/833).
5. *Papers in Poggendorf*: Notably Peter Riess, 'Bemerkung über die electrische Influenz' [Remarks on electric induction], *Poggend. Annal.,* 93 (1854), pp. 626–27; 'Ueber den

Einfluss der Leitung eines electrischen Stromes auf die Act seiner Einladung' [On the effect of conduction of an electric current on its discharge], *Poggend. Annal.,* 98 (1856), pp. 524–27; and 'Ueber die electrischen Pausen' [On electric discharge], *Poggend. Annal.,* 99 (1856), pp. 571–98.

6. *as given by De la Rive*: Harris's remark must refer to the first volume of A. De la Rive, *Traité d'Electricité Théorique et Appliqué* (Paris: J. B. Ballière, 1854).

To Thomas Archer Hirst 19 April 1856 1211

Royal Institution, Saturday | 19th. April, 1856.

My dear Tom

I wrote to you a long letter[1] some days ago and had it duly posted, since then I have heard nothing from you until your note this morning[2] fell into my hands: I reply to it at once.

I cannot do better than forward the note to Col. Wynne. Of course my dear Tom I know there is no mixture of selfishness in your advice; in fact I rather feel that it is pure self sacrifice on your part. Col. Wynne called on me a few days ago he at that time expressed a strong disinclination to let Eddy come forward in July,[3] and an equally strong desire that he should remain near you for 6 months longer. I have little doubt that this will be his decision when he reads your excellent note. I know nothing regarding the continuation of the examinations, but he I doubt not will ascertain this at the War Office in Pall Mall. I promised to ask you whether it was not from you that Eddy was to have his knowledge of electricity, magnetism, and heat. I take it for granted that this is the case. You will I doubt not hear from Col. Wynne soon after he receives your note. I will post it to him at once or perhaps take it to him.

I am greatly concerned to hear of the state of Anna's health.[4] I have however little fear of her lungs. But Fox's[5] advice is undoubtedly the best. have London advice if matters dont soon mend. My faith is that when this east wind which now searches the bones and marrow of us all vanishes she will sprout out again like one of the spring flowers.

Ever affectionately | John

RI MS JT/1/HTYP/443
LT Transcript Only

1. *long letter*: letter 1208.
2. *your note this morning*: note missing.

3. *come forward in July*: Wynne thought it was too soon for his son, Edward, to take the examinations for entry to Woolwich. In his Journal for 20 April 1856 Hirst noted that Eddy Wynne had heard from his father that the next Woolwich examinations were in July. Hirst advised him to wait six further months and wrote to Tyndall accordingly (RI MS JT/2/32c/ 1192).
4. *Anna's health*: Anna Hirst was displaying symptoms of tuberculosis.
5. *Fox's*: Hirst's Journal for 9 April 1856 says that his wife was examined by the Queenwood physician, Dr Luther Fox (RI MS JT/2/32c/ 1191).

From Joseph Allen Galbraith 23 April 1856 1212

23rd April 1856 | Dublin

My dear Sir

There is one feature in Mr Moseley's report[1] of the Burlington House Examination which has not a little astonished me, and to which I beg to direct your attention, as I think you will agree with me in opinion. In noticing the various departments of the examination held in Burlington House reporting progress in some and regreting its absence in others he has carefully ignored the branch of which you had charge, a very cursory glance indeed at the numbers will prove the very great superiority of the answering in this over every other part of the examination. Mr Haughton and I have been largely concerned in bringing about this excellent result and we wish to spare no pains in producing still better results, and as we are hard and I believe successful labourers I think you will allow it is only natural to expect a little encouragement. A cold silence will have a depressing effect on future preparation in this I believe most important branch of the examination

I remain | Yours faithfully | Joseph A. Galbraith.

RI MS JT/2/7/306–7

1. *Mr Moseley's report*: *Report on the Examinations for Appointments to the Royal Artillery and Practical Class of the Military Academy at Woolwich Held . . . on the 21st of January, 1856* (London: Harrison, 1856), pp. 22–23.

To the Editor of the Times 26 April 1856[1] 1213

Sir,—In justice to the candidates for Commissions in the Royal Artillery and Engineers examined by me in Natural Philosophy and Chemistry you will, perhaps, permit me to state that the general level of the answers in the last examination was much higher than that attained in the first. Some of the papers[2] returned to me gave evidence of rare ability; and if, during their future career,[3] the authors of those papers continue to cultivate the powers which they have shown themselves to possess they will, I doubt not, justify by their deeds the high opinion entertained of them by

Your obedient servant, | John Tyndall

Royal Institution | 26th April 1856.

RI MS JT/2/13c/836

1. Published in *The Times*, 28 April 1856, p. 12. The published version differs slightly from the manuscript version as noted.
2. *Some of the papers*: The editor substituted 'many' for 'some', a significant alteration by the editor.
3. *during their future careers*: phrase deleted by the editor.

From the War Department 28 April 1856 1214

War Department | Pall Mall | 28th. April 1856.

Sir

A letter with your signature[1] under date 26th Inst having appeared in today's impression of 'the Times' newspaper relative to the proficiency displayed by the candidates for appointments to the artillery &c. at the examination conducted by you in the subjects of natural philosophical and chymistry in January last., I am directed by the Secretary of State for War[2] to acquaint you that however glad his Lordship may be to receive such favourable testimony to the attainments of the Candidates from a judge so competent as yourself, yet he is of opinion that an examiner appointed by his Lordship should not take such a step as to address himself to the public prints[3] without in the first instance communicating with, and receiving the sanction of, the authorities of the War Department.

I have the honour to be | Sir | Your obedient servant | H. R. Drewry.

RI MS JT/2/7/310–11

1. *A letter with your signature*: letter 1213.
2. *Secretary of State for War*: Fox Maule, second Baron Panmure (1801–1874), Scottish politician, Secretary of State for War (1855–58).
3. *the public prints*: the newspaper press.

From Henry Moseley 28 April 1856 1215

Olveston | 28 April

My dear Sir,

I am very glad to find that Mr. Galbraith is giving attention[1] to instruction of Candidates for the Artillery examination in Experimental science.

I do not know whether in forwarding to me the results of the examination you expressed any decided opinion as to the improvement which had been made in that subject. If so I regret to have overlooked it. Mr Galbraith has however on the whole no reason to be dissatisfied with the examination.

Yours my dear Sir | very truly | H. Moseley.

RI MS JT/3/903
LT Transcript Only

1. *Mr. Galbraith is giving attention*: see letter 1212.

From Gustav Magnus [April 1856] 1216

Mein theurer Tyndall

Sie haben mir eine sehr große Freude durch Ihren lieben Brief vom 20 v.M. bereitet, denn er hat auf mich den Eindruck gemacht als ob Ihnen wirklich daran gelegen sei direkte Nachricht von mir zu erhalten. Zwar habe ich nie daran gezweifelt daß Sie mir aufrichtig zugethan sind, aber eine Aeusserung, oder eine Handlung eines Freundes, durch welche die Anhänglichkeit eine neue Bestätigung findet, ist stets wohlthuend, besonders für jemand wie ich bin. Indem ich dies fühle und erkenne, mache ich mir große Vorwürfe dass ich Ihren lieben Brief solange unbeantwortet gelassen habe, und Ihnen nicht durch schnelle Beantwortung zu erkennen gegeben habe, wie sehr mich Ihr Brief erfreut hat. Aber Sie wissen wie es zugeht. Ich hatte den Brief auf meinen Schreibtisch gelegt um ihn gleich zu beantworten; und bin heut erstaunt daß nun schon 4 Wochen vergangen sind seitdem ich ihn erhalten habe.

Für die Uebersetzung meiner Abhandlung sage ich Ihnen, dem Mr. Atkinson und besonders Dr. Hirst meinen aufrichtigen Dank.

In Bezug auf die Explosionen von Locomotiv Kesseln habe ich mich erkundigt, und gehört daß solche gar nicht vorkommen. Es ist gewiß daß die Cohaesion des Wassers zunimmt wenn dasselbe frei von Luft ist. Sie können dies auch ersehen aus Versuchen die ich bekannt gemacht habe, unter dem Titel „Ueber die Kraft zur Erzeugung von Dämpfen". Diese Versuche sprechen sehr dafür daß durch Wasser, das frei von Luft ist, Explosionen entstehen können, dennoch glaube ich daß der Grund ein anderer ist. Sollten Sie ein besonderes Interesse an dieser Sache haben so theile ich Ihnen meine Ansicht gern mit.

Es ist mir nicht gelungen die Stelle über Thomson's Bakerian lecture in Ihrem Briefe zu lesen, und bin ich deßhalb außer Stande darauf zu antworten. —Verzeihung!

Sir Ch. Lyel hatte mir schon, als er hier war, gesagt, daß Sie sich mit der Schichtung oder Absondrung der Gebirgsmassen beschäftigen. Ich freue mich wenn es Ihnen gelungen ist durch Versuche diesen Gegenstand aufzuklären.

Ich bin im vergangenen Winter mit einem Gegenstand beschäftigt gewesen an den ich früher nie herangegangen war. Nämlich mit Electrolyse. So vollständig dieses Capitel durch Faraday bearbeitet ist, so war doch noch immer einiges unklar, namentlich die wunderbare doppelte Zersetzung die Daniell und Miller bei den Salzen beobachtet haben. Und wenn man die Abhandl. liest, die Kohlrausch jetzt im 3^{n} Heft von Pogg. hat abdrucken lassen, so überzeugt man sich welche Confusion hierüber herrscht. Einen ersten Theil dieser Untersuchung habe ich bereits der Academie mitgetheilt, allein ich will denselben nicht veröffentlichen bevor nicht auch der zweite fertig ist; was wie ich hoffe nicht lange währen soll. Was ich in den Monatsberichten unserer Academie mitgetheilt habe ist nicht der Mühe werth.

Ausserdem bin ich mit der Redaction einer kleinen Note beschäftigt die mit den Versuchen über die Abweichung der Geschosse in Verbindung steht.

Habe ich Ihren Brief richtig gelesen, so ist Faraday mit der Durchsichtigkeit der Metalle beschäftigt. Das klingt sehr wunderbar, aber in seinen Händen und durch seinen Geist wird das unglaublichste entziffert. Bitte grüßen Sie ihm und drücken Sie im meine Verehrung aus. Es giebt wenig Menschen die ich so hoch schätze, nicht sein Talent allein macht mich zu seinem Verehrer sondern seine Hingebung für die Wissenschaft, die alles andre unberücksichtigt läßt und nur dieser lebt. -

Wie glücklich sind Sie der Sie berufen sind in seine Fußtapfen zu treten. -

In Liebe und Freundschaft | Ihr | GMagnus.

Meine Frau und die Kinder grüssen Sie bestens. -

Ob ich nach Wien gehe kann ich noch nicht bestimmen. Wenn Sie dorthin gehn so könnte mich dies wohl veranlassen auch hinzu gehn, obgleich ich sonst keine große Neigung verspüre. Jedenfalls hoffe ich Sie im Herbst zu sehn, in Wien oder in Berlin oder sonst in Deutschland, nur thun Sie mir den Gefallen und geben Sie mir Nachricht über das was Sie thun werden.

My dear Tyndall,

You have given me very great pleasure through your kind letter[1] of the 20th of last month, for it has given me the impression that it is really important to you to receive direct news from me. Admittedly I have never doubted that you have a genuine liking for me, but a comment or a deed of a friend through which devotion finds new confirmation is always agreeable, especially for someone like me. As I feel and recognize this, I reproach myself greatly for having left your kind letter unanswered for so long and for not having indicated to you by means of a prompt reply how much your letter pleased me. But you know how things are. I had put the letter on my desk in order to answer it straight away and today I am astonished that now 4 weeks have passed since I received it.

I sincerely thank you, Mr Atkinson[2] and especially Dr Hirst for the translation of my article.[3]

I have made enquiries regarding the explosions of locomotive boilers[4] and heard that they do not occur at all. It is certain that the cohesion of water increases when it is devoid of air. You can also see this from experiments that I have published under the title 'On the force for generating steam'. These experiments strongly suggest that water that is devoid of air can produce explosions; nevertheless, I believe that the reason is a different one. Should you have a special interest this matter, then I shall gladly tell you my opinion.

I did not succeed in reading[5] the passage in your letter about Thomson's Bakerian lecture[6] and am not therefore in a position to comment on it. —Sorry!

Sir Ch. Lyel had already told me when he was here,[7] that you are busy with the stratification or structure of mountain masses. I shall be delighted if you have succeeded in explaining this subject through experiments.

The past winter, I have been busy with a subject that I had never worked on before: namely electrolysis. As thoroughly as this matter has been treated by Faraday, some things were still unclear, particularly the wonderful double dissociation that Daniell[8] and Miller[9] observed with salts. And when one reads the paper that Kohlrausch[10] has now had printed in the 3rd issue of Pogg., one sees for oneself what prevails about this. I have already notified the Academy[11] about a first part of this investigation but I do not want to publish it until I have also finished the second part, something that I hope will not take too long. What I have reported in the monthly communiqué of our Academy is not worth the trouble.

In addition to this, I am busy with the editing of a short note relating to the experiments on the deviation of projectiles.

If I have read your letter correctly, then Faraday is busy with the transparency of metals. That sounds very wonderful, but in his hands and through his mind the most incredible things are deciphered. Please give him my regards and express my admiration to him. There are few people whom I regard so highly; it is not his talent alone that makes me his admirer, but also his dedication to science that disregards everything else and lives only for this. —

How lucky you are to be called to follow in his footsteps. —

In affection and friendship | Your | GMagnus.

My wife and children[12] send you their best regards. -

I still cannot decide whether I shall go to Vienna.[13] If you are going there, then this could well prompt me to go there as well, although I otherwise feel no great inclination to do so. In any case I hope to see you in the autumn—in Vienna, or in Berlin, or elsewhere in Germany. Just do me the favour of letting me know what you are going to do.

RI MS JT/1/M/17

---

1. *your kind letter*: letter missing.
2. *Mr Atkinson*: Edmund Atkinson; see letter 1157, n. 5.
3. *the translation of my article*: G. Magnus, 'Hydraulic Researches', *Phil. Mag.,* 12 (1856), pp. 89–107, 178–96.
4. *explosions of locomotive boilers*: the subject of Tyndall's report to Colonel Wynne. See letter 1045.
5. *I did not succeed in reading*: either Magnus could not understand Tyndall's English or the handwriting was too poor. Tyndall's letter to Magnus is missing.
6. *Thomson's Bakerian lecture*: W. Thomson, 'On the Electrodynamic Qualities of Metals', *Phil. Trans.,* 146 (1856), pp. 649–751, given at the RS on 28 February 1856.
7. *Sir Ch.Lyel . . . when he was here*: Sir Charles Lyell had visited Berlin in March 1855.
8. *Daniell*: John Frederic Daniell (1790–1845; FRS 1814), Professor of Chemistry at King's College, London. He had collaborated with Miller (n. 9) on 'Additional Researches on the Electrolysis of Secondary Compounds', *Phil. Trans.,* 134 (1844), pp. 1–19. This extended previous work by Faraday and explained electrolytic decomposition products in terms of the different mobilities of anions and cations.
9. *Miller*: William Allen Miller (1817–70; FRS 1845), a pupil and assistant to Daniell, he succeeded him as Professor of Chemistry at King's College, London, in 1845.
10. *Kohlrausch*: Rudolph Kohlrausch (1809–58), German physicist, Professor of Physics at the University of Marburg (1853–57). Tyndall had worked with him in 1849. Magnus was referring to W. Weber and R. Kohlrausch, 'Ueber die Elektricitätsmenge, welche bei galvanischen Strömen durch den Querschnitt der Kette fließt', [On the quantity of electricity which flows through the cross-section of the circuit during electrolysis] *Poggend. Annal.,* 99 (1856), pp. 10–25.
11. *I have already notified the Academy*: The Akademie der Wissenschaften zu Berlin (since 1992 The Berlin-Brandenberg Academy of Sciences and Humanities) was founded by Gottfried Leibniz (1646–1716) in 1700.
12. *My wife and children*: Tyndall had met Magnus's family several times when he worked in Berlin in 1851. See letter 1084, n. 11.
13. *I shall go to Vienna*: A meeting of the Gesellschaft Deutscher Naturforscher und Ärtze was to be held in Vienna in September 1856. Tyndall attended but failed to meet Magnus.

To the War Department 1 May 1856 1217

Royal Institution 1st. May 1856

My Lord.[1]

While fulfilling the engagements of the last three days I have been accompanied at all hours by the painful consciousness that I was the object of your lordship's censure. I at once acknowledge the justice of that censure, and had I reflected sufficiently before acting I think I should have seen the irregularity of the course which the feelings of the moment induced me to pursue.

And now would your lordship's kindness permit me to state (though not as any extenuation of my error) the serious considerations which resulted in the hasty publication of the note[2] to which your lordship has referred. If the statement serve no higher purpose it will at all events satisfy the desire which I feel to appear before you in my true character.

It is doubtless well known to your Lordship that among a highly educated, earnest, and influential class of men the opinion is daily gaining ground that experimental science is neglected to a deplorable extent in this country. Sharing this opinion I accepted the examinership in Physics[3] with the strongest desire to render the department efficient, and with the hope that the subject would at length have a position assigned to it in the service commensurate with its great importance. I believe that no other portion of science could appeal to and exercise to the same extent the mental qualities required by an engineer or artillery officer. Such an officer on the field of battle, or before the beleaguered city cannot proceed from postulates and axioms. The process of thought under such circumstances is not solely <u>deductive</u>. He has to observe and compare facts, to draw conclusions from them and to convert these conclusions into events. The qualities of mind here necessary are precisely those which experimental science is calculated to exercise and develop; and yet in the values assigned to the various subjects in which candidates have been examined, pure and mixed mathematics, which are wholly <u>deductive</u>, and thus appeal to only one half of the powers which an officer of artillery or engineers must call into action are rated at more than three times the value of those sciences which appeal to all the intellectual powers: namely natural philosophy and chemistry.

Regarded simply as a means of general intellectual development I would respectfully submit to your lordship whether the experimental sciences ought not to occupy a higher place. But if their possible practical applications be taken into account the argument in their favour is surely greatly strengthened. There are only a few who are aware of what might be done with the agencies of Light, heat, magnetism, and electricity, if their principles and treatment were understood by military men. I will not detain your lordship by dwelling upon particular illustrations. Daily and hourly the laws of physics might, and would, find application in the hands of an artillery or engineer officer who had mastered them, and become familiar with the experimental combinations by which nature is questioned and her forces directed and controlled. To one like myself, who sees in experimental science a powerful accessary to the military strength of the country, it must be a source of pain to find in the examinations instituted under your lordship's authority, natural philosophy and chemistry rated at precisely half the value of Greek and Latin, and considered of less importance in a soldier's practical life than English literature.

Feeling that the sciences referred to did not occupy a position at all corresponding to their importance, I ventured at an early period to represent this to Mr Moseley. I have reason to believe that though Mr Moseley's own achievements have been more in the domain of deductive than of inductive science, still his opinion on the point in question was substantially the same as my own. In reply to my representation he however stated to me, that as the values had been fixed by a committee appointed for the purpose we had no alternative but to accept them. After the second examination,[4] and more through a feeling of duty than from any hope of effecting a change, I again stated my convictions upon this subject; but, as I anticipated, the statement was unavailing.

Though I could not subscribe to the estimates of the committee referred to, and though I knew I should be backed in my dissent by some of the most eminent individuals in this country, still I resolved to work on in accordance with the decisions of the committee until time should give me an opportunity of pleading the cause of experimental science with some prospect of success.

This was my state of mind when the enclosed note[5] from Professor Galbraith of Trinity College Dublin came into my hands. Professor Galbraith complains that the department under my charge has been ignored in the report submitted to your Lordship,[6] 'though a very cursory glance at the numbers would prove the very great superiority of the answers in this over every other part of the examination'.[7] It was evident to me from the tenor of the report referred to,[8] that Mr Moseley had restricted himself to the publication of the incidental remarks which accompanied the returns furnished by the examiner. I recollected that I had made no such remarks, as I trusted that the creditable array of figures which I sent in would tell their own tale. I saw that in this omission the complaint of Professor Galbraith really originated, and at the moment I supplied the omission in the manner which has invoked your Lordship's reproval.

Having thus both by the acknowledgement of my irregularity, and the performance of what I believe to be a duty to your lordship, to the service and to science, relieved my mind from a painful burthen, I place myself entirely in your lordships hands; prepared to acquiesce in any decision which the circumstances of the case may render necessary.

I have the honour to be | Your lordship's most humble servan^t^ | John Tyndall

RS MS JT 2/7/313–19

1. *My Lord*: Lord Panmure. See letter 1214, n. 2.
2. *The note*: letter 1213 to *The Times*.
3. *I accepted the examinership in Physics*: Tyndall had accepted the examinership in July 1855; see letter 1098.
4. *After the second examination*: January 1856; see letter 1177.
5. *the enclosed note*: letter 1212.
6. *The report . . . to your Lordship*: Moseley's report; see letter 1212, n. 1.
7. *'though a cursory glance . . . part of the examination'*: Tyndall is quoting from Galbraith's letter 1212.
8. *the tenor of the report referred to*: see letter 1212, n. 1.

To Thomas Archer Hirst 4 May 1856 1218

My dear Tom

Before I start out upon my Sunday walk[1] I will write to you. I had a letter from Magnus[2] on Friday in which he desired me to return you his special thanks for having looked over the translation of his paper.

I am at a little crisis of my life[3] at present and were I disposed to place more weight in externals and less on internals than I am accustomed to do I should doubtless be in a small fever. You saw a short note of mine[4] perhaps some time ago in the Times referring to the examinations in which I have been engaged. Since the commencement of these examinations I have endeavoured to get a higher recognition for Natural Philosophy & Chemistry but in vain, they rate as less than one third of pure & mixed mathematics and are below English literature. on the 25th of last month I had a letter from Prof. Galbraith[5] of Trinity Coll. Dublin complaining that in the report submitted to Lord Panmure my department was totally ignored, 'although a very cursory glance at the numbers would prove the past superiority of the answering in it over every other part of the examination'. He complained because he & Haughton had worked hard to produce this result. Well I saw from the turn of the Report he referred to[6] that Mr. Mosely had restricted himself to the publication of the remarks which accompanied the examiners returns. I made no such remarks as I trusted to the figures speaking for themselves; in this omission I saw the complaint of Galbraith originated. Well at the moment feeling the step could do no injury to anybody but myself I supplied the omission by writing that short note to the Times.

As I expected, a letter from the War office[7] at once reached me telling me that I had no right to appear thus without sanction in the public prints. I felt that the War office was perfectly right from its own point of view. Still as I had besides the War office the general cause of science in view I felt that I was not altogether wrong. I thought therefore that an opportunity was thus given me to lay the whole subject before Lord Panmure. I set to work and after much thoughtful deliberation managed to say nearly what I intended to say, and sent the letter into his Lordship.[8] before sending it I set clearly before my mind the possibility, nay the probability, that the whole matter would result in my removal from the post of examiner, and thus having accustomed myself to look at the worst, I became tranquil. What the issue is to be I do not know. I have shot a bomb which may recoil upon myself. But the great gods be thanked my existence or my happiness does not not hang upon the result. If it give them a better notion of science with reference to the practical life of a soldier my end is joined; if not it only proves that I have been too early, for a day will yet come, where natural philosophy & Chemistry will not in a soldier's education be ranked at half the value of Greek & Latin, and occupy a lower place than English literature.

Affectionately Yours John.

Tell me how Anna is getting on this weather must try her very much. I had your brothers card[9] and have written to him. Today it is cold & unfriendly here, a gray cloudy heaven & north east wind.

RI MS JT/1/T/897

---

1. *my Sunday walk*: On Sunday 4 May he walked over Hampstead Heath (RI MS JT/2/13c/840).
2. *a letter from Magnus*: letter 1216.
3. *a little crisis of my life*: Although Tyndall had thought the War Department's reprimand (letter 1214) 'not wholly unjustifiable' (RI MS JT/13c/837), he was deeply worried that he would be dismissed from the examinership and publicly disgraced.
4. *a short note of mine*: letter 1213.
5. *a letter from Prof. Galbraith*: letter 1212.
6. *the Report referred to*: H. Moseley, Report of Examinations for Appointments to the Royal Artillery . . . on 21 January 1856 (London: Harrison, London, 1856).
7. *a letter from the War office*: letter 1214.
8. *the letter into his Lordship*: letter 1217.
9. *your brothers card*: a visiting card left at the RI or at Tyndall's lodgings by John Henry

Hirst (1826–82). John Hirst entered the Royal Military Academy at Woolwich in 1841 and subsequently practised as an architect. His most spectacular building was the Byzantine Gothic 'Towerhirst' (1860) on the edge of Clifton Gorge, Bristol.

From César Despretz 6 May 1856 1219

Mon cher Mons. Tyndal

Je me propose de partir demain soir pour Londres. Je demeurerai d'abord dans un hotel, puis je chercherai un logement garni dans les environs de l'Athénée de l'Institution ou de Sommerset House. C'est là je crois le centre des établissements que je suis intéressé à visiter. Je connais bien une dame, Grosvenor Square, chez laquelle j'ai déjà logé, mais je serais là trop loin du quartier central. J'aurai l'honneur d'aller vous voir à mon arrivée, pour que vous m'aidiez un instant à trouver un logement ce qui ne sera pas difficile ni long. Notre ami Mr Grove voulait m'arrêter un logement dans les environs de sa maison. Je le prie de n'en rien faire. Je serai trop loin de ce que je considère comme le centre et je serai trop exposé à le déranger souvent.

Rappelez-moi, je vous prie au souvenir de Mr Faraday.

| Votre dévoué | C. Despretz | Le 6 mai 1856

[Paris] Le 6 mai | 1856

My dear Mr Tyndal,

I intend to come to London tomorrow evening. First I will stay in a hotel, and then I will seek furnished accommodation in the vicinity of the Athenaeum of the Institution,[1] or of Somerset House.[2] I believe that is the centre of the institutions that I am interested in visiting. I know a lady in Grosvenor Square[3] with whom I have stayed before, but I would be too far away from the central area. I will have the honour of seeing you upon my arrival, so that you might help me to find lodgings, which should not be difficult nor take long. Our friend, Mr Grove, wanted me to stop at a place close to his home.[4] I declined his suggestion. I would be too far from what I consider the centre, and I would be too prone to bother him often.

Please remind Mr Faraday of me. | Your devoted | C. Despretz | May 6th 1856

RI MS JT/1/D/124

1. *the Athenaeum of the Institution*: 'of the Institution' is unclear. In letter 1159 Despretz had talked about being admitted to the Athenaeum Club.
2. *Somerset House*: A government building on the Strand and overlooking the River Thames. In 1856 it housed, among other offices, the Royal Academy, the Royal Society, the Society of Antiquaries, and the administrative offices of the University of London.

3. *Grosvenor Square*: Clearly, Despretz was unfamiliar with London since Grosvenor Square, a fashionable garden square in Mayfair, was closer to the RI and Tyndall's lodgings in Mount Street than the Athenaeum.
4. *close to his home*: William Robert Grove lived at 6 Queen Anne Street in Marylebone to the north of Oxford Street. The family later moved to neighbouring Harley Street.

From Thomas Archer Hirst 6 May 1856 1220

Broughton, | May 6th/56

My dear John,

I heard of your letter in the Times and read your letter[1] today with great interest. I do not think you have much reason to fear the consequences of what you have done—on the contrary I have much faith that some good will come of it. I shall be anxious to see how it ends. The last week has been one of the most distressing ones to me since I came to Queenwood. We have been compelled to send two boys away on account of the evil habit[2] with which we have so long been fighting. One of them was Taylor's son (of Romsey),[3] the other Le Quesne of Jersey[4]. Le Quesne was a puny, weak-minded boy. He was only detected about a month ago, but it proved that he had been a slave to the habit for some years. I spoke to him very earnestly and he made me many solemn promises to amend. In less than a week after he was detected again and I thrashed him, but not very severely. I spoke to him especially about contaminating others, or speaking to others in a light manner about the habit. I thought I had frightened him but in less than a fortnight he was detected again. On being accused he stoutly denied it and I gave him a very severe thrashing. This broke him down and he made some revelations which astounded and afflicted me greatly. He and Taylor went out together, talked in the most obscene manner to one another and finally committed the act in one another's presence. Now Taylor was a companion of George Williams,[5] he was spoken to six months ago and by his manner he had led us to believe that our warnings had been effectual. As to encouraging others he had given us a solemn promise never to do so. Le Quesne's evidence however proved that we had produced no impression upon him whatever. Of course our only resource now was to have both boys removed. We did so very promptly but very quietly. Le Quesne was put on board the steamer at Southampton, and I went to break the news to Taylor's father at Romsey. He received my information as you may guess with deep grief. He is a good and I believe a high-minded man, but of course the shame brought upon his family obscured his calmer judgment, and he pleaded pitifully for his son, causing me terrible pain for I had to be firm. His son was at the moment at home on pleasure;

after I left him he examined his son who denied the charge. I had scarcely returned to Queenwood when he arrived with his son, and a distressing scene occurred. The father thinking his son had been unjustly accused was indignant, requested me to repeat the charge in his son's presence. I did so, stating all particulars, and turning to the son I said is all this true or not. The boy dared not deny it again, and the confession smote the father terribly. Again the father pleaded; Edmondson was present and gladly would I have left him to settle the matter but I was afraid of his giving way & was compelled to take it all upon myself though my heart ached for the poor smitten down man. It was melancholy to notice how the fear of shame and of what the world would say clouded his otherwise clear and high principles. Had it been another boy I saw he would have been on our side and would have insisted upon our doing what we did. The poor fellow at last roused me by descending to sophistry which he evidently felt to be such. He pleaded the prevalence of the habit and the differences of opinion of the medical men,[6] some asserting (said he) the physiological necessity of it. To eradicate the vice was impossible he urged, and seeing that the lad was in other respects not bad we ought not to inflict upon him so terrible a punishment and bring his innocent family also to shame. This and much more he said sufficient to shake the firmest resolution but his sophistry roused me, as I said, and I told him firmly: 'Mr Taylor, your present distress which I feel acutely blinds you to what is your own duty and ours. It is evident you do not believe in the physiological necessity of which you speak and the argument is unworthy of you. No right minded man will deny that it is a solemn duty in us to check this vice by every means in our power. Had your son himself fallen ninety nine times we could have forgiven him if we had been convinced that he was trying his best and was at any rate ashamed of his vice. But you forget it is for contaminating others that he must be removed. You will find no one to plead a physiological necessity for this. Let me beg you for a moment to put yourself in our position. We are entrusted by parents with the charge of many sons who may yet be ignorant of this degrading habit, we have lost all trust in your son who has proved himself incapable of appreciating its degrading character. Should we be fulfilling our duty were we to let them associate with your son any longer when we know that in all probability they would learn from him a beastly habit which would ruin them for ever, cling to them through life, and choke whatever was high and noble in them. Whatever I can do to mitigate your distress I will—a public expulsion would do more harm than good—it is not to punish him but to protect others that he must be removed and the more quietly it can be done the better. As to ruining him for life, that depends in a great measure upon yourself; your responsibility is great, but not so great as ours, and I trust that both you and ourselves will, when the first pangs of grief have departed, see our duties clearly, and do them.' This produced some impression upon him but not much. I extended my hand and repeated my willingness at all times to do what I conscientiously could to assist him or his, but although he shook me by the hand he shook his head and said, 'no, you can do nothing, you have damned him for life.' These hard words hurt me much, I felt inclined to resent them but thought better of it, and merely looked him steadily in the face; he did not recall them but modified them merely by saying—'Perhaps not intentionally but really you have damned him for life.' I left the room, and shortly after he took his boy away.

Both boys have gone now and the place is rid of two sources of contamination at any rate. But really, John, I am sick at heart with the prospect before me. Week by week we find more and more implicated and our preventative measures grow more severe. Continual fighting with such an ugly enemy has its reaction upon me and I am weary of finding my thoughts so incessantly dragged down to the subject. Many many times I have had to go through scenes that have quite unmanned me and I ask in vain—when will the end of all this come, or is there to be no end? I often ask myself have I found my life's labour in this ugly thing? Am I to struggle with the effects of the negligence of a hundred fathers and take their responsibilities upon me? Sometimes I am prompted to reply 'by Heaven, I will not' and then again I see the selfishness of this and how it would be in me a shirking of duty to wash my hands of the whole profession of schoolmaster which has half lost its interest for me.

Well, the end of it all is that I reconcile myself to the duty which is nearest to me, and leave the future to itself.

Anna recovers slowly, the terrible weather keeps her back. For her sake and for my own I long for midsummer and the vacation. Never before did I long for it so much.

Pardon my complaining and say nothing to me but what will encourage me to persevere. | Yours affectionately, T.A. Hirst.

P.S. Of course like ourselves you will, both now and always, keep silence as to the cause of the expulsion of these two lads—indeed, as to the expulsion itself.

RI MS JT/1/H/223

---

1. *your letter today*: letter 1218 dated 4 May.
2. *the evil habit*: masturbation. The events described in the letter occurred while Hirst was trying to translate one of Clausius's papers for *Phil. Mag.*
3. *Taylor's son (of Romsey)*: Frank Taylor (1844–1904), son of Francis Taylor (1812–71), a medical practitioner and Mayor of Romsey 1854–56, a town 8 miles south of Broughton. The father was also acutely embarrassed because he had four daughters and was at a loss

how to explain their brother's expulsion. Expulsion evidently did Taylor no harm since he later qualified as a doctor and inherited his father's practice in Romsey.

4. *Le Quesne of Jersey*: Philip Winter Lequesne (b. *c.* 1839) according to the 1851 Census. Hirst thrashed him 'for lying and sensual practices' on 1 May 1856 (RI MS JT/2/32c/1193). On his return to Jersey he became a corn miller at St Helier.
5. *George Williams*: not identified.
6. *the differences . . . of the medical men*: As a medical doctor himself, Taylor would have been aware of the contemporary controversy over masturbation.

To Thomas Archer Hirst 7 May 1856 1221

7th May 1856

My dear Tom

In the midst of my preparations your letter[1] has broken in upon me, absorbing all else by its intense interest. I do not recollect to have read of any thing more truly painful than the position of poor Taylor;[2] I do not know any man who has been subjected to a stronger temptation and trial than you have been, and my admiration of the manner in which you have borne yourself through it corresponds fully to the difficulty of your task. Had it been the boy himself—he confined his sin in his own heart we might have yielded under the circumstances, but I can see nothing but the most disastrous consequences had you given way as matters really were. You have done your duty in a truly noble manner: time heals the necessary lacerations which accompany such an act, but time only brings out into fairer relief the merit of such a deed. I would not thrash if I were you. that is, I would not thrash for the act itself, though for a lie a thrashing may be the best thing possible. Try and make yourself acquainted with the length breadth and depth of this problem, become to a certain extent devoid of feeling, investigate the thing as you would a physical question. God only knows to what valuable account your experience may be turned at a future day. You must not be a sacrifice to the neglect of fathers and teachers, but you can and ought now to make yourself acquainted with the full depth of their negligence, your word upon this subject at a future day may have a weight and influence which will be a blessed recompense for your present labours—Meanwhile, act like a physician who has to arm himself with inhumanity in order to attack a difficult case aright. Relax a little now and then—deviate from your routine, and think a little now & then upon the subject. —New modes of action may thus suggest themselves, and you may try these new methods with fresh hope & fresh vigour. You ought I think to procure that French work of which we spoke when we were last together—Saland[3] I think it was. You are now low Tom, and things appear dark to you—this will not continue—At all events I never regarded your present position in any other light than that of a transitory one—but while you are there sound all the depths of your position—make yourself acquainted with its joy & its sorrow. Store your journal[4] with the account of your experience—it may yet be a blessing to humanity. Besides Tom. hard as these boys seem now a day will come when the consciousness of right and wrong has ripened, and when they can estimate the magnitude of your efforts to save them, when the very memory of these efforts will be an inspiration and a safeguard—encouraging them against temptation, and thus bring forth effectual though tardy fruit. God bless you my dear fellow and strengthen your hands & heart in the task that lies before you.

affectionately yours | John Tyndall

I am greatly pleased to find that Anna is improving. I hope by the time that the mid summer concert[5] takes place that I shall see the roses on her cheeks once more.

As yet I have not received my dismissal.[6]

RI MS JT/1/T/621

---

1. *your letter*: letter 1220.
2. *poor Taylor*: the Queenwood parent discussed in letter 1220, n. 3.
3. *that French work . . . —Saland*: Tyndall misremembered the name of Claude François Lallemand (1790–1854), French pathologist, Professor of Surgery at the University of Montpellier (1819–45). He published *Des Pertes Séminales Involontaires*, 3 vols. (Paris: Béchet Jeune, 1836–42), a work which attributed male physical and mental degeneration to masturbation. The work, which appeared in English as *A Practical Treatise on the Causes, Symptoms and Treatment of Spermatorrhea* (London: J. Churchill, 1847), influenced the English physician William Acton in his *The Functions and Disorders of the Reproductive Organs in Childhood, Youth, Adult Age* (London: J. Churchi ll, 1857), a work that led to further masturbation hysteria.
4. *Store your journal*: In fact Hirst had already written a long account of the affair on 4 May (RI MS JT/2/32c/ 1193–95).
5. *mid summer concert*: a concert held annually at Queenwood College to which parents and sponsors were invited.
6. *my dismissal*: Tyndall's concern that he would be sacked for writing letters 1213 and 1217.

From the War Department 8 May 1856 1222

War Department Pall Mall | 8th May 1856

Sir,

I am directed by the Secretary of State for War[1] to acknowledge the receipt of your letter[2] of the 1st Inst., and to inform you that his Lordship considers the Explanation you have furnished to be quite satisfactory.

I have the honour to be | Sir | Your obedient servant | H.R.Drewry | J.Tyndall Esqr^{e}

RI MS JT/2/7/322
LT Transcript Only

---

1. *Secretary of State for War*: Lord Panmure.
2. *your letter*: letter 1217.

To Michael Faraday 9 May 1856 1223

9 May 1856

My dear Mr. Faraday,

The following note from the War Department reached me last night, it contains my doom.

> War Department Pall Mall | 8th May 1856
>
> Sir,
>
> I am directed by the secretary of State for War to acknowledge the receipt of your letter of the 1$_{st}$ instant, and to inform you that his Lordship considers the explanation you have furnished to be quite satisfactory.
>
> I have the honour &c | H. R. Drewry

I opened this letter with the neutrality of a man who has really made up his mind for the worst, but I confess it gives me great pleasure to think that the matter has thus terminated.

I hope you are improving,[1] and that you will soon return quite restored. This weather is awfully trying—I was at a very low ebb in my lecture of yesterday, but barring the failure of a difficult experiment or two, managed to drag through pretty well. Frankland is in town for a few days and Despretz—who desires me to present his respects to you[2]—is coming. He purposed starting on Tuesday evening,[3] but he has not yet appeared.

I hope Mrs Faraday is well. I should think that Folkstone cliff is very cold for her at present.[4]

Last Wednesday evening[5] I was at Lord Ashburton's and met Wheatstone there. We had a conversation about the stereoscope and I was pained to see the desire manifested so strongly to deprive Brewster of what he really merits—or to reduce his merits to zero.[6] Wheatstone says he must give me a lesson on the subject; that is, he must instruct my muddy understanding—the matter appears to me as plain as a proposition of Euclid:[7] but it does not seem as if he had really mastered any theory but his own.

Good bye | Ever Yours most truly | John Tyndall

RI MS JT/12/8
Faraday Correspondence, 5:3139

---

1. *I hope you are improving*: Faraday and his wife had borrowed Bence Jones's holiday home in Folkestone in order to restore his health. Faraday to Bence Jones, 9 May 1856 (*Faraday Correspondence*, 5:3138).
2. *his respects to you*: letter 1219.
3. *Tuesday evening*: 6 May 1856.
4. *very cold for her at present*: The weather was so bad that Mrs Faraday did not venture out at all (*Faraday Correspondence*, 5:3138).
5. *Last Wednesday evening*: 7 May 1856.
6. *reduce his merits to zero*: A very 'bitter dispute' had broken out when Brewster publically claimed priority over Wheatstone for the invention of the stereoscope in his *The Stereoscope: its History, Theory and Construction* (London: J. Murray, 1856). See letter 1109, n. 13.
7. *a proposition of Euclid*: Euclid (fl. 295 BC), Greek mathematician whose geometry was the epitome of deductive logic.

From Michael Faraday 10 May 1856 1224

Royal Institution | 10th. May. 1856.

My dear Tyndall,

I am very glad of your letter[1] I was hoping to hear of the end I could not doubt that it would be satisfactory but still wished to know that it was so as it is I think you stand in a good independent position and have been very useful in the great matter, whether it progresses now or has to wait a little. That it must progress is I think certain.

I am better: in fact all the matter is that years do their work and why should they not if they have done it in one way i.e. for my comfort and happiness they ought also to do it in another as for instance in wear and tear. Who would wish for an ever-enduring life in this world, above all if a better hope appears before him?

I hope we shall be home on Thursday night[2] and to find you right. I have heard well—very well of your last lecture.[3]

Ever Truly Yours | M. Faraday
Dr. Tyndall

RI MS JT/12/4140
Faraday Correspondence, 5:3140

LT Transcript Only

1. *your letter*: letter 1223.
2. *Thursday night*: 15 May 1856.
3. *your last lecture*: That was at the RI on 7 May. Tyndall's own opinion was different. 'Lectured: felt very poorly: so low that I was careless how the lecture went. This very carelessness however enabled me to drag through pretty well. An experiment or two failed but the lecture as a whole was not a bad one' (RI MS JT/2/13c/ 841).

To Thomas Archer Hirst 10 May 1856 1225

May 10th 1856

My dear Tom.

Behold my doom! it[1] reached me last night.

War Department | Pall Mall 8th May 1856

Sir

I am directed by the Secretary of State for War to acknowledge the receipt of your letter of the 1st. Inst. and to inform you that his lordship considers the explanation you have furnished to be quite satisfactory

I have the honour to be | H.R. Dewry | J. Tyndall Esqre

I am very glad that it has terminated thus, though I had fully made up my mind for another result: I am glad that a person in a high official position is able to listen to reason, and to allow it to sway him in opposition to the prudence of routine. This Lord Panmure has done in the present instance.

I know you will be glad to hear this & therefore I communicate it to you without delay.

Yours affectionately | John.

Again I would say to you don't allow your feelings to get too far entangled in that matter[2]—a Surgeon on the battle field would be of little use if he suffered the painful scenes around him to unman him. By this coldness as a general rule you will husband your strength, and cases will arise when you can bring the lever of feeling to bear with effect.

1. *it*: letter 1222.
2. *in that matter*: the outbreak of masturbation at Queenwood College. See letters 1220 and 1221.

From Thomas Archer Hirst 11 May 1856 1226

Broughton, | May 11th/56

My dear John,

Thanks for your cheering letter[1] which has done more than all else to encourage me to perseverance and cause me to forget the pain of the preceding week. To every word my better self could respond and by responding grew stronger. I wish you would ask Huxley for the title of the book by Lallemand[2] to which he referred, for it will unquestionably be of use to us. I was also glad to hear the result of your letter to Lord Panmure.[3] I expected it and feel sure your letter will prove of service.

This is the first fine sunny day we have had for some time and I am going to give Anna a drive. You will have seen from the Times that John Martin is at last at liberty.[4] We are expecting him daily and if he comes I have resolved to send poor Anna with him to Ireland to try what native air will do towards restoring her health.

At the break up[5] Debus and I intend to exhibit a few experiments to amuse the company for an hour and give a scientific tone to the assembly. We are very poorly off for apparatus since our fire [6] (Frictional electricity being a too capricious subject[7] for the occasion) Ruhmkorffs coil[8] is one of our most available instruments and I want to know if you can suggest a suitable experiment or two with it besides the one I usually shew with the light falling over the edges of a beaker of mercury in a vacuum.[9] Debus is going to produce some chemical blazes and I only want two or three good experiments to serve as introduction. I shall not make it a lecture but rather a series of striking experiments accompanied with very brief descriptions. If you could suggest one or two I should be thankful for I am sorely pressed with other things which unfortunately are not connected with Physical experiments.

With love | Yours affectionately | T. A. Hirst.

RI MS JT/1/H/224

---

1. *your cheering letter*: letter 1225.
2. *the book by Lallemand*: letter 1221, n. 3.
3. *letter to Lord Panmure*: letter 1217.
4. *John Martin is at last at liberty*: Reported in the *Times* on Saturday 10 May 1856 in its 'Parliamentary Report', p. 6. Martin had been transported to Tasmania in November 1849 to serve a ten year sentence, but was pardoned in 1854 on condition that he did not settle in Britain or Ireland. He settled in Paris in October 1854 and received an unconditional pardon in May 1856 following which he returned to his estates in Ireland.

5. *At the break up*: the last day of Queenwood College's summer term.
6. *since our fire*: There appears to be no record of this incident which probably occurred in the laboratory. The school was to be destroyed by fire on 10 June 1902.
7. *a too capricious subject*: spectacular experiments with frictional electricity machines had been a common entertainment since the 18th century, but can easily fail because of damp atmospheric conditions.
8. *Ruhmkorff's coil*: an induction coil capable of producing spectacular sparks of over a foot in length devised by the Franco-German instrument maker, Heinrich Daniel Rühmkorff (1803–77) in 1851. Tyndall had previously borrowed the coil in April 1855; see letters 1080 and 1081, n. 5.
9. *mercury in a vacuum*: Hirst was here describing the stratification of light later more commonly reproduced in a discharge tube.

To Thomas Archer Hirst 12 May 1856 1227

12th. May. 1856

My dear Tom

Ruhmkorff's coil[1] is a difficult subject for an assembly such as yours. but you might lead them up to it by experiments on induction with the spiral of your magnet. You might shock a large eel in this way. the experiment always produces a fund of amusement—a flounder is also good, but the eel is best. Could you not get a large live eel (say 18 inches long) The extra current might also be shewn, by shewing the influence of the spiral on the strength of the spark when broken in a cup of mercury. The dancing of the spirals through their own attraction is a beautiful experiment—In fact here you have a field open to you of the greatest interest at last—the mutual action of currents and magnets. With regard to the spark of Ruhmkorff a beautiful experiment may be made by connecting the ends with wires which you attach to a handle, and move quickly through the air—the sparks at first continuous are separated into wide intervals by the motion of the wire.

The diminution of the extra current by closing the secondary wire is also a very striking experiment. I don't think you can be at a loss for material.

I send you a letter from Faraday[2] and another from Tommy Grieve.[3]

Sadly pushed | Yours affectionately | John

I was extremely glad to her of John Martin's release.[4] I saw it as you judged in the Times

RI MS JT/1/T/898

---

1. *Ruhmkorff's coil*: see letter 1226, n. 8.

2. *A letter from Faraday*: probably letter 1224.
3. *Tommy Grieve*: a former pupil of Tyndall's at Queenwood in 1853. The letter is missing. Grieve was possibly Thomas Grieve (b. 1845), son of William Grieve (b. 1796), a widowed sheep farmer owning 4000 acres at Branxholm Park, Hawick, Roxburghshire, Scotland. See volume 4, letters 0817, 0832, 0887, 0913, 0968, 0977, 0989 and 1011.
4. *John Martin's release*: see letter 1226, n. 4.

From Carlo Matteucci 15 May 1856 1228

Pise 15 mai 56

Mon cher monsieur,

Je suis bien content d'avoir une occasion pour me rappeller à vous et pour revenir un peu à la [Science[1]]. J'ai addresse au General Sabine un memoire manuscrit qui serait pour le Phil Transac et que fait suite à toute la serie de mes Memoires d'Electrophysiologie qui ont déjà eu l'honneur de paraitre dans les memoires de la S.R.[Royal Society]. J'espere que celui-ci aussi [aura][2] la meme distinction. C'est un sujet nouveau et c'est la première fois qu'on explique par un [fait][3] physique bien connu un phenomene d'Electrophysiologie tres obscur. Je serai bien recompensé de tout ce que j'ai du faire et travailler pour [y] reussir, si je parviens a [savoir/faire][4] qu'il est aprouvé par la [S.R.].

Rappelez aux souvenirs de M. Faraday et des autres amis de [la S.R.].

[Croyes a ma/en m?][5]

C. Matteucci

Pisa, 15th May 1854

My dear Sir,

I am pleased to have an opportunity to remind you of me and to come back to science for a while.[6] I have sent a manuscript to General Sabine which is aimed at the Phil. Trans.[7] and follows the series of my essays on Electro-physiology previously honoured by publication in the proceedings of the R.S. I hope that this one too will receive the same distinction. It is on a new subject and for the first time it explains a phenomenon of Electro-physiology by a well known piece of physics.[8] I would be amply rewarded for all the actions and work I had to do in order to succeed, if I obtained the approval of the R.S.

Send my regards to Mr Faraday and the other friends at the R.S.

Believe in [?].

C. Matteucci

RI MS JT/1/M/64

1. *[Science]*: The word, which is almost impossible to decipher, might also be 'Lumière'.
2. *[aura]*: word missing in manuscript.
3. *[fait]*: the word appears to begin with a 'T', but 'fait' is the obvious meaning.
4. *[savoir/faire]*: it is unclear which verb was intended.
5. *Croyes a ma/ en m ?]*: the closing phrase is indecipherable.
6. *to come back to science for a while*: an alternative reading might be: 'to return into the limelight'. Matteucci had been heavily involved in lecturing and electrophysiological research during 1855–56 but was now able to publish new findings.
7. *Phil. Trans.* C. Matteucci, 'Physical-chemical Phenomena of Muscular Contraction', *Phil. Trans.*, 147 (1857), pp. 129–44, received in 1856.
8. *piece of physics*: the dynamical theory of heat. Matteucci argued (n. 7, pp. 143–44) that during muscular contraction chemical action produced an electric current that correlated with the amount of heat produced by the same contraction.

To Thomas Archer Hirst 18 May 1856 1229

Sunday 18th. May. [1856]

My dear Tom.

Rain & pelting hail prevent me from going out, and as I am a prisoner against my will and have done a capital mornings work I will rest and amuse myself at the same time by writing to you. I have rummaged through my books and found an essay upon the Induction Coil[1] which may be of service to you and which I will therefore send to you by the same post that carries this. During the week my health has been low: on Thursday it was however in good condition and I managed to give them a lecture accordingly.[2] I yearn for the time when these lectures will be over and I can lift my liberated forehead to heaven, snuff the air without thinking of chemistry, and enjoy the blue heaven without breaking my head with natural philosophy. I saw yesterday what pleased me and will please you—a full length translation of your paper on the magnetic medium in L'Institut.[3] They have translated it all which is rather unusual and which therefore proves their appreciation of the essay. By the way Tom I have a few things here[4] belonging to the Royal Society; and which I am competent to lend—among the number is a magnet, but I believe nothing else that you could use—If they would be of use to you when you make your experiments at Queenwood they are quite at your service. The Royal Institution apparatus I cannot lend. I was at the Duchess of Northumberlands[5] on Wednesday even[in]g and was received with great kindness by both Duke and Duchess. It was a glorious assembly—that is to say women's eyes made it glorious. Hoffman gave us a capital lecture on Friday night.[6] I dined yesterday[7] with a fine fellow, an Irishman, Sir Emerson Tennent of Castle Tempo.[8] Thackeray[9] sat opposite to me at dinner. Whiteside the barrister[10] who made the recent motion against the Government was also there, and a son of Dallas the American Minister.[11] My host gave me his fair daughter[12] to take down to dinner and we had much pleasant talk over Tennyson, Longfellow[13] and a hundred other things. By George no man can catch your heart quicker than a noble-minded warm souled Irishman. Tennant talked to me of his son:[14] a youth he said who has just come of age and never gave him 10 minutes uneasiness in the course of his life. He was in Ceylon,[15] and he sent his son home to Rugby to be educated. They returned him drilled in the preter pluperfect tense,[16] but ignorant of every thing else—His time was completely lost there. He made up for this by tutors and then had some thoughts of sending him to the university. But, said he, shall I expose this fine nature which has developed itself at my side to the temptations of the University—shall I send my son among those whose highest idea of human life is horseracing & such things—He did not do it—He took time to himself in Whitehall, and when he gets a little older he will send him into parliament & let him make his way as a public man. There will be a seat open for him when he is ready for it. We talked of Faraday and of other things—Tyndall said he I admire a man of eminence in science, but when I see an Irishman making his way upward in spite of opposition and this damned prejudice that exists against him I love him, and that is my feeling towards you. The man who spoke there has a fine high intellectual brow, an eye clear enough to see his soul of kindness showing through it and altogether by the Lord one of the finest fellows I have met. I had a conversation with Playfair[17] about Queenwood at the Duchess of Northumberland's. He asked me would I recommend him to send a boy 10 years old there. I said that if I had a boy 10 years old I would keep him under my own roof. Ah said he that is supposing you could look after him, but that I cannot do with my son.[18] If, said I, you ask me what school I would choose in comparison with other schools, I should without a moments hesitation say Queenwood. How are Mathematics there first rate, Classics, a man from Oxford. He talked much of the matter and I gave him a good picture of the noble efforts that were making to keep the moral as well as the intellectual tone of the school high. I believe he will send his son there—meanwhile I think you might send him a prospectus enclosed to Marborough House | Pall Mall[19]—

He said he would visit Queenwood previous to deciding.

And now I think I have said all I intended to say—a sunbeam smites my shoulder, and though before I get shaved it will probably be quenched in darkness still I must contrive to get a little fresh air. Love to Anna—I hope she continues to improve—The native air I doubt not will be better than all medicines to her. I may be at Queenwood before you lecture and if so will assist you with your experiments.

Ever yours affectiona[te]ly | John | over

I have opened this to say that your letter has just reached me.[20] I shall be glad to see both you and Anna, though a thousand times gladder if she had come in the bloom of health instead of to see a doctor. | J.T

RI MS JT/1/T/900

1. *essay on the Induction Coil*: essay missing. Hirst had asked for advice on physics demonstrations to entertain pupils and boys at the end of the Queenwood term. See letter 1226.
2. *to give them a lecture accordingly*: Tyndall's weekly lecture to RI on the subject of light.
3. *your paper on the magnetic medium*: T. A. Hirst, 'On the Existence of the Magnetic Medium', *Roy. Soc. Proc.*, 7 (1854–55), pp. 448–54, appeared in French in the *L'Institut (Journal des Académies et Sociétés Scientifiques de la France et de l'Étrangers),* 1856. The reference has not been traced.
4. *a few things here*: Tyndall had bought some apparatus with a grant from the RS and was proposing to lend items to Hirst for his lecture demonstration.
5. *Duchess of Northumberlands*: Lady Eleanor Grosvenor (1820–1911), wife of Algernon Percy, fourth Duke of Northumberland (1792–1865), the President of the RI.
6. *a capital lecture*: A. W. Hofmann, 'On the Chemical Type: Ammonia' was delivered as a Friday Evening Discourse at the RI on 16 May 1856, and published in *Roy. Inst. Proc.,* 2 (1854–58), pp. 274–82.
7. *I dined yesterday*: Saturday 17 April.
8. *Sir Emerson Tennent of Castle Tempo*: Sir Emerson Tennent (1804–69; FRS 1862), Irish politician, barrister and author. Secretary to the India Board (1841–43); Colonial Secretary to Ceylon (1845–50); Secretary to Board of Trade (1852–67); and MP for Lisburn from 1852. Born James Emerson, he married the heiress Letitia Tennent in 1831 and took the name Tennent and inherited Castle Tempo in Co. Fermanagh, Ireland. Knighted (1845). Tyndall had first met Tennent at Peter Barlow's home on 30 April 1856 (RI MS JT/2/13c/637–38).
9. *Thackeray*: William Makepeace Thackeray (1811–63), English novelist and satirist.
10. *Whiteside the barrister*: James Whiteside (1804–76), Irish barrister who made his name defending Daniel O'Connor and William Smith O'Brian. Solicitor-General for Ireland. In a parliamentary motion on 28 April 1856 he blamed the government for its lack of foresight and energy in allowing the capture of the Turkish garrison at Kars by the Russians in November 1855.
11. *a son of Dallas*: George Mifflin Dallas (1792–1864), American lawyer and Democrat, 11th Vice-President of USA (1844–49). Between 1856 and 1861 he was Minister (Ambassador) to Great Britain. He had two sons, Philip Mifflin Dallas (1825–66) and Alexander James Dallas (1826–?); it is unclear which son was being discussed.
12. *his fair daughter*: She was one of Tennent's two daughters, Sarah or Edith Laetitia Tennent.
13. *Longfellow*: Henry Wadsworth Longfellow (1807–82), American poet and author of *Hiawatha.* See letter 1188, n. 12.

14. *his son*: William Emerson Tennent (1835–76) was a Clerk in the Board of Trade. He became second Baronet in 1869.
15. *He was in Ceylon*: Former British island colony to south of India which became independent in 1948 and renamed Sri Lanka. Tennent had been in Ceylon from 1845 to 1850 and wrote several books on the island, including *Christianity in Ceylon* (London: John Murray, 1850); *Ceylon, Physical, Historical and Topographical*, 2 vols. (London: Longman, 1859); and *Sketches of the Natural History of Ceylon* (London: Longman, 1861).
16. *The preter pluperfect tense*: The praetor pluperfect is the grammarian's term for the past perfect tense.
17. *a conversation with Playfair*: see letter 1170, n. 6.
18. *I cannot do with my son*: George Playfair (1849–1939) who succeeded to Playfair's baronetcy after a career in the Royal Artillery as a Brigadier-General. It is not known whether he attended Queenwood.
19. *Marlborough House, Pall Mall*: Marlborough House had been originally built for Sarah Churchill, the Duchess of Marlborough, in 1711. It was acquired by the Crown in 1817 and used by members of the Royal Family until 1853, when Prince Albert adapted it for use as a National Art Training School (later the Royal College of Art). The building was also the headquarters of the Department of Science and Art, for which Playfair was the Science Secretary.
20. *your letter has just reached me*: This appears to be a missing letter from Hirst saying that he was coming to London in order for Anna to consult a doctor.

From Gustav Magnus 18 May 1856 1230

Berlin 18 Mai 1856.

Verehrter Freund

Entschuldigen Sie daß ich Ihren lieben Brief vom 3ten d. erst heut beantworte. Ich war in dieser Zeit sehr beschäftigt und konnte meine Thätigkeit nicht gut unterbrechen, um so mehr als ich die Festtage auf dem Lande zuzubringen versprochen hatte. —Sie wünschen meine Ansicht über die Ursache des Zerspringens der Kessel von Dampfmaschinen.

Springen der Kessel, sowohl der Locomotiv- wie der übrigen Kessel tritt, nach meiner Ansicht besonders dann ein, wenn es an Wasser in dem Kessel fehlt. Bei den Locomotiven kommt ein Mangel an Wasser bestimmt viel häufiger vor als bei stehenden Maschinen. Denn denkt man sich daß eine Locomotive anhält, so verbraucht sie doch Dämpfe, weil sie gewöhnlich so stark geheizt wird um Dämpfe in hinreichender Menge zu liefern wenn der Zug wieder in Bewegung gesetzt werden soll; auch öffnet sich fast jedes Mal das Ventil während der Zug anhält. Die Locomotiven sind aber so eingerichtet daß sie nur während ihrer Bewegung das Wasser aus dem Tender pumpen. Bei uns pflegt man auf der Locomotive noch eine besondere kleine

Dampfmaschine anzubringen, in die man, nur wenn der Zug steht, Dämpfe einströmen läßt und die dann Wasser in den Locomotivkessel pumpt. Wo eine solche Vorrichtung fehlt, ist es unvermeidlich daß das Wasser im Kessel ein Mal unter seinen normalen Stand hinabsinkt. —Nehmen wir an daß dieser Fall eintritt, so nehmen die Wände des Kessels und die Heizröhren, welche vom Feuer umspühlt, und im Innern nicht mehr von Wasser bedeckt sind, eine sehr hohe, der Glühhitze nahe Temperatur an. Das Wasser wird an seiner Oberfläche die heißen Wände gar nicht berühren, sondern den sogenannten sphaeroidalen Zustand annehmen; es wird deßhalb auch nur eine sehr mäßige Dampferzeugung stattfinden. Wenn aber durch irgend einen Umstand der Druck den die Dämpfe im Kessel gegen die Oberfläche des Wassers ausüben, sich vermindern sollte, so würde ein Aufkochen eintreten. Dann werden die Wassermassen gegen die heißen Wände geworfen, diese kühlen sich ab, es entsteht eine plötzliche Dampfentwicklung und der Kessel springt. Solcher Umstände, durch die eine plötzliche Verminderung des Drucks der Dämpfe entsteht, können sehr verschiedene eintreten. Zunächst eine Abkühlung des oberen mit Dampf gefüllten Raumes von aussen, durch einen kalten Wind oder dergleichen. Diese Abkühlung wird aber selten genügen um ein Aufkochen zu erzeugen. Öfter möchte dieses eintreten durch Oeffnung des Sicherheitsventils. Steht die Locomotive still, wird kein Dampf verbraucht, so steigert sich die Spannkraft der Dämpfe doch allmälig, obgleich das Wasser sich im sphaeroidalen Zustand befindet, endlich wird seine Spannkraft so groß, daß das Sicherheitsventil sich öffnet. So bald dies geschieht vermindert sich plötzlich die Spannkraft und das Wasser kocht auf. Eine andere Veranlassung zum Aufkochen ist die, daß ein Theil des Dampfes in die Cylinder gelassen wird, damit die Locomotive sich bewege. Sobald dies geschieht vermindert sich die Spannkraft des noch vorhandenen Dampfes, und das Aufkochen tritt ein.

Auch bei feststehenden Maschinen kommt nach meiner Ansicht ein Springen der Kessel nur vor, wenn es an Wasser in denselben fehlt. Leider ereignen sich mit solchen Kesseln auch bei uns bisweilen, jedoch nur sehr selten Unglücksfälle. Man bringt aber immer mehr, auch bei den stehenden Kesseln, die vorhin erwähnten kleinen Maschinen an, die nur dazu dienen Wasser in den Kessel zu pumpen. Denn wenn aus irgend einem Grunde eine Maschine für einige Zeit still stehen muß, so nimmt das Wasser in dem Kessel ab, indem der Dampf durch das Sicherheitsventil entweicht ohne daß neues Wasser zugepumpt wird, weil die Pumpen immer mit der übrigen Maschine in Verbindung sind. Durch die getrennte kleine Maschine lässt sich dieser Uebelstand leicht beseitigen.

So viel über diesen jedenfalls sehr interessanten Gegenstand. Sollten Sie darüber etwas veröffentlichen, oder einen Bericht darüber erstatten, so wäre

es mir lieb wenn Sie ihn mir mittheilten. Ich verspreche Ihnen von hier aus zu senden was ich über denselben erhalten oder erfahren kann. Uebrigens glaube ich daß eine Veröffentlichung ganz gut wäre damit das Publicum wüßte worauf es zu achten hat. Vielleicht würde dieselbe auch die Veranlassung für die Auffindung einer mechanischen Vorrichtung durch welche das Sinken des Wassers unter dem normalen Stand unmöglich würde.

Was Sie mir von Plücker schreiben ist mir sehr interessant. Zu mir sagte Wheatstone in Paris daß Plücker nur als Mathematiker Mitglied der Royal Society geworden sei.

Faraday ist zu bescheiden. Nicht seine Arbeiten sondern sein vortrefflicher Character binden seine Verehrer so fest an ihn, und dieser Character, verbunden mit seinen colossalen Leistungen stellt ihn in meinen Augen so hoch. Denn leider findet man nur zu häufig wo bedeutende Leistungen vorliegen, ein [einen] Mangel eines tüchtigen und liebenswürdigen Characters. (Sie brauchen Faraday übrigens nicht immer vorzulesen was ich von ihm schreibe, denn sonst kann ich nicht mehr schreiben wie ich möchte.) Aber grüßen mögen Sie ihn von mir.

Wegen der Reise nach Wien weiß ich noch nichts zu bestimmen, aber einen Vorschlag will ich machen. —Wenn Sie nach Deutschland kommen, werden Sie nicht allein nach Wien gehn wollen. Ich hatte mir vorgenommen im August mit meiner Frau in die Schweiz zu gehn, kommen Sie mit Frankland und Bence Jones auch dort hin, und gehn Sie von dort nach Wien. Wir wollen dann in der Schweiz die schöne Natur genießen und dabei uns von Physik unterhalten. Schreiben Sie mir bald ob und was Sie in dieser Beziehung beschließen, und seien Sie auch so gut mir zu sagen was Sie gegen meine Theorie von dem Springen der Kessel auszusetzen haben.

Bence Jones, Frankland, Wheatstone, Francis, und alle die sich meiner erinnern bitte ich Sie zu grüßen.

Von Herzen | Ihr | GMagnus.

Meine Frau und Kinder tragen mir die freundlichsten Grüße auf.

Berlin 18 May 1856.

Esteemed friend

Excuse the fact that I am answering your kind letter[1] of the 3rd of this month only today. I have been very busy lately and could not easily interrupt my activities, all the more so as I had promised to spend the holidays in the countryside. —You wish to hear my view about the cause of the explosion of boilers of steam engines.

In my view the bursting of boilers, both of locomotives as well as of other boilers,[2] occurs especially when there is a lack of water in the boiler. A deficiency of water definitely occurs much more often with locomotives than with stationary machines. If one imagines a locomotive stopping, then it nevertheless consumes steam, because it is usually heated so intensely as to provide steam in sufficient quantity if the train should be set in motion again; also, the valve opens almost every time while the train is stopping. However, locomotives are constructed in such a way that they only pump water from the tender while in motion. Here, it is customary to attach on top of the locomotive another special little steam engine,[3] into which—only when the train is stationary—one lets the steam flow and which then pumps water into the locomotive's boiler. Where such a device is wanting, it is inevitable that the water in the boiler falls below its normal level at some stage. —If we suppose that this event occurs, then the walls of the boiler and the heating pipes, which are lapped by the fire and are no longer covered inside by water, attain a very high temperature, close to red hot. On its surface, the water will not touch the hot walls at all, but will assume the so-called spheroidal state;[4] only a very moderate generation of steam will therefore occur. If, however, the pressure that the steam in the boiler exerts against the surface of the water should decrease through some circumstance or other, then boiling would occur. The masses of water will then be thrown against the hot walls, these will cool down, a sudden generation of steam will occur, and the boiler will burst. Some very different circumstances can occur by which a sudden decrease of pressure of the steam will happen. First of all, a cooling of the upper steam-filled space from the outside, through a cold wind or the like. This cooling, however, will rarely suffice to cause boiling. From time to time this might occur through the opening of the safety valve. If the locomotive is standing still, if no steam is consumed, then the steam's expansibility[5] will still increase gradually even though the water is in the spheroidal state; finally, the tension force will become so great that the safety valve opens. As soon as this happens, the tension force decreases suddenly and the water boils. Another cause of boiling is that a part of the steam is led into the cylinders so that the locomotive may move. As soon as this happens, the tension force of the steam that is still present decreases and boiling begins.

In my opinion, in stationary machines as well, bursting of the boilers only happens if there is a lack of water in them. Unfortunately, accidents involving such boilers also occurs here from time to time, but only very rarely. However, the little machines that I mentioned before, and that serve solely to pump water into the boiler, are being installed more and more, even with stationary boilers. For whenever a machine has to remain inactive for a while for some reason or other, the water in the boiler will decrease as steam escapes through the safety valve without new water being pumped in, because the pumps are always connected to the rest of the machine. By means of the separate little machine, this problem can be easily eliminated.

So much for this at any rate very interesting subject. If you happen to publish something about it, or report on it, I should be glad if you told me about it. I promise to send you from here what I can obtain or learn about it. By the way, I think that a publication would be quite good, so that the public would know what it has to pay attention to. Perhaps the publication would also be an inducement to discover a mechanical device by means of which the sinking of water below its normal level would become impossible.

I am very interested in what you tell me about Plücker. Wheatstone told me in Paris that Plücker became a member of the Royal Society only as a mathematician.[6]

Faraday is too modest. Not his works, but his splendid character keep his admirers so close to him, and this character, combined with his colossal achievements, raises him so high in my eyes. Unfortunately one finds only too frequently, where important achievements exist, a lack of a good and amiable character. (By the way, you do not always need to read out to Faraday what I write about him, for otherwise I shall no longer be able to write how I would like.) But you can give him my regards.

I still cannot decide about the trip to Vienna,[7] but I shall make a suggestion. —If you come to Germany, you will not want to go just to Vienna. I had intended to go to Switzerland in August with my wife; come there as well, together with Frankland and Bence Jones, and go to Vienna from there. We shall then enjoy the beautiful natural scenery in Switzerland and talk about physics at the same time. Write soon whether and what you decide in this respect, and also be so good to tell me what objections you have about my theory of the bursting of boilers.

I ask you to give my regards to Bence Jones, Frankland, Wheatstone, Francis, and everyone who remembers me.

Sincerely | Your | GMagnus.

My wife and children[8] ask me to give you the kindest regards.

RI MS JT/1/M/18

---

1. *your kind letter*: Tyndall's letter of the 3 May 1856 is missing.
2. *as well as of other boilers*: stationary steam engines in factories and pumping stations.
3. *special little steam engine*: In early steam locomotives the injector to replenish water that had evaporated in the boiler was linked to the locomotive's mechanism. It therefore only worked when the engine was in motion. Magnus is referring to a special steam engine, possibly mounted on a truck, which operated a mechanical water pump to replenish water in the boiler when the engine was stationary. The accessory engine did not become redundant until after 1864 when the British engineer, James Gresham (1836–1914) developed the steam injector.
4. *the so-called spheroidal state*: Magnus uses the phrase as a specific scientific term rather than using the common German words for bubbling or boiling.

5. *Expansibility*: 'die Spannkraft' literally means the 'tension force'. The modern engineer would call this thermal energy and the physicist, 'potential energy'.
6. *only as a mathematician*: Wheatstone had been one of Plücker's sponsors for his election as a Foreign Member of the RS on 21 June 1855. Magnus's comment was not entirely correct since the RS certificate says: 'distinguished for his investigations in analytical Geometry, and his researches in various branches of physical science'.
7. *the trip to Vienna*: Tyndall must have told Magnus that he was planning to attend the meeting of the Gesellschaft Deutscher Naturforscher und Ärzte in Vienna in September 1857.
8. *My wife and children*: see letter 1216, n. 12.

From Mary Anne Coxe 19 May 1856 1231

9 Athol Crescent | [Edinburgh] |Monday | May 19th.

Yesterday in church in the middle of the sermon (Oh! fie Mrs Coxe!) it suddenly flashed across me to wonder if Dr Tyndall had ever read Aytoun's Lays![1]—on the instant I determined to send them to him--Unlike (too) many Sunday resolutions this had not, when Monday came—melted into thin air. So here they are—read them alone—and if you do not, while so employed, feel 'the old blood bounding free, through pulse and heart and vein'[2]—you are not the John Tyndall I take you for, and never more shall you be officer of mine! With a high tone of chivalry, there is a simple pathos—that even yet,—(countless as are the times that I have read and spouted them to myself.) gives me a choking sensation in my—throat! Then they are so absolutely without the morbid maudling sickly kind of sentimentality—for it is not sentiment—that so many/modern poets delight in—and for all that I know almost every one of them by heart. I have as much pleasure in them yet, as in the first time of their perusal.

You will be sorry you are not a Scot!

I shall mark them according to my idea of their excellence and it will please me to know us d'accord.[3]

The other day I saw in the paper an anecdote, which took my fancy so much, I shall run the risk of your having seen it, rather than let you miss it—It was headed 'How to put a thing'[4]

In a theatre in Paris, while Rubini[5] was in the act of singing, a Frenchman began to hum the same song. 'Che Bestia'![6] wrathfully exclaimed an Italian—on which the Frenchman turned upon him with 'qui ça Monsieur'?[7]

'C'est Rubini—qui m'empêche d'entendre Monsieur'[8] blandly replied the Italian! Confess that se non è vero, e ben trovato![9]

My husband[10] desires me to thank you for him—he says you will know how to make allowance for a man full of his own idea, and slaving to death. Really I do wish this Lunacy Commission[11] were at an end, for save at meals and at my constitutional, I never now set eyes on him till 10 and after 11 at night. He sits in his den with an amount of papers and Documents that make one shudder to look at. I have the conviction that I shall end by becoming a "fatuous Pauper" myself![12]

Now good bye—| Yours always | Mary Anne Coxe. |

RI MS JT/1/1/272
LT Transcript Only

---

1. *Aytoun's Lays*: William Edmondstoune Aytoun (1813–65), Scottish barrister and poet who contributed to *Blackwood's Magazine*. Professor of Rhetoric and Belles-Lettres at the University of Edinburgh from 1845. Lady Coxe was referring to his *Lays of the Scottish Cavalliers and other Poems* (Edinburgh: Blackwood, 1849) which appeared in a new edition in 1856. According to an annotation by LT, Lady Coxe wrote a dedication on the flyleaf of the copy she sent to Tyndall. The copy has not survived at the RI.
2. *The old blood . . . heart and vein*:

   'Come brothers; let me name a spell
   Shall rouse your souls again,
   And send the old blood bounding free
   Through pulse, and heart, and vein!'

   Ayrtoun, *Lays of the Scottish Cavalliers,* stanza VI of the poem 'The Island of the Scots'.
3. *know us d'accord*: know us to be in agreement (French).
4. *How to put a thing*: quotation not traced.
5. *Rubini*: Giovanni Baptista Rubini (1794–1854), Italian tenor famous for his bel canto and tremulo.
6. *Che Bestia*: Italian, literally, 'How beastly', but in Italian slang also means 'a load of bollocks'.
7. *Qui ça, Monsieur*: (French) 'What's that you said?'
8. *C'est Rubini—qui m'empêche d'entendre Monsieur*: (French) 'It's Rubini, who prevents me from hearing you.'
9. *se non è vero, e ben trovato*: an Italian proverb, 'even if it is not true, it is a good story.'
10. *My husband*: James Coxe.
11. *Lunacy Commission*: Sir James Coxe was a Commissioner in Lunacy for Scotland. See letter 1077, n. 1.
12. *fatuous pauper*: In the phraseology of the law of Scotland, 'persons who, by reason of mental unsoundness are unfit for the management of themselves or their property, are termed "furious or fatuous persons and lunatics", the first being applied to maniacs, the second to imbecile persons or idiots, and the last to insane persons generally.' *Report by Her Majesty's*

Commissioner to Inquire into the State of Lunatic Asylums in Scotland (HMSO: Edinburgh, 1857), p. 3.

From Rudolph Kohlrausch[1] 19 May 1856 1232

Marburg den 19 Mai 1856.

Mein lieber Tyndall!

Beikommend übersende ich Ihnen eine Arbeit, welche vor etwa zwei Jahren vollendet war, deren Redaction und Druck indessen sich verzögert hatten. Wenn Sie sich wundern, daß mein Name vor dem des Prof. Weber steht, so kann ich darüber nur sagen, daß es ohne mein Wissen und gegen meinen Wunsch von Weber so eingerichtet worden ist.

Zugleich lege ich ein Exemplar von Webers Schrift über den Diamagnetismus bei. Er sagte mir [zu] Ostern, daß er Ihnen ein solches zu überschicken dächte, da er vermuthe, daß Sie es vielleicht nicht besäßen. Nun hatte er, als er die Exemplare unserer Abhandlung herschickte, kein Exemplar jener Schrift zur Hand, weshalb ich das meinige beilege, indem ich mir von ihm ein anderes verschaffen kann, sobald sich Weber mit einigen neuen wird versehen haben. Sie dürfen sich nicht daran stoßen, daß von seiner Hand mein Name vorn hineingeschrieben ist.

Drittens begehe ich die Unverschämtheit, Sie zu bitten, die beiden einliegenden Exemplare an Prof. Thomson und Stokes weiterbefördern zu wollen. Ich kenne nämlich deren Adreßen nicht. Sollte es mehrere Thomson geben, so meinen wir den, welcher im Februar Heft des Philos. Magazin „On the theory of the electric telegraph" geschrieben hat, und schon viele werthvolle Untersuchungen lieferte. Ich werde stets zu allen Gegendiensten bereit sein.

Indem ich Ihnen meine schönsten Grüße sende und die von Ihren hiesigen Freunden dazugeselle, bemerke ich noch, daß vor einigen Tagen die Prof. Bunsen und Kirchhoff uns einen Besuch abstatteten, letzterer, um mit den electrischen Apparaten zu arbeiten, welche ich hier für ihn hatte anfertigen laßen und daß wir in alter Weise einige sehr vergnügliche Zusammenkünfte feierten. Von Knoblauch habe ich länger nichts gehört.

In alter Freundschaft | der Ihrige | Kohlrausch

Da die Post mir Schwierigkeiten bei der Uebersendung der Bücher machte, laße ich diese Zeilen allein mit ihr abgehn, während Herr Elwent die Bücher Ihnen über Leipzig in etwa 14 Tagen zu senden übernommen hat.

Marburg 19 May 1856

My dear Tyndall!

I am enclosing a paper[2] for you which was completed about two years ago whose editing and printing had been delayed in the meantime. If you are surprised that my name precedes that of Prof. Weber, I can only say that this was arranged like this by Weber without my knowledge and against my wish.

At the same time I enclose a copy of Weber's work on diamagnetism.[3] At Easter he told me that he was thinking of sending you one, as he suspected you may perhaps not have it. Well, when he sent the copies of our paper,[4] he did not have a copy of that work at hand which is why I enclose my own, as I can obtain another one from him for myself as soon as Weber has provided himself with some new ones. You must not take exception to the fact that my name is inscribed at the front by his hand.[5]

Thirdly, I am being so impertinent to ask you to forward two enclosed copies to Professors Thomson and Stokes. I do not know their addresses. Should there be several Thomsons, then we mean the one who wrote 'On the theory of the electric telegraph'[6] in the February issue of the Philos,[ophical] Magazine and who has already provided many valuable investigations. I shall always be ready to do any favours in return.

In sending my finest regards to you and adding those of your friends here, I shall mention in addition that a few days ago Professors Bunsen and Kirchhoff[7] paid us a visit—the latter in order to work with the electrical apparatus which I had had made for him here—and that we celebrated with some very enjoyable gatherings in our old fashion. I have heard nothing from Knoblauch for some time.

In old friendship | Yours | Kohlrausch

Since the postal service was creating difficulties for me to forward the books, I shall have these lines sent on their own, while Herr Elwent[8] has undertaken to send you the books via Leipzig in about a fortnight's time.

RI MS JT/1/K/31

---

1. Rudolph Kohlrausch (1809–58), German physicist, Professor of Physics at the University of Marburg (1853–57).
2. *I am enclosing a paper*: W. Weber und R. Kohlrausch, 'Ueber die Elekricitätsmenge, welche bei galvanischen Strömen durch den Querschnitt der Kette fließt' [On the quantity of electricity which flows through the cross-section of the circuit during electrolysis], *Poggend. Annal.,* 99 (1856), pp. 10–25. No English translation was made. In the printed and bound version of the paper, Weber's name precedes that of Kohlrausch; possibly the names were reversed in the offprints that preceded formal publication.
3. *Weber's work on diamagnetism*: Weber's work on diamagnetism using a moving cylinder of bismuth inside a solenoid was reported in W. Weber, *Elektrodynamische Maasbestimmungen* (Leipzig: S. Hirzel, 1852).
4. *the copies of our paper*: paper cited in n. 2.
5. *inscribed by his hand*: Kohlrausch was sending the copy of *Elektrodynamische Maasbestimmungen* whose flyleaf Weber had signed.

6. *On the theory of the electric telegraph*: W. Thomson, 'On the Theory of the Electric Telegraph'. This appeared as part of the report of the proceedings of the RS, *Phil. Mag.,* 11 (1856), pp. 146–60.
7. *Kirchhoff*: Gustav Kirchhoff (1824–87), German physicist. Professor of Physics at the University of Heidelberg where he joined Bunsen in 1854.
8. *Mr Elwent*: an unidentified Leipzig bookseller.

From Thomas Archer Hirst 1 June 1856 1233

Broughton | June 1st 1856

My dear John.

I have never had a spare half hour to devote to writing to you since I left you in London.[1] I have had a hard time of it, with a heavy heart I have had to assume a cheerful face. I scarcely ever allow myself to look far forward into the future for it would be to anticipate (probably) great sorrow[.] I do not allow myself to look back on the past and dwell on the hopes which I once indulged for to do so would be to give way to useless regrets and all would unfit me for the duty close at hand to keep Anna cheerful and to do everything possible to ward off the consequences of her malady. So far I have succeeded very well she has reconciled herself to leaving our little home to which we have become greatly attached and I have managed to bring this about without alarming her. John Martin came a week ago and on Thursday[2] he took her with him to Ireland. I am hoping to hear tomorrow of their safe arrival at Newry.[3] As soon as possible I shall follow her for in her present state of health I cannot bear to be away from her very long. On Monday week I shall commence my examinations and get them over on the Wednesday[.] On the Thursday morning I shall leave for Ireland. When you come I want you to take up your quarters here and keep house for me.[4] Wynne[5] will remain here of course until after the concert. Our servant will remain in the house for a fortnight at least after I am gone and she will make you comfortable. Thus you will have a quiet and independent lodging for yourself and it will give us great pleasure to think that you are making our house your home. I have of course told Mr E.[6] about my leaving and <u>of course</u> he has not a word of any kind to say about it or about my successor. There is one plan of supplying my place which Wright, Debus and I simultaneously entertained that is to get as good a mathematician and Physicist (if possible) as we can find and then ask you to spend your long vacation here and help by your advice and presence as much as you can. Much as this would be to the advantage of Queenwood I must say I hesitate greatly to ask you to make such a sacrifice knowing as I know what you <u>ought</u> to be doing; and what terrible anxiety of mind your residence at Queenwood will involve, I can scarcely ask or even advise you to do any such thing after your long toil;

peace of mind is essential to you; at any rate you should have such peace as will allow you to give your whole attention to your own investigations, and to harrass you with such matters as will be brought before you at Queenwood appears to me cruel indeed. Yet on the other hand if you do not come the prospect for Queenwood will be very dark. Wright as far as he goes is an excellent & pure minded fellow, but there are some things he cannot do, he and Haas alone will never pull well together. Then again there is Debus who could be of immense service, but he wants practical tact and with Haas he will never be able to cooperate without giving rise to absurd jealousies. As for a new man there are 9999 chances to 1 that he will be <u>merely</u> a mathematician besides at the best he would require time, a long time to acquaint himself even with the length and breadth of his position, and meanwhile much mischief may be done. What is to be done you alone will be able to say. I must speak to Edmondson this week, but before doing so I should like to hear from you, for upon your opinion would depend altogether what I should say to him. As to a new mathematician and physicist I have been thinking I would ask you to give me a letter of introduction to Haughton and I would call upon him and see if he knows anybody suitable. As far as physics & mathematics are concerned Dublin is the best field I know.

We shall remain in Ireland 6 or 8 weeks, then return here to settle our affairs and in August or September we will start for the Continent.

When I attempt to look calmly at my position and prospects I feel a coward. When I remember what were my hopes and intentions only 6 months ago, the contrast upsets me completely[.] I feel, well enough, the danger and uselessness of dwelling on such things but I cannot always avoid it, and it requires all my courage to face what is before me and in doing the best at this moment to trust the rest to God. But never fear I will do so.

Yours as ever | <u>TA Hirst</u>

RI MS JT/1/H/225

---

1. *I left you in London*: On the recommendation of Dr Fox, Hirst took his wife to London on Tuesday 20 May 1856 for a consultation with a Dr Walsh. This was probably the physician W. H. Walshe, 40 Queen Anne Street, Cavendish Square. The Hirsts must have taken the opportunity to see Tyndall who did not record the visit. See Hirst's Journal for 18 May 1856 (RI MS JT/232c/1195).
2. *Thursday*: 29 May 1856. Hirst recorded in his Journal: 'For the first time since we were man and wife Anna has left me. For the sake of her own health I have had to send her away from me to Ireland, whither I am to follow in a fortnight' (RI MS JT/2/32c/1195).
3. *Newry*: Newry, a town in what is now Northern Ireland, 35 miles from Belfast and 67 miles from Dublin. Anna Hirst's family lived in Newry. See letter 1027, n. 2.
4. *keep house for me*: Tyndall lodged at Vine Cottage from 14 June 1856.

5. *Wynne*: Eddy Wynne, who had been studying with Hirst.
6. *Mr E.*: George Edmondson, the headmaster.

To Michael Faraday 2 June 1856 1234

Royal Institution of Great Britain

Dear Mr. Faraday,

Sir Roderick Murchison[1] I am informed has kindly expressed his readiness to grant us specimens from the Museum of Practical Geology[2] provided an official application be made to him on the part of the managers. Fortunately a meeting occurs today, and I should feel obliged if you would bring the subject forward and obtain for me the application which Sir Roderick considers necessary.[3]

Believe me | most truly yours | John Tyndall
2nd June

for certain geological specimens illustrative of slaty cleavage[4]

RI MS AD/03/D/04/03
Faraday Correspondence, 5:3154

---

1. *Sir Roderick Murchison*: the geologist; see letter 1170, n. 10.
2. *Museum of Practical Geology*: The Museum of Practical Geology (originally Economic Geology) opened in Craig's Court, Whitehall, in 1835, the brain-child of the Director of the Geological Survey, Henry De la Beche (1796–1855). It moved to a purpose-built building in Jermyn Street in May 1851 under the direction of Sir Roderick Murchison. Besides the Museum, it was the headquarters of the Geological Survey, the Mining Records Office, and the School of Mines.
3. *considers necessary*: Tyndall required the specimens for his Friday Evening Discourse of 6 June 1856, 'Comparative View of the Cleavage of Crystals and Slate Rocks', *Roy. Inst. Proc.*, 2 (1854–58), pp. 295-308.
4. *for certain . . . slaty cleavage*: endorsement in Faraday's hand.

To Thomas Archer Hirst 2 June 1856 1235

2nd June 1856

My dear Tom

I have just received your letter,[1] and were it not that sorrow is like magnetism which communicates itself without loss your sorrow would be already half borne. I wish I could transfer a portion of it from you to myself,—I wish

I could gather that cloud which you think overhangs your future before my own vision and so relieve you of at least a portion of it. And yet this may be, and I would hope & trust probably is, the very darkest phase of things. Before taking any decided step I would have additional medical advice. I am strongly inclined to believe that the Isle of Wight would be as good for an invalid as Italy—but take my word for nothing if unsupported by suitable medical testimony. I saw Bevington yesterday & he thinks the same. He has a sisterinlaw[2] who has been far worse than Anna, one of whose lungs is indeed entirely gone, and she is now in good health. I will do as you say—take up my quarters at Vine Cottage,[3] I will come to you the moment I can get away from London and will talk over all. With regard to going to Queenwood I have a mortal dread of it in my present state of health. Believe me I need rest—deep rest—and I know not how this would be attainable at Queenwood. Besides there are many other grave demands upon me here. But I will think of all and talk over all when we meet. God bless you & support you Tom

Your affectionate | John

RI MS JT/1/T/623

1. *your letter*: letter 1233.
2. *He has a sisterinlaw*: Mary Bevington's sister has not been identified.
3. *Vine Cottage*: the Hirsts' home at Broughton which Tyndall was to use as a vacation home for the week of 14–21 June 1856 while the Hirsts were in Ireland.

To Thomas Archer Hirst 7 June 1856 1236

7th June 1856

My dear Tom.

I ordered one or two little things bearing upon your proposed experiments at the breaking up meeting[1]—they are now beside me, and consist simply of a Delarue's pile,[2] a steel cylinder magnet along which it moves, a simple contrivance for shewing the attraction, & repulsion of electric currents. Shall I send the things to you by rail or wait until I go myself and take them along with me

affectionately John

RI MS JT/1/T/624

1. *breaking up meeting*: the end of term Open Day for Queenwood parents and investors on 16 June 1856. See letters 1226 and 1227. Hirst never used the apparatus since he left Queenwood abruptly on Wednesday 11 June.

2. *DelaRue pile*: an improved version of Daniell's constant voltage battery which Warren de la Rue devised in 1836. 'On Voltaic Electricity, and the Effects of a Battery Charged with Sulphate of Copper', *Phil. Mag.*, 9 (1836), pp. 484–87 and 10 (1837), pp. 244–46.

From Geraldine Jewsbury[1] 7 June 1856 1237

3 Oakly Street King's Road | Chelsea | June seventh 1856

dear Sir. Mrs Carlyle is gone out of town until Monday so it falls to me to write to you to thank you for the tickets you sent us.[2] We quite promised ourselves the pleasure of going but when it came to the time Mrs Carlyle's cold was not well enough to make going out at night prudent; of course I c^{d} not go alone & I sh$^{d.}$ have regretted my disappointment still more if I had not gone & tired myself to death at the Crystal Palace Concert.[3] As we hope you will not be disgusted at our apparent want of zeal but send us other opportunities of hearing you—will you kindly let it be in a forenoon if possible as we can attend far better than in the evenings—going into crowded places at night generally makes her ill.

Do you never come to Chelsea when you are taking yr walks abroad? if so I wd suggest that Mrs Carlyle & I both live in Chelsea.

I am dear Sir | yrs truly & respectfully | Geraldine E Jewsbury | To—Tyndall Esq.

RI MS JT/1/J/27

1. *Jewsbury*: Geraldine Endsor Jewsbury (1812–80), a popular novelist, prolific reviewer and publisher's reader. She was a close friend of Thomas and Jane Carlyle. The latter referred to her as 'Miss Gooseberry'. Tyndall had first met Jewsbury on 13 February 1856. Following a dinner at the Huxleys on 31 July, Tyndall recorded in his Journal: 'Walked all the way to Piccadilly with Miss Jewsbury: we talked of the stars and of their science, then of other matters, finally of science again, and I muddled both her and myself: went a little astray, and found a Chelsea omnibus at length, and saw her safely on board. The account she gave of Carlyle's present selfworship gave me great concern. When she knew him first this had no existence, but now it cuts away all sympathy with anything but himself' (RI MS JT/2/13c/824 and 863).
2. *tickets you sent us*: tickets for one of Tyndall's RI lectures.
3. *Crystal Palace Concert*: Following the success of the Great Exhibition in 1851, Paxton's Crystal Palace was dismantled and re-erected on grounds at Sydenham. Under the management of George Grove (1820–1900), the engineer and musician, it became a popular venue for concerts and exhibitions.

To Thomas Archer Hirst 9 June 1856 1238

9th June 1856

My dear Tom

My lectures are not yet finished: my last lecture[1] will be given on Thursday next. Faraday's concluding lecture[2] takes place on Friday and before Saturday or Sunday it will be impossible for me to move—I thought you intended remaining until the Concert week.[3] I quite hoped to meet you and talk with you at Queenwood. but perhaps it will do equally well here. Remember that a little reflection on your future movements may eventually be a saving of time. Francis and I have had a long conversation about you,[4] and he knowing the continent so well is able to give much useful advice; but it is needless for me to dwell on this—you will I doubt not take your measures with due deliberation. Poor Francis: as we talked of you yesterday he almost wept.[5] The more I see of him the more I admire his sincere, warm nature. 'If the thing could be transferred to myself' he said 'I would bear it for Tom's sake, and this is more founded on my instincts regarding Tom than on any thing thing I absolutely know concerning him.' I am very weary & very confused this morning. Another misfortune now looms upon me. For Wednesday I have entered into an engagement from which I do not know how to extricate myself.[6] I wish I had known your intentions & I would have gone to see you yesterday. Can you not come up on Thursday? I shall then have thrown my great weight of my shoulders and we can have an evening together. I would ask Francis to meet you, and he I am sure could give you many valuable hints regarding climate, cost of living and other matters. You will get this note tomorrow morn[in]g just send me a line by return stating what your decision is. I put 50 pounds in bank a few days ago—if you need money they are at your service; and any you may need on the Continent I shall be able to send to you. Your proposal regarding Haughton[7] I think a very good one, and I will write to him as you suggest: Give my loving remembrance to Anna when you write.

Ever | John

RI MS JT/1/T/625

1. *my last lecture*: Tyndall's final lecture on light on 12 June 1856.
2. *Faraday's concluding lecture*: 'On M. Petitjean's Process for Silvering Glass', delivered on 13 June 1856 and published in *Roy. Inst. Proc.*, 2 (1854–58), pp. 308–12.
3. *until the Concert week*: Hirst had abandoned Queenwood on 11 June, such was his concern to join his sick wife in Ireland.

4. *Francis and I . . . about you*: The conversation took place during a weekend hike through the Chiltern Hills from Tring (Hertfordshire) to Windsor on 7–8 June (RI MS JT/2/13c/848).
5. *he almost wept*: Francis was upset that Hirst had had to abandon his teaching career to look after his dying wife, Anna.
6. *I do not know how to extricate myself*: Tyndall had been invited to the Lord Mayor of London's banquet at the Mansion House on Wednesday 11 June (RI MS JT/2/13c/ 848).
7. *Haughton*: In letter 1243, Hirst asked Tyndall for a formal introduction to Haughton so that while he was in Ireland with his ailing wife, he might ask Haughton's advice regarding someone to replace him as a mathematics teacher at Queenwood. However, when he called on Haughton at Trinity College Dublin on 29 June he found Haughton had gone to London (RI MS JT/2/32c/1203).

From William Snow Harris 9 June 1856 1239

Monday 9th. June 1856[1]

Dear Dr. Tyndall,

I am to read a Paper at the Royal Society[2] on Thursday and have sent there a large amount of Electrical apparatus—which I am very desirous to show you quietly as between ourselves. Faraday is coming to me at 3 this afternoon and I should be happy to see you at the same time at the Royal Society's apt.[3] If however this does not suit your engagements we will arrange for some other day. I have been more than once about to write to you since we last met[4] but have not had sufficient leisure for the purpose of writing all I had to say.

believe me to be most truly yours | W. Snow Harris

Dr. Tyndall. F.R.S. | &c. &c.

RI MS JT/1/2/470
LT Transcript Only

1. *Monday 9th June 1856*: The letter was probably sent from Harris's home at 6 Windsor Villas, Plymouth.
2. *a Paper at the Royal Society*: W. S. Harris, 'On Quantitative Measurement of Statical electricity, and on Some Phenomena of Electrical Force', *Proc. Roy. Soc.*, 8 (1856–57), pp. 166–71, read on 12 June 1856.
3. *the Royal Society apt*: The apartments of the Royal Society were at Somerset House in the Strand. Tyndall did join Faraday to view the apparatus (RS MS JT/2/13c/848).
4. *since we last met*: Tyndall had last met Snow Harris on 10 April 1856 when they discussed the draft of the paper mentioned in n. 2 (RI MS JT/2/13c/833).

From César Despretz 15 June 1856 1240

Mon cher Mr Tyndall

J'ai tardé à vous écrire. J'ai eu des cloux à Londres. Un nouveau clou (Boil) qui ne faisait que poindre à mon départ, s'est enflammé par le frottement. Bref, il a fallu l'inciser en quatre. Il était placé à la nuque. Je suis à peu près guéri. Il faut espérer qu'il n'en reviendra pas d'autres. J'aurais pu vous écrire dès mon retour. Mais comme depuis plus de trente ans que je suis professeur (depuis 20 ans à la Sorbonne), je n'ai jamais dû manquer ni à une leçon, ni à un examen, par suite d'une indisposition. Je me trouve tout déconcentré, quand j'ai quelque chose j'oublie mes lettres, &c.

J'ai lu le travail de Mr Wiedeman. Je n'ai rien à changer au plan de travail dont j'ai donné une partie dans ma réponse à Mr Langberg. Je ne crois pas que je me sois trompé dans mon premier travail sur l'effet du contact. Cependant c'est à revoir. Je dirai, comme dans toutes les circonstances, ce que j'aurai trouvé, que j'aie raison en tout ou seulement en partie ou que j'aie même tort en tout ou en partie. Nous verrons. Comme ce travail, à la manière dont je le conçois exigera assez d'expériences et occasionnera assez de dépenses, je vous prie de n'en pas parler. Quoiqu'il soit très naturel de se défendre quand on est plus ou moins contredit et même c'est un devoir pour celui qui recherche la vérité avant tout.

Je vous réitère mes remercîments pour toutes les aimables dispositions que vous m'avez montrées à Londres. Présentez, je vous prie, mes compliments à vos deux jeunes collègues Hofmann et Malone et entretenez moi dans la bienveillante amitié de Mr Faraday.

15 juin 1856 | Votre dévoué | C. Despretz

P.S. J'ai écrit à Mr Grove. Je vais écrire à Mr Babbage.

Je vais faire venir Noad que j'ai parcouru à Royal Institution. Non pas pour y voir mes expériences, mais celles qui ont été faites en Angleterre sur l'électricité car je suppose que s'il oublie quelques étrangers, il n'oubliera pas les compatriotes.

My dear Mr Tyndall,

I have delayed writing to you.[1] I had boils in London.[2] A new boil appeared as soon as I left, and it flared up due to friction. In short, it had to be lanced into four. It was on the nape of the neck and is almost healed. Let us hope that it will not return. I could have written you a letter as soon as I returned. But as I have been a professor for more than 30 years (for 20 years at the Sorbonne[3]), I have never had to miss either one lesson or one examination due to illness. I find myself confused when I have something;[4] I forget my letters, etc.

I have read Mr Wiedeman's work.[5] I have nothing to alter in terms of the work that I partly gave in my answer to Mr Langberg.[6] I do not think I was wrong in my first investigation on the effect of contact. However, it can be criticised. As I have found on all occasions: either I was wholly or partially right, or I was wholly or partially wrong. We will see. Since this work, as I see it, will require quite a few experiments and will involve quite a bit of expense, I ask you not to speak of it. Although it is very natural to defend oneself when more or less contradicted, and even though it is a duty for those who seek the truth above all.

I would like to repeat my gratitude for all the kindness that you have shown me in London. Please send my compliments to your two young colleagues, Hofmann and Malone,[7] and give my friendly respects to Mr Faraday.

15 June 1856 | Your devoted | C. Despretz

P.S. I wrote to Mr Grove. I will write to Mr Babbage.

I will get hold of Noad,[8] the book I skimmed through at the Royal Institution. Not to see whether my experiments were reported, but to learn about those electrical experiments which were made in England, since I assume that if Noad ignores some foreigners, he will not forget hsi compatriots.

RI MS JT/1/D/125

---

1. *writing to you*: Despretz's staccato style with sentences lacking subordinate clauses was probably deliberate in a kind attempt to make his French easier for Tyndall to understand.
2. *I had boils in London*: Despretz had been in London in May 1856. See letter 1219.
3. *at the Sorbonne*: Despretz was Profeesor of Physics at the École Polytechnique, the College Henri IV, and at the Sorbonne in Paris.
4. *when I have something*: whenever he felt unwell.
5. *Mr Wiedeman's work*: Gustav Heinrich Wiedemann (1826–99), German physicist and physical chemist. A life-long friend of Helmholtz, he was appointed Professor of Physics at the University of Basel in 1854. There he worked chiefly on endomosis and the magnetization of iron and steel. Tyndall had first met Wiedemann in Berlin on 2 May 1851 (RI MS JT/2/13b/541). After another meeting on 15 July 1851 he recorded: 'Wiedermann is a clever fellow also, but somehow you feel that there is no real manly bottom to the man. "He smiles and whispers low" but the smile is not one you would trust very far' (RI MS JT/2/13b/611). Despretz was referring to Wiedemann's paper, 'Ueber die Fortplanzung der Wärme in den Metallen', *Poggend. Annal.*, 45 (1855), pp. 337–47. This was translated as 'On the Transmission of Heat in Metals', *Phil. Mag.*, 10 (1855), pp. 393–400.
6. *My answer to Mr Langberg*: Christian Langberg (1810–57), a Norwegian pupil of Gustav Magnus in Berlin who had been Professor of Physics at the University of Christiana since 1847. In 1845 he had published 'Versuche über das Wärmeleitungsvermögen fester Körper', *Poggend. Annal.*, 66 (1845), pp. 1–30; this appeared as 'On the Determination of the Temperature and Conducting Power of Solid Bodies' in *Phil. Mag.*, 28 (1846), pp.

161–66. The paper suggested Despretz's observations were inaccurate. Twelve years later, Despretz challenged Langberg's findings in 'Note en réponse à M. Langberg, de Christiana, nouvaux nombres sur la propagation de la chaleur dans les corps', *Paris, Comptes Rendus,* 35 (1852), pp. 540–48.

7. *Malone*: Thomas Malone (1823–67), photographic chemist and partner in the firm of Henneman and Malone from 1848 to 1851. He lectured on chemistry and photography at the London Institution.
8. *I will bring Noad*: Henry Minchin Noad (1815–77; FRS 1856), chemist and electrician. Professor of Chemistry at St George's Hospital in London since 1847. He published an influential text, *A Course of Eight Lectures on Electricity, Galvanism, Magnetism and Electro-Magnetism* (London: Scott, Webster & Geary, 1839). There were three editions before he republished it as *A Manual of Electricity,* 2 vols. (London: George Knight, 1855).

To Thomas Archer Hirst 15 June 1856 1241

[Vine Cottage, Broughton] |15th. June, 1856|

My dear Tom

Here I sit in your little study all alone except the accompaniment of an oblique sunbeam which comes athwart the door and smites your bookcase. The birds are singing and the grass shining like an emerald while butterflies are wafted to and fro across your little paddock—all these, with inexpressible softness, tell me that [I] am far from the smoke and roar of London. I had a good nights rest in the charming room you left prepared for me, and feel better this morning than I have done for a long time. If this mortal coil of mine would only remain in order I think nature intended me to be a happy joyous fellow; accepting of course the sorrow as well as the gladness which accrues from the changeableness of external things, but untroubled by vapours or spleen which arise either from dyspepsia or the confusion and clash of the moral powers. I walked into your study after breakfast; It was hushed and peaceful; and turning to your bookcase I saw there 'Plutarch's Lives.'[1] I remember our evenings in Marburg[2] when, after the accomplishment of a hard day's work, we regaled ourselves at night with the companionship of this glorious old soul, and this memory lent an increment to the pleasure with which I read this morning the life of Marieus Coriolanus.[3] Plutarch at the end compares Alcibiades[4] and Coriolanus; almost all my nature sympathizes with the latter, only a fraction of it with the former. Both perhaps were equally competent, but Alcibiades loved riches and honour and would win them by craft and flattery of the mob. Coriolanus erred in his contempt of the populace, and in an obstinacy which Plutarch himself analyzes in an admirable manner, and strongly condemns. The remarks of the old biographer are full of wisdom, which smacks of the practice of life, and is not that merely which the student gathers in woods and solitudes. But a truce to Plutarch. I have been out in your garden and seen the boundary of it all purple with the bloom of flowers; the bells of the church are now chiming merrily and Mr Lee is preparing for the delivery of his weekly lesson on religious mechanics.[5]

I was up at Queenwood last night: I came by the last train, and sending my portmanteau on to Broughton turned off from Rose Hill to Queenwood: I saw Edmondson; he wished to appear frank, friendly, and hospitable; but there was a dryness in his words, which seemed to me to shew a hospitality reaching from the larynx upwards. I wonder has the man ever had a serious earnest moment. I hardly think it. The few moments conversation I had with him last night, although on matters as light as a salad supper, and a knife and fork at dinner, was not at all pleasing to me, and when I bade him good night I found it impossible to look with frank and friendly trustfulness into his eyes, but made the act of shaking hands as mechanical as his own nature.[6] Debus asked me some questions relative to my coming to Queenwood. If I thought my vocation lay here, that I had not work before me which has a still graver demand upon me I would come. But in aiming at too much I should I fear accomplish very little. I set the question thus before me—either you give up your present vocation and turn once more to Queenwood heart and soul, or you cling to your present course with an undivided spirit—both you cannot accomplish, and by endeavoring to accomplish both you will do neither as it ought to be done. This is the inexorable fact Tom, and I have no doubt as to which course you would counsel. In the matter of education I see much to be done in another field—in a field where I can be more useful than at Queenwood. All this I stated in substance to Debus: he regrets it, but still could not advise me alter my determination. I see danger, great danger to Queenwood in advance. I do not in the least see how your place here is to be filled, and I fear the gods which have thus far befriended Edmondson are about to forsake him, though he from man can discover nothing of the kind. He cannot look beyond the present hour and I would venture to say would be a little surprised if he hears me express my estimate of the loss he has suffered in losing you. It would be a weary work for me Tom to endeavor to couple with the tranquil & earnest pursuit of my researches the propping up of such a shaky edifice, and to keep a man whose nature is as quicksilver from running into error and mischief. I could not do it Tom, so I see nothing for it but to permit Queenwood to roll along with the momentum which it has already acquired, and to hope that the forces now within it may not be exhausted simply in neutralizing one another. I think your suggestion regarding Haughton[7] a good one: if you like I will write to him direct, but by far the best way would be to give you the introduction you spoke of so that you might see with you own eyes, and converse with the person he might recommend. We had some bitter hail showers yesterday. I thought of Anna as I came along in the train: a handsome lady sat opposite to me who appeared to be in weak health: she was married and her husband I think is an officer at Aldershot.[8] To keep out the cold draught a gentleman had closed the main door of the compartment. I saw the lady become paler & paler and finally almost white. I opened the window, gave her air[,] tried to heat her up and finally she became restored and animated. She told me all about her health—She got chilled after dancing last winter and the cold has clung to her since. She takes cod liver oil, which she abominates, not being able to take any food for hours after each dose; but she is getting stronger and hopes soon to be quite well again. I united Anna and her in one enormous wish, hoping however that Anna would get well first and she afterwards, or more correctly that both might be restored within the shortest compass of nature's power of effecting such a restoration.

And now my dear Tom I think I have said all I intended to say, I hope you will have patience to read the letter. Have you made any arrangements about the letting of your house? I think you ought to look to this, and if necessary advertise that it is to be let. Have you thought of my means of occupying your leisure on the Continent—You know the subject of education is one of great & general interest in England at present and I know that few are able to look at it in a truer light than yourself. Could you not turn your experience to account and give the world something on this subject. Let you be Emerson and I will be Tom Carlyle, write an introduction to the book and get it quick access to the pundits of London. I think this subject well worthy of your consideration—You will do the thing with a freshness now which may not be attainable when your present impressions become dim.

Ever your affectionate | John

RI MS JT/1/T/626

---

1. *Plutarch's Lives*: Plutarch's *Lives of the Noble Grecians and Romans* was compiled at the start of the 2nd century AD. It consists of 23 pairs of lives (one Greek with one Roman) that share certain moral characteristics. The first English translation, made by Thomas North in 1579, was used by Shakespeare.
2. *Our evenings in Marburg*: Tyndall was probably confusing discussions during Hirst's brief visit to the University of Marburg in August 1849 with evenings reading Plutarch together at Queenwood. See Tyndall's Journal entry for 13 January 1851 (RI MS JT/2/13b/520).
3. *Marieus Coriolanus*: Gaius Marcius Coriolanus, a probably fictitious Roman general of the 5th century BC. Early historians, such as Plutarch, treated him as a real person. Shakespeare's play, *Coriolanus,* was based upon Plutarch.
4. *Alcibiades*: Alciabiades (c. 450–404 BC), an Athenian orator and statesman.

5. *Mr Lee . . . on religious mechanics*: The Revd Stanlake Lee (1815–96), Rector of Broughton. 'religious mechanics' referred to Bible discussion classes rather than science lectures.
6. *I found it impossible . . . as his own nature*: Tyndall (as well as Hirst and Frankland) harboured a deep dislike of Edmondson. In the absence of any alternative contemporary discussion of his character, it is hard to judge whether theirs was a balanced view.
7. *your suggestion regarding Haughton*: While he was in Ireland Hirst was to consult Haughton regarding a replacement teacher at Queenwood. See letter 1233.
8. *Aldershot*: As a result of the Crimean War, a military garrison was established at Aldershot in Hampshire in 1854.

To Thomas Archer Hirst 17 June 1856 1242

[Vine Cottage] Broughton, 17th. June, 1856.

My dear Tom

The concert[1] is already numbered among the things that were. Heaven smiled upon Queenwood and gave them a glorious day. After the concert a terribly long winded gentleman—Spooner's father[2]—addressed the boys, and was eventually coughed down by Fox and others. Mr. Edmondson in his wisdom had given this gentleman permission (on his own application some days previous to the concert) to address to the boys some words which were to be for their good. He came charged with the usual platitudes and wearied every body. Any body but Edmondson would have inferred that the man who would make such an application was just the man likely to accomplish the result which he accomplished. However the thing passed away, and the tea arrangements on the lawn were excellent. A tent of home manufacture embraced by its white walls those who wished for an extra shadiness; but the day did all that was required. Light clouds drew themselves over the sun and thus tempered his glare. Many talked of you. I did not know indeed that you were so much beloved and respected by the Broughton people. Mrs. Lee[3] was fervent in her expressions of regret. So was Lee[4] as far as his wooden nature would permit him. Various suggestions were made. Mr Best,[5] who was here with his fine daughter,[6] feared the changing winds of Italy, and thought the Isle of Wight would be more suitable for Anna. Lee proposed an experimental trip with one of Anna's own relatives, leaving you to work at home. My answer to all was that you would do as your feelings prompted and in time you would do right. The gymnastics passed off well and afterwards Debus entertained us for an hour with his experiments. They all proceeded, though I was in terror half the time lest some of his fire should sputter over the light dresses of the ladies who were two feet in front of him. He also forgets that ladies are no lovers of smoke and smells; but on the whole his proceeding was very successful.

I brought down the apparatus which I had ordered for you it is simple and would have given Edmondson an opportunity of shewing his scientific knowledge, but he let the opportunity slip. I told him on Sunday that I had brought the things—'Well,' said he, 'Richard Wright will pay thee for them'—and this in a tone as if my sole object in telling him of them was to obtain payment. I said I took it for granted that they would be paid for, that I had not paid for them, but that the maker of them would doubtless send in his bill in due time. So they were not used but are still in my bedroom upstairs. After Debus's lecture as it was approaching 10 o clock I turned my steps towards Broughton thinking an early bed would be the best thing for me, but on reaching the bottom of the hill I said when shall I witness another day of this kind. Tom is gone and thus the chief tie which linked me to the place is cut in two. I will turn and see the day's proceedings out. I came back, and sat alone in the tent upon the lawn, watching the moon sailing through a mottled heaven and the people at supper moving to and fro across the window blinds. Some camping boys came near me and declaring there was a ghost in the tent vanished again. Here I managed to get some of the chlorine of the lecture out of my lungs which was really a source of unpleasant irritation. I watched the proceedings, mixing them with other thoughts, for an hour. With its light and its warmth, and its friendliness I still felt Queenwood becoming strange to me, and with a calmness which had something of sadness in it, yet without a trace of any thing approaching to unhappiness I resigned myself to the thought that a day was coming when I should have to strike the roots of my affections elsewhere. The appearance of the boys was not such as to drive this feeling away. I have looked at them once or twice in line, and seen them pass me one by one. To my eye they are on the whole far below par; many of them with rude and rather animal countenances; but among the number Charley Cumming[7] shone like a blessed little star which still kept a man's hopes aloft. There are some other nice looking boys, but they appear to be in sad minority.

I know not how matters will tend. Wright is fearful—He sees most clearly the shock Queenwood has sustained by your departure. When I think of God's goodness to me he once said, and see in Tom a far nobler and braver man so smitten and afflicted a kind of shame fills my heart. Had he the means he declares that he would go along with you to Italy, and Fanny[8] should nurse Anna with the loving tenderness of a sister. He sees the danger of a collision between himself and Haas. Debus thinks that Queenwood is about to become a Swiss colony.[9] I have had no communication with Haas. He is a hard worker, and if his other qualities were commensurate with his zeal, Queenwood would be safe in his hands. But they are not, he is narrow, ungentlemanly and egotistical, and the more the pity.

To my terror I find it is near 10 o clock. I thought I could not do better than spend this morning in writing to you—when the spirit moves me I will do so again—an engagement calls me now to Queenwood. Affectionate remembrance to Anna.

As ever Your | John

RI MS JT/1/H/459–60

1. *The concert*: The Queenwood summer concert had taken place on Monday 16 June 1856.
2. *Spooner's father*: the father of a boy at the school, otherwise unidentified.
3. *Mrs Lee*: Elizabeth Lee, the Rector of Broughton's wife.
4. *So was Lee*: The Rector of Broughton. See letter 1242, n. 5.
5. *Mr Best*: The Revd Samuel Best (1802–73), Rector of Abbotts Ann, near Andover, Hampshire. Educated at Eton and Cambridge, Best started a non-denominational primary school at Abbotts Ann in 1831 as well as a Friendly Society for the villagers. Despite his liberalism he was conservative in biblical interpretation. See his *After Thoughts on Reading Dr Buckland's Bridgewater Treatise* (London: J. Hatchard, 1837).
6. *fine daughter*: Best's second wife was Emma Duke (b. 1811). They had a son and two daughters. Tyndall was probably referring to the eldest daughter, Grace Best (b. 1841), aged 15.
7. *Charley Cumming*: Charles Cumming was the ward and nephew of Sir James Coxe.
8. *Fanny*: Fanny Wright, wife of the Queenwood mathematics teacher Richard Wright.
9. *A Swiss colony*: Haas was Swiss. In the absence of Tyndall and Hirst, and the timid direction of Edmondson's headmastership, Debus evidently feared that Haas, a Pestolozzian-trained teacher, would come to dominate the school's policy.

From Thomas Archer Hirst 18 June 1856 1243

Newry | June 18th 1856

My dear John

I was glad to get your long letter[1] dated from my study and to find from it that you are enjoying yourself at our house. I hope you will stop there as long as you can enjoy it for it gives us both pleasure to think that its quiet beauty is not unappreciated in our absence but is in its own way contributing to your enjoyment peace of mind and bodily health. I am sorry to say the change of air has not yet had so favorable an effect upon Anna as I hoped. Although she is certainly a little stronger yet the cough and expectoration are just as usual. In a few days now we shall leave for Dublin[2] and if the weather will but change and become warm and sunny I hope she will improve there. If you will send me a letter to Haughton[3] I will call upon him as soon as I arrive in Dublin and see what can be done for Queenwood. Of course looking at it from the Queenwood point of view (not forgetting Debus and Wright) I regret the conclusion to which you have come, but I perfectly agree with your reasons for coming to such a conclusion. The only doubt I have is that in considering the matter over you may have been devoting more time to Queenwood than would be absolutely necessary. It would be a great benefit to Queenwood to have your council and presence now and then even if you took no active part in its management. Could you not pursue your own work in the neighborhood and do as you did last year?[4] Recollect what an immense advantage to Queenwood was your presence there last year. After what you have said I could not advise you to take any prominent part in tuition or connect your name with Queenwood publicly but if you could visit there for a month or so it would be an admirable thing. Before this time Wright will have talked to you on the subject and will have probably urged what I am now enquiring. As to the moral charge of the lads at Queenwood I know that Wright having once undertaken it will never be able to continue at Queenwood and neglect it. With Haas and Edmondson it would be otherwise. Now you could stir them up and your earnestness on the subject would stimulate them for many months. This is all I would ask you to consider again. Besides if I am fortunate enough to meet with an earnest young fellow to take my place he would require a certain initiation into the duties I have above alluded to and no one could impress upon him the importance of those duties as well as you could, no one could influence him as much or encourage him as well to accept such responsibilities.

Debus is well capable of lending a helpful hand in the government of the place if you could but give him a little more practical tact. If you could be present when he commences to do so, watch him and tell him candidly (no body else dare) what things he should do and avoid doing it would be well. I too think that there are forces still in Queenwood which will carry it on for some time to come especially if they are now and then controlled. Your proposal about occupying my leisure time has occurred to me too and should I find that I could say what would be of use I would try the experiment. But I do not think that the world would well receive what I should have to say about our schools. What leisure time I have after securing bread & lodging I shall devote to my mathematics. Let me know your intended movements for the next month or two, are you not going to take a trip somewhere? The weather here is very changeable and very unfavourable for Anna. Tell me how the Queenwood Concert &c went off and what you did.

Yours affection[ate]ly T.A. Hirst

Do you know John I often think that I shall be better able to say something on the subject of Education after a little while than just at present. After working so long on the darkest side of education I am rather afraid that I should take too gloomy and despondent a view of the subject. At present my theory of education consists almost entirely in the prevention and care of one fatal habit.[5] Do you think the time has come to tell the world this and to assert what is taking place in every school in the kingdom? Known it must be at sometime or other but whenever the exposure is made one ought to be able to follow it up and to give all necessary evidence. In fact a regular war against the thing must be raised and to be successful ones forces should be properly organized. Queenwood would probably have to suffer. Such questions as these require mature consideration and make me doubt sometimes whether the time has arrived to act.

RI MS JT/1/H/226

1. *your long letter*: letter 1242.
2. *we shall leave for Dublin*: The Hirsts left Newry for Dublin on 20 June 1856 (RI MS JT/2/32c/1200).
3. *send me a letter to Haughton*: This formal introduction to Haughton is missing.
4. *do as you did last year*: Tyndall had stayed at Queenwood twice during 1855: 9–14 June prior to travelling to Paris with Debus via Jersey and Normandy; and 18 August to 5 September when he worked in the school's Lodge House and slept in the main school (RI MS JT/2/13c/752–53, 783–86).
5. *one fatal habit*: masturbation. See letters 1220, 1221, and 1226.

From Gustav Magnus 20 June 1856 1244

Berlin 20 Juni 1856

Theurer Freund

Es hat mich sehr gefreut aus Ihrem lieben Brief vom 10^{n}. d. zu ersehen daß meine Mittheilungen über die Explosion der Dampfkessel Ihnen genügt haben. Dieselben sind so hingeschrieben wie sie mir in die Feder gekommen sind, ich bin deßhalb nicht sicher, ob sie sich zu einer Veröffentlichung eignen, doch will ich dies Ihrer Beurtheilung anheim geben, nur muß ich gestehn daß ich eine Erklärung, die nur aus theoretischen Ansichten abgeleitet ist, die nicht auf eigenen neuen Versuchen beruht (und daher auch falsch sein kann) nicht so abgedruckt wissen möchte, daß man glauben könnte, ich hätte sie für die Veröffentlichung bestimmt. Wenn Sie hingegen so gut sein wollen einige Worte vorauszuschicken, durch welche der Leser erfährt, daß dies von mir als meine Vermuthung über die Ursache der Explosionen ihnen auf Ihre Anfrage mitgetheilt worden sei, so habe ich nichts gegen die Veröffentlichung. Die bloße Ueberschrift „From a letter of Prf. M." würde mir nicht genügen, weil man dann glauben könnte ich hätte diesen Brief zur Veröffentlichung bestimmt. —Sie müssen mir diese Weitläufigkeit nicht übel nehmen, ich glaube aber man muß sich hüten etwas bekannt zu machen, wovon andre sagen können: das sind ja ganz bekannte Sachen! Und ich bin überzeugt daß es eine Anzahl von Personen, namentlich in Paris geben würde, die, sobald sie die Erklärung lesen sagen „Dazu brauchen wir nicht die Weisheit von Hr. M."—Daß dies vermieden werde darauf kommt es mir an. Ich denke Sie werden dies nicht mißbilligen, und wenn das der Fall ist, so überlasse ich es Ihnen gern, ob und wie Sie die Veröffentlichung einrichten wollen. Ihr richtiger Tackt wird leicht eine geeignete Form finden.

Wenn Sie glauben daß die Physik in England in einer andern Lage sei als bei uns, so thun Sie Ihrem Vaterlande Unrecht. Auch bei uns, und selbst in Frankreich sind die Physiker entweder Mathematiker oder Chemiker, oder richtiger zu sagen: nur wo chemische Kenntnisse mit mathematischem Sinn verbunden sind, kommt ein guter Physiker zu Stande, z.B. Regnault. Aber das ist selten, und daher mag es auch wohl kommen, daß es so schwierig ist junge Leute zu finden, die sich zu physicalischen Untersuchungen eignen. Sie wissen wie gern ich dergleichen bei mir aufnehme, aber es ist gegenwärtig niemand vorhanden der Zeit, Lust und Geschicklichkeit hätte. Ihr Plan ein physicalisches Laboratorium für die Militairschule zu errichten, ist gewiß vortrefflich, und ich zweifle auch nicht daß wenn ein solches existirt, sich auch junge Leute finden werden die es benutzen, aber ob das viele sein werden, ist ein Frage, die ich nicht beantworten kann da ich die Vorkenntnisse der jungen Militairs nicht zu beurtheilen vermag, mir will indeß scheinen, daß die Errichtung eines physicalischen Laboratoriums nur dann von rechtem Nutzen ist, wenn zugleich ein chemisches vorhanden ist, indem die practischen Uebungen beginnen. Doch alles kommt auf die jungen Leute, namentlich auf ihre Vorbildung an, und die mag in England besser sein als bei uns. Sie haben ohne Zweifel die Einrichtungen in der Ecole polytechnique in Paris gesehn, die bestimmt sind die Zöglinge in physicalischen Experimenten zu üben. Es wäre sehr interessant zu wissen wie sich diese Einrichtungen bewähren.

Sie schreiben mir in Ihrem letzten Briefe nichts über Ihre Reise. Ich rechne um so bestimmter darauf Sie in diesem Jahre auf dem Continent zu sehn, als ich im vergangenen in meinen Hoffnungen mit Ihnen in Paris zusammen zu treffen getäuscht worden bin. Was ich unternehmen werde, weiß ich noch nicht. Aber lassen Sie mich nur wissen wie Sie Ihre Reise einrichten, ich werde dann schon sorgen daß wir uns treffen. Sobald ich einen Plan gemacht theile ich Ihnen denselben mit, vielleicht schliessen Sie sich uns an, und wir reisen zusammen. Doch ich hoffe wir wechseln noch, bevor es zum Reisen kommt, einige Briefe. Viele Grüße für alle Freunde, wenn Bence Jones und Frankland mit uns reisten wäre es vortrefflich.

An Faraday meinen Respect | von Herzen | Ihr GMagnus.

Berlin 20 June 1856

Dear friend

I was very pleased to see from your kind letter[1] of the 10th of this month that my notes on the explosion of steam boilers were sufficient for you.[2] They are written down off the top of my head[3] and I am therefore not sure whether they are suitable for a publication,[4] but I shall leave this to your judgment. I have to admit that I would not like to know that an explanation deduced only from theoretical views and that is not based on new experiments of my own (and can therefore also be wrong) has been printed in such a way that a reader could believe I had intended it for publication. If you, on the other hand, will be so good as to preface a few words to inform the reader that this has been communicated by me to you in reply to your enquiry as my assumption about the cause of the explosions, then I shall have no objection to publication. The mere title, 'From a letter of Prf. M.' would not be sufficient for me, because someone could then think that I had intended this letter to be published. —You mustn't hold this long-windedness against me, but I think one must be on one's guard about publishing something of which others can say: these are all well-known things! And I am convinced that there would be a number of persons, especially in Paris, who will say as soon as they read the explanation: 'That is why we do not need the wisdom of Herrr M.'—That this may be avoided is what matters to me. I think that you will not disapprove of this, and if that is the case, then I will gladly leave it up to you whether and how you want to arrange publication. Your correct sense of tact will easily find a suitable form.

If you think that physics in England is in a different position than here, then you are doing your fatherland an injustice. Here as well, and even in France, physicists are either mathematicians or chemists, or to put it more correctly: only where chemical knowledge is combined with a mathematical mind will a good physicist come about, e.g., Regnault. But this is rare and therefore it may also probably happen that it is so difficult to find young people who are suited for physical investigations. You know how gladly I take in those kind of people, but there is currently nobody available who would have the time, inclination and skill. Your plan[5] to set up a physical laboratory for the military school is certainly excellent and I do not doubt that once one exists there will be young people who will use it, but whether there will be many of them is a question that I cannot answer, as I cannot judge the young officers' previous knowledge; however, it seems to me that the setting up of a physical laboratory will only then be of proper benefit if, at the same time, a chemical one is available in which the practical exercises are begun. But, everything depends on the young people, particularly on their previous education, and this may be better in England than here. You have doubtless seen the facilities at the École polytechnique[6] in Paris, which are intended to train students in physical experimentation. It would be very interesting to know how these facilities prove their worth.

You did not write anything in your last letter about your journey.[7] I am counting all the more resolutely on seeing you on the continent this year, my hopes of meeting you in Paris were disappointed last year.[8] What I shall do, I still do not know. But just let me know how you are arranging your journey, and I shall then make sure that we meet. As soon as I have made a plan, I shall inform you about it, perhaps you will join us and we shall travel together. But I still hope that we shall exchange some letters before it comes to travelling. Many greetings to all my friends; if Bence Jones and Frankland were to travel with us, it would be excellent.

My respect to Faraday | sincerely | Your GMagnus.

RI MS JT/1/M/19

---

1. *your kind letter*: Tyndall's letter of 10 June 1856 is missing.
2. *my notes on . . . steam boilers were sufficient for you*: see letter 1230.
3. *the top of my head*: literally, 'the way they came off the knib.'
4. *suitable for a publication*: In view of Magnus's hesitation, not surprisingly Tyndall did not publish Magnus's reflections on steam boiler explosions in *Phil. Mag.*
5. *Your plan*: Unfortunately, nothing is known of Tyndall's proposal to set up a physics laboratory at a military school—presumably at Woolwich, where Faraday taught chemistry.
6. *the facilities at the École polytechnique*: There is no evidence that Tyndall had visited the École Polytechnique during his first visit to Paris with Frankland in July 1848, or during the Exposition Universelle in July 1855.
7. *your journey*: Magnus was assuming that Tyndall would be attending a meeting of the Gesellschaft Deutsche Naturforscher und Ärtze in Vienna in September 1856 and was wondering what his plans were for crossing the continent.
8. *were disappointed last year*: Magnus had failed to meet Tyndall in Paris when Tyndall was a juror for the Expsoition Universelle in July 1855.

To Thomas Archer Hirst 21 June 1856 1245

[Vine Cottage, Broughton] | 21s[t]. June, 1856.

My dear Tom

My sojourn in Broughton draws near a close. This is Saturday and tomorrow morning I start for London. I have got my questions nicely together,[1] so that I have nothing to do but to hand them in at our meeting on Monday. My health except a touch of indigestion which makes me feel that I have a stomach is good, and to overcome the enemy alluded to I have only to use ordinary care in my diet. I have lived almost exclusively on your substance throughout the week. Have swallowed your gooseberries, fed on your rhubarb, tasted your honey, smacked my lips on your currant jam, eat your bread and drank your beer. I have besides have had sundry chops and cutlets, a pull or two at your brandy bottle, and today when I attempt to form an estimate of the amount of my consumption I find difficulties in the way which neither Elizabeth[2] nor myself can remove. Today I have been hammering in the chalk pits of the neighborhood and have learned a great deal about flint, concerning which the most vague, and I believe erroneous notions are entertained by geologists. First I spent half an hour in the pit beside the Houghton road,[3] thence I directed my steps to that large cliff near Bossington[4] where the defunct lime kiln stands, and then I picked and smashed for a couple of hours. I shall have something to say upon the matter at a future day—My lecture[5] at the Royal Institution on Slate Cleavage is printed by this time. I will send you a copy soon. It was heard with great attention and I believe produced general—not universal—conviction. I have no doubt upon my own mind as to the subject, so that I expect those who still differ from me to come round to my way of thinking by and by. I have had two or three rides on Fanny.[6] Once she and I had to contend with an angry thunder cloud which seemed bent on reaching a summit before us—On reaching the said summit Fan would halt, but I expostulated with the uninstructed beast, and told her plainly that she did not understand the subject[,] for if she did she would not make [,] as her deportment was calculated to do[,] herself and me a lightning conductor. Fox was with me for a couple of hours last night[.] the day weather was cold and we had to lay a fire in your study. Fox smoked and we talked of you. He remarked that there was a good deal of pallor about your cheeks, which I must say I have noticed myself for some time—and attributed it, without suggestion from me I protest, to your smoking too much. I do believe that damned tobacco has an evil influence on you, and I would ask you to observe its influence, and if you conscientiously believe it bad or doubt its goodness to moderate it. Take two pipes a day. That will be sufficient and wont harm, and by their superior sweetness will compensate for the reduction of your present superfluity. This is Fox's advice, and I merely commend it to you as a thing you ought to attend to. What are you doing with yourself? I hope you have plenty of wholesome work at your elbow, and do not permit that brain of yours to break and ruin itself in bootless brooding. There is a sorrow which with all its pain has some thing purifying and ennobling in it. God forbid that I should wish to quench this high feeling in your soul, but you must not let it degenerate into dark repining. What do you purpose doing with your house. I have been told that you intend keep it in hand till September. Why not try to let it now—it surely is the best season, and to keep it in your own hands besides keeping a paying tenant out will add considerably to your expense. I think you ought to let it without delay if you have made up your mind to go to Italy. Your crops are beautiful, your potatoes, beans, peas [,] cabbage—all promise abundance. Your grass is fit for mowing. I hope you will turn your thoughts to these things among others, and give distinct orders how every thing is to be done. Nobody here appears to know distinctly what your wishes or intentions are. Debus & Tribe[7] are off to the Tyrol. They have a glorious trip before them and constitute pleasant companions. Lee has offered Tribe this curacy but I believe he has not accepted it, fearing that he could not attend to it and to Queenwood. The school is quite a wilderness at present, with one single rose of Sharon[8] viz. Miss Edmondson,[9] blooming in its wastes. The boys are all gone & the playground is as silent as the ruins of Nineveh.[10] What are your thoughts with regard to Queenwood: do you purpose striking into a new line when you return from Italy. I should like to have at least a shadow of your feelings on this subject, so that I may know how to deport myself. Edmondson, mentioned that matter to me a few days ago proposing as terms 50 pounds until December.[11] He does not see the difficulty of his position, and professes to Richard Wright that although things apparently make against him, Providence has something good in store for him behind it all. My fear is that the said Providence is now very likely to demand payment of a debt which has been long due to her by the aforesaid George Edmondson. George William[12] is at home; a good account I hear has been sent with him from his master—he is miserably pale—his cheeks and eyes sunken, and the whole expression of his countenance unusual. Fox I believe is plying him with beer to improve his health. I'm told he has a serious cough also. Mrs Edmondson I am informed has also forsaken hydropathy,[13] her medical preceptor Dr Martin[14] having proved apostate to the cold water cure.[15] Of Haas I have seen very little since I came. We have had lately a specimen of the collective wisdom of parliament as regards the system of National Education in Ireland.[16] Were it not that I like to reserve myself for other matters I would certainly rush into print. And still what need—those fellows are hastening the doom of the church of England in Ireland[17] more effectually than the most direct attacks of their enemies could do. God will shower judgment upon them by and by and all I have to say is 'may it come quickly'.

I have been thinking for sometime of making another onset on Lord Panmure as regards the question of physics.[18] The case is so sound that I think I cannot fail: But I will wait and mature my thoughts upon the matter. Write to me Tom & tell me what you are doing—tell me how is Anna. Her note to Mrs Wright was bright & cheery I am told. Remember Tom no brooding—action of some kind becomes you now. Goodbye & God bless you.

Ever Your | John

RI MS JT/1/T/628

---

1. *my questions nicely together*: examination questions for the Royal Artillery and Royal Engineers to be sat in July 1856. An examiners' meeting to discuss these questions was held on 23 June (RI MS JT/2/13c/850).
2. *Elizabeth*: the Hirsts' servant. Surname not known.
3. *the Houghton Road*: Road linking Broughton to the village of Houghton on the River Test two miles to the east.
4. *Bossington*: A village at the junction of the rivers Anton and Test, one mile south-east of Houghton.
5. *My lecture*: The RI discourse of 6 June 1856, 'Comparative View of the Cleavage of Crystals and Slate Rocks', *Roy. Inst. Proc.*, 2 (1854–58), pp. 295–308; *Phil. Mag.*, 12 (1856), pp. 35–48.
6. *Fanny*: the Hirsts' horse.
7. *Tribe*: Walter Harry Tribe (1832–1909) had been appointed to teach classics at Queenwood in July 1855. Subsequently, he took holy order in 1857 and was Rector of Stockbridge (1860–67). He served on the north-west frontier of India (1867–85) and was Archdeacon of Lahore (1885–92). He ended his days as Vicar of Awliscombe in Devon.
8. *rose of Sharon*: 'I am the rose of Sharon, the lily of the valleys', Song of Solomon, 2:1.
9. *Miss Edmondson*: Jane Edmondson (1823–1906), the Edmondson's only daughter. She had been born in St Petersburg, and was to marry a Lancashire business man, David Benson (1820–92) in 1859.
10. *the ruins of Nineveh*: Nineveh, the former Assyrian city on the River Tigris whose archaeology had been investigated by Austen Henry Layard between 1847 and 1849. Tyndall had heard Colonel Henry Rawlinson's account of the excavations in an RI lecture on 15 June 1855, *Roy. Inst.Proc.*, 2 (1854–58), pp. 143–44.
11. *50 pounds until December*: Edmondson was offering Tyndall a fee of £50 to teach at Queenwood in place of Hirst while a new master was sought.
12. *George William*: George William Edmondson (1837–1913), the Edmondson's only son. He became a professional photographer in Cleveland, Ohio. See letter 1086.
13. *forsaken hydropathy*: a popular Victorian water treatment.
14. *Dr Martin*: not identified.
15. *cold water cure*: a drastic form of hydrotherapy that involved wrapping a patient in ice–cold wet sheets.
16. *National System of Education in Ireland*: A system of non-denominational elementary

education set up by the British government in 1831 and administered by a National Board of Education (2 Catholics, 2 Protestants and 2 Presbyterians). The government paid for building schools and teachers' salaries. All the lessons, apart from religion and morals, were secular, but religious lessons could be denominational. In practice, as the century wore on, the schools assumed a denominational character at local level. The government provided cheap textbooks, including primers on scientific topics; these were also widely adopted in English schools and this led to protests from English publishers of school textbooks.

17. *the doom of the church of England in Ireland*: On 17 June 1856, the House of Commons discussed the national education system in Ireland. A proposal by the former Home Secretary, Spencer Horatio Walpole (1806–98) to extend grants to 'unvested schools', provided they removed religious instruction from the curriculum, was defeated. He argued it was unfair to exclude some schools (the 'unvested') from government grant when everyone, whatever their religion, paid their taxes. Although education ought to be based on religion (in a moral sense) it was not right that the Protestant Church of England and Ireland schools were ineligible for grant aid because they insisted on teaching their doctrines. Despite some opposition, the proposal was passed by a majority of ten MPs. Tyndall was angry because he disliked the idea of protestant schools being forced into the same mould as Catholic schools with an entirely secular curriculum.
18. *the question of physics*: This remark may have related to the unfair weighting of marks for scientific questions in the Royal Artillery examinations; or to the plan for practical physics teaching at Woolwich mentioned in letter 1244, n. 5.

To Miss Julia Moore 24 June 1856 1246

24th June 1856

My dear Miss Moore

If I am in town I will gladly come on Saturday.[1] I should be more definite were it not that my head admonishes me to day to be off for a week's rest. I may take the hint; but if I despise it and remain you are sure to see me early at Wimbledon.[2] The whole distance is only a pleasant walk.

Yours most sincerely | John Tyndall

RI MS JT/1/T/1071

1. *Saturday*: 28 June 1856.
2. *Wimbledon*: The Moore family lived at Wimbledon, then a village some 6 miles south west from central London.

To Thomas Archer Hirst 25 June 1855 1247

June 25 1856

My dear Tom

I have just time to send you the enclosed,[1] and to say little of any thing else. When I have an opportunity of laying my thoughts before you I am sure you will by no means counsel my going to Queenwood. It would be a terrible invasion of my time & thoughts. If they[2] thought it worth while and would pay my fare I should not mind attending their Saturday Conference[3] say once a fortnight, but more than this I could not promise today. I am going to run away from Town today & to have a week's real idleness[4]—I am very weak and need this rest.

Dubois[5] has sent you a beautiful lithograph of himself, it came to Bence Jones in company with one to Mr Faraday and another to me.

ever yours aff[ectionate]l[y] | John | 25th June 1856

RI MS JT/1/T/629

1. *the enclosed*: enclosure missing; probably the lithograph of Du Bois-Reymond mentioned.
2. *they*: the Edmondsons and teachers at Queenwood.
3. *Saturday Conference*: presumably a weekly meeting of Queenwood staff to discuss pupils' progress and other matters.
4. *a week's real idleness*: Tyndall joined the Wrights for a walking holiday in Hampshire and the Isle of Wight that lasted from 25 June to 2 July 1856. He described much of it in verse in his Journal (RI MS JT/2/13c/850–61).
5. *Dubois*: Emil Du Bois-Reymond. The lithograph has not been found in the RI collections.

To Miss Julia Moore 26 June 1856 1248

26th June 1856

Dear Miss Moore

I have felt myself compelled to move from London to give myself a week's rest, so that I shall not be able to grant myself the great pleasure of meeting you and your guests on Saturday.[1] The deep quiet of this place[2] contrasted with the roar of London is extremely refreshing. I hope to return in a week a new man and then to attack my work with tough muscles and a clear brain.

Believe me | Ever yours | John Tyndall.

RI MS JT/3/866c
LT Transcript Only

1. *Saturday*: 28 June 1856. See letter 1246.
2. *this place*: The letter is without an address, but Tyndall's Journal reveals that he was staying at Mudeford in Dorset on 26 June (RI MS JT/2/13c/850).

To Emil Du Bois-Reymond 27 June 1856 1249

[Mudeford, Dorset] | 27th June 1856.

My dear DuBois

I am greatly obliged to you for the portrait[1] you have sent me, and Mr Faraday also thanks you for the one you sent to him. It is a most excellent likeness, a fine, healthy, handsome countenance of which your wife[2] has every reason to be proud. It brings you very vividly to my remembrance, and I mean what I say, when I state that there are but few men whose images it will give me a greater pleasure to look upon.

You also sent Hirst a portrait.[3] The poor fellow is now in Ireland sadly afflicted. You know he got married. Half a year ago his wife caught cold; it clung to her through the winter, but still he thought it was not serious, and that the fine weather would cause it at once to disappear. He however was obliged to have a London physician's advice some time ago, and to poor Tom's dismay it was discovered that his wife was far gone in a consumption. He is now with her in Ireland. The change of scene and air appears to have done her good, but the cough and expectoration continue unchanged. He has given up his post in Hampshire and purposes going with his wife to Italy during the winter. Poor fellow he is sadly smitten down.

My session has just ended and I am now giving myself a week's rest. Faraday is very well—he purposes going over to Paris this year to thank them for the honour[4] they have done him. Bence Jones is in capital health. I never saw him look better.

Believe me | ever yours sincerely | John Tyndall

RI MS JT/1/T/400

1. *the portrait*: see letter 1247.
2. *your wife*: Du Bois-Reymond had married Jeannette Claude (1822–1911), an Englishwoman, at her family home in the English Lake District in 1855.
3. *Hirst a portrait*: see letter 1247.
4. *thank them for the honour*: Faraday had been made a Foreign Member of the Académie des Sciences and attended a meeting in Paris on Monday 4 August 1856. *Faraday Correspondence*, 5:3172.

To Juliet Pollock 28 June 1856 1250

Mudeford near Christ's Church. | 28th. June 1856.

My dear Mrs Pollock,

Miss Herries's note, but not her book,[1] has reached me this morning. I shall be in London on Wednesday next and my first act after I return will be to attend to her request. I am trying to be idle here for a week, and thus to permit my brain and muscles to regain their normal toughness. At my feet is the sea, splashing and booming and causing the pebbles to rattle up and down along the beach. Right opposite are the white cliffs of Alum Bay,[2] it seems as if I could swim to them with ease, and yet I am told they are eight miles off. What glorious weather this is, the central day is hot, but the morning and evening are delicious; the deep quiet, and the perfume of the honeysuckle and beanflowers are a slight improvement upon the growls and gully holes of London. I wonder does science really injure a man's imagination?* Sometimes when looking at the sprouting flowers I endeavour to follow the <u>mechanical</u> action of the sunbeams upon the atoms of matter whereby they are caused to arrange themselves into these beautiful structures, and to dye their petals in those splendid colours. You remember Mr Faraday's experiment with the iron filings[3] which you liked so much. These iron filings when acted on by the magnet, moved towards each other and clung together in a peculiar manner. The sunbeams appear to act in a somewhat similar manner: they shake up the particles of matter and cause them to arrange themselves according to certain laws, the result being the formation of grass, gooseberries and flowers. We eat the gooseberries; but the grass is too much for us and so we cunningly submit it first to the digestive apparatus of the ox and sheep; but the sun is still the original grand engineer to whom we owe the architecture of our bodies. Behind these of course are the questions which beset us now, but which are perhaps to be answered by a race of beings bearing the same relation to us that we do to the tadpole and iguanadon. The vision which is rudimentary in us may in them be developed into perfect day; and thus

> 'The something in this world amiss
> May be unravelled[4] by and by.'[5]

I will not afflict you longer—goodbye
Yours sincerely | John Tyndall
*I wish I had Walter[6] here to instruct in the noble art of swimming.

RI MS JT/1/T/1128

1. *Miss Herries's note but not her book*: Maria Julia Herries (1820–57), the daughter of the politician John C. Herries, was an aunt of Frederick Herries Pollock, who had a brother, Edward Herries (1821–1911), in the consular service in Berne, where Tyndall met him in 1855 (RI MS JT/2/13c/869). Tyndall did not meet her personally until 9 May 1857 when he noted in his Journal: 'Dined at [Peter] Barlow's, and sat beside Miss Herries at dinner, a lady of remarkable intelligence much resembling Mrs Pollock' (RI MS JT/2/13c/ 933). Her note is missing, and the book not identified.
2. *Alum Bay*: a picturesque beach at the western tip of the Isle of Wight within sight of the Needles rocks. It is renowned for its coloured sands.
3. *You remember . . . the iron filings*: Faraday's famous demonstration of a magnetic field with iron filings was first shown at the RI on 23 January 1852 in 'On the Lines of Magnetic Force', *Roy. Inst. Proc.* 1 (1851–54), pp. 105–8; *ERE,* vol. 3, pp. 402–6. Plates showing the distribution of iron filings in a magnetic field were printed in 'On the Physical Character of the Lines of Magnetic Force', *Phil. Mag.,* 3 (1852), pp. 401–28.
4. *unravelled*: The correct word was *unriddled.*
5. *The something in this world amiss*: Tennyson, *The Miller's Daughter* was first published in A. Tennyson, *Poems* (London: Moxon, 1833), pp. 33–47. The quoted stanza continues: 'There's somewhat flows to us in life, But more is taken away.'
6. *Walter*: Walter Herries Pollock (1850–1926), the Pollocks' youngest son. He later attended Eton and Trinity College, Cambridge, where he studied classics before embarking on a literary career. Editor of the *Saturday Review* (1884–94).

To Joseph Allen Galbraith [June-October 1856][1] 1251

Dear Mr Galbraith

This letter will be handed to you by my friend Col. Wynne of the Royal Engineers. He has a son preparing for our next examination and as I have reason to think that under the auspices of yourself and Mr Haughton Trinity College is well [regarded? organised?] with reference to the wants of your candidate[s]. I have recommended it to Col. Wynne as the place most likely to forward the objects which his son has in view. I should take it as a favour impressed upon myself if you would kindly afford Col Wynne your advice & assistance in this matter. His son has already made considerable progress [*word illeg*] his four subjects of study which are Mathematics—English—French Expressions [*words illeg*][2]

If he has the opportunity of [*word illeg*] this you will soon learn that young Mr Wynne possesses a most estimable character.

RI MS JT/3/1/-

1. Draft of an undated letter scribbled into Tyndall's pocket book for 1856 (RI MS JT/3/1).

The notebook otherwise contained pencilled scraps on apparatus, trips and expenses. When Hirst left Queenwood College abruptly on 11 June 1856 he was still coaching Eddy Wynne for the Royal Artillery examinations which Wynne wished to take the following month. When the results were published in September 1856 he had failed. Eddy's father, Col. George Wynne, must then have sought Tyndall's advice on how his son could be coached for the entrance examinations in July 1857. Because Haughton and Galbraith had enjoyed success in passing Trinity College students through the Royal Artillery examinations, Tyndall evidently approached Galbraith for advice. This would date the draft letter to between June and no later than October 1856. Wynne passed and entered Woolwich in September 1857 and graduated as a Cadet in the summer of 1858.

2. *French Expressions*: The words here should refer to Eddy's fourth examination paper.

To Julia Moore [5 July 1856][1] 1252

My dear Miss Moore

I am happy to say that my week's idleness[2] has been of immense service to me—I returned with the strength of a giant condensed into my little body. Could you not name either Monday Tuesday Friday or Saturday? The examination of candidates for the Artillery and Engineers takes place next week and I am down for the 9th and 10th[.][3] Any other day during the week it will give me very great pleasure to come—Indeed I was almost tempted to make a descent upon you yesterday—Mrs Pollock had put your garden in my mind and I had a strong desire to test the scent of its flowers.

Believe me | ever yours sincerely | John Tyndall | Saturday night

Wellcome MS 7777/82–82a

1. Undated, but since Tyndall recorded in his Journal that he visited the Moores on Saturday 12 July the internal evidence dates the letter as 5 July (MS RI JT/2/13c/861).
2. *my week's idleness*: his trip to Dorset and the Isle of Wight. See letter 1250.
3. *I am down for the 9th and 10th*: Examinations were held during the week 7–11 July 1856, and Tyndall invigilated 9–10 July.

From Carlo Matteucci 7 July 1856 1253

Pise 7 juillet

Mon cher monsieur,

Je vous remercie bien de votre lettre: je desire que votre sejour en Suisse retablisse bientot votre santé. Je voudrais bien etre avec vous aux glaciers, mais je ne puis pas et ma santé aussi n'est pas bonne.

Vous verrez donc M. de la Rive et toute sa famille et je vous prie de me rappeler a tous.Vous me parlez d'experiences sur des tons obtenus dans la condensation des gaz, quand vous en aurez le temps, je vous prie de m'en ecrire plus largement ainsi que sur les glaciers. Cela me promet d'abord une lettre de vous et puis une nouvelle Scientifique pour le Nuovo Cimento.

Savez-vous si l'eau des glaciers est pure, comme l'eau de la neige ou usé ou bien si elle contient des sels?

Adieu. Croyez-moi [*word illeg*] | C. Matteucci

Pisa 7 July

My dear sir,

I thank you for your letter.[1] I wish your stay in Switzerland will rapidly restore your health—I would love to be with you on the glaciers, but I cannot as my health is not good either.

You will be bound to meet Mr. de la Rive and all of his family, and I ask you to send them my regards. You talk about experiments on the tints obtained in the condensation of gas;[2] when you find some time, please write more extensively about it as well as about the glaciers—I shall expect a letter from you and then a scientific paper for the Nuovo Cimento.[3]

Do you know if glacial water is as pure as water from the snow or spoiled or if it contains salts?

Goodbye. Believe me, yours truly | C. Matteucci

RI MS JT/1/M/66

1. *your letter*: letter missing.
2. *tints obtained . . . condensation of gas*: Since we do not have Tyndall's letter, it is unclear what observation of Tyndall's Matteucci is referring to. Tyndall's work on absorption of heat by gases did not begin until 1858; instead Matteucci may be referring to Tyndall's experiments on pond water at Wandsworth. When he sprinkled olive oil on pond surfaces, ripples caused by the breeze disappeared and coloured diffraction patterns (Newton's rings) spread over the surface. Tyndall attributed the effect to trapped gas bubbles rising to the surface. See Journal for 15 February 1855 (RI MS JT/2/13c/825).
3. *Nuovo Cimento*: see letter 1123, n. 5.

To Charles Babbage 10 July 1856 1254

10th July 1856

Dear Mr Babbage

I have long been under a promise to go to Wimbledon,[1] and Saturday (if it be fine) is the only available day this week. Do you think you can come? If so I should be glad to wait upon you at any hour or place you may appoint, and accompany you by foot, omnibus, cab, or railway as it may suit your pleasure.

Most truly yours | John Tyndall

BL Add MS 37197–70

---

1. *Wimbledon*: The home of the Moore family. See letters 1248 and 1252. In his Journal for Saturday 12 July Tyndall made no mention of Babbage as having been present (RI MS JT/2/13c/861).

To Thomas Archer Hirst 11 July 1856 1255

11th July, 1856.

My dear Tom

I saw this day Messrs Galbraith & Haughton.[1] They feared that they would not be able to [do] any thing, as their best men get posts immediately in India & elsewhere, in which the possibility of advancement is before them. But two they mentioned as possible, these they promised to consult and to let me know the result of their enquiries in a week. They are themselves solid men, and will I think recommend none but a competent man for the post. If therefore Mr Edmondson is willing to wait a week he will have the benefit of their researches. I have written to him letting him know what I now tell you.

I think I told you that I had a week's idleness with Wright at Christchurch. It was of amazing advantage to me as you will glean from the journal of one day which I will send you soon, and which is written in blank doggrel.[2] This London has something vicious in its air—my stock of strength has already ebbed, and sometimes I almost feel tempted to give up science for a time and turn to the doggrel or something else. Though it is a rigmarole, still I felt a pleasure in writing it, and it showed me that with a little discipline I could throw some power into a thing of the kind. I dare say 3 months, spent at some folly of this kind would be very good for my health physical & mental.

But I talk while in a crisis. Goethe says each transition is a crisis[3] & I have ever found it so. I am now floundering in the dark with my experiments; no light of hope to cheer me on, nothing but that dumb stubborn faith that there is always something to be gained by seeking. This however requires a physical substratum of sound health, and hence it is in my case at present not as strong as it might be. But away with complaining—There is still a balm in Gilead,[4] there is still a physician there.

Affectionate remembrance to Anna | Ever Yours | John

RI MS JT/1/T/630

---

1. *Galbraith & Haughton*: It is not known what brought these Irish scientists to London. Tyndall consulted them on a replacement for Hirst's teaching position at Queenwood. The meeting was not recorded in Tyndall's Journal.
2. *blank doggrel*: The long piece of doggerel describing his adventures with the Wrights at Alum Bay on the Isle of Wight is found in Tyndall's Journal for Monday 30 June 1856 (RI MS JT/2/13c/853–60); there is also a fair version in RI MS JT/2/7/355-373.
3. *each transition is a crisis*: 'Every transition is a crisis; and a crisis presupposes sickness'. The quotation is from Goethe's novel *Wilhelm Meisters Lehrjahre* (1795–96) which Tyndall had read in its English translation by Thomas Carlyle, *Wilhelm Meister's Apprenticeship and Travels* (Edinburgh: Oliver & Boyd, 1824). Quotation at p. 260.
4. *a balm in Giliad*: a rare perfume mentioned in the Bible figuratively to indicate a perfect cure (Jer. 8:22).

From Gabriel Stokes 19 July 1856 1256

Pembroke College, Cambridge | July 19, 1856

My dear Sir,

You did not get my letter[1] till some days after it was written, and the Friday referred to was last Friday,[2] so there is no particular hurry about Mr Claudet's paper,[3] though it is pleasant to get such things off one's stomach.

There will be no more meetings of the Govt. Grant Committee[4] until after the recess.[5] I don't think it likely they would wish to recall the grant[6] if you are likely to make use of it, though not immediately. It is just possible there might be some particular pressure on the grant this year which might make them wish to recall it for the present, to be renewed again, but I don't think this likely.

Yours very truly | G.G. Stokes
Dr Tyndall F.R.S.

RI MS JT/1/S/218

---

1. *my letter*: letter missing.
2. *last Friday*: Unclear; Stokes may have meant either Friday 11 or 18 July 1856.

3. *Mr Claudet's paper*: Tyndall had been asked to referee a paper submitted to the RS by the French photographer Antoine Claudet (1797–1867; FRS 1853). Tyndall submitted the 'Report on A. F. J. Claudet on Various Phenomena of Refraction through Semi-lenses or Prisms, Producing Anomalies in the Illusion of Stereoscopic Images' on 30 July 1856. See letter 1264. An abstract of Claudet's paper was published in *Roy. Soc. Proc.,* 8 (1856), pp. 104–10.
4. *the Govt. Grant Committee*: See letter 1091, n. 9. Tyndall had applied successfully for £50 in 1852 and for £100 in 1855. He had applied again in 1856.
5. *the recess*: the RS did not meet during the summer months and resumed meeting on 29 November 1856.
6. *to recall the grant*: Tyndall had applied for a grant of £100 for the purchase of apparatus. Thinking that the request would be refused, Bence Jones had given Tyndall the money so that the magnetic research could continue without interruption (RI MS JT/2/13c/850 and RS CM, 11 July 1856, 2: 363–64).

From Warren De la Rue — 21 July 1856 — 1257

| 7 St. Mary's Road, Canonbury, | July 21, 1856.

MY DEAR TYNDALL,

In an account of your lecture 'On the Cleavage of Crystals and Slate Rocks' (Phil. Mag. vol. xii. p. 46),[1] you have noticed my experience in the consolidation of precipitated white lead, and have correctly stated the unfavourable results of my first attempts with conical moulds.[2] It may be of interest to the readers of the Philosophical Magazine to learn that the difficulty was entirely overcome by substituting cylindrical moulds. By their employment the lateral spread of the mass was prevented, and cakes of white lead[3] were produced which broke with a conchoidal fracture,[4] and which consequently could readily be ground to a granular powder.

I am, | Yours very truly, | WARREN DE LA RUE.

Phil. Mag., 12 (1856), p. 157[5]

1. *Phil. Mag. vol. xii, p. 46*: J. Tyndall, 'Comparative View of the Cleavage of Crystals and Slate Rocks', *Phil. Mag.*, 12 (1856), pp. 35–48.
2. *conical moulds*: In his paper (n. 1), at p. 46, Tyndall had stated: 'As time passed on cases multiplied, illustrating the influence of pressure in producing lamination. Mr. Warren De la Rue informs me that he once wished to obtain white-lead in a fine granular state, and to accomplish this he first compressed the mass: the mould was conical, and permitted the mass to spread a little laterally under the pressure. The lamination

was as perfect as that of slate, and quite defeated him in his effort to obtain a granular powder.'

3. *cakes of white lead*: basic lead carbonate used as a white pigment in oil painting and as the basis of domestic paints.
4. *conchoidal fracture*: The fracture of minerals or rocks such that they have smooth convex or concave surfaces.
5. Published under the title, 'CLEAVAGE OF COMPRESSED WHITE LEAD'.

From John Frederick William Herschel 21 July 1856 1258

Collingwood | July 21, 1856

Dear Sir

If you think anything in my letter[1] worth printing you are quite at liberty to do so. I would merely add a word or two in a note to the effect that in forcing down the strata[2] which still in some degree yielding down the general slope of the ocean bed, they must often be driven over its larger inequalities

in a way very likely to contort and corrugate them and this too, equally whether the softness be that remaining in them previous to their consolidation—or be produced anew by invasion of heat from below producing semifusion and weakening their support.

As the Glacier in its descent will have to be forced over occasional protuberances in its bed its structure (as modified by your causes of lamination) may be expected to bear the traces of this action. The ice of a glacier however does not seem to be a substance very favourable for exhibiting your class of phenomena.

Wishing you much enjoyment in Switzerland.

I remain | Dear Sir | Yours very truly | J. F. W. Herschel.

J. Tyndall Esq

RI MS JT/1/H/93

1. *my letter*: letter missing; but see letter 1261.
2. *in forcing down the strata*: the phrase is preceded by the symbol ' [*' in Herschel's handwritten copy, but not in his copyist's version at the Royal Society (Royal Society Letters 23_175). It is unclear whether Herschel intended to include a footnote, or whether the mark was Tyndall's to remind him of this striking phrase.

To Michael Faraday 22 July 1856 1259

R. I. | Tuesday

My dear Mr. Faraday,

I have often thought of writing to you, and should have done so before now had I not been tossed about in body and soul more than I anticipated.

The examinations[1] took me a long time. It is dreary work plodding out these answers, hunted by the consciousness that you may possibly do a candidate wrong. Then this time the examiners had to accompany their returns with a report, which to draw up properly also took time. I have a suspicion that this is one result of <u>that</u> correspondence.[2] I have put my notations again strongly, but have done it by arranging facts, rather than by expressing an opinion. This I shall continue to do whenever I have an opportunity, as long as I find natural philosophy, taken together rated equal to French alone, Latin alone, and at 250 marks less than the number awarded to English literature. Such an absurdity I think cannot live long in the world, if there be only somebody to hold it up to the attention of men.

I have made a number of experiments connected with the passage of a current across the place of junction of two metals—Peltier generalised[3]—Here there is a conversion of heat into electricity; I fear the knowledge is a long way off still one does long to know something more of the intimate nature of their conversion. As yet I am in perfect darkness.

Monday I was swept suddenly down to Hampshire to see my young friend[4] who, in accordance with the advice of Dr. Stokes of Dublin,[5] goes off to the South of France immediately with his wife. She manifestly has no idea of her danger[6] which makes it all the sadder. She talks about coming back as a matter of course, when it is the opinion of some that she hardly ever will return.

I have had rather a strong letter from Mr. Sorby[7] the geologist complaining of my misapprehension of his views, in my late lecture[8] at the R.I. He has also sent a paper on the subject to Francis.[9] In his eagerness to make out a strong case against me, he has left one little more to do than to quote his own words. This I have done[10] without any show of feeling—indeed before I wrote I got all feeling under, for the longer I live the more I feel that these personal discussions are a perfect nuisance to science. There may be an avarice of the intellect as well as of the pocket, and one, in quality is not very much superior to the other. Snow Harris has sent up a violent response to Riess's last short communication,[11] which also might have been far milder and more courteous. Francis has written to him advising him not to print it—I do not know what the result will be.[12]

One morning soon after you left I received a note from Stokes[13] informing me that the grant committee had acceded to my application for £100. I wrote to him[14] requesting him to return my thanks to the committee for this proof of their confidence; I said however that having been led to conclude that my application had been considered ineligible I had resorted to private sources and obtained from them all the funds I needed. If therefore the grant committee thought the money voted to me could be otherwise usefully employed, I should be glad if they would consider my application as extinct, and the £100 again at their disposal. I concluded by saying that at some future time, if permitted, I should probably be glad to avail myself of the assistance which the grant offers.* Stokes has written to me again,[15] the upshot of his note is that he does not think it likely that it will be recalled, so the matter rests at present.

The morning on which Sorby's letter reached me I met Prof. Haughton of Trinity Coll. Dublin. He is an eminent mathematician and geologist; Professor indeed of Geology in Dublin, and he told me that his interpretation of Sorby was precisely the same as mine.

Now a truce to science. I hope you are well, I hope Mrs. Faraday and Miss Barnard are well. This is glorious weather in the country—but at Newcastle[16] with its artificial atmosphere it may be the reverse. I associate with the name visions of foundries, smoke, and grimy miners—perhaps the reality is not quite as bad as my picture of it.

I am half inclined to write this letter again, for it is very confusedly written but I remember that you are accustomed to Matteucci,[17] and after him you will not find me difficult to decipher.

Believe me ever | faithfully yours | John Tyndall

22nd. July 1856

*Bence Jones agreed with this view of the subject.

RI MS JT/12/4049–51
Faraday Correspondence, 5:3166.
LT Transcript Only

---

1. *The examinations*: the Artillery and Royal Engineers examinations for entry to Woolwich.
2. *that correspondence*: see letters 1213 and 1217.
3. *Peltier generalised*: The French physicist and clockmaker Jean Charles Peltier (1785–1845). In 1834, while constructing a galvanometer to measure the different conductivities of antimony and bismuth, he discovered the so-called 'Peltier effect' in which the junctions of two metals in an electric circuit display heating or cooling effects. J. C. A. Peltier, 'Nouvelles Experiences sur la Caloricité des Courans Électriques', *Annal.de Chimie.*, 56 (1834), pp. 371–86. See also letter 1155, n. 2.

4. *my young friend*: Hirst, who had returned with his wife from Ireland on 16 July, was back in Broughton on 18 July 1856. Tyndall had visited them on Monday 21 July.
5. *Dr Stokes of Dublin*: William Stokes (1804–78), Regius Professor of Medicine at Trinity College Dublin and author of *A Treatise on the Diagnosis and Treatment of Diseases of the Chest* (Dublin: Hodges & Smith, 1837).
6. *her danger*: Tyndall implies here that Anna Hirst had no idea that she was dying from tuberculosis.
7. *a strong letter from Mr Sorby*: letter missing, though recorded in Tyndall's Journal as received in July (RI MS JT/2/13c/862).
8. *my late lecture*: J. Tyndall, 'Comparative View of the Cleavage of Crystals and Slate Rocks' delivered on 6 June 1856. Sorby had read the version of the lecture in the July issue of *Phil. Mag.*, 12 (1856), pp. 35–48.
9. *a paper on the subject to Francis*: Henry Clifton Sorby, 'On the Theory of the Origin of Slaty Cleavage', *Phil. Mag.*, 12 (1856), pp. 127–29.
10. *This I have done*: J. Tyndall, 'Observations on the Preceding Paper', *Phil. Mag.*, 12 (1856), pp. 129–35. The paper immediately followed Sorby's paper.
11. *Riess's last short communication*: P. Riess, 'On the Law of Electric Discharge', *Phil. Mag.*, 11 (1856), pp. 524–27. This was published in a Supplementary issue in July 1856.
12. *what the result will be*: It was published as W. Snow Harris, 'On Certain Phenomena of Electric Discharge', *Phil. Mag.*, 12 (1856), pp. 136–40.
13. *I received a note from Stokes*: letter missing.
14. *I wrote to him*: letter missing.
15. *Stokes has written to me*: See letter 1256.
16. *at Newcastle*: The Faradays had gone to Newcastle to visit the Sandemanian community there. They also visited the mining engineer Thomas Sopworth at Haydon Bridge. *Faraday Correspondence*, 5:3163.
17. *you are accustomed to Matteucci*: a reference to Matteucci's staccato style rather than to his atrocious handwriting.

To Julia Moore 24 July 1856 1260

<u>24th July 1856</u>

Dear Miss Moore.

If Saturday[1] be fine I will gladly come in the afternoon. I do not however expect that even sunshine can make the day more pleasant than the last one that I enjoyed at Wimbledon

Yours ever truly | <u>John Tyndall</u>

---

1. *Saturday*: 26 July 1856. In the event, Tyndall was unable to visit because of bad weather. He wrote a letter of apology on Tuesday 29 July which will be published in the final volume of *Tyndall Correspondence*.

From John Frederick William Herschel 7 July 1856 1261

Collingwood | July 27 1856

Dear Sir

Let me thank you for several instances of obliging attention in forwarding me copies of your scientific papers, on Diamagnetism, and quite recently of your lecture on the Slaty cleavage,[1] which appears to me a singularly happy explanation of that odd Phaenomenon. I may mention an illustration of the same nature with some others you have introduced, which occurs in the manufacture of the Coin of The Realm.[2] One part of the process consists in 'Ringing'[3] each piece of Gold or Silver Coin (after receiving the impression) every piece which will not ring on dashing it against a stone is rejected and re-melted, and such are numerous. Now the cause of this can be no other than minute air-blobs in the Bars out of which the lamina of metal are rolled by a flattening mill—each blob being flattened out into an infinitesimally thin film of air parallel to the face of the coin and so producing a solution of continuity and of course destroying the harmonic vibrations. —

There are one or two points suggested by your views of the slaty cleavage which harmonise so entirely with certain ideas I have long held about geological stratification in general that I cannot forbear mentioning them. I refer to the contortions of strata which are usually accounted for (as Poulett Scrope[4] has done) by the Protrusive violence of igneous rocks crushing together and doubling up the strata. To me it has always seemed that this feature might be more tranquilly produced thus[5]

Suppose A the ocean's surface and B B B its bed forming a basin having irregularities C C in it. Let sediment be deposited over the whole—uniformly—till the irregularities are filled in and the bottom is reduced to a generally speaking basin-like depression.

Then up to a certain point the friction of the bed will retain the deposited mud on its slope—and in this way strata nearly of uniform thickness (and parallel) will be formed.

But as the thickness of the whole increases—the weight will overcome the friction and the still soft strata will yield inwards from all sides toward the central portion of the bed—increasing the thickness of the strata there, and at the same time crumpling up the strata into contortions &c.

The whole affair may go on with any degree of slowness. Now as this process goes on the middle thickness will increase—and that by a general lateral compression and vertical dilatation of the whole central mass. And here we have the condition which your theory requires for the Slaty cleavage which (be it observed) does not absolutely require violent force (for mere force cannot produce any rearrangement of molecules)—but only intermolecular movement in a given fixed direction. Thus we see the blobs in slow moving lava currents drawn out in length and thus we see liquids in which excessively fine spicular crystals are floating, assume a silky appearance when stirred or agitated (I have a most beautiful case of the kind under my eye at this moment) by the tendency of the particles to arrange themselves when in motion all in one direction according to the laws not of pressure but of friction a distinction which is quite necessary to be borne in mind. Infinitesimal films of mica disseminated through amorphous mud would afford a precise parallel to this case.

Your account of the fibrous structure in tough Iron is quite beautiful and perfectly accounts for the exceeding toughness and tenacity of bar Iron formed by welding together smaller bars &c. —In fact such a bar is a rope of Iron wire held together as the strands of a hempen rope are by lateral friction & each portion each potion being per se of trivial length compared with the whole.

I am sure you will excuse me for writing to you in this desultory manner which perhaps I am not entitled to do but the fact is that since reading a great many years ago Wells's little tract on dew[6] I have not met with anything which seems to bring so many loose-lying phœnomena under a general and very simple principle.

I remain | D[ea]r Sir | Yours truly | JFW Herschel

J. Tyndall, FRS

RI MS JT/1/H/92[7]

1. *lecture on the Slaty cleavage*: J. Tyndall, 'Comparative View of the Cleavage of Crystals and Slate Rocks', *Phil. Mag.*, 12 (1856), pp. 35–48. See letter 1259.
2. *The Coin of The Realm*: Herschel had been Master of the Mint between 1850 and 1855.
3. *Ringing*: When silver coins are tapped they produce a bell-like sound; mixed metal coins produce a duller tone.
4. *Poulett Scrope*: George Julius Poulett Scrope (1797–1876; FRS 1826), English geologist and MP for Stroud (1833–68). Born George Poulett Thomson, he assumed the surname Scrope on marrying an heiress in 1821. Herschel was referring to Scrope's *Memoir on the Geology of Central France* (London: Longmans, 1827) in which he argued that valleys surrounded by striated larval rocks had been produced by slow river erosion, and not by a

universal flood. The foliation of the stratified rocks had been caused by differential movements at the time of deposit.

5. *thus*: the diagram was redrawn for publication in *Phil. Mag.*, 12 (1856), p. 197.
6. *Wells's little tract on Dew*: William Charles Wells (1757–1817; FRS 1793), Edinburgh-trained physician who made meteorological studies. His *Essay on Dew* (London: Taylor & Hessey, 1814) meticulously explained the formation of dew as caused by the condensation of air on plants that had cooled from loss of radiant heat.
7. Printed under the heading, 'Remarks on Slaty Cleavage, and the Contortions of Rocks' in *Phil. Mag.,* 12 (1856), pp. 197–98.

To John Frederick William Herschel 30 July 1856 1262

Royal Institution of Great Britain | 30th July 1856

Dear Sir

I write to ask your permission to publish in the next number of the Philosophical Magazine the interesting letter to which you have favoured me.[1] I placed the letter today before the proprietor[2] of that journal and he agrees with me that it would be most desirable to have it published. Apart from the intrinsic interest of the letter there are, I know, many who would be delighted to hail it as an evidence of the improvement of your health.

I quite agree with you that there is no necessity whatever for assuming the action of violent or sudden forces. I have at present before me a portion of a rail sent to me a few days ago from Basingstoke[3] and the laminar structure of which is striking in the extreme. This was brought about by the accumulated displacements due to the passage of many thousand trains, the mass having been subjected to this action for a period of 18 years.

In a few days I intend setting off to Switzerland & the Tyrol and shall make it a principal object of my journey to see with my own eyes some of the phenomena of glacial descent. At a future day I hope to be able lay before you the result of my observations on this much disputed question.

Believe me dear Sir | most faithfully & respectfully yours | John Tyndall

Royal Society Letters, Herschel 384

1. *letter to which you have favoured me*: letter 1261.
2. *the proprietor*: William Francis.
3. *Basingstoke*: a town in north-east Hampshire some 19 miles from Winchester. It was a junction for the Great Western and the London & South West Railway Companies. Tyndall's remark implies that he was still being consulted by railway companies following his 1855 report on boiler explosions.

To William Sharpey[1] 30 July 1856 1263

30th July 1856

My dear Sir

Your own request,[2] without any further stimulus, would be quite sufficient to induce me to give due attention to the paper of M. Matteucci. I have read it through, and although touching upon a controversial topic[3] it is I think very temperately written. You doubtless remember the experimental fact demonstrated to the satisfaction of the French Commission,[4] that during contraction the muscular current is enfeebled, the needle of the galvanometer not simply approaching zero, but actually swinging to the negative side of it. Du Bois[5] states that this reversion of the deflection is no proof that the direction of the current in the muscle is reversed, that such is really not the case, but that the polarization of the platinum plates by the current in its first direction, tending to establish a current in the opposite direction, becomes triumphant, where the muscular current is enfeebled by contraction, and produces the observed effect. Matteucci resorts to plates where no such polarization takes place, and finds that the contraction produces the same effect; hence he concludes that the effect is due to a real discharge of electricity of the muscle during the act of contraction. His experiments and reasoning are marked by great ingenuity: for his facts he must of course be responsible, but the whole character of the paper is such as to entitle it, in my opinion, to a place in the Philosophical Transactions.

Believe me dear Sir | very sincerely yours | John Tyndall
Dr. Sharpey | Sec. R.S.

RS MS RR 3/193

---

1. *William Sharpey*: William Sharpey (1802–80; FRS 1839), Professor of Anatomy and Physiology at University College, London. He was Biological Secretary of the Royal Society (1853–72). Tyndall was acting as referee for Matteucci's paper, 'Physical and Chemical Phenomena of Muscular Contraction'. It was subsequently published as 'Electro-Physiological Researches. Physical and Chemical Phenomena of Muscular Contraction—Tenth Series. Part 1', *Phil. Trans.* 147 (1857), pp. 129–44. Matteucci was upset by its delayed publication and complained to Faraday on 8 May 1856 (*Faraday Correspondence,* 5:3137).
2. *Your own request*: request missing.
3. *a controversial topic*: Matteucci had long complained that he had priority over Emil Du Bois-Reymond for the discovery and demonstration that nerve fibres conducted an electric current. (For a detailed even-handed account of the controversy, see Giuseppe

Moruzzi, 'The Electrophysiological Work of Carlo Matteucci', *Brain Research Bulletin,* 40 (1996), pp. 69–91; first published in Italian in *Physis,* 6 (1964), pp. 101–40).

4. *French Commission*: It is unclear whether this refers to a Committee of the French Academy of Science in 1837 or to a more recent commission.
5. *Du Bois*: In his paper (n. 1) Matteucci cited *On Animal Electricity: Being an Abstract of the Discoveries of Emil Du Bois Reymond,* edited by H. Bence Jones (London: John Churchill, 1852), p. 132.

To George Gabriel Stokes[1] 30 July 1856 1264

Royal Institution | 30th July 1856

My dear Sir,

If a straight line, or the edge of a white sheet of paper placed upon a black ground be looked at through a glass prism with a small refracting angle, the refracting edge being parallel to the straight line; the latter, instead of appearing straight, will appear curved, turning its concave outline towards the edge of the prism. Hence, when two corresponding straight lines on two pictures intended for the stereoscope are brought into union by two such prisms, the coalescence of the lines, (rendered curved by the prisms) will give the appearance of curvity to the surface upon which the stereoscope image rests. What is true of the prisms is also true of the semilenses of the ordinary lenticular stereososcope,[2] and the statement and explanation of the fact constitutes the major part of Mr Claudet's paper. The circumstances are ingeniously varied, and the author also proposes stereoscopes by which the curvity of surface above referred to is abolished and a truthful flat surface obtained. The paper, in my opinion, is well worthy of a place in the Philosophical Transactions. The only qualifying remarks I have to make are the following: —

The principles of the stereoscope are already well known to those who have given due attention to what has been published on the subject, and to such persons, who do not need elementary instruction, Mr Claudet's paper may appear prolix. The explanations are also given without any reference to the real authors of them, and it is a question whether some such reference may not be necessary to remove all just complaint on the part of those to whom we are really indebted for the theory of the stereoscope. The council will however have to consider whether the unfortunate personal differences to which the stereoscope has given rise, may not justify Mr Claudet's reservation.[3]

I am dear Sir | very sincerely yours | John Tyndall

Prof. Stokes | Sec. R. S.

RS MS RR/3/79

1. The letter is annotated 'Dr Tyndall's Report on M. Claudet's Paper'. Tyndall had delayed refereeing Antoine Claudet's paper according to letter 1256. The paper was published as A. J. F. Claudet, 'On Various Phenomena of Refraction through Semi-Lenses or Prisms, Producing Anomalies in the Illusion of Stereoscopic Images', *Roy. Soc. Proc.,* 8 (1856), pp. 104–10.
2. *the semilenses of the ordinary lenticular stereoscope*: The stereoscope is an optical device for combining photographs taken from points a little way apart to give the illusion of three dimensions. This can be done with mirrors (reflecting stereoscope) or lenses. In the latter refracting or lenticular stereoscope, the viewing lenses or eye glasses are called semilenses from the fact that they are segments (cut portions) of spherical convex lenses. Effectively they act as prisms.
3. *Mr Claudet's reservation*: Tyndall was implying that Claudet might have deliberately avoided giving references to Wheatstone's and Brewster's publications on the stereoscope because of the feud between the two men over priority of discovery. In the paper eventually published, Claudet made no reference to the controversy, but this may have been because only an abstract of his paper was accepted.

To Charles Babbage 1 August 1856 1265

My dear Sir

It will give me very great pleasure to join you on Sunday.[1] Both Sylvester[2] and the mechanical notion conspire to draw me towards you early, so I will come to you at about 1 o'C P.M.

very sincerely yours | John Tyndall | 1s[t]. Aug. 1856

BL Add MS 37197–48

1. *Sunday*: 3 August 1856. In his Journal Tyndall referred to visiting Babbage to discuss mechanical rotation (RI MS JT/2/13c/864).
2. *Sylvester*: see letter 1088, n. 1.

To Edward Frankland 1 August 1856 1266

1s[t] Aug 1856

My dear Frankland

Your note[1] lies before me, and I almost fear you will damn my impudence when you hear the extent to which I propose to modify it. I have the strongest reason for visiting some of the glaciers of the alps. I did think of going to Monte Rosa[2] but that is too far and too expensive. But the glacier of the Rhone, and of the Aar[3] I must see, and make a few observations upon them. The scenery there is magnificent some of the finest in Switzerland. The Furka pass[4] is most noble, and the Grimsel,[5] where I should have to take up my quarters for a few days—perhaps three—stands at the head of the magnificent Hasli Thal,[6] with the falls of Handeck[7] an hour or two distant. Having accomplished the survey of the glaciers, I should be able to go forward to the Tyrol.[8] But here you have omitted the wild and splendid scenery of the Gross Glockner[9] and this I think we should by all means see. Indeed here also I want to make a few observations on the glaciers. Thence I would go whither you might be pleased to command. Francis assures me that Schaffhausen Bregenz and Feldkirsch[10] are comparatively uninteresting. Now then I make the following proposal. Hooker & Huxley, two excellent fellows, who know you by renown, will be at Bale[11] on the 13th of August. Is it not possible for us to join them there go forward to Interlaken,[12] and thence to the Grimsel. You will be glorified in the splendid region and while I am cracking my shins & imperilling my neck in the glacier you might make the acquaintances of the splendors round. We then go forward to the Tyrol. Write to me at once and say whether this is not feasible. I must as I said before go to Paris to order the necessary instruments for my next course. Could we not go by Dieppe? It is only a pound 2nd class to Paris. I hope you have made up your mind to travel third class whenever that is possible. Now say yes my dear Tris[13] to what I have proposed. Say that you will start with me to Paris on the 9th leave me a day there for my instruments and get along to Bale in time to meet Hooker & Huxley. If you strongly oppose this then I will place one or two general facts in your hands, and conform to any plan in which you can incorporate them.

1st fact: give me two or three days upon the glaciers of the Rhone & Aar.

2nd fact. Give me an opportunity of observing the glaciers of the Gross Glockner.

This I believe is all.

You have devoted 7 mortal days to reach Zurich. Far less than this would carry us to Interlaken.

If you have any proposal to make respecting my meeting you at Chur (Coire)[14] or elsewhere I am ready to fall in with it. But the simplest and sweetest plan would be for us to start to Paris on the 9th. Do dear Tris, sweet Tris, my own Tris, say yes and cut the matter short.

Ever yours | John Tyndall

The Boden See is all comparative rot![15]

JRL Frankland, 245
RI MS JT/1/12/3975

---

1. *Your note*: note missing.
2. *Monta Rosa*: The western side of Monta Rosa, the highest mountain (15,203 ft) in Switzerland, contains the Gorner glacier which descends to Zermatt. Tyndall did not climb the actual mountain until August 1858.
3. *the glacier of the Rhone and of the Aar*: Glaciers some 5 miles from Grimsel that descend from Mt Galenstock (11,765 ft) and feed the rivers Rhône and Aar. They are approached from the alpine passes of Grimsel and Furka.
4. *Furka pass*: The highest mountain pass (7,969 ft) in Switzerland; it connects the villages of Gletsch and Realp in the Upper Rhône valley.
5. *Grimsel*: The site of the Hospice for travellers undertaking the Furka Pass.
6. *Hasli Thal*: The Hasli Valley, from the Old High German, *hasal,* meaning 'hazel'. The valley stretches from Meiringen to Innerkirchen in the Bernese Oberland.
7. *The falls of Handeck*: The site of a spectacular waterfall at Guttannen where the waters of the Aar and Ärlenbach meet.
8. *the Tyrol*: Before the First World War this referred to the Alpine, wine-growing region of Austria. The South Tyrol was ceded to Italy in 1919.
9. *Gross Glockner*: an Alpine peak of 12,461 ft. which is now reached easily by the Grossglockner High Alpine Highway.
10. *Schaffhausen Bergersz and Feldkirsch*: Schaffhausen is a city in North Switzerland in the Canton of the same name. Here both Swiss German and Alemannic German dialect were spoken; Bregenz is a town on the eastern shore of Lake Constance; Feldkirsch or Feldkirch is a medieval Austrian city on the borders of Switzerland and Liechtenstein.
11. *Bale*: the French spelling of Basle or Basel, the Swiss city situated on the borders between France, Germany and Switzerland.
12. *Interlaken*: a town in the Canton of Berne situated on a tongue of land between the lakes Thun and Brienz through which the River Aar flows.
13. *Tris*: Tyndall's nickname for Frankland, from Trismegestus, a legendary alchemist.
14. *Chur (Coire)*: reputedly the oldest town in Switzerland, lies on the River Rhine; it is the capital of the Canton of Graubünden
15. *The Boden See is all comparative rot!*: The Bodensee is the German name for Lake Constance. Frankland may have mentioned a prospective visit in a missing letter and presumably Tyndall had been told (by Francis?) that is was not worth visiting.

From Thomas Archer Hirst 2 August 1856 1267

Biarrits.[1] | August 2nd, 1856

My dear John,

At length our long journey is finished[2] and we have arrived safely at our destination. We started from Broughton yesterday week,[3] arrived in Paris on the Saturday afternoon, rested there until Monday, when we travelled as far as Orleans. After remaining there 30 hours we continued our journey and stopped the next night at Tours. From Tours we went to Bordeaux, where we halted again for 24 hours, and lastly we travelled to Bayonne, and reached Biarrits, which is only 5 miles distant, on Thursday evening last.[4] By means of these short stages Anna has borne the journey wonderfully well. From beginning to end the weather has been magnificent, with the exception of a thunderstorm in Paris. I do not know the temperature here, but it must be very high, for from morning to night I am always bathed in perspiration. I have thrown off drawers and flannel and by and bye I shall cease wearing a shirt I think. The greatest luxury I have is a bath every morning at 6 in this glorious bay of Biscay.[5] At present it is exceedingly calm and yet the waves, as they break into foam on the beautiful beach, are larger and stronger than I have ever seen them in England. What they will be in a storm I can scarcely imagine. I am greatly disappointed to find Biarrits a very expensive place. After great trouble I failed to find lodgings in the village that were cheap enough for us. For a single room with two small beds in it they wanted 5 francs a day (this of course includes attendance) and would not take less even if we promised to remain for six months. At length about a mile from Biarrits I met with two small rooms, adjoining one another, and having a bed in each, for 4 francs a day and these we have concluded to take. The situation is certainly beautiful and superior to any in the village of Biarrits. It is at the top of a cliff overhanging the bay, and from our windows we have a magnificent view of the bay at the feet of the Pyrenees.[6] But only fancy 23s/4^{d} a week for two small rooms and attendance, and yours which are far superior only cost 17s. Nevertheless I think I shall be able to live on £150 a year.

Immediately on our arrival we sought out the best physician and I got him to examine Anna. His opinion coincides precisely with that of all the other physicians we have consulted. He thinks Biarrits will be of some service to her, and that Pau[7] will be the best place to winter at. I consulted him thus early in order to have his opinion of the progress Anna makes. He appears to be a very decent fellow and has already been of great service to us in finding lodgings. When I found them so expensive I went to him and told him frankly what income I had, and that I wished to procure for my wife the best of everything that such an income could afford. He was evidently pleased with this frankness and we understood each other better on account of it. He made a few calculations and then recommended us to go to the place I have described. At present we are still at the hotel our lodgings will be ready to receive us tomorrow and then the doctor is to submit Anna to some new kind of treatment the precise nature of which I do not yet know, but I trust it will be beneficial. Please send this letter to Wright; when I have ascertained more about money matters I will write to him. Meantime should he or you have occasion to write please address Poste Restante,[8] Biarrits. If Francis would send my journals[9] here I should be obliged to him. Remember me kindly to him please, and tell him of our safe arrival. I wish too you would let me know the result of your letter to Roby Pridie.[10] Remember me kindly to Debus when you see him and accept our kind love. Mr & Mrs Wright I hope will do the same.

Yours affectionately, | T.A. Hirst

What about Wynne?[11]

RI MS JT/1/H/227

---

1. *Biarrits*: Biarritz, a resort in the Bay of Biscay on the west coast of southern France. Hirst spells it Biarrits (as pronounced) throughout the letter.
2. *our long journey is finished*: Hirst was taking his wife to Biarritz in the hope of curing her tuberculosis. He provided a detailed account of the journey to Spain in his Journal, 27 July and August (RI MS JT/32c/1214–17).
3. *yesterday week*: Friday 25 July 1856.
4. *Thursday evening last*: 31 July 1856.
5. *bay of Biscay*: The stretch of Atlantic ocean from Brest in northwest France to Cabo de Perias in the northwest of Spain.
6. *the Pyrenees*: The mountainous range that forms the border between France and Spain and extends from the Bay of Biscay to the Mediterranean Sea.
7. *Pau*: a French town 60 miles from the Atlantic coast and 30 miles from the Spanish border. It had become a popular winter resort for Victorian tourists.
8. *Poste Restante*: post remaining (French); an international arrangement whereby travellers could collect their post from a town's central post office.
9. *send my journals*: monthly copies of the *Phil. Mag.* to which Hirst presumably subscribed.
10. *Roby Pridie*: William Robert Pridie (1825–74), the son of an Independent Minister in Halifax who was a close friend of Hirst's. He spent his life as a humble bank clerk in Halifax. Hirst's other Halifax friend, the dyer's clerk, Francis Booth (1826–53), had died on 13 March 1853 and Hirst had promised to support his widowed mother Sarah Booth with an annuity. When Hirst left Queenwood he could no longer afford to support Mrs Booth and Tyndall had volunteered to pay the annuity. This arrangement continued until November 1856 when Mrs Booth died. See the letter from Pridie to Tyndall, 8 November 1856, in volume 6 (RI MS JT/1/P/258).
11. *What about Wynne?*: Hirst was asking whether Eddy Wynne had passed the Woolwich entrance examinations. He had not.

To Edward Frankland 4 August 1856 1268

4th Aug 1856

My dear Tris

I am ready to go down upon my marrow bones and implore the forgiveness of Mrs. Frankland.[1] But she does not know what a man I intended to send you home to her. She does not know how I would have taken care of you, and prevented your lank carcase from been[2] squashed in the crevasses of the ice[.] She does not know how I would have fed you in the choice mountain milk, and kept everything dyspeptic out of your stomach's way—If she knew all this—if she could contemplate the prospect of your returning to her a compact mass of bone and muscle weighing at least thirteen stone, with vigour in your heart sufficient to carry you triumphantly through the mist & soot of ugly Manchester[3] during the coming winter—She would I am sure forgive the effort I made to confer all these benefits upon you and her. But it was ever thus—This was always the treatment which the world's benefactors received at the world's hand, and I am content like them to suffer for righteousness sake—Mrs Frankland! Mrs Frankland! Your black ingratitude brings tears to my eyes: but I will pray for my enemy and say God bless her! You can have half a dozen boxes[4] if you like at the R.I. I will meet you as you propose, but Chur (Coire)[5] will be better than Feldkirch.[6] I will write to you again upon this point—Perhaps I may take up my quarters upon the Finstermünz Glacier[7] and await your arrival there. I will write to you again before I start.

Faraday's picture[8] I have received. Thank you twice over for it. The face is Faraday's face without the soul of Faraday shining through it. Give my kind regards to Williamson and respects to his wife.[9]

Ever yours | John Tyndall

JRL Frankland, 246

---

1. *the forgiveness of Mrs Frankland*: Tyndall was responding to a missing letter from Frankland. It would appear that Sophie Frankland did not want her husband to go to Switzerland. The Franklands had had their third child, Sophie Frankland, on 3 November 1855.
2. *been*: being.
3. *Manchester*: Frankland continued to teach chemistry at Owens College, Manchester until October 1857.
4. *half a dozen boxes*: Tyndall was employing hyperbole. The lecture theatre seating at the RI did not include boxes.
5. *Chur (Coire)*: see letter 1266, n. 14.

6. *Feldkirch*: see letter 1266, n. 10.
7. *Finstermünz Glacier*: the glaciers surrounding the Finstermünz Pass on the borders between Austria and Italy and between the towns of Landeck and Nauders. In his Journal for 10 August 1856, Tyndall stated that he had 'Fixed to meet Frankland at Landeck in the Tyrol on the 1st September' (RI MS JT/2/13c/865). In the event the two men met in the Finstermünz Pass on 2 September 1856 (RI MS JT/2/13c/897).
8. *Faraday's picture*: Frankland was a keen photographer and had probably taken a photograph of Faraday which he sent to Tyndall.
9. *Williamson and... his wife*: William Crawford Williamson (1816–95; FRS 1854), medical practitioner in Manchester and, from 1851, inaugural Professor of Natural History, Anatomy and Philosophy at Owens College. His publications mainly concerned palaeontology. Tyndall had first met him in London on 23 November 1854 (RI MS JT/2/13c/699).

From Heinrich Debus 5 August 1856 1269

Queenwood College | Aug. 5 | 1856

Mein lieber John!

Das Reisebuch war schon eingepackt; ich hatte nur vergessen es gestern abzuschicken. Hierbei erhältst Du dasselbe und auch das gewünschte Thermometer. Es thut mir leid, dass wir kein besseres Thermometer besitzen, da es aber das Einzige ist, dass wir haben, so kann ich kein anderes schicken. Der Gefrierpunct des Wassers liegt bei 0° und der Kochpunct bei 75° so dass also 1° auf dem Thermometer = $1\ {}^{1}/_{3}$° Celsius ist.

Um es also noch einmal zu wiederholen: 0° auf dem beifolgenden Thermometer zeigt den Gefrierpunct des Wassers und 75° den Kochpunct des Wassers an. Wahrscheinlich ist diese sonderbare Eintheilung durch einen Irrthum entstanden.

Wann wirst Du abreisen? Ich habe nur noch einige wenige Worte hinzuzufügen. Reise glücklich! Der Himmel gebe Dir freundliche Wirthinnen, reinliche Betten, schönes Wetter und angenehme Gesellschaft. Doch dies ist zu gewöhnlich und jeder gemeine Mensch kann die Dinge verstehen. Jeder Pavian, jeder Zeitdieb, jeder leere Kerl wünscht & kann dasselbe wünschen. Nein! Gott gebe Dir einen offenen kindlichen Sinn um die Pracht seiner grossen Werke mit vollen Zügen mit all ihrer Herrlichkeit in Dich aufzunehmen. Jede Pore Deiner Haut sei ein Telescop, und Microscop, jedes Haar ein Fühlfaden, & jeder Sinn verdoppelt. Deine Brust sei offen und alle Fasern Deines Herzens seien nach aussen gekehrt. Und wenn Du so vorbereitet in die Welt gehst und von ihr Alles Grosse, Erhabene, und Schöne eingesogen hast, wenn Du und die Welt Eins geworden sind, dann komme zurück nach Queenwood.

Dies war der erste Theil meiner Rede. Der zweite ist kürzer und lautet also:

"Bleibe nicht zu lange aus"

Dein | H. Debus.

Queenwood College | Aug. 5 | 1856

My dear John!

The travel guide[1] was already packed; I had just forgotten to send it off yesterday. You[2] receive it herewith and also the thermometer you requested. I am sorry that we do not possess a better thermometer, but as it is the only one we have, I cannot send another. The freezing point of water lies at 0° and the boiling point at 75° so that 1° on the thermometer = 1⅓° Celsius.

Just to repeat it one more time: 0° on the enclosed thermometer indicates the freezing point of water and 75° the boiling point of water. This strange calibration has probably arisen through an error.[3]

When are you going to depart? I have only a few more words to add: Have a happy journey! May heaven give you friendly landladies, clean beds, fine weather and pleasant company. But this is too conventional and every common person can understand such things. Every baboon, every time-waster, every vapid fellow wishes & can wish this. No! May God give you an open childlike mind in order to take in the splendour of His great works in all their glory to the full. May every pore of your skin be a telescope, and microscope, every hair a sensor, & every sense doubled. May your breast be open and all the fibres of your heart be turned outwards. And when you go thus prepared into the world, and have absorbed everything great, sublime and beautiful from it, when you and the world have become one, then come back to Queenwood.

That was the first part of my speech. The second is shorter and simply says:

'Don't stay away for too long'[4]

Your | H. Debus.

RI MS JT/1/7/2415
LT Transcript Only

1. *The travel guide*: This was probably J. Murray, ed., *Handbook for Travellers in Switzerland and the Alps of Savoy and Piedmont* (John Murray: London, 1838); 7th ed. 1855 and 1856.
2. *You*: informal address (Du).
3. *through an error*: It would normally be 100 degrees; possibly the thermometer was made in the Queenwood laboratory.
4. *Don't stay away for too long*: It is unclear why Debus has put this fairly conventional remark in quotation marks.

To Sarah Tyndall 8 August 1856 1270

London, 8th, Aug, 1856.

My dear mother

I have been very much pressed by work for many days: now I am thankful to say a good portion of it is thrown off my shoulders and I have time to write to you. Early next week I start for the Continent and purpose spending the next six weeks among the glaciers of Switzerland and Southern Germany. This journey is very needful for me at present for my brain wants rest and my body exercise. In the healthy mountain air I shall soon get very strong and hope to return with a stock of good health sufficient to last for the next half year.

The reason I send you money so soon,[1] and so little, is that I just wish you to have what will keep you going until I return. I will make the rest of my debt good. You need not answer this letter, as I shall be away before your reply could reach me. I will write to you soon after I commence my operations upon the mountains, and let you know from time to time where I am and how I am getting on. I think I used to hear a song long ago in Ireland commencing

> 'Green hills of Tyrol again I see &c'[2]

Well in a very short time I shall be among these green hills, and snow clad mountains. A portion of the time I shall be alone, and the other portion I shall have an old companion and fellow student at my side.[3] It is a pleasant thing to be able to make one's self at home in this distant place, to speak the language of the people and feel less a stranger than a Leighlin Bridge man would in the County Kilkenny.[4] I hope to be able to do some scientific work while I am away, and thus turn my journey to useful account. You are now I have no doubt fast asleep in bed, while I am here writing to you in the midst of great London. It is just on the stroke of midnight and I am getting sleepy.

Good night | Your affectionate son | John

RI MS JT/5/16/159–60
LT Transcript Only

---

1. *money so soon*: Tyndall had been sending occasional sums of money to his mother since his surveying days. See, for example, *Tyndall Correspondence*, vol. 1, letter 0006, 8 June 1840.

2. *Green hills of Tyrol again I see*: The first line of a widely-circulated ballad that was used by Gioachino Rossini in Act 3 of his opera *Guglielmo Tell* (William Tell) in 1829.
3. *an old companion . . . at my side*: Edward Frankland.
4. *County Kilkenny*: Because of dialect variations and the possibility of Gaelic being spoken, a man from County Carlow might find difficulty in communicating with someone from Kilkenny. Tyndall had experienced this while working as a surveyor in rural Ireland. See *Tyndall Correspondence*, vol. 1, letter 0008, n. 3, and letter 0043.

To Edward Frankland 9 August 1856 1271

Royal Institution of Great Britain | Saturday night

My dear Tris

Here goes to give you my last word and Testament: I start on Tuesday morning, and meet Hooker & Huxley at Basle. Thence I get on with all speed to the Grimsel[1] where I shall remain 5 or 6 days and then my thoughts will turn towards you. Like Wellington longing for Blücher[2] on the field of Waterloo, I shall exclaim 'Would to Heaven that Frankland were come!' I shall be dismal without you my beautiful Tris! Well then here goes to fix you. You propose that I should meet you at Feldkirch[3] on the 1s[t] of September. This I wont do but I will do something very near it. While you are on your way I shall be sauntering among the glaciers about the Finstermünz,[4] but to enjoy the prospect with you at my side I will descend from my glory and meet you either at the Goldner Adler Landek,[5] or at the Inn in the little town of Ried[6] a few miles further on. I shall aim at the former, but if you do not find me there you will be sure to find me at Ried. I will endeavour to leave some intelligence regarding my movements at the Goldner Adler; so you had better call there. I hope to meet you on the 1s[t] of September; but lest you should not be able to accomplish this I will remain over the 2nd at Ried. And now my dear boy good bye: may the gods guard you till I get you under my protecting care. Tell Mrs. Frankland that in treading the glaciers I shall tie you with a rope to my girdle so that if I fall into a crevasse, which is quite likely, I will carry you to heaven along with me, & thus make you happy for ever & ever Amen!

Ever Yours John Tyndall

JRL Frankland, 247

---

1. *Grimsel*: letter 1266, n. 5.
2. *Like Wellington waiting for Blücher*: The Battle of Waterloo in what was then the Nether-

lands on Sunday 18 June 1815 saw the end of the Napoleonic Wars when Napoleon was defeated by the combined forces of an allied army led by the Duke of Wellington and a Prussian force commanded by Prince Gebherd Leberecht von Blücher (1742–1819). While waiting for the Prussian reinforcements, Wellington reputedly said: 'Would to heaven that Blücher were here.' Tyndall had made a close study of the battle during his first visit to the Continent as recorded in his Journal for 19–20 July 1850 (RI MS JT/2/13b/362–68).

3. *Feldkirch*: letter 1266, n. 10.
4. *Finstermünz*: the snow-covered Finstermüntz Pass which marked the boundary between Italy and the Austrian Tyrol.
5. *Goldener Adler Landek*: Golden Eagle Inn at Grindelwald near Landeck (*Murray's Handbook for Travellers in Switzerland* (John Murray: London, 1838), p. 77).
6. *Ried*: a village in Oberinntal eleven miles south of Landeck.

To Thomas Archer Hirst 9 August 1856 1272

[London] | 9th. August, 1856.

My dear Tom

I have received your letter,[1] read it with pleasure, sent it to Wright, and also sent your brother's letter[2] to him. I was not surprised to hear your account of the dearness of things in Biarrits. Such places are always dear, simply because they are the resort of people who can afford to pay. Every body in London, that is every physician, knows Biarrits. Your plan would be, if you intended to make a long stay to take up your quarters elsewhere. Francis has no doubt that in some of the places he had in his eye you might have excellent lodgings for ten shillings a week: you would of course have to dispense with fashionable neighbours, and this I take you would consider no great evil. But on the whole I think you have acted the wisest part under the circumstances. I am glad that you have established yourself with the physician and your method of doing so was certainly, the shortest, the frankest and the best route possible. Our weather here for a long time has been excessively warm; for many years I have seen nothing like it. This morning, the 9th, the weather has changed and the rain is falling gently but persistently. I wish I could join you in your plunge bath in that 'glorious bay of Biscay' those waves would make me a poet, and probably a better natural philosopher. I have derived much instruction from watching waves. But I am on the point of commencing a trip of my own which promises well: four days hence will find me in Switzerland. I have been reading much of later regarding the descent of glaciers, a subject on which a great deal has been written, but on which, I believe a great deal remains to be done. I have read the 'travels' and 'letters' of Forbes,[3] and made myself thorougly acquainted with his theory. The impression left upon my mind by his writing is that the matter needs closer looking after. So this I intend to combine with my journey. I will spend a week or ten days at the Grimsel,[4] and proceed from there to the glaciers of the Aar, the Rhone and Grindelwald,[5] and hope to learn wisdom from the iceblocks. Huxley will be with me part of the time. Then I break away to the Tyrol and meet Frankland on the Summit of the Stelvio,[6] where there are other glorious glaciers. I could content myself with the glaciers of the Tyrol, but in order to appreciate Forbes's theory I must see what he has seen. Afterwards Frankland and I push on to Vienna. The journey is too long, or rather it will be too expensive, and were it not that the double motive of the Naturforscher Versammlung[7] and the fact of my never having seen Vienna I would not go there.

I saw Atkinson[8] two days ago, on his way from Oxford to Scotland. One of his first questions was regarding you. Debus has returned in good health from the Tyrol, where he made but a brief journey, being chiefly in Munich. I had a note from George Edmondson a few days ago in reply to a question of mine regarding the acoustic bellows.[9] He said at the end they were hourly expecting Mr Smith;[10] and that they would be fuller and stronger this half year than last. I wish he may be able to say the same two years hence.

For the last week or two I have been terribly worn down with work. I am clear in head thank heaven but my body needs the freshness of the mountain air which is now in prospect. This caused me to put off writing to Priddie[11] longer than I otherwise should have done. I have written to him, but shall not receive his reply before I start. I have told him simply that he must substitute me for you, and that I will settle with him when I return. Fox was with me here a few days ago. I accompanied him to University College hospital[12] and saw one awful operation performed. I never before had so lively a sense of the blessings of chloroform: the sufferer did not feel anything. My work[13] has been thus far wholly unsuccessful, save the augmentation of my experience, which will come in useful some day. I have obtained totally negative results where others (Thomson for example) have obtained positive ones with means which are perhaps a hundred times less perfect than mine. But I pause, and instead of science send both you and Anna the very XXX of my affection.

Ever yours | John Tyndall

RI MS JT/1/T/631

1. *your letter*: letter 1267.
2. *your brother's letter*: This letter from John Henry Hirst to Tyndall is missing.
3. *'travels' and 'letters' of Forbes*: J. D. Forbes, *Travels Through the Alps of Savoy and other Parts*

of the Pennine Chain, with Observations on the Phenomena of Glaciers (Edinburgh: Adam & Black, 1843; 2nd ed., 1845). Forbes had also published a series of 16 'Letters on Glaciers' in the *Edinb. New Phil. Journ.* between 1843 and 1851.

4. *the Grimsel*: see letter 1266, n. 5.
5. *the glaciers of the Aar, the Rhone and Grindelwald*: See letters 1266 and 1268.
6. *the summit of the Stelvio*: The Stelvio Pass is some 9000 ft above sea level and connects Stilfs (Stelvio in Italian) in the South Tyrol with Bormio in northern Italy. The statement appears to contradict the arrangement suggested in letter 1268.
7. *Naturforscher Versammlung*: Tyndall was to attend the meeting of the Deutscher Naturforscher Versammlung in Vienna from 16 to 22 September 1856.
8. *Atkinson*: see letter 1216, n. 2.
9. *acoustic bellows*: Presumably an apparatus whereby air is driven through a pipe, as in an organ or siren. It is unclear why Tyndall had asked Edmondson about such a device.
10. *Mr Smith*: Not identified; he was evidently Hirst's replacement at Queenwood College.
11. *Priddie*: see letter 1267, n. 10.
12. *University College hospital*: The medical school and hospital in Gower Street attached to University College.
13. *My work*: Prior to his Alpine vacation in August and September 1856 Tyndall was concluding his research on diamagnetism.

To Michael Faraday 27 August 1856 1273

27 Aug 1856

My dear Mr. Faraday,[1]

I have bidden Huxley goodbye[2] and watched him descending the slopes of the Furca till his gray coat was lost amongst the gray rocks. I am now all alone in a snug little pen, which may perhaps be justly dignified by the name of bedroom; and my thoughts find refuge in writing to you. I left London on the 12th, and had pleasant intelligent companions to Dieppe. I was not sick in the passage, but came in sight at least of sickness. I reached Paris a little after midnight, had my luggage examined, the act being facilitated by promptness and civility on my part towards the officials. Took a voiture,[3] and while I rumbled through the forsaken streets amused myself by watching the meteors vaulting through the sky. Had three hours bed, but no sleep at the hotel de Strasbourg, and next morning at 6 o clock was on the road again. Part of the way an intelligent Parisian was my companion, and from him I learned that Regnault was better,[4] but that great anxiety was still felt regarding him. Since then I have heard nothing regarding Regnault. Put up at the Maison Rouge in Strasbourg.[5] I dreamt that I was in a besieged fortress, with the cannon rattling about my ears. As my consciousness became clarified I found I had transformed the knocking of the 'boots' into the explosion of cannon. The man had already become irascible on account of my want of attention. this was at half past 3 o clock in the morning and a little after 4 I started for Basle, hoping almost against hope to catch Huxley and Hooker there. I arrived just in time to see Huxley in a corner of the diligence,[6] to apprize him, by a thump on the shoulder of my arrival, to take my seat in an extra omnibus and set off to the railway station. At Berne we were joined by Dr. Hooker and a party, and then we went on to Thun. From Thun by steamer to Interlaken, the weather being magnificent. On the day of our arrival, we took donkeys, one for the luggage, and the other to carry Mrs. Huxley,[7] and reached the Jungfrau hotel[8] upon the Wengern Alp. Here we spent the night and next day we examined a glacier formed by the avelanches which thunder down the sides of the Jungfrau. We derived much instruction regarding both glaciers and avelanches. Though many prophets of evil told us we should have bad weather we ventured forward to Grindelwald, and as far as the weather is concerned nothing could be more glorious. Drifting cloud masses smote the sides of the Jungfrau. Along the slopes of the Alp the black pine crowded sometimes into forests, and again dispersed in clusters, was interleaved with patches of the most lively green; the sun poured his glory over all. Right before us a rainbow planted one end of its curve in the valley and leaning over the mountain summits to the right, seemed to clasp the savage crags of the Wetterhorn in its embrace. We reached Grindelwald just in time to secure bedrooms at the Bear Hotel.[9]

Next day we engaged a guide and ascended the slope which bounds the glacier. It was most beautiful, and most instructive; the snowy minarets shining in the sun and relieved against the black slope of the opposite mountain, were undescribably beautiful. But there was a scientific pleasure superadded. We examined the conformation of the glacier, the nature and direction of the fissions, the varying inclinations of the surface, and after a little time a mechanical problem of great beauty revealed itself. The structure and modifications of the entire mass were such as might be predicted a priori from mechanical laws. A certain relationship was thus established between our own minds and the mass before us, which rendered our subsequent journey upon it pleasanter. I will not bother you with this further now, as I hope to be able to make a Friday evening[10] out of the subject. We ascended to what is called the Eis mer 'mer de glace' in some places the breaking up of the mass into chasms, rendered our progress difficult and not quite without danger. In one place we had to cross an edge 10 inches wide, with a chasm 60 feet deep yawning at each side of us: but our progress since has caused us to think little of this. In some places deep shafts are formed into which the surface water of the glacier tumble with the noise of thunder. One of these shafts we found by letting a stone fall to be at least 400 feet deep. High above us on the mountain slope was a flat mass of brown crag, surmounted by a crown of ice cliff. Summer or winter on this rock no snow rests; and it is thence called the Heisseplatte. (the hot slab). Down this while we were on the glacier at least a dozen avalanches darted downwards: no wonder that the noise of these equal the sound of thunder. They are composed in part of solid blocks of ice: one of these blocks which we measured, and which was cast to a distance of 1000 yards weighed at least 5 tons. We derived great instruction from this day's journey, and I trust we shall be able to put certain glacial phenomena in an entirely new point of view.

Next day Mrs. Huxley was ill, and Huxley had to return with her to Interlaken. I was joined by Hooker and we again ascended the glacier together. We followed one great tributary of the Grindelwald glacier as far as our guide (a powerful dark looking Swiss) dared to take us. When further progress on the ice was impossible we ascended the steep slope of a mountain side to a height sufficient to give us a view of the entire glacier. The dirt bands of Prof Forbes[11] were here strikingly manifest; but it did not appear quite so manifest that he had solved them. I will reserve all further expression of opinion until my thoughts are riper on the subject. While on the mountain the rain came down in torrents, and we returned to Grindelwald well drenched. I went to bed until my trousers (for I had not a second pair) were dried.

Next morning I left Grindelwald,[12] and had a glorious excursion over the Sheideck to Meyringen, on the following day I walked to the Grimsel, and saw the Handeck falls[13] to great perfection. The Aar had a bountiful supply of water; and the Aerlen, a river which darts down a precipice to the left, was also well supplied. The Aar plunges into a chasm 200 feet deep. The smaller river first falls upon a rock ledge and rebounding from this darts at the Aar, and both plunge together like a pair of fighting demons to the bottom of the gorge. The spray produced a vivid rainbow; which was beated about hither and thither, as the water smoke shifted its position.

A day has flown[14] by since the last word was written, and I have left 12 hours of hard walking, during seven of which I carried my heavy pack, behind me, I am now in a little town in the Canton of Graubünden, surrounded by Roman catholics. As I write the bells of a large abbey are summoning the Benedictine monks to their duties; but I must quit the present and recur the past. I reached the Grimsel and found Huxley there before me. His wife became better, and he came straight from Interlaken to Meyringen, and thence to the Grimsel. I had read previous to leaving London that a certain M. Dollfus of Mielhausen[15] had built a hut by the side of the glacier of the Aar, for the purpose of making observations upon the motion of the mass. All furniture had been removed from the hut, but there was still a litter of old hay there, and this Huxley and I proposed to make our bed, and thus enable ourselves to go to the very summit of the glacier, which would be impossible in one day from the Grimsel. We had some blankets packed together; a couple of stout fellows carried up some food and firewood, and on the evening of the 22nd, after a hard day's work, during which I visited the mass of rock, which in former years sheltered Agassiz,[16] we returned to our hut. After cold mutton, sour bread and a glass of beer we betook ourselves to our hay and spent the night, if not comfortably, at least philosophically. Next morning at day dawn we found the snow falling heavily. We waited until noon, and finding it still persistent, and deriving no comfort from the predictions of our guide, we resolved to return—Made on our way as many observations as the fresh fallen snow permitted us to make. As we descended the snow flakes became smaller and finally merged into a heavy rain which drenched both of us. Notwithstanding this defeat we found our visit to the glacier extremely instructive, the structure of the ice, the formation of the moraines, the great blocks of white granite which came slowly sailing down the glacier—all were objects of the highest interest. To form the lower glacier two others combine, and their confluence reminds me strikingly of the union of two rivers. Indeed the idea of viscosity, or plasticity, or semi fluidity, is that which must first suggest itself to any reflective mind. Forbes' comparison with an overturned pail of mortar[17] is in many cases admirable; but whether the comparison will stand philosophic analysis is quite another question. I believe myself that your old experiments[18] will help us greatly here.

Next morning we paid our bill in good time at the Grimsel and sallied forth towards the glacier of the Rhone. I sought for a cloud in heaven as we started, but could not find one. Clouds and vapours had all swept themselves away during the preceding night, and we had the unspotted firmament above us. Two hours brought us to a point from which we had a capital view of the lower portion of the glacier of the Rhone. Our view upwards was bounded by ice cliffs, and beyond these it was our purpose to penetrate towards the origin of the glacier. Having procured some food at the Auberge and a bottle of wine we set out, conducted by our athletic guide. Having examined the lower portion of the glacier, and observed some remarkable and beautiful phenomena, we crossed, and ascending a mountain at the opposite side, we came right over the upper portion of the glacier: Here on the crags we had our frugal dinner, in a manner which made the pomp of emperors poor. A scene of indescribable magnificence opened before us. Right in front was the mighty mass of the Finister Aar horn, further towards the horizon was the grand peak of the Weisshorn, more to the left we had the snowy summits of Monte Rosa by the side of which the cone of the Matterhorn (Mont Cervin) rose like a black savage tattooed with streaks of snow. Still further to the left the chain of the

Furca, with shoulders of snow as smooth as chiselled carrara marble, completed the picture; and over all this the glorious sun poured his undimmed radiance. It was a scene calculated to stir the heart of man, and to carve for itself an everlasting resting place in his memory. We went upon the glacier and tramped for hours upwards over the newly fallen snow. We learned much, and our bump of caution was in perpetual action, for the crevasses were terrific. As we ascended, the blue of the sky became deeper: and at the highest point there was something awful in the blueblackness it assumed. Once the sun got above a thin veil of frozen particles, and we had a splendid ring like a circular rainbow round him. The veil gradually melted away to the westward, and we had simply the eastern semicircle: by degrees this also melted away and the circle wholly disappeared. As we returned the shadow of the Finster Aar horn caused the vapours to curdle up, and to flow with great velocity into the valley of the Rhone. Here however the sun still shone, and the vapours were licked up as fast as they came: hence though the supply from behind was incessant, the cloud river made no progress. Huxley accompanied me to the Furca, at the summit of which stands a little inn. The valley was full of vapours when we arrived, and standing on a mountain ridge, with the sun behind us each of us suddenly observed his head surrounded by a coloured halo, and his shadow projected on the vapour mass in front. We raised our hands, the gigantic specture before us did the same, and imitated all our other actions—we had in fact 'the spirit of the Brocken'[19] in all its splendour. We closed this day of wonders over a chop and a bottle of wine in the Furca Hotel, whence I descended to see Huxley on his way to the Rhone Auberge. The marmots were piping on the rocks as we bade each other good bye, and I was soon on my way back to my lonely lodging.

One's pleasure is sometimes marred by the cupidity of the Swiss guides: but on this I will not dwell at present, I bought a salad spoon and fork at Grindelwald for Mrs. Faraday but cannot undertake to deliver it to her safely. To her and to Miss Barnard pray present my kind remembrances. I am now at Feldkirch having just crossed the Austrian frontier. Tomorrow morning at 4 o clock, I start for Landeck. Meanwhile I will post this. Would you have the kindness to say to Anderson that I received my letters safely at Chur. Would you add to this kindness by handing him the enclosed letters[20] and asking him to have them prepaid and posted to the gentlemen to whom they are addressed? Two of them require envelopes and postage stamps, I had nearly forgotten an important point. Knowing that Mr. Addams[21] formerly possessed some acoustic apparatus and thinking that he might be willing to dispose of it cheaply, I called upon him previous to leaving London but he was absent. I wrote to him but his people did not forward my letter. Last night at

Chur I had a note from him saying that he had not disposed of his apparatus. Would it not be worth while to see it, and ascertain whether it, or part of it could not on account of being second hand be more cheaply purchased than new instruments at Paris. I should be greatly obliged by your advice upon this point. If you would write to me Poste Restante Vienna I should know how to act there and in Paris on my return. I shall not reach Vienna[22] until the 16th of September. I would thank you to ask Anderson to direct my letters there up to the 15th of September. The meeting will continue 6 days and that will afford sufficient time for all letters posted in London up to the 15th to reach me. And now I must beg of you to excuse all the trouble which I so impudently lay upon your shoulders, to wish you patience through this tedious ill written letter and to believe me

ever most faithfully Yours | John Tyndall

RI MS JT/12/4052–9
Faraday Correspondence, 5:3185
LT Transcript Only

---

1. *My dear Mr Faraday*: Most of this long letter was extracted by Tyndall from his Journal (RS JT 2/13c/866–89).
2. *Huxley goodbye*: The letter was posted from Feldkirch in Austria.
3. *voiture*: a carriage (French).
4. *Regnault was better*: A catastrophic explosion in his laboratory in 1856 had left Regnault unconscious for a week and with serious head injuries.
5. *Maison Rouge in Strasbourg*: an historic hotel in the Place Kléber.
6. *diligence*: a stage coach for public use.
7. *to carry Mrs. Huxley*: Nettie Huxley was pregnant, though Tyndall was probably not aware of this.
8. *The Jungfrau hotel upon the Wengern Alp*: The rich meadowland of the Wengen Alp lies at a height of 6148 ft between the village of Wengen and the Kleine Scheidegg Pass. The Jungfrau Hotel at Wengen was opened in 1841 to cater for tourists.
9. *Bear Hotel*: The Hotel Bär (Bear) was one of two hotels erected at Grindelwald. in the 18th century. During Tyndall's lifetime it was very considerably enlarged, but was demolished in 1979.
10. *to make a Friday evening*: On 23 January 1857 Tyndall gave a Friday Evening Discourse at the RI, 'Observations on Glaciers', and published in *Roy. Inst. Proc.*, 2 (1854–58), pp. 320–27.
11. *The dirt bands of Prof. Forbes*: J. D. Forbes, 'Account of his Recent Observations on Glaciers", *Edinb. New Phil. Journ.*, 33 (1842), pp. 338–52. These concentric light and dark marks in glacial ice that resemble a rattlesnake's bands are caused by different rates of

glacial flow in summer and winter. Now termed 'Forbes Bands', Forbes had attributed the bands to interior fissures caused by tensions in the plastic mass of the glacier as its parts moved down a slope with different velocities—the fastest being in the centre.

12. *I left Grindelwald*: In what follows, Tyndall basically describes his exploration of the Bernese Oberland, as recommended as Route 25 in the 5th edition of Murray's guide to Switzerland (J. Murray, Murray's *Handbook for Travellers in Switzerland* [London: J. Murray, 1852], pp. 67–82).
13. *Handeck falls*: a spectacular waterfall at Handeck where the river Aar tumbles into a chasm 100ft below.
14. *A day has flown*: Tyndall evidently began this letter on 26 August 1856.
15. *M. Dollfus of Mielhausen*: Daniel Dollfus-Ausset (1797–1870), a Swiss industrialist and glaciologist from Mulhouse who had pioneered glacial photography.
16. *in former years sheltered Agassiz*: see letter 1274, n. 10.
17. *an overturned pail of mortar*: Forbes had made the comparison in his *Travels through the Alps of Savoy* (Edinburgh: Adam & Black, 1843), pp. 378–82.
18. *your old experiments*: These experiments had been performed by Faraday in a Friday Evening Discourse entitled 'Certain Conditions of Freezing Water' on 7 June 1850 and reported in *The Athenæum*, 15 June 1850, pp. 640–41.
19. *the 'spirit of the Brocken'*: Brockengespenst. An enlarged shadow of the observer cast upon clouds opposite the sun and the observer in misty conditions. The optical illusion, which is caused by refraction, was first noticed on the Brocken, the highest peak in the Hartz mountains of Germany. As Huxley and Tyndall observed, the illusion can occur in any mountainous region when the conditions are right.
20. *the enclosed letters*: letters missing.
21. *Mr Addams*: Robert Addams (1790–1875), a London scientific instrument maker. Tyndall was planning to give a series of eleven lectures on sound at the RI in the Spring of 1857.
22. *Vienna*: Tyndall was travelling to the meeting of the Gesellschaft Deutscher Naturforscher und Ärzte in Vienna.

To Hector Tyndale 27 August 1856 1274

Feldkirch Austria | 27th Aug. 1856.

My dear Hector

It was with a kind of pang—a sting of selfreproach, that I recognized your hand in the address of a letter last night[1] in a town near the Swiss frontier. You have been good enough not to complain of my neglect; but nevertheless I feel that I have been guilty of this in not replying long ago to your last kind letter.[2] Your books—for I knew they were yours—reached me safely; at least one part of the translation of Heine,[3] and that wonderful book of Walt Whitman's.[4] I welcomed them warmly and shall be glad to welcome your friends with equal warmth, if I be in London when they arrive. I have been for the last fortnight among the glaciers of Switzerland, and during the next fortnight I hope to visit some of those of the Tyrol. Afterwards my object is to join the German Savans[5] at their meeting in Vienna, which commences on the 17th of September and continues for 6 days. I hope you have given your friends letters to Chapman.[6] He is in connexion with so many eminent literary men that his acquaintance would be peculiarly valuable. I believe he is thinking of giving up the publishing business and of turning his thoughts to medicine: he will still carry on the Westminster Review,[7] but all else he purposes resigning. I sometimes, when I have sufficient leisure, write an article for the Review myself.[8]

I sent you a little paper[9] some time ago on the cleavage of Slate rocks, and one object I had in view in coming to Switzerland was to see whether the simple principle there enunciated did not make itself manifest in the structure of glacier ice. I am greatly pleased to find that it does so. Much, as you are aware, has been written about glaciers. Agassiz once took up a position[10] for a considerable time upon the middle of one of them, for the purpose of making observations upon them; and I and a friend of mine Huxley—whose name is already distinguished and will one day be celebrated in Natural History, slept one night in a hut[11] on a litter of hay, 8000 feet above the level of the sea, for the purpose of being able to carry on our observations at a sufficient height. We were beaten back by a heavy snow storm, but nevertheless learned a great deal. This was on the glacier of the Aar,[12] in sight of the position of Agassiz: but we were more fortunate upon two glaciers of Grindelwald and of the Rhone. You know the glacier slowly moves downwards, and various theories are now afloat to account for this motion. That most generally accepted in England is the so called viscous theory of Prof. Forbes.[13] He supposes ice to possess a quality like that possessed by warmed pitch or sealing wax. A power of flowing to some extent, and to this power he ascribes the descent of the mass. It is the idea which must first suggest itself, for nothing can be more like a gigantic pail of mortar overturned than some of these glaciers. But whether the notion will bear philosophic analysis is quite another question. Our ramblings upon the ice have been sometimes attended with great difficulty and certainly not without danger. I was very near losing Huxley upon the Rhone glacier last Sunday.[14] The chasms, or crevasses as they are called, are in some places terrific and we sometimes have had to walk between two of them upon an edge of ice not more than ten inches in width. We have been blessed, on the whole, with splendid weather, and some of the glimpses which we have had of the mountains summits have been indescribably magnificent. Last Sunday from an eminence about 9000 feet above the level of the sea, we could see at once the gigantic Finsteraarhorn[15] just beside us, the Weisshorn,[16] Monte Rosa,[17] with the black cone of the Matterhorn (Mont Cervin)[18] rising between them, like a savage tattooed with streaks of snow. Round further to the left were the snow summits of the Furca chain, and over all this the glorious sun shone from a sky the blackness of whose blue was almost awful. We had also the good fortune to see many phenomena of which we had previously but read, a splendid coloured circle drawn like a rainbow all round the sun; our own shadows cast upon clouds with a glory round the head of each: in fact we saw the 'spirit of the Brocken'[19] in all its splendour. You may infer from all this that I am in good health; I hope to return to London with the true muscles of a Tyndale—lithe and tough as the willows of the Barrow![20]

I hardly know what to think of Walt Whitman, though I have read many parts of his book three or four times. He seems to have occupied his mind in drawing out a catalogue of natural objects, which might, were sufficient power at hand,[21] be fused into a glorious poem. But whether it is that I demand too distinct a purpose from the poet, or that his purpose lies too deep for my discernment, I cannot see the drift of his poem. His perpetual reference to natural objects, and his wonderfully naked statements regarding them, prove that he possesses what the Germans call the objective faculty; but the whole is so mixed up with his own subjective thought, with the hues and tinges of a mind rank by nature, and unprimed by discipline, that I am rather bewildered when I seek real nourishment from this poem. The fault may lie in me but if you like frankness in Walt Whitman, you will not be inclined to blame it in me if I say that I believe he sometimes writes what he himself will not understand seven years hence. The world to him is a kind of spirit of the Brocken, a gigantic fog—maps with a halo round its head. I hope the discovery of this, which he will one day probably make, will not impair his strength, and that when the illusion vanishes, the power which made it its medium of expression will still remain.

Did I tell you that I had the great privelege and pleasure of meeting Carlyle? I spent[22] a few days at the seat of Lord Ashburton in Hampshire last January,[23] and met there Carlyle, his wife, and many other people of note. Tennyson had left on the morning of the day on which I arrived. When I next meet the old Titan I will ask him what he thinks of Walt Whitman. He once told me that if he were director of the British Museum library he would select the bad books from the good ones—a bad book he said was like a bad man, whose acquaintance was to be shunned for fear of injury. The bad books he would not be permitted to destroy, but he would put them out of the daylight. In three cellars of the building he would stow them away, so that they might be read by those who admired them, and on the doors of those cellars he would write Damned N° 1! Damned N° 2! Damned N° 3! I am curious to know into which of these he will thrust the Leaves of Grass![24] Mrs Carlyle is a very remarkable woman and her power of conversation quite wonderful.

The secret of her power appears to consist in a clear insight, and a terrible sincerity in announcing what she sees. I have also lately made the acquaintance of another estimable lady, Miss Jewsbury,[25] with whose writings you are no doubt better acquainted than I am.

And now Hector I think I must bid you goodbye. Give my love to all my cousins; and when your occupations let you, remember the drop of your blood which circulates in the laboratory of the Royal Institution.[26]

Ever affectionately yours | John Tyndall

[Postscript][27]

If the account which I read of the matter be correct the act to which you refer[28] was one of the most cowardly and brutal ever committed. It would be a national scandal to let it go unpunished.

I met a young American[29] a few days ago at the Grimsel, he was in mortal terror of death from fever. I comforted him and bathed his feet with hot water: my friend gave him some pills and he was well the next day. He was altogether helpless: had plenty of money but no culture—could not speak a word of French or German, and was terrified at the idea of being left among those who did not understand him. He told me his father had an office in Philadelphia, but I did not ask his name.

RI MS JT/1/T/1444

---

1. *a letter last night*: letter missing.
2. *your last kind letter*: letter missing.
3. *Heine*: Hector had evidently sent Tyndall a translation of the poems of the German poet Heinrich Heine. See letter 1140, n. 25.
4. *book of Walt Whitman's*: Walt Whitman (1819–92), American poet. The reference is to Whitman's *Leaves of Grass* (Brooklyn, N.Y.: priv. pr., 1855). Initially published anonymously with a portrait of the poet in workman's clothes, the 95-page book opened with the untitled poem that was later known as 'Song of Myself'. Tyndale probably sent Tyndall the 2nd edition published in 1856.
5. *German Savans*: Savants (French). Tyndall, Hooker and Frankland attended the Gesellschaft Deutscher Naturforscher und Ärtze in Vienna from 17–22 September 1856.
6. *friend's letters to Chapman*: Tyndall had evidently recommended that Hector's unspecified friend should send his literary efforts to the London publisher John Chapman.
7. *Westminster Review*: Chapman had edited the *Westminster Review* (founded 1824) since buying it in 1851, but did not resign as editor until 1887. He did, however, relinquish ownership in 1860 when he was declared bankrupt. He had studied medicine in Paris in the 1840s.
8. *article for the Review myself*: Tyndall had contributed reviews of scientific books to the *Westminster Review,* 63 (1855), pp. 551–58; 64 (1855), pp. 255–63; 64 (1855), pp. 557–65;

65 (1856), pp. 254–61; and 65 (1856), pp. 596–601. He made no further contributions.

9. *a little paper*: J. Tyndall, 'Comparative View of the Cleavage of Crystals and Slate Rocks', *Phil. Mag.*, 12 (1856), pp. 35–48.
10. *Agassiz once took up a position*: Agassiz had set up an observation hut (the 'Hotel Neuchatellois') on the Finsteraar glacier in 1838. See illustration in the Atlas accompanying his *Études sur les Glaciers* (Neuchâtel: Jent et Gassmann, 1840). Tyndall had visited the spot on 22 August 1855 (RI MS JT/213c/882–84). 'The rock is still there, bearing traces of names now nearly obliterated by the weather, while fragments around also bear inscriptions.... It was a temple of science now in ruins' (J. Tyndall, *The Glaciers of the Alps* [London: J. Murray, 1860], p. 19).
11. *slept one night in a hut*: This occurred on 22 August 1855 (RI MS JT/2/13c/883–84).
12. *the glacier of the Aar*: A system of glaciers in the Bernese Alps that drain into the river Aar.
13. *viscous theory of Prof. Forbes*: According to James D. Forbes, ice flowed slowly like a viscous liquid. See the editors' introduction.
14. *last Sunday*: 24 August 1856 (RI MS JT/2/13c/886).
15. *Finsteraarhorn*: The highest mountain (14,000 ft) in the Bernese Alps.
16. *Weisshorn*: White Mountain (German), a mountain (14,783 ft) in the Pennine Alps that Tyndall was to be the first to climb in August 1861.
17. *Monta Rosa*: see letter 1266, n. 2.
18. *Matterhorn (Mount Cervin)*: A pyramidal mountain (14,692 ft) on the borders between Switzerland and Italy and overlooking Zermatt. Mont Cervin is its French name.
19. *'the spirit of the Brocken'*: see letter 1273, n. 18.
20. *as the willows of the Barrow*: This appears to be a reference to the willows that were a regular feature of the course of the River Barrow in County Carlow.
21. *hand*: at this point a quarter of the original letter has been torn away. The letter continues in LT's transcription (RI MS JT/1/1614–18) until it resumes at the word 'spent'.
22. *spent*: the original holograph letter resumes here.
23. *last January*: see letter 1187. This suggests that Tyndall had not written to his cousin since September 1855, letter 1140.
24. *the Leaves of Grass*: see n. 4 above. Although Emerson alerted Carlyle to the publication of *Leaves of Grass* on 6 May 1856, Carlyle made no comment in his reply to Emerson on 20 July 1856 (*The Correspondence of Thomas Carlyle and Ralph Waldo Emerson, 1834*–1872, ed. C. E. Norton, 2 vols. (Boston: James R. Osgood, 1883), vol. 2, pp. 282, 285).
25. *Miss Jewsbury*: letter 1237, n. 1.
26. *remember the drop of your blood . . . the Royal Institution*: For Hector Tyndall's reply on 29 December 1856, see vol. 6.
27. *Postscript*: Only in LT transcript (RI MS JT/1/1618).
28. *the act to which you refer*: This probably related to John Brown (1800–59), the white abolitionist who was responsible for the murder in Kansas in May 1856 of five men who supported slavery (the 'Pottawatomie Massacre').

29. *a young American*: The episode was recorded in Tyndall's Journal (RI MS JT/2/13c/822). The American was apparently restored to health after Huxley gave him some pills.

From Henry Clifton Sorby 1 September 1856 1275

Broomfield | Sheffield, Sep. 1/56.

My dear Sir,

I was very sorry you were not at Cheltenham,[1] for I wished particularly to see you, and took with me a number of specimens and microscope objects on purpose to show to you. However as that could not be, I must content myself with thus writing. In the first place I must beg to thank you for the plumbago which I got all safe, and also the copy of your paper.[2]

I fear we have been calling by the term 'slaty cleavage' different things. In my paper on the microscopical structure of mica-schist, read at Cheltenham,[3] I described a structure which is seen to the greatest perfection in that of the Loch Lomond district.[4] For this I proposed the term 'crumple or contortion joints.' If the rock was of such a nature as not to yield as clay, but was a metamorphosed schist, and was compressed laterally, sharp contortions of small size are produced thus

If the force was not exactly evenly along the beds there is a lateral movement of parts with or without actual small breaks thus—

In some cases these joints are ½ inch apart, but in others only 1/100, and then it appears very much like cleavage. However I have not called it so, because you cannot cleave between joints—you can only fracture along them. It is not an ultimate structure, as in genuine solid slates, but only finite, like joints; though it may be on a very small scale. This structure explains the double cleavage of such slate as that at Shap,[5] in which one is of the ordinary ultimate kind, and the other is like that just described, in which the joints are often only 1/4000 of an inch apart. None of the beds nearly coincide with such a structure, there occurs exactly what you suggest in your last paper. The various parts do not move over one another with the same velocity but slip over each other and produce what I have always classed with [*word illeg*] sided joints, running nearly in the plane of cleavage. Such indeed to a small extent exist in very many slates, and must materially influence its splitting; but in my paper I have not alluded to this, since my object was to explain that ultimate structure which is found in the slates and which I have defined to myself as genuine slaty cleavage, as distinct from close jointing. Now you will see from this that if by pressure of any substance you gave rise to a close system of joints, I should demur to its being called cleavage; though it might indeed cause its material to cleave up into very thin laminæ. I may be scarcely correct in your opinion in doing this, but the microscopical structure in the two cases is so different, that some difference in times is absolutely requisite. You will understand me better by this drawing,

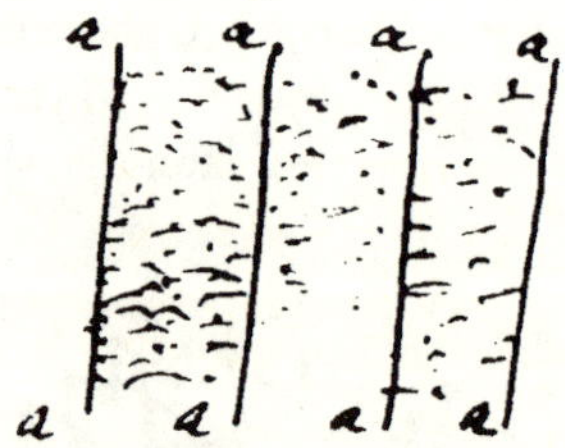

where we have for instance close joints at (a a) and you could not cleave the material in the intermediate space, whilst there is a structural arrangement of the ultimate particles in quite a different direction as you see drawn, along which you could cleave.

Thinking if you did not thus understand my views respecting what I would define as cleavage and close jointing, we might be led in our publications to call by the same term things that are different, I thought it best to write thus to you: for it would give me much pain if anything should occur to make us squabble about mere words. I wish you could contrive to come and see me some time or other, and inspect my collection to illustrate chemical and physical geology. I do not know whether you ever go into the country to lecture; but, if you would, I think I could arrange for you to deliver a course before our Sheffield Society.[6] If you would will you inform me.

Trusting you are well and enjoyed your trip

I remain | Yours very truly | Henry Clifton Sorby.

RI MS JT/1/4/1345–46.
LT Transcript Only

1. *Cheltenham*: The BAAS had met at Cheltenham 6–13 August 1856.
2. *the copy of your paper*: J. Tyndall, 'Comparative View of the Cleavage of Crystals and Slate Rocks', *Phil. Mag.*, 12 (1856), pp. 35–48. It is not known why Tyndall had sent Sorby a specimen of plumbago (an alternative name for graphite).
3. *read at Cheltenham*: Sorby read three papers at Cheltenham, notably 'On the Microscopical Structure of Mica-schist', but only an abstract is given in the *Brit. Ass. Reports,* 1856, p. 78.
4. *Loch Lomand district*: see letter 1137, n. 9.
5. *Shap*: a village in Cumbria near Penrith noted for its pink granite.
6. *our Sheffield Society*: The Sheffield Literary and Philosophical Society, in which Sorby was extremely active, had been founded in 1822.

To Michael Faraday 11 September 1856 1276

Innsbruck. 11th Sep. 1856.

My dear Mr. Faraday,

I believe my last letter[1] was posted to you at Feldkirch. Yesterday I reached this place[2] and had one letter from the post office. It was from Magnus,[3] dated August 17th, and had reference to a plan of a journey into which I had already plunged. He wanted to arrange matters so that we might meet somewhere but this will now be impossible. On crossing the frontier[4] I was happy to find better people and better German. From Feldkirch I proceeded to Landeck: the first instance of the piety of the Tyrolese struck me here, on passing the kitchen of the hotel I heard a confused murmur of voices and looking in saw the whole household upon their knees praying aloud before going to bed. From a conversation with some guests I learned that the nearest glaciers were in Kaunserthal[5] and up this valley I wandered next day. In the valley is a place of pilgrimage called Kaltenbrunnen[6] which is much resorted to by the devout. Near the chapel is a figure of Christ with a tube in his side from which is projected a stream of limpid water. The walls of the chapel are covered with offerings to the virgin, for here it is said she loves especially to dwell, and to shower her benefits upon her adorers. Tablets containing rude drawings of men and women in all possible positions of peril are hung along the walls; some half crushed under fallen rocks, others falling into rivers, women with carts overturned tumbling down precipices; most however were of men on mountain slopes lying under fallen pine trees. At the bottom of these tablets it was sometimes coolly stated that the person represented was saved by the manifest and immediate interposition of the Mother of God at Kaltenbrunnen. I like the piety of the people, but cannot say that this form of it pleased me. Onward to the highest inhabited part of the valley and here I engaged a chamois hunter to be my guide on the glaciers for two days. About 10 miles higher up was the Gebatsch Alp to which the dwellers in the lower part of the valley send their herds in summer; and in a cow house on the Alp I proposed to take up my nights quarters. There were two huts, one of which was occupied by 4 herdsmen, who milked the cows and made cheese and butter. A fine tall fellow with a wild countenance, his stockings, without feet, being drawn upon his legs, leaving a black zone of dirty skin between the upper rim of his stockings and the lower extremity of his breeches, asked me what he could cook for me. I asked for some milk and attacked the fare of the place, viz brown bread a month old, with the relish of hunger. The milk was rich and sweet, and I fared sumptuously. Went with my guide upon the glacier afterwards and remained upon it for several hours. The mass of ice was all in gray shadow, and a fringe of dying sunshine lay all along the opposite mountain summits as we finished our day's work. On coming again to the chalet, my cook asked me what I wished to eat: my reply was 'boiled eggs'. The four herds were there, all brownly sunburnt, the brown deepening into black with beard and dirt. The grouping of these men round the pine fire, which gave us our only light, was sometimes extremely picturesque. All of them smoked, and now and then one of them took a blazing torch from the fire to light his pipe, thereby casting a ruddy glare upon his wild face. A quantity of flour was placed in a wooden dish; milk was gradually added and the mass stirred up, all being done by the gleam of the pine logs. A saucepan was taken and some eggs were broken into it, sometimes the yolk clung to the shell and on such occasions the cook, who was the chief of the establishment, scooped out the contents with his finger, shaking dexterously the portions which attached themselves to the latter into the general mass. I had seen his fingers before and only dreaded that through a misunderstanding of my request the eggs were intended for me. A small broom of stiff twigs was taken from two pegs which supported it, and with this the eggs were pricked and whipped into a liquid, which, to my unspeakable satisfaction, was poured into the wooden dish already referred to. Meanwhile my eggs were selected and boiled—two of them were rotten—I could not help admiring at times the fine straight figure of the cook as he stood erect in the firelight and talked in the intervals of his vocation with those around him. The fire was made on a platform of stone elevated three feet above the floor, there was no chimney but the smoke found ample egress through the fissions in the sides and roof. At one end of the chalet was a hook which turned a small wheel, which imparted an up and down motion to a churn dish, which in its turn again made the butter. The beams and rafters were covered with cheeses drying in the warm wood smoke. The men had their supper and were the images of tranquil content afterwards. They all gathered round the fire, lighted their pipes and talked, with the gravity of philosophers, about those things which concerned them. Finding my interest alive as to his manner of making cheese the cook took me into his stores and explained the process to me. In one room were three gigantic masses of butter, and I amused my instructor very much by calling them butter glaciers.

As I have said there was a cowhouse near the chalet and above the ground floor a kind of cockloft was supported by pillars of pine, on the loft hay was strewn, and we reached it from the outside by means of a broken ladder. My guide shook up the hay and broke its nodules, piling up an eminence for my head. I lay down in my clothes with my scotch plaid as a blanket: my valet said I should find my feet cold before morning, and to prevent this piled hay upon me up to the hips. He suggested the propriety of continuing the process till my head was covered, but this I declined. Before morning however the biting coldness of the air which, sometimes blew in upon us proved that there was some wisdom in the suggestion. Having set me right my chamois hunter prepared a place for himself, and soon his heavy breathing told me that he was in a state of blessedness that I could only envy. During the long hours of the night I found a melancholy amusement in watching the keen gleaming stars travelling across the apertures in the roof above me. Once as I opened my eye the Pleiades[7] were there, twinkling in beauty: I strove to admire them, but an hour's sleep would have been worth a score of constellations. Sometimes I did approach a doze, which, when on the verge of deepening into slumber, was rudely broken by the clamour of an unamiable group of pigs which occupied the ground floor of our dwelling. The commotion of these animals usually commenced by a small grunt; this, like the first rattle of a pebble which announces the incipient motions of an avalanche, ascended to a grumble, and broke out at length in vociferous and angry expostulation. Rose at 5oclock greatly unrefreshed by my night's attempt at sleeping, still strong enough to bear the fatigues of the coming day.

Spent five hours upon the glaciers of Gebatsch[8] and learned a good deal: saw some marmots scampering over the rocks: returned to our hut, fortified ourselves, as far as it was possible to do with bread and milk, and faced the mountain, the summit of which we had to cross to reach Launtaufers Thal. After a rough ascent through the mossy pine clad Alp we came to the bare rocks. The weather has made havoc with the mountains broken them up into ruinous masses and showered them down the slopes. Among these we picked our way. Reached the end of a steep glacier whose surface was coated with sloppy snow; had this been frozen it would have been impossible to ascend. Once indeed I found my footing insecure and knowing by experience that if I fell so as to sit upon the ice I should be shot like an avalanche to the bottom, I took hold of the hand of my guide, to check this motion in case I slipped. He was a fearless man, far more reckless than the Swiss guides and not at all accustomed to act the cautious part of such. Whenever I made a proposal regarding the attainment of a perilous point, the word 'impossible', which I often heard in Switzerland, appeared not to be in his vocabulary. 'I dont know' was his usual reply 'it will certainly be dangerous but we can try it'. Near the summit of the slope up which we now toiled was an islet of stones and debris upon the glacier. Here we sat down to rest; right in front of us, and far above all vegetable life, surrounded only by bare sharp crags and dazzling snow, stood a beautiful chamois watching all our motions. The association of animal life and beauty with dead savagery around was exceedingly striking. Onward again for a time over a gentle ascent, till we reached the base of an ice slope steeper than any we had yet encountered. The chamois hunter faced it with confidence, though I should have regarded the ascent as impossible. I plodded after him through the snow slush fitting my feet into the tracks which he left behind him, almost afraid to look upwards or downwards. After a desperate struggle, which was felt as much by my guide as by myself, we reached the narrow row of crags which formed the summit and looked into the world of mountains beyond. The keen breeze here smote us but it put new vigour in our muscles and by jumping springing and sliding we descended through the mass of debris which here forms the mountain side. We reached the glacial ice once more. The mass had melted away from the mountain side which rose perpendicular to our right; and to our left an ice wall rose to a height of 50 or 60 feet. Through the gorge these formed we walked. From what I had seen of other glaciers I inferred that the mass on which we trod merely formed the roof of a cavern underneath; and I soon found that this conjection was right; for on turning an angle of rock my guide muttered with an expression of surprise and concern upon his countenance—'I did not expect this.' The ice had wholly melted in one place thus revealing a green tarn of unknown depth between the ice and mountain side. My guide inspected the place and desiring me to stand still, crept like one of his own marmots along the ice slope above the tarn. I endeavoured to round it by clambering over the rocks, but my guide warned me back. Returning to me he said 'we must try it together'–'good' I replied 'only do you take hold of this' handing him my plaid. Thus prepared for the cold plunge bath I expected I took his hand, and his glacier stick. Striking the latter into the ice I disposed of a portion of my weight and following cautiously reached the land safely. The sketch[9] will give you some idea of the gorge in section we first walked along the roof a until we reached a place where the ice from a to b had melted away, and the cavern c was exposed. We had to edge our way along the slope above a and this, contrary to my expectations, we accomplished in safety. I met Frankland next day.[10] He is now beside me and desires his kind remembrances. I think I must tell you the rest of my adventures by word of mouth. Kind remembrances to Mrs. Faraday and Miss Barnard. good bye—

As ever Yours | John Tyndall

RI MS JT/12/4060–6
Faraday Correspondence, 5:3186
LT Transcript Only

1. *my last letter*: letter 1273. Most of the contents of the present letter were taken from Tyndall's Journal (RI MS JT/2/13c/890–907).
2. *this place*: Innsbruck, in Austria, lies about 300 miles south west of Vienna.
3. *from Magnus*: letter missing.
4. *the frontier*: the frontier between Switzerland and Austria; Tyndall was now travelling in the Austrian Tyrol.
5. *Kaunserthal*: (modern) Kaunertal in Austria is a region 12 miles to the south east of Landeck containing many small towns and villages that ascend to up to 9000 ft above sea level. The Kaunertal Glacier is the highest in the Tyrol.
6. *Kaltenbrunnen*: a village to the south of Prutz about 8 miles south of Landeck at the head of the Kaunertal valley. It was renowned for its chapel.
7. *Pleiades*: a prominent cluster of seven stars in the northern hemisphere located in the constellation of Taurus and sometimes called the Seven Sisters.
8. *The glaciers of the Gebatsch*: (modern) Gepatsch. Now known as the Kaunertal Glacier which drains into the modern (1980) Gepatsch reservoir.
9. *The sketch*: not in transcript.
10. *Frankland next day*: Tyndall met Frankland on the Finstermünz Pass on Tuesday 2 September 1856.

From Carlo Matteucci[1] 13 September 1856 1277

Mon cher Monsieur

Il y a bien longtemps que j'aurais désiré de vous ecrire d'autant plus que je devais vous remercier des different mémoires et brochures que vous avez eu la bonté de m'envoyer. J'avais espéré de vous rencontrer quelque part en Suisse mais il parait que vous n'y êtes pas venus, au moins dans les deux mois d'été que j'y ai passé. J'ai eu le plaisir de voir à Pise M. et M[de] Sabine et nous avons beaucoup causé de vous et de vos travaux. J'ai donné à M. Sabine un petit appareil que j'ai fait faire dernierement pour montrer et mesurer la difference de conductibilité du bismuth axial et equatorial. Avec cela il y a quatre tiges de bismuth, deux de chaque qualité. Il ne manque plus pour faire l'expérience que d'avoir un galvanomêtre différentiel et il suffit qu'il soit fait d'un double fil de cuivre qui fait 15 a 20 tours autour d'un bon système astatique. J'ai prié

M. Sabine de vous donner tout celà. Mais en cas que vous vouliez faire l'experience je vous prie d'avoir soin de bien amalgamer les extrémités des tiges en plongeant ces extremités dans du mercure pur et legerement chauffé. Il est aussi essentiel que ces extremités plongent egalement dans le mercure contenu dans les petites cavités de l'appareil. Enfin, une fois qu'on s'est assuré qu'avec les deux tiges de cuivre et les deux petits axes du même métal, les deux circuits du galvanometre sont egaux, il faut enlever les tiges de cuivre et mettre à leur place celles de bismuth [sans rien bouger au reste].[2] Il faut encore eviter de tenir fermé le circuit de la pile pour plus qu'il n'est necessaire pour voir le sens de la déviation. C'est à peu près un tiers ou un demi seconde. En essayant les deux tiges de bismuth de la même espece, il est presque impossible de ne pas avoir quelque deviation. Si les tiges ont été bien réduites des mêmes dimensions, les différences sont très petites et autrement celà serà du à quelque defaut d'expérience. En comparant une tige axiale et une equatoriale la différence est grande et constante. Je vous demande pardon d'avoir insisté sur ces petits détails car j'ai été obligé de quitter Pise [avant le départ de M. Sabine][3] et le jour même que mon Mécanicien a préparé les tiges de bismuth, de sorte que je n'ai pu m'assurer de leur exactitude. J'espère qu'elles seront bien, car ce Mécanicien a acquis une grande habitude et il travaille avec un bon comparateur de Froment.

Je viens d'achever mon memoire sur les courants du disque d'Arago et j'espere bientot pouvoir vous en envoyer un exemplaire. En attendant je m'occupe de reunir mes experiences sur le diamagnetisme que je poursuis depuis trois ans presque sans interruption. Malheureusement mes resultats ne peuvent pas se dire en deux mots.

Croyez-moi | tres affectuesement | C. Matteucci | Pise. 13 Sept. 1856.

My dear Sir,

I have been wanting to write to you for a long time, especially as I had to thank you for the various memoirs and brochures[4] that you were kind enough to send me. I had hoped to meet you somewhere in Switzerland, but I heard you did not visit, at least not during the two summer months that I spent there. I had the pleasure of meeting with Mr and Mrs Sabine[5] in Pisa, and we talked a lot about you and your work. I gave Mr Sabine a small device that I had recently made to demonstrate and to measure the difference in the conductivity of axial and equatorial bismuth. With this there are four rods of bismuth, two of each quality.[6] The only thing missing to conduct the experiment is having a differential galvanometer and it suffices if this is made out of a double copper wire that wraps 15 to 20 times around a good astatic system.[7] I asked Mr Sabine to give you all of it. But if you decide to conduct the experiment, I urge you to take care to fuse together the ends of the rods by dipping them in pure and slightly heated mercury. It is also essential that these ends are plunged in the mercury while held into the small cavities of the device as well. Finally, once it is certain the two circuits of the galvanometer are equal with the two copper rods and the two small axes of the same metal, the copper rods must be removed and replaced with those made of bismuth, without changing the rest [of the device][8]. Then, the circuit of the battery should not be closed for longer than is necessary to see the direction of the deviation. [It is about a third or a half a second.][9] While trying the two bismuth rods of the same type it is almost impossible not to have some deviation. If the rods were properly shortened to equal size,[10] the differences are very small; otherwise it will be due to some fault in the experiment. The difference is greatest when comparing an axial rod and an equatorial one. I beg your pardon for insisting on these minute details, for I had to leave Pisa before Mr Sabine's departure and on the very day my mechanic[11] had prepared the bismuth rods, so I was not able to ensure[12] their exactness. I hope that they will be good, since this mechanic has acquired great skill and he works with a good comparative device[13] designed by Froment.[14]

I have just completed my memoir on the currents of Arago's disk,[15] and I hope to be able to send you a copy of it soon. In the meantime, I am busy putting together my experiments on diamagnetism, which I have worked on for three years almost without any interruptions. Unfortunately my results cannot be described in two words.

Sincerely | My best regards | C. Matteucci | Pisa 13.Sept 1856

RI MS JT/1/M/59

---

1. *Matteucci*: Unlike the other Matteucci letters in the collection, this letter is written in a neat clear hand of either his wife or a secretary. Two interpolations in Matteucci's own handwriting are noted below.
2. *Sans rien bouger au reste*: interpolation in Matteucci's hand.
3. *avant le depart de M. Sabine et*: interpolation in Matteucci's hand.
4. *various memoirs and brochures*: presumably offprints of Tyndall's papers, though what 'brochures' refers to, is unclear.
5. *Mrs Sabine*: Elizabeth Juliana Sabine (née Leeves) (1807–79), whom Sabine married in 1826. She translated Humboldt's *Cosmos* and several other scientific works into English.
6. *two of each quality*: that is, two samples of bismuth axial in conductivity, and two samples equatorial in conductivity.
7. *a good astatic system*: Matteucci was referring to the astatic galvanometer in which the disturbing effect of the earth's magnetic field is eliminated by fusing two magnetic needles together with their poles reversed. This produced a net magnetic field of zero and no preferred direction to the earth's magnetic field.
8. *of the device*: interpolation in Matteucci's hand.
9. *It is about a third or a half second*: interpolation in Matteucci's hand.

10. *in equal size*: or, to the same size.
11. *my mechanic*: not identified.
12. *I was not able to ensure*: or, I was unable to check up on.
13. *a good comparative device*: or, measuring device.
14. *Froment*: Paul-Gustav Froment (1815–65), Parisian instrument-maker and electrical engineer. The specific instrument mentioned has not been identified.
15. *Arago's disk*: When a copper disk carrying a centrally–mounted compass needle is rotated by means of a crank and pulley system, the compass needle begins to rotate due to eddy currents induced by the disk. The effect was first noted by Dominique Arago in 1824. Matteucci's investigation of the device was reported as: 'Mémoire sur l'État Électrique Induit dans un Disque Métallique Tournant en Presence de l'Aimant', *Paris, Comptes Rendus,* 43 (1856), pp. 286–88.

From Michael Faraday 15 September 1856 1278

Hornsey[1] | 15 Sep 1856

My dear Tyndall,

It is too bad of me to think of putting you off with a short letter, and yet I think I must do so, to save the post, for the 15th[2] has come most unawares on me. Being here in the country for my wife's benefit, the dates are beguiled away from me. Your letter was most welcome, and many have enjoyed it. I return you our most sincere thanks for the great favour. On our part we have no events to describe; we are rusticating. I try to experiment and write, but I forget so fast that I have the pleasure of doing both over two or three times. Never mind. A time for all things.[3]

I called and saw Mr. Addams[4] the other day. I found him disinclined to separate his acoustical apparatus[5] one part from another. He has told you, I believe, that the price of the whole would be £80. I have not seen them, and indeed if I had, I could not have formed any judgment, for I do not know the present state of acoustical apparatus and should be unable to make any comparison with that which exists elsewhere, or in the collection of lectures. He said he would send me a bit of them, but I have not received it.

I suppose before you receive this you will meet with Dr. Bence Jones.[6] If so, do not let him forget me. Very kindest thoughts to both him and you from all of us here. My wife is really better for the country, and we have had very fine weather. As to the house in Albemarle St., Anderson is well, Mr. Vincent[7] is well, Miss Savage[8] is somewhere in France, and has sent a letter, but no address. I do not know whether you expect to meet with any of our German friends at Vienna. Schrotter[9] remember me to, and the others if you see them. I think you told me something about Magnus and the transparency of gold [foil]. If so, tell him it works on pretty well—very curiously in my mind—but as yet with no broad distinct result which will strike scientific men, though I hope it may help the working of some.

Your account of your wanderings in the Oberland and also by Strasburg Bale, &c &c &c was most delightful to us, having been over the same ground.[10] We also were stopped at the Grimsel glacier, by illness. We had agreed to meet Agassiz, and could not go up to him, and so it goes in intentions and acts: not always well accordant, but let us hope well.

I must not stop | Ever, dear Tyndall, | Yours very truly | M. Faraday.

Dr. Tyndall | Post Restante | Vienna | Austria[11]

RI MS JT/12/4141
Faraday Correspondence, 5:3189
LT Transcript Only

---

1. *Hornsey*: The Faradays were spending the summer at Hornsey, which was still a village to the north of London near Highgate and Crouch End, for the sake of Sarah Faraday's health.
2. *the 15th*: In letter 1273 Tyndall had told Faraday that the 15 September would be the last date by which letters posted to him would reach him in Vienna.
3. *A time for all things*: Shakespeare, *The Comedy of Errors*, II, 2, 65.
4. *Mr Addams*: see letter 1273, n. 20.
5. *his acoustical apparatus*: see letter 1273, n. 20.
6. *you will meet with Dr. Bence Jones*: It is not known whether Jones attended the Vienna meeting.
7. *Mr Vincent*: Benjamin Vincent (1819–99), Sandemanian Librarian of the RI who had married Ellen Barnard.
8. *Miss Savage*: Sarah Savage (1808–65) had been the RI's housekeeper since 1835.
9. *Schrotter*: Anton Schrötter (1802–75), Austrian chemist who discovered the red form of phosphorus in 1848. He was Professor of Chemistry at the Vienna Polytechnic from 1845 and Secretary of the Vienna Academy since 1851.
10. *over the same ground*: The Faradays had travelled in Switzerland in the summer of 1841 and seen the upper glacier of Grindelwald.
11. *Post Restante | Vienna | Austria*: Address transferred from cover.

From Auguste Arthur De La Rive 24 September 1856 1279

Florence le 24 9bre 1856

Mon cher Monsieur,

Je reçois dans cet instant votre bonne & intéressante lettre dont je vous remercie excessivement. Toutes les nouvelles que vous me donnez m'ont vivement intéressé. Vous comprenez combien je suis de votre avis en ce qui concerne notre excellent ami Faraday. Il y a encore plus longtemps que vous que j'ai pu apprécier la noblesse de son caractère qui ne le cède en rien à la hauteur de son intelligence. Si vous y pensez, rappelez moi à son bon souvenir quand vous le verrez, & faites lui nos compliments les plus affectueux de ma part & de la part de ma femme. Ce que vous me dites du travail qu'il vient d'achever, a piqué vivement ma curiosité, & je suis bien impatient d'en prendre connaissance. Vous ne me dites rien de vos propres recherché; je suis sûr que vous avez quelque chose d'intéressant sur le tapis. —Quant à moi maintenant que mon troisième volume est achevé, je vais, si Dieu le permet, revenir à des travaux de recherche. Je suis vivement impatient, comme vous le comprenez. Vous ne vous faites pas une idée de tous les querenda que la confection de mon ouvrage m'a suggéré. Je voudrais seulement avoir la capacité physique & intellectuelle nécessaires pour attaquer toutes les questions qui me semblent mériter qu'on les traite. Heureusement que j'aurai incessamment une aide précieuse dans un fils âgé de 23 ans qui, après être sorti honorablement de l'École Polytechnique Française (il est sorti le 25eme sur 120) vient de travailler un an dans le laboratoire de M. Deville à Paris & ira probablement passer un an à Berlin, comme je vous l'ai dit. Vous savez même que vous m'avez offert vos bons offices pour l'introduire chez M. Magnus, je ne l'oublie pas & j'y aurai recours pour l'année prochaine.

Je n'ai point encore vu Matteucci; j'irai incessamment à Pise passer deux ou trois jours avec lui. C'est un homme d'esprit & d'un bon caractère; mais il n'est pas toujours bien sur en science, quoiqu'il soit très ingénieux & perspicace. —Il est bien pourtant parmi les hommes scientifiques de l'Italie l'un des meilleurs.

Cette lettre vous sera transmise par Mr Walker à qui j'écris de vous envoyer les épreuves. C'est vous dire que j'accepte votre coopération malgré le dérangement que cela vous cause. Aussi croyez bien que c'est un service que je n'oublierai jamais, & j'espère avoir dans ma vie plus d'une occasion de vous prouver ma reconnaissance. —Il va sans dire que vous prendrez pour lire les épreuves tout le temps que vous voudrez; rien ne presse. J'écris à Mr Walker de vous laisser toute liberté à cet égard. —Quant aux corrections, ce n'est pas seulement pour le style, mais pour le fonds que je voudrais vous en voir faire. Vous avez toute liberté à cet égard. —Ma seule crainte, c'est le temps que cela vous prendra, vous à qui il est si précieux sous tant de rapports. —Encore un mot sur ce sujet : si, après quelques essais, vous apercevez que ce travail vous prend trop de temps, dites le moi franchement, je recourrai alors à cet ami non-scientifique dont je vous ai parlé. —Ce n'est qu'à condition que vous userez avec moi de cette franchise qui me montrera que vous êtes véritablement mon ami, que j'accepte. —C'est donc entendu.

—Vous pouvez m'écrire encore à Florence poste restante. —À partir du 10 Décembre c'est à Rome qu'il faudra m'adresser les lettres chez Mess. Schlotter frères.

Votre tout dévoué & affectionné | A. de la Rive.

Florence, November 24th 1856

My dear Sir,

I just received your pleasant and interesting letter,[1] for which I am extremely grateful. I have been interested in all the news you gave me. You realise how much I agree with you regarding our excellent friend Faraday. For much longer than you, I have been able to appreciate the nobility of his character that is in no way inferior to the level of his intelligence. If you remember please remember him to me when you see him, & send him our most affectionate regards from me & my wife. What you are telling me about the work that he has just completed greatly piqued my curiosity, & I am eager to become familiar with it. You have not told me a thing about your own research, and I am sure you have something interesting to bring to the table. For my part, now that my third volume is complete,[2] God willing, I will return to research work. I am very much longing for it, as you can imagine. You have no idea about the querenda [complaints][3] that the making of my book has brought on me. I only wish I had the physical & mental capacity needed to tackle all the questions that I deem deserving to be dealt with. Fortunately I will shortly receive the precious help of my 23-year-old son,[4] who after finishing the course with honours at the École Polytechnique Française (he finished 25th out of 120) has just worked for a year in Mr Deville's laboratory in Paris[5] & should probably spend a year in Berlin, as I have told you. You also know that you had offered your assistance to find him a position with Mr Magnus;[6] I am not forgetting it & I will return to this next year.

I have not yet seen Matteucci; I will soon spend two or three days in Pisa with him. He is a witty man & has a good character, but his science is not always very reliable, although he is very ingenious & insightful. However, he remains one of the best amongst the men of science in Italy.

This letter will be delivered to you by Mr Walker,[7] and I am writing to him to send you the advanced copies.[8] That is to say that I accept your cooperation despite the inconvenience it causes you. Hence, take it as a favour, & I hope to have in my lifetime more than one opportunity to show you my gratitude. It goes without saying that you can take as much time as you would like to read the proofs; there is no hurry. I am writing to Mr Walker to grant you complete freedom in this matter. As for the correction, it is not merely the style, but also the content that I wish you would rectify. You will have complete freedom in this matter. My only concern is in regards to the time it will take you, your time is so invaluable to you in every respect. One more word on this subject: if after a few attempts you realise that this work takes up too much of your time, be honest with me about it and I will turn to this non-scientific friend I have mentioned to you. I will only accept providing you will treat me with this openness demonstrating that you are my true friend. It is agreed then. You can still write to me by post restante in Florence. From December 10th you will have to address your letters to Messrs Schlotter brothers in Rome.[9]

Your devoted & affectionate | A. de la Rive.

RI MS JT/1/D/90

1. *interesting letter*: letter missing.
2. *my third volume is complete*: De la Rive had just finished writing the third volume of his *Traité d'Électricité Théorique et Appliqué,* but it was not published until 1858. No acknowledgment to Tyndall was made in the Preface.
3. *complaints*: or laments. De La Rive uses the Latin, *querenda,* for emphasis.
4. *my 23 year old son*: Lucien de la Rive (1834–1924) was the only son of Auguste de la Rive's first marriage to the writer Jeanne-Mathilde Duppa (1808–50). He published a large number of mathematical papers on electromagnetism and gravitation, as well as a translation of Tennyson's poetry into French. See also letter 1105, n. 10.
5. *Mr Deville's laboratory in Paris*: Henri Étienne Sainte-Clair Deville (1818–81) had been Professor of Chemistry at the École Normale Supérieure in Paris since 1851.
6. *introducing him to Mr Magnus*: Formal notices of introduction and recommendation were etiquette amongst Europeans.
7. *Mr Walker*: The electrician Charles Vincent Walker (1812–82) who was translating De la Rive's *Traité d'Électricité* into English.
8. *to send you the advanced copies*: i.e. of De la Rive's treatise.
9. *Messrs Schlotter brothers in Rome*: not identified.

To Thomas Archer Hirst 1 October 1856 1280

Paris, Wednesday night

My dear Tom

When I left London I purposed writing to you from Switzerland or the Tyrol, but the resolution faded before other demands. To save my purpose from total shipwreck I sit down on the last night of my journey,[1] with the rumble of the Paris omnibuses in my ears, and the day's traffic in Paris intellect in my brain, to write to you. I am weary and muddy and hence cannot offer you the fare you merit, but still I know that any stupid lines will not be altogether unacceptable to you. I have not written, but on the mountains and in the warm sheltered valleys of the Tyrol I have thought of Anna and yourself.

I thought of Anna at Meran[2] while looking at the vineyards laughing in the sunshine, and wished both of you at my side. The thought of you came with all the greater force because Meran is resorted to by patients suffering from chest ailments both for the mildness of its air, and for the purpose of feeding on the ripe grapes in due season. In company with Huxley I visited many of the glaciers of Switzerland, and afterwards alone and in company with Frankland I clambered over the icehills of the Tyrol. While in a church at Innsbruck my friend Hooker his wife and a couple of friends made their appearance: I attached myself to their party, journeyed with them through many delightful scenes, reached Vienna along with them, remained there for 9 days, and lived through the meeting of the German Naturforscher:[3] and now we[4] are all housed together at the Hotel des Étrangers in the Rue Vivienne. We had in general glorious weather during our journey in Switzerland. We managed, that is Huxley and myself to combine a little work with our excursion. The subject of glaciers is one that has engaged the attention of a great number of men. In England the views of James Forbes of Edinburgh appear to be almost universally accepted. He has written on the structure of the ice, which is very curious, and proposed a theory to account for this structure.[5] Both Huxley and myself imagine that precisely the same principle which I have applied to the explanation of slaty cleavage applies also to the structure of glacial ice. We spent some days upon the Grindelwald glacier. In a hut 8000 feet above the level of the sea we lay one night wrapped up in hay in order to be able start early next morning and reach the origin of the Aar glacier: we were beaten down again by a snow storm. On the glacier of the Rhone[6] we had a glorious day. Soon after entering the Tyrol I engaged a chamois hunter[7] for my guide. The first night we passed together in a cow house on the Gebatsch alp: A kind of cockloft supported by pillars of pine was raised above the lower floor, and on this a litter of hay was placed. My chamois hunter shook up the mass to break its nodules and render it soft. I lay down with my plaid wrapped about me. He told me my feet would be cold before morning and to prevent this piled hay upon me up to the hips. He proposed to continue the operation until my head also should be covered but this I declined; nevertheless the bitter air of the morning, for we lay beside a glacier, told me that there was good sense in his suggestion. He next prepared a place for himself and his heavy breathing soon spoke to me of a state of blessedness which I could only envy. Sleep kept away from me during the night and I had a melancholy satisfaction in watching the keen eyed stars travelling across the aperture in the roof above me. Once as I opened my eyes the Pleiades were there twinkling in beauty. I watched them as they moved away, and sought to admire them, but an hour's sleep would have been worth a dozen constellations. Next day my daring guide led me over the glacier, and across a mountain 10000 feet in height; that day was one of the greatest toil, and perhaps of the greatest risk that I ever passed. My guide was almost as much cut up as myself, and he doubted his ability to return to his home without devoting a day to the recovery of his strength. I have spoken of risk. The clefts and chasms of the glaciers are in some places terrific, and you have sometimes to cross a knife edge for a distance of 10 or 15 yards, with about ten inches in width to walk upon and a chasm 60 feet deep yawning like the jaws of a ferocious whale on the right hand and on the left. Huxley was once within a hair's bredth of falling into a gulf of unknown depth, for neither of us could see the bottom of it. Some of the shafts in the ice are undoubtedly 400 feet deep at least. This I found by calculation. I think my greatest peril was occasioned by a small chasm which had been wholly bridged over and hidden by soft snow. My right leg went right into it, but with the purchase of the other I pushed myself vigorously forward and thus threw myself to the opposite side of the fissure. I have a good mind to make a small book of my journal;[8] I think it might be made instructive and would help perhaps to pay the expenses of my two months journey.

In Vienna I made many acquaintances and had every reason to be gratified by the cordial welcome and good treatment we received. I met Plücker there. He was polite and cold, and I reconciled myself to the fact. I saw him afterwards at Ettingshausen and I thought he seemed to relent as Ettinghausen and myself conversed together. I was induced to make a communication on the ice structure,[9] and among the audience I could see the smiling countenance of our good friend Gerling,[10] as if he seemed content with his old pupil. Here in Paris also I have met with much civility and kindness, so that I return to London with good health and good feeling towards mankind in general. And now Tom let me hear from you soon. I long to hear about Anna; give my best love to her. Her feelings towards me would be of a very curious nature if I told her all the experiences I have had during the last 3 weeks. God bless you and her is the prayer of your

affectionate | John

RS MS JT/H/470–1
LT Transcript Only

1. *on the last night of my journey*: This remark dates the letter as Wednesday 1 October 1856.
2. *Meran*: Meran or Merano is a town in the Italian part of the South Tyrol where Tyndall stayed on 6 September 1856 (RI MS JT/2/13c/900–901).
3. *German Naturforscher*: Tyndall had attended the Vienna meeting of the Deutscher Naturforscher Gesellschaft from 16 to 22 September 1856.
4. *and now we*: Tyndall was staying in Paris with Hooker and Frankland before returning to England.

5. *proposed a theory . . . for this structure*: For Forbes' viscous theory of glaciers, see our Introduction.
6. *Rhone*: The Rhône glacier is the source of the river Rhône and is situated between the Grimsel and Furka passes.
7. *A chamois hunter*: Tyndall only named him as 'Johann' in his Journal, but in *Glaciers of the Alps* (London: J. Murray, 1860), p. 29, he was named as Johann Auer. He was hired between 31 August and 2 September to guide Tyndall over the Gepatsch Alp (RI MS JT/2/13c/891–97).
8. *a small book of my journal*: This materialized (after several more visits to the Alps) as *The Glaciers of the Alps*. See n. 7.
9. *communication on the ice structure*: On 18 September 1856 Tyndall recorded in his Journal that 'talked to them for half an hour: a hearty round of applause at the conclusion, which was the first thing of the kind that I had heard in the [physical] section' (RI MS JT/13c/913). The extempore talk was not published.
10. *Gerling*: Christian Ludwig Gerling (1788–1864), Professor of Physics at the University of Marburg from 1817 to 1864. Tyndall had attended his lectures and joined him in astronomical observations in the university's observatory in the autumn of 1848 and spring of 1849.

From Thomas Archer Hirst 5 October 1856 1281

14, Rue de la Mairie | à Pau. | Oct. 5th, 1856

My dear John,

I was rejoiced to receive your letter[1] this morning for Anna and I had long wondered at your long silence and we began to fear that you had disappeared in one of those horrible four-hundred feet deep fissures. I often recalled your former adventures, and the fascination that bodily peril always had for you, until I began to feel some anxiety at your silence. I consoled myself generally, however, by reflecting that there always was a certain higher prudence even in your recklessness, and that there were many other assignable reasons for your silence. Thank God however you are safe home again, and I shall hear from you more frequently. I approved highly of your glacier journey the moment you mentioned it to me; it was calculated to satisfy all your wants, scientific, mental and physical. I trust you have returned home with bodily strength enough to carry you through the winter, & that the scientific fruits of your journey will be at any rate sufficient in your own eyes to banish the thought that any time was lost. I have no doubt a description of your journey would sell;[2] I know one who would read it with pleasure whether in print or manuscript. Indeed you will, I hope, give me many accounts of your rambles by letter as it will be long perhaps before we can talk together about them. Tell me above all some particulars of your scientific observations and your intentions respecting them.

We left Biarritts about a fortnight ago, and are now quietly settled in our winter residence.[3] I cannot say that Biarrits was beneficial to Anna; she never liked the place, and was glad to leave it. Her health fluctuates very much, sometimes she is better, sometimes worse than she was in England; on the average I think we may say she is about the same. For the last few days there has been a decided improvement in her strength, spirits and appetite, and this leads us to hope that Pau may be beneficial to her. We have taken a pretty little 'apartment meuble, au premier'.[4] It consist of a salon, sale-à-manger, and kitchen, and three very small bedrooms which are, in fact, mere alcoves two of which open into the salon and are occupied by myself and Anna; and the third opens into the dining room and is intended for John Martin whom we expect daily. For this accommodation we pay 900 francs (£36) for the season of 8 months; and although far more expensive than I anticipated when in England, I have the satisfaction to know that we could not have obtained lodgings in Pau for less. Besides this we have a servant to whom we pay £1 a month; she takes her meals with us, and sleeps 'chez elle'.[5] Here too we have been fortunate; we took the first servant who presented herself and so far she has proved herself to be an honest, faithful girl. Moreover she has taken a great liking to Anna and nurses her affectionately. She is a Bearnese girl,[6] and she amuses herself often by teaching Anna to say a few words in her own patois. When she leaves us at night she bids me 'Bon soir, Monsieur,' in good French, but to Anna she says 'ally shat',[7] and smiles pleasantly when Anna returns the greeting in her own words. Pau is a lovely place as far as its situation is concerned. In five minutes we can be in the Place Royal, which is a fine square planted with trees and abundantly provided with wooden seats. In the centre is a fine statue of old Henri IV.[8] The view from the south side of it is as fine as any I ever saw. Two hundred feet below you is the rapid river Gave, which flows westward to the sea through a beautiful and fertile valley. Immediately on the other side of this river the Pyrenees may be said to begin. In the foreground are bold but not lofty hills, well wooded to their tops, and here and there amongst the trees may be perceived quaint picturesque old chateaux, like gems set in emerald; beyond these rise other hills, each growing more barren, more rugged and more blue in succession, until at last high up in the bright blue sky are the glittering while peaks of the summits of the Pyrenees. Not satisfied with this magnificent panorama visible to the bodily eye, ever-craving fancy descends beyond the snowy peaks to imaginary vineyards of Spain. Immediately to the South, when looking from the Place Royal, the eye rests with delight upon the boldest peak which is detached from the rest and looks like a mighty sugar loaf chopped down the middle with a mightier axe; it is the Pic du Midi d'Ossan—the highest mountain but one of the range. The highest, the Pic du Midi de Bigorre, closes the view eastward, and from being surrounded by other high mountains is less bold and striking. This is enough about the beauties of Pau though I could easily write more. Now and then, Anna being unable to walk far, we hire a gig and horse and trot up some of the lovely branch valleys that lead from the broader valley of the Gave towards the mountains. These were treats indeed to both of us but our last trip had nearly been fraught with serious consequences. The horses they let out to hire are miserable creatures. I have scarcely seen one without broken knees. On Friday last I observed as usual that our horse had been accustomed 'to praying' on the highroads, and consequently I was, as usual, very careful, so much so that we accomplished our journey outwards without a stumble. On our return, however, when about 5 miles from Pau on a beautiful level piece of road (the horse going at a jog trot) I looked for a moment on one side to admire some object or other, and in a second down came the brute, not merely on his knees but on his side. The gig did not overturn but we were both pitched out over the dash-board. I miraculously alighted of my feet—like Tidmarsh in the celebrated crash at Northowram[9]—and to my horror I saw Anna thrown with her head on the horse and made to turn a complete somersault. I had her in my arms almost as soon as she reached the ground and would you believe it, she was not in the least hurt, her bonnet even was scarcely bruised, and she was calm and collected in a moment. An umbrella which she had had to protect her from the glaring sun she still held in her hand. She insisted she was not hurt, but I could not at the time believe her; I thought her calmness and presence of mind unnatural, and dreaded a relapse might come. All this happened in far less time than you can read it, and a struggle of the prostrate horse reminded me that I must seize his head and keep him down, or with his struggles he would dash the gig to pieces. This I did, and Anna took up a gig cushion, and seated herself on some stones on the roadside and watched me and a good-natured farmer, who had just passed us and came galloping back to our assistance, extricate the poor horse. This done, we mounted again and drove very slowly and very carefully home. Anna continued quite well and even lovely; my nerves appeared to have sustained a far severer shock than hers. For a day or two I looked anxiously for bad consequences of our upset, but no, they did not appear, and as, indeed, she has continued to improve ever since I can look back upon our accident as an adventure merely and from it learn caution for the future. It was the umbrella, I believe, which saved Anna; it was bruised terribly, it had an iron stem which was bent into a right angle, and altogether it looked as little like an umbrella after its fall as one could well conceive. Our good Bearnese was horrified when we told her the story, she too could not believe that Anna had sustained no injury. After assisting her to take off her shawl and bonnet she looked at her with wonder and kindness and ejaculated, '<u>Vraiment Dieu a vous aimé</u>'.[10]

And now a word about myself. I have, as you will readily conceive, been much occupied with Anna night and day since we left England. In the few hours per day that I could call my own I have worked at my mathematics. I have taken up my old work on attracting bodies,[11] completed much that was unfinished, improved upon what was done, and pushed my investigations a little further. I have plenty of work before me, to which I have cut out the road. I could at any time complete a little notice for publication, but I postpone doing so from a wish not to let what I have done go from me until I have worked more at kindred subjects so as to feel sure that my treatment of the problem is good. As long as I can make what I have done more complete or improve upon it I will keep it by me. Since I left Biarrits I have occupied myself exclusively with the subject of determinants,[12] a subject which I must master. I am doing so though with great labour. The only treatise I possess is one by Spottiswoode,[13] published by Crelle.[14] It is one of the most confused things I ever read in mathematical literature, full of misprints, of the gravest kind, and full of looseness on the author's part. I expect an English edition by the same author shortly which I hope will lighten my labour. Nevertheless I already feel that my great labour will be beneficial. I have had absolutely to discover theorems for myself so that out of a bad book one can get great good. Probably I should not have had my present practical insight into the subject had the treatise been more correct and lucid. 'It is an ill wind that blows good to nobody.' [15]

John Martin will be here soon, and I shall have more time at my disposal. But I must devote an hour a day to lucrative employment. I do not forget your advice on the subject and still less your brotherly offer of pecuniary help. But with no definite prospect of repayment and with the probability before me that my expences may increase, I should be far more at ease if I were earning a little, and more especially as there is an opportunity of doing so here in a manner congenial to my tastes. Several people have assured me I may obtain a private pupil or two in mathematics who will be glad to pay me well. It very often happens that young fellows, intended for the army, or University, are compelled by circumstances to accompany an invalid relative hither, and such would be glad to continue their studies. You may tell me it would be better to retrench a little than spend some of my hours thus. Believe me, I have considered that subject carefully. I am economizing as much as I consider prudent, under the circumstances—I will not enter into particulars—and still my income will not suffice. Now I want you like a good fellow to send me a testimonial stating my capability to give lessons in Mathematics, and Natural Philosophy, for being a perfect stranger here some certificate of the kind will be necessary. I will ask Edmondson for one also, but I will not promise to use it. As stating my position at Queenwood it might be of some service, but if it bore traces of the weakness of the Principal of Queenwood[16] it would in the end be useless and I would prefer putting it into the fire to using it.

From Queenwood I have good news. Smith[17] appears to give satisfaction, and everything to be going on as usual. Wright writes to me often, and his letters are like himself, plain, straightforward, kind and good. Wright is a worthy fellow, indeed, his tendencies are always to the honest, the beautiful, and the noble. Debus too wrote me an excellent letter a short time ago. He is longing to see you, and I hope you will run down and see how matters are going on. I should like your report of it all. Remember me kindly to Heinrich, and tell him I will answer his kind letter shortly. To Wright and Mrs Wright remember us also affectionately, and give them the latest news of us and the assurance that they will hear directly from us before long. To Bevington and to Francis do likewise.

To yourself Anna sends her love with mine. She read your letter with pleasure and would gladly read more of your adventures on the glaciers. With our best wishes for your health and experimental prosperity, I remain,

Yours affectionately, | T.A. Hirst

P.S. Wynne,[18] I was sorry to learn, did not pass his examination. I hope he will gain experience from his failure and succeed next time. He must not think too slightly of the difficulties before him. Remember me kindly to him and to the Colonel.[19]

RI MS JT/1/H/228

1. *your letter*: letter 1280.
2. *would sell*: see letter 1280. After four further visits to the Alps 1857, 1858, 1859 (twice), Tyndall published the popular *The Glaciers of the Alps. A Narrative of Excursions and Ascents* (London: J. Murray, 1860). This drew heavily on his Journals and RI lectures.
3. *winter residence*: The Hirsts were lodging at Pau, a French town near the border with Spain and lying within 60 miles of the Atlantic Ocean. It had become a popular resort for European visitors because of its views of the Pyrenees.
4. *apartment meuble, au premier*: a first-floor furnished apartment (French).
5. *chez elle*: at her home (French).
6. *a Bearnese girl*: a native from Bréan in the Pyranese mountains who spoke a unique dialect.
7. *ally shat*: presumably Hirst's phonetic rendering of 'cheerio' in the servant's dialect.
8. *old Henri IV*: Henry IV (1553–1610), King of Navarre (1572–1610) and King of France (1589–1610), head of the House of Bourbon. Raised as a Protestant, he became a Catholic as King of France. He is remembered for giving French Protestants their liberty in the Edict of Nantes (1598).

9. *Tidmarsh in the celebrated crash at Northowram*: Not mentioned in Hirst's Journal, but presumably a railway accident near the village of Northowram in West Yorkshire involving Tyndall and Hirst's fellow railway surveyor, John Francis Tidmarsh (1824–1906). See Biographical Register for vol. 1.
10. *Vraiment Dieu a vous aimé*: Truly God loves you! (French).
11. *old work on attracting bodies*: Eventually published as T. A. Hirst, 'On Equally Attracting Bodies', *Phil. Mag.*, 13 (1857), pp. 305–24; 'On Equally Attracting Surfaces', *Phil. Mag.*, 6 (1858), pp. 160–77, 266–84.
12. *determinants*: A method of solving simultaneous equations in algebra by arranging the coefficients in a matrix of lines and columns and applying specific rules of multiplication and subtraction.
13. *Spottiswoode*: William Spottiswoode (1825–83, FRS 1853; PRS 1878–83), Oxford mathematician and physicist. His *Elementary Theorems Relating to Determinants* (London: Longman, 1851) rapidly fell out of print. He prepared a new edition for *Crelle's Journal* (note 14).
14. *Crelle*: August Leopold Crelle (1780–1855), German mathematician and engineer. He was employed as a civil servant in the Prussian building administration. Founded *Journal für die reine und angewandte Mathematik* in 1826 (*Crelle's Journal*). Spottiswoode's revised book on determinants appeared as 'Elementary Theorems Relating to Determinants', *Crelle, Journ.*, 51 (1856), pp. 209–71, 328–81.
15. *'It is an ill wind . . . to nobody'*: English proverb. Hirst is reflecting that he has benefited from studying a poor treatise on determinants.
16. *the Principal of Queenwood*: Hirst is doubting the ability of George Edmondson to write an honest reference.
17. *Smith*: A Dr Smith replaced Hirst at Queenwood, but he has not been further identified. He was sacked by Edmondson in September 1857. See Hirst's Journal for 13 September 1857 (RI MS JT//2/32c/1313).
18. *Wynne*: Eddy Wynne, who had been tutored by Hirst. See letter 1194.
19. *the Colonel*: George Wynne, Eddy's father.

To Edward Frankland [6 October?] 1856 1282

Royal Institution of Great Britain | Monday

My dear Frankland

To Cassel it was impossible to come.[1] Hooker had made many deviations in his plans on my account and finally agreed to accompany me to Paris. The most I could ask him to agree to was to stay a single night in Halle. we travelled day and night till we reached Paris and remained there 4 or 5 days. Thence over Calais to England where I terminated the happiest excursion which I ever made.

We were like a little family in Vienna, always in harmony, always helping each other and contributing to each others pleasure. We went to Reichenbach[2] and saw all his paraphernalia, his magnets his crystals and his electrical apparatus—He described his processes minutely[3] but could shew us nothing as he had no sensitive person to operate on. I asked him how he knew that among the number there present including Frankenheim[,][4] Gustav Rose, Dr. Daubeny his two nieces,[5] Hooker & Mrs. Hooker, Grailich[6] and others, there was no one sensitive: he replied that we had all walked up the hill to his house, and if we were sensitive the state of our health would not permit us to do this. Columns of light are seen by his patients issuing the poles of magnets—one pole also gives a sensation of cold, the other of heat. Crystals emit differently coloured lights from their ends—Indeed some of the people there present imagined they felt something issuing from the crystals they required only a little encouragement to make them quite sure of it. An hysterical woman surrounded by crystals & magnets & dense darkness could be made I think to see almost any thing. However I think the whole matter may be readily reduced to an experimentum crucis,[7] and if I have time I will try it before the close of the year.

I sent your letter to the Hookers.[8] They were much pleased with it, and desire your presence at Kew[9] when you come to London. I shall feel very thankful to you for Adolf's Book[10] and will probably write a review of it[11] if I have the time. Happy man to be able to travel 3rd class,[12] but we had the luxury of 1st or 2nd all through—We bribed the guards and they always left us to the peaceful enjoyment of a carriage to ourselves. Really it was delightful, and I shall always remember the trip with pleasure.

I must try and put my observations on the glaciers into some form. Huxley is not yet in Town and I must consult with him on the way in which they ought to appear. We must be guarded; for Forbes is sure to swoop down upon us—the subject is the great foundation of his reputation. Present my kindest regards to Mrs. Frankland.

Ever yours | J. Tyndall

I will send you your cash shortly[13]

JRL Frankland, 248

---

1. *To Cassel it was impossible to come*: Tyndall had probably suggested meeting Frankland in Kassel where he had gone to visit his wife's parents on his way back from Vienna to London.
2. *Reichenbach*: See letter 1153, n. 22. Reichenbach owned a castle at Reisenberg, near Vienna, where he conducted the magnetic experiments on 'sensitive' people that were recorded in his 'Untersuchung über den Magnetismus und damit verwandte Gegenstäde'

that Justus Liebig published as a special issue of *Annalen der Chemie,* 53 (1845), pp. 1–270. Reichenbach subsequently published *Physikalisch-physiologische Untersuchungen über die Dynamide des Magnetismus,* 2 vols. (Braunschweig: Vieweg, 1850) which appeared in William Gregory's English translation as *Researches on Magnetism* (London: Taylor, Walton & Moberly, 1850).

3. *He described . . . processes minutely*: Reichenbach attributed the ability of mediums to detect hidden magnets and crystals in darkened rooms to their possession of an 'odylic force', as outlined in *Untersuchungen* in n. 3.
4. *Frankenheim*: Moritz Ludwig Frankenheim (1801–69), German physicist. Professor of Physics at University of Breslau who had published on thermoelectricity. His interest in crystallography no doubt prompted his curiosity concerning Reichenbach's claims.
5. *Dr Daubeny his two nieces*: Charles Daubeny (1795–1867, FRS 1822), Anglican chemist and botanist. Professor of Chemistry at Oxford since 1822 to which he added the chair of Botany in 1834 and a chair of Rural Economy in 1840. He had been President of the BAAS at Cheltenham in 1856. Nieces not identified.
6. *Grailich*: Josef Grailich (1829–59), Austrian physicist and crystallographer. Professor of Physics at the University of Vienna.
7. *Experimentum crucis*: a crucial experiment (Latin); an experiment that decides between two alternative hypotheses.
8. *letter to the Hookers*: letter missing.
9. *desire your presence at Kew*: presumably a social visit.
10. *Adolf's Book*: Frankland's brother-in-law was Adolphe Fick (1829–1901), Marburg-educated physiologist and ophthalmologist. Professor of Physiology at the University of Würtzburg. He had just published *Medizinische Physik* (Braunschweig: Vieweg, 1856).
11. *a review of it*: Tyndall never published such a review.
12. *travel 3rd class*: With the introduction of railways in the 1840s most European countries deployed three classes of compartments and comfort that reflected the existing class structure of the upper, middle and working classes. 3rd-class travel offered the cheapest fares and least comfort. Frankland (a middle-class scientist) had evidently returned to Paris from Kassel by cheap 3rd class train.
13. *send you your cash shortly*: Tyndall must have borrowed some money from Frankland while they were on the Continent.

To Thomas Archer Hirst 9 October 1856 1283

My dear Tom

One of the best letters[1] you ever wrote to me reached me today. I have enclosed it to Debus, requesting him to hand it over to Wright, and him again to send it back to me as I intended it for still further circulation among your friends. To avoid all delay I sent, as soon as <u>your</u> letter reached me, the certificate[2] which you requested—I hope it will be what you wished it to be.

Your account of your Houyhnhnm[3] made me wish to be at your side there, and the description of your escape from the effects of that brute's unstable equilibrium had something of romance in it. It is singular how such things befal. Regnault, as you are doubtless aware, fell some time ago[4] upon his feet through a height of 90 feet and his life was dispaired of for a time. He is now recovering with extreme slowness. About the same time a little Irish boy fell[5] from the top of one of the high houses in Hyde Park and escaped without a bruise. It would seem as if the gods extended an especial protection to some people under such circumstances. This is your second freak escape.[6] I have not yet got properly to work as the molecules of my brain seem to require time to subside into London occupations. Before going to the Continent I wrote to Priddie[7] concerning Mrs Booth,[8] I have written to him again since my return, so that in my next letter I hope to be able to give you some information upon the subject. As Anna wishes it I will give you from time to time scraps from my journal. If I published any thing the journal of course would merely form the nucleus of my disquisition. I left London on the 12th of August and went by Dieppe to Paris. I had agreed to meet Huxley at Basel on the 14th, so I travelled as swiftly as possible and arrived just in time to see him in the corner of the diligence and to follow him in a beiwagen.[9] At Berne we were joined by Hooker Mrs. Hooker and some of their friends, and we all travelled in jolly fashion to Interlaken together. Here we left Hooker (the Himmelayan traveller)[10] and party, and Huxley Mrs Huxley and myself ascended the Wengern Alp.[11] We stayed the night at the hotel opposite the Jungfrau. My window was open and sometimes during the night the wind rose from perfect silence to a sigh, ascended to a moan of unearthly wilderness, and finished by blowing my hair into confusion, sinking afterwards into the same perfect silence as before. At 5 o'C I rose. The sun was hidden behind the Monk,[12] and his first beams had already smitten the peaked snow of the Silberhorn, converting it on one side into a pane of gold. I never saw any thing more beautiful. The white cone so high up in heaven, covered with snow of perfect purity, unsullied by a particle of dust or debris, rising higher & higher into the golden sunlight—, so silent and so solemn, helped by the contrast of the savage brown crags around it, formed an object never to be forgotten. Descended the valley towards the base of the mountain. An avelanche descended with the sound of thunder, sometimes continuous and sometimes broken into rounded explosions which seemed to assert a passionate predominance over the general tenor of the sound. The latter was like the magnified clamour of a railway train passing through a tunnel. Reached finally a secondary glacier formed from the snow which fell from the higher slopes of the Jungfrau. The view from this place down the valley has a savage magnificence which I have never seen surpassed. Clambered on the glacier with a feeling of timidity being ignorant of its structure and altogether unprepared with the weird and musical loneliness of the place. The ghostly rattle of the debris as it fell from the melting ice into the fissures made me start. One object of my journey was to examine the structure of the ice. Much has been written upon this subject especially by Forbes who first noted that it was striated and resembled slate rock. I observed the structure here for the first time, but as yet was much puzzled to account for it. Forbes' explanation seemed to me wholly inadequate,[13] but to supply a correct one was still a difficulty. The true nature of the avelanches first revealed itself to me this morning. It had puzzled me to conceive how mere snow could produce the thunder peals which announce the descent of an avelanche. This morning I learned that it was not snow but masses of solid ice which produced the noise. Once I noticed what seemed to me a sudden and enormous augmentation of the waters of a cascade which fell from the heights. In a moment I was informed by the sound that the increase was due to the descent of an avelanche which had chosen the track of the stream. From afterwards my eyes were fixed upon the snow resting on a slope perhaps 5000 feet above me: a kind of zig zag channel was cut into the almost perpendicular sides of the mountain and along this I saw the rush of an enormous avelanche. The thunder came some time after the motion was first seen: the moving mass was sometimes revealed, sometimes concealed by the sides of the channel through which it ran—sometimes it appeared shooting like a cataract down a height of 1000 feet. Some of the masses darted like inverted rockets with snowy trails in advance of the general mass: These were the solid larger lumps which had thus far escaped being shattered, and which in the resisting atmosphere moved forward. The thunder of this avalanche endured for several minutes.

Rain and thunder caught me in the valley, but it cleared up with magical speed; I returned to the hotel and although many prophets of evil promised us bad weather we pursued our journey. Crossed the summit and here our difficulties commenced. The road was steep at intervals, indeed sometimes so steep as to be formed into stairs down which the little donkey which bore Mrs Huxley had to pick its steps: she was totally unable to walk herself. Anxiety for her impaired our enjoyment but could not totally quench it. Wild masses of clouds smote the flanks of the great Eiger;[14] before us one end of a splendid rainbow rested in the valley and beaming over the crags of the Matterhorn,[15] seemed to clasp the mountain in its embrace. The sun which painted thus formed a flood of golden light a down the slopes of the Wengern Alp. Black pines, crowded here and there into the forest, anon broken into clusters, were interleaved with sunny slopes of the most lively green. The cows browsed and shook their mellow bells; chalets and patches of golden corn spotted the landscape. The ragged crags of the Eiger rose 10000 feet to the right. The peaks of the Matterhorn were in front with light on their summits, and gloom on their sides. The Jungfrau[16] thundered in the rear his avalanches being brought down by the rain which still fell there. Thus sunlight, shadow, colours, sound and vastness conspired to render the scene a miserable one. The journey was too much for Mrs Huxley, and instead of pursuing her way to Meyringen as was first proposed she had to wend her way back by coach along a level road to Interlaken.

Having taken up our quarters at Grindelwald our object was to examine the glacier. We engaged a powerful guide and with him faced a swift ascent which brought us to the side of this glacier. Our arrival was greeted by a thundering crash, the mass was split into chasms and ridges, transverse masses which sometimes fell together and produced the noise that we had heard. At first the ice seemed a mass of confusion, long fissures with their corresponding ridges; broken blocks, and minarets of dazzling whiteness. Reached a spot which commanded a long view and here a mechanical problem of exquisite beauty revealed itself, and the whole structure of the mass was such as might be predicted à priori, from mechanical principles. I cannot of course enter into it here. We stood above a place where the glacier had changed in gradient, falling from a more level to a steeper incline. In the centre the mass was broken right across by parallel fissures of great regularity. The glacier here was narrow, but a little lower down it widened, owing to the lateral movement thus permitted the transverse ridges broke across and became intersected by fissures transverse to them. At first the square blocks thus formed bore the dirt which was strewn over the surface of the glacier. But as they slowly descended (for you know the glaciers are always in motion) their angles were melted away, the dirt disappeared and the blocks became converted into those beautiful minarets. Coming down the glacier with those white peaks shining in the evening sun and projected against the black pines upon the opposite mountain their appearance is indescribably lovely. As the mass crossed the crown of the slope above which we stood, it was as I have said broken into transverse chasms. The force was here a force of tension. Higher up the surface was more level, the force was pressure, and here the chasms formed at a more elevated point were pushed together and quite closed up. The line below represents the surface of the glacier. At a wide fissures yawn,

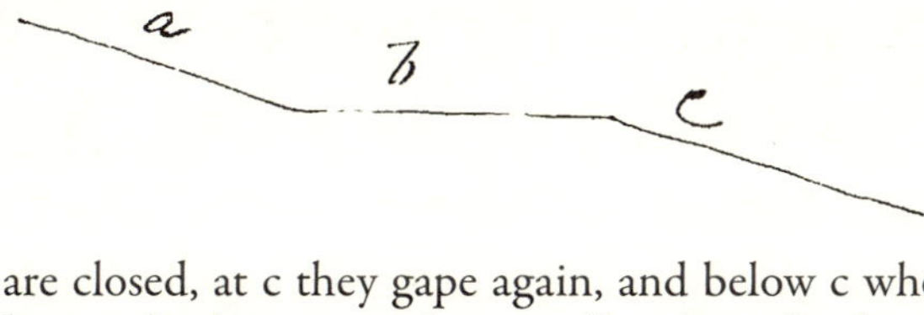

at b they are closed, at c they gape again, and below c where the glacier widens the ridges are broken across: lower still we have the shiny minarets. To a superficial observer the sides appear to be dragged downwards so that it is difficult to avoid the conclusion that the sides of the mass move more quickly than the centre; indeed this is stated in the writings of Agassiz.[17] I will here stop—more in my next. affectionately—John.

RI MS JT/1/T/1005

1. *best letters*: letter 1281.
2. *the certificate*: letter 1284.
3. *Your account of your Houyhnhnm*: Hirst's account of the accident with the horse recounted in letter 1281 had reminded Tyndall of Jonathan Swift's account of the horse-like Houyhnhnms in Book 3 of *Gulliver's Travels* (1726).
4. *Regnault . . . fell some time ago*: See letter 1273, n. 4.
5. *A little Irish boy fell*: The accident has not been identified.
6. *Your second freak escape*: Hirst's first lucky escape has not been identified.
7. *Priddie*: Roby Pridie, Hirst's Halifax friend. See letters 1267, n. 10 and 1272, n. 11.
8. *Mrs Booth*: the widow of Hirst's Halifax friend, Francis Booth, who had been financially supported by Hirst and Tyndall. See letter 1267 n. 10.
9. *beiwagen*: literally a side-car (German). This was a horse and cart vehicle for postal deliveries that also carried passengers.
10. *The Himmeleyan traveller*: Joseph Hooker had travelled through the Himalayas between 1847 and 1849 collecting plants (including rhododendra).
11. *Wengern Alp*: Now Wengen. An area of alpine meadows beneath the Eiger in the valley of Lauterbrunnen.
12. *the Monk*: the Mönch is a 13,474 ft. mountain in the Bernese Alps.
13. *Forbes' explanation . . . wholly inadequate*: as letter 1272 suggests, Tyndall was already of this opinion before he went to the Alps.
14. *The flanks of the great Eiger*: The 13,020 ft Eiger overlooks Grindelwald in the Bernese Alps.
15. *Matterhorn*: see letter 1274, n. 18.
16. *the Jungfrau*: One of the principal mountains (13,642 ft) in the Bernese Alps.
17. *the writings of Agassiz*: L. Agassiz, *Études sur les Glaciers* (Neuchâtel: Jean et Gassmann, 1840).

To Thomas Archer Hirst 9 October 1856 1284

London 9th Oct. 1856.

My dear Friend,[1]

You ask me for a certificate of your abilities as a Teacher of Mathematics and Natural Philosophy. The request reminds me of the fact that it is not always the world's best men who are most widely known in the world; for assuredly if your merits were known as they deserve to be no certificate from me would be needed to attest them. Both your acquirements as a mathematician and physicist, as also your power of communicating knowledge, are well known to me; and the result of this knowledge is that if I had a son to instruct in the above sciences I would place him in your hands sooner than in those of any other living man. To natural endowments of the first order you have been able to add a personal intercourse for years with the most distinguished scientific men of Germany, and I know that their opinion of your qualifications entirely coincides with my own. You have had three years experience as teacher and lecturer in a most important school in the south of England; and to your ability and the salutary influence which you exercised upon the pupils, the present prosperity of Queenwood College is, in my opinion, mainly due. If at any future time you should deem the expression of these opinions useful to you it will give me happiness to repeat them to Faraday, Sabine, Grove, Lyell and other men of distinction with whom I have the pleasure of being acquainted.

Believe me | ever sincerely yours | John Tyndall F.R.S. &c.

Professor of Natural Philosophy in the Royal Institution of Great Britain.

RI MS JT/1/H/476

1. *My dear Friend*: This is Tyndall's formal testimonial in response to Hirst's request in letter 1280.

From Henry Moseley 9 October 1856 1285

Olveston | 9 Oct. 1856.

My dear Sir,

My report on the last exam.[1] for the Engineers has been printed and I trust that you will in a few days receive a copy of it, if you have not already done so.

I am very glad that your examination of the glaciers is favourable to my theory[2] on some of the points to which Prof. Forbes takes objection. I have given such consideration to it lately as my other occupations have permitted and prepare [propose?] to put my thoughts on paper[3] when I can find a few days leisure. Will you permit me to ask your opinion on one point. Prof. Forbes describes an experiment in which a thermometer was placed in a hole in the glacier made on purpose, and another (I think) in which it was let down into a 'crevasse', and showed in both cases a temperature of <u>zero</u>, and he concludes that the <u>whole mass</u> of the glacier is therefore of that temperature.[4] Is this to be relied upon? Does not the presence of the air in the crevasse (or in any other hole he might make) when the external temperature is above zero imply a thawing condition of the surface ice of the crevasse and therefore a temperature of zero in the thermometer brought in contact with it?

Might not the solid depths of the glacier be of a temperature below zero notwithstanding that the walls of the crevasse (or of a hole made artificially) showed that temperature? Would not—in point of fact—a mass of ice 200 or 300 feet thick and as wide as the Mer de Glace[5]—of necessity be of various temperatures from its surface to the parts furthest removed from it; supposing it originally to have been formed (or at any time to have existed) at a lower temperature? Does not the heating of a mass take place under conditions analogous—in this respect—to its cooling? and may the glacier not keep far into the summer in the abysses of the solid ice some of the cold of winter? May there not be depths in it to which the summer heat scarcely reaches—or rather layers in it which that heat reaches in succession, and to which it is communicated in different degrees—raising none however above zero—and thus expanding each differently and causing that 'striated' appearance on which Forbes so much insists?

Pardon me for troubling you with this.

Yours truly | H. Moseley.

P.S. Do you know of any facts bearing upon the question of the distribution of temperature at different depths of a solid body in the process of cooling and heating? I never can too much lament my own imperfect knowledge of these things. Is there not something analogous in the temperature of earth's surface itself?

RI MS JT/3/897
LT Transcript Only

1. *My report on the last exam*: report not found.
2. *my theory*: Moseley had published an explanation in 'On the Descent of Glaciers', *Roy. Soc. Proc.*, 7 (1854–55), pp. 333–42.
3. *to put my thoughts on paper*: Moseley did not do this until 1860: 'On the Cause of the Descent of Glaciers', *Brit. Assoc. Rep.,* 1860, pp. 48–50. This was subsequently incorporated into 'On the Motion of a Plate of Metal on an Inclined Plane, when Dilated and Contracted; and on the Descent of Glaciers', *Roy. Soc. Proc.,* 11 (1860–62), pp. 168–77.
4. *Prof. Forbes describes . . . of that temperature*: No such experiments were described in J. D. Forbes, *Travels Through the Alps of Savoy ad Other Parts of the Pennine Chain* (Edinburgh: Adam & Black, 1843)—though see, pp. 36, 358. Moseley must have been referring to one of Forbes's many papers on glaciers.

5. *Mer de Glace*: A valley glacier in the Chamonix valley of France. At the time of Tyndall's visit (and until 1860), it was part of the kingdom of Sardinia.

From Carlo Matteucci[1] 10 October 1856 1286

Pise 10 octobre 1856

Mon cher Monsieur,

J'avais bien appris à mon grand regret par le Gen[al]. Sabine votre voyage en Suisse presque en même temps que nous; ce qui m'a faire penser lorsque les savants font des voyages de plaisir pendant les vacances ils devraient bien faire connaitre d'avance aux amis, leur itineraire. Je n'ai pas beaucoup vu, ni medité cette fois sur les glaciers: ce que me frappe le plus maintenant dans ce grand et obscure phenomène de Physique terrestre, c'est le <u>mouvement de masse</u> de tout un glacier, depuis son commencement a sa fin au milieu de tant de tortuosités et de variations de la section et de la forme du lit. Cela ferait bien penser que Forbes à raison et l'état de la glace des glaciers qui est je crois imbibée d'eau pourrait bien se concilier avec le principe fondamental de sa theorie. Je lirai avec grand plaisir votre mémoire sur ce sujet.

Je viens d'achever mon Cours d'Electro Physiologie qui est ainsi la nouvelle Edition depuis si longtemps annoncée du Traité des Phenomènes Electrophysiologiques. J'avais désiré beaucoup une occasion pour rendre justice aux travaux de M. du Bois Reymond et je remercie la Providence de me l'avoir accordée, tout en faisant un ouvrage qui est plein de nouvelles expériences coordonnées entre elles. J'espère que vous recevez régulierement à la Royal Institution le <u>Nuovo Cimento</u>: ce Cours se compose de cinq longues Leçons qui ont parues successivement dans ce journal. Je serais bien content de pouvoir vous envoyer des Extraits des parties nouvelles rédigés en français pour le Philosophical Magazine; mais absolument je n'ai, du moins dans le moment, ni le temps ni la patience pour cela. D'ailleurs ce cours doit paraitre traduit en français en 1857. Quant aux Memoires sur le disque tournant d'Arago je crois qu'il est impossible d'en donner une analyse ou d'en faire un Extrait. C'est un memoire complet dont la traduction dans les Annales de Phys. et de Chim. Doit paraitre dans le cahier de Novembre ou de Decembre prochain; je suis maintenant comme je vous le disais, sur le diamagnétisme, et j'espere bien que si nous ne tombons pas d'accord sur les conclusions, celà n'empecherà pas à tous les deux de continuer a y travailler sans troubler nos rapports d'amitié auxquels j'attache une grande importance.

Veuillez me rappeler aux bons souvenirs de M. Faraday /et M. Sabine /[2] et me croire avec estime et attachement. | C.Matteucci

P.S. J'ai demandé un appareil semblable au vôtre a M. Leyser de Leipsik. Auriez-vous la bonté de lui écrire et de l'envoyer aussi parfait que possible et dans le moindre delai possible?[3]

Pisa 10th October 1856.

My dear Sir,

I had indeed learnt from Genal Sabine to my greatest regret that your trip to Switzerland was almost at the same time as ours; it made me think that when scientists take leisure trips during the holidays, they ought to let their friends know in advance of their itinerary. I have neither seen much, nor thought much about glaciers:[4] now what strikes me the most about this grand and obscure phenomenon of terrestrial physics is the mass movement of an entire glacier, from its origin to its end amidst so many tortuosities and variations of the section and the shape of the bedrock. It seems to me that Forbes could be right and that the condition of ice in the glaciers, which is, I believe drenched in water,[5] could well be reconciled with the fundamental principle of his theory. I will read with much pleasure your memoir on this subject.[6]

I have just finished my 'Course on Electro-Physiology' which thereby forms the new edition of the long announced Treatise on Electro-Physiological' phenomena.[7] I had much desired an opportunity to do justice to Mr DuBois Reymond's work and thank Providence for allowing me to do so while publishing a work that is full of new experiments, coordinated together. I hope that you regularly receive Nuovo Cimento at the Royal Institution:[8] this Course is composed of five long lectures that were published in this journal[9] one after the other. I would be very happy to be able to send you excerpts of the new parts written in French for the Philosophical Magazine; but by all means I have not, at least not at the moment, the time nor the patience for this. Also this Course will be published in French in 1857.[10] As for my memoir on Arago's turning disk,[11] I believe it is impossible to give an analysis of it or to summarize it. It is a complete memoir from which the translation in the Annales de Phys. et de Chim. must be published in the issue of the next November or next December; I am now, like I told you, working on diamagnetism, and I hope that if we end up disagreeing about the conclusions, this will not prevent the two of us from continuing to work on it without troubling our friendly relationship on which I attach much value.

Would you mind remembering me to Mr Faraday and Mr Sabine and receive my sincere esteem and affection. | C.Matteucci

P.S. I have asked for a device similar to yours from M. Leyser in Leipzig.[12] Would you have the kindness to write to him and send it to me as perfect as possible and as soon as possible?

RI MS JT/1/M/60

1. The letter is not in Matteucci's handwriting, but in his wife's or a secretary's hand. The postscript and signature are, however, in Matteucci's usual scrawl.
2. *et de M. Sabine*: interpolation in Matteucci's hand.
3. *P.S: J'ai demandé . . . delai possible*: This postscript is in Matteucci's handwriting.
4. *thought much about glaciers*: This might also be translated as 'meditated while on the glaciers', an image reminiscent of Caspar David Friedrich's painting, 'Wanderer above the sea of mist' (1818).
5. *drenched in water*: or, 'full of water'.
6. *your memoir on the subject*: presumably Matteucci means a paper on glaciers that Tyndall was yet to write.
7. *Treatise on Electro-Physiological phenomena*: Matteucci was referring to his just-completed lectures at the University of Pisa: *Lezioni di Electrofisiologia* (Torino: Paravia, 1856), and to his later work *Corso di Electrofisiogia in sei Lezioni* (Torino: Castellano e Vercellino, 1861).
8. *at the Royal Institution*: see letter 1123, n. 3.
9. *published in this journal*: The Pisan lectures cited in note 7 also appeared in *Nuovo Cimento,* 3 (1856), pp. 359–85; and 4 (1856), pp. 5–34, 177–92.
10. *French in 1857*: C. Matteucci, *Cours de Électro-physiologie* (Paris: Mallet-Bachelier, 1858).
11. *Arago's turning disk*: See letter 1277, n. 15.
12. *M. Leyser of Leipzig*: Georg Leyser, the University of Leipzig's instrument maker. See letter 1139, n. 2.

From Heinrich Debus 12 October 1856 1287

Queenwood, am 12 Oct. 1856.

Mein lieber John!

Obgleich ich heute sehr beschäftigt bin, so will ich doch ein Paar Zeilen schreiben. Tom's Brief habe ich Deinem Wunsch gemäss Mr. Wright übergeben. Komme nur nach Queenwood. Edmondson ohne Zweifel wird Dir mit grossem Vergnügen Alles geben das er einem Gaste bieten kann. Ich machte das Arrangement, worüber ich Dir schrieb, in der Absicht um mehr frei und unabhängig zu sein und ein Recht zu haben meine Freunde hierher einzuladen. Doch in Deinem Fall glaube ich war eine solche Massregel unnöthig. Jedermann in Queenwood wird sich freuen wenn Du hierherkommst. Uebrigens hast Du Dich um solche Fragen gar nicht zu bekümmern. Ich habe Dich eingeladen und wenn Du überzeugt bist dass ich Dich gern sehe dann solltest Du nicht weiter fragen und sobald als möglich kommen.

Emmersons neustes Buch habe ich nicht gelesen, werde es aber bald thun. Die Art der Gegenstände über die Emmerson schreibt gehört zu den wichtigsten die den Menschen interessiren können. Seine Denkungsweise über diese Sache ist gross und wahr. Er dringt zum Grund der Fragen. Das Urtheil Emmersons über Geschichte, Freundschaft, Liebe oder Natur verbreitet sich wie Donner durch die Seelenwelt. Die Wahrheit und die Kraft sind die Impulse die Alles erschüttern und die Sympathie der Menschen unter seiner Fahne versammeln. Dies Alles erkenne ich und es ist die Hauptsache. Jedoch etwas Anderes ist die Illustration, die Mittheilung dieser Ideen. Immer und mit leichter Mühe kann ich die fundamentalen Gedanken in einem Essay finden. Wie der Faden in dem Knäuel folge ich seiner Grundidee. Da begegnet es mir denn oft dass ich ein gutes Theil Abstraction in Thätigkeit setzen muss um Fremdes abzusondern um die Sache zu entdecken die eigentlich klar gemacht werden soll. Eine Gedankenreihe ist eine Perlenschnur die in der unendlichen Zeit fortrollt deren Anfang und Ende Niemand kennt und deren Gesetz in der Geschichte liegt. Nun ist es ein erstes Erforderniss dass je höher der Gegenstand um so einfacher klarer und kräftiger auch die Sprache sein sollte. In jenen Tiefen oder Höhen, (wie man es nennen will) zu denen nur wenige schwingen können, sollte der reine Gedanke herrschen und das Fremde so viel als möglich ausgeschlossen sein. Emmerson hat eine mächtige Einbildungskraft. So kommt es denn also dass oft an den Faden seiner Ideen sich Krystalle ansetzten die eigentlich incongruent zum Ganzen sind, die mehr zur Verschönerung, zur Vervollkommnung der Form also zur Bildung der Sache gehören, rein die als ausländische Waare sogar den Bau seiner Gedanken stören. Wenn die Kraft seiner Spontaneität die Ideen wie einen reissenden Strom in die Welt treibt fliessen häufig fremde Tropfen mit fort die durch ihre brilliante Farbe den Anblick des Ganzen wohl verschönern und eindruckreicher machen aber vor einer scharfen Analyse doch zurückweichen müssen. Doch dies thut wenig. Das Hauptwasser ist rein und himmlisch und kommt aus jenem ewigen Brunnen dessen Ursprung in einem unbekannten Lande liegt. Es ist eine sonderbare Sache um den Menschen. Immer habe ich gefunden dass je solider, kräftiger, gediegener ein Buch geschrieben ist um so weniger wird es Allgemeingut d. h. je gedrängter, je reiner die Wahrheit ausgetheilt wird um so weniger schmeckt es den Leuten. Eine gute Novelle die ein sittliches Subject behandelt findet tausend Leser während ein Buch über Moral obgleich sein Inhalt tausendmal reicher ist doch kaum einen Leser findet. In dieser Beziehung gleichen wir den Pflanzen. In der besten Manure wenn rein, gedeiht kein Pflänzchen. Um aber Effect hervorzubringen muss die Manure mit Dreck und Steinen ohne Ende gemischt werden und dann vegetirt Alles darin mit der grössten Gemüthlichkeit. Ja Kalkpflanzen wollen nicht auf Kalkboden wachsen wenn ein paar pro cent Kalk zu viel im Boden sind.

Wilhelm Meister enthält hier und dort prächtige Ideen. Jeder schätzt dieselben. Viele lesen das Buch. Nun nehme das Schlechte hinweg, stelle die Kernsachen alle zusammen, sie werden vielleicht keinen Bogen bedecken, und gebe die letztern Jemanden in die Hand. Werden sie von so vielen gelesen werden als das Buch aus dem sie entnommen werden? Die Wahrheit ist die Leute haben schwachen Magen darum muss Ihnen der Spiritus in homöopathischen Dosen verabreicht werden sonst werden sie besoffen und bekommen einen Abscheu davor. Jeder bewundert einen Heiligen aber keiner will ein Heiliger werden, nein Niemand hat so viel als den Gedanken ein heiliges Vorbild nachzunahmen.

Diese Bemerkungen mache ich nur nebenbei und sollen natürlich nicht Emmerson treffen. Er ist das Gegentheil. Doch diese Eigenthümlichkeit unserer geliebten Mitmenschen ist eine wichtige psychologische Thatsache die wohl ernstes Nachdenken verdient. Ja sie hat ihre Analogie in der ganzen Natur und in der Geschichte. Daran muss entweder eine innere Nothwendigkeit dafür dasein oder äussere Umstände haben es so herbeigeführt.

Mit herzlichem Gruss | Dein | H. Debus.

Queenwood, 12 Oct. 1856

My dear John!

Although I am very busy today I shall nevertheless write a couple of lines. I have given Tom's letter[1] to Mr Wright in accordance with your wish.[2] Just come to Queenwood. Edmondson will no doubt give you with great pleasure everything that he can offer a guest. I made the arrangement about which I wrote to you[3] in the intention of being more free and independent and having a right to invite my friends here. But in your case, I believe, such a rule was unnecessary. Everyone in Queenwood will be pleased if you come here. Incidentally, you do not have to worry about such questions at all. I have invited you and if you are convinced that I shall be glad to see you, then you should not ask further and come as soon as possible.

I have not read Emerson's latest book[4] but shall do so soon. The kind of subjects about which Emerson writes belongs to the most important ones that can interest men. His way of thinking about this matter is grand and true. He gets to the bottom of questions. Emerson's opinion about history, friendship, love or nature spreads like thunder through the world of the soul. Truth and power are the impulses that shake everything and gather the sympathy of men beneath his banner. I recognize all this, and it is the main thing. However, something else is the illustration, the impartation of these ideas. I can always find the fundamental thoughts in an essay, and with little trouble too. Like the thread in a knot, I follow his basic idea. In doing so, it often happens to me then that I must put a good deal of abstraction into action in order to separate out the foreign, in order to discover the matter that should really be made clear. A line of thought is a string of pearls that rolls away in endless time whose beginnings and end no one knows and whose law lies in history. Now, it is a foremost requirement that the higher the subject, the simpler, clearer and more powerful language should be. In those depths or heights (whatever one wants to call it) to which only few can attain, pure thought should rule and the foreign be excluded as much as possible. Emerson has a mighty power of imagination. Thus it comes about then that to the thread of his ideas crystals often attach themselves that are actually incongruent to the whole and that belong more to the embellishment and to the perfection of the form, that is, to the shaping of the matter—and clearly that, as foreign goods, disturb the construction of his thoughts. If the power of his spontaneity drives his ideas like a raging stream into the world, then frequently foreign droplets flow away with them, that through their brilliant colour surely embellish the view of the whole and make it more impressive, but must nevertheless shrink from sharp analysis. But this matters little. The headwater is pure and heavenly and comes from that eternal spring whose source lies in an unknown land. It is a strange thing about men. I have always found that the more solidly, powerfully and tastefully a book is written, the less it becomes common property, i.e. the more tersely and purely the truth is dispensed, the less it appeals to people. A good novella that deals with a moral subject finds a thousand readers while a book about morality finds hardly one reader, even though its contents are a thousand times richer. In this respect we resemble plants. In the best manure, when pure, no little plant thrives. In order to produce any effect, however, the manure must be endlessly mixed with dirt and stones, and then everything vegetates in it with the greatest ease. Indeed, calcareous plants do not like growing on calcareous soil if there is a few percent too much chalk in the soil.

Wilhelm Meister[5] contains some splendid ideas here and there. Everyone treasures them. Many read the book. Just take away the bad stuff, put all the crucial points together—they probably will not cover one sheet of paper—and hand the latter to someone. Will they be read by as many as the book from which they are being taken? The truth is that people have weak stomachs and so the spirit must be administered to them in homeopathic doses; otherwise they will get pickled and develop a loathing for it. Everyone admires a saint but no-one wants to become one, nay, no one has as much as the thought of imitating a holy example.

I make these observations only in passing, and they should naturally not concern Emerson. He is the opposite. But this peculiarity of our beloved fellow men is an important psychological fact that probably deserves serious reflection. Indeed it has its analogy in the whole of nature and in history. There must either be an inner necessity for it about it all, or else outside circumstances have brought it about this way.

With warm greetings | Your | H. Debus.

RI MS JT/1/7/2416–17
LT Transcript Only

1. *Tom's letter*: Hirst's letter 1281.
2. *your*: informal address (Du).
3. *arrangement about which I had written to you*: Debus's letter is missing and it is not clear what arrangement he was making, though it seems that Debus's school lodgings at Queenwood did not allow him to accommodate guests.
4. *Emerson's latest book*: Ralph Waldo Emerson had visited England in 1833 and 1847. He summed up his impressions in *English Traits and Representative Men* (London: Routledge, 1856).
5. *Wilhelm Meister*: Goethe's autobiographical novel which was first published in 1796. Debus would have read this in German and not in the English translation (1824) by Carlyle. See also Debus's discussion of Goethe in letter 1200.

From Samuel Haughton 20 October 1856 1288

Trin: Coll: Dub: | 20 October 1856.

My dear Sir,

I sent to Messrs Taylor and Francis,[1] by this day's post, for the Phil. Mag., my paper on the Cleavage of rocks and Distribution of Fossils[2]—I fear it is almost too late for the November number, but I should like it to appear as soon as may be convenient, as I think I have obtained a few steps in advance on this interesting subject.

I have hit the Geological surveyors a little hard in the closing words of my paper,[3] but in truth they richly deserve it, at least some of them and those who do not feel the justice in my remarks as much as I do myself. A new doctrine has sprung up among us lately, to the effect that the Government, and in especial his Royal Highness Prince Albert,[4] is the fountain of wisdom and knowledge, and consequently that all Government officials, ex termini necessitate,[5] are competent, wise and good.

This is a perversion of the old Stoical Doctrine[6] with a vengeance—This pernicious dogma, which is equivalent to Scientific Popery,[7] should be resisted by every one who has the interests of Science really at heart, and not the advancement of cliques aping the manners and pretending to speak the sentiments of men of real merit.

I could mention many instances in Ireland of grievous wrong done to modest worth, because it wanted an official puffer;[8] and a contrâ[9] of successful quackery promoted and applauded because it was loud, noisy and forward to attract attention—I beg your pardon for having written such a rambling note, but I am never tired of this subject which has not received yet the attention it deserves.

I you agree with me, and your experience accords with mine, you will understand my warmth; if not, please burn my letter and forget my ebullition of spleen.

I am yours sincerely | Saml. Haughton.

RI MS JT/1/2/482
LT Transcript Only

1. *Messrs Taylor & Francis*: The publishers and owners of *Phil. Mag.*
2. *my paper on . . . Distribution of Fossils*: Houghton's paper 'On the Cleavage of Rocks and the Distribution of Fossils' was published in *Phil. Mag.,* 12 (1856), pp. 409–21.
3. *the closing words of my paper*: Haughton's paper included a mathematical analysis of the dips of cleavage and concluded: 'An opportunity of obtaining the information requisite for a perfect mechanical theory of cleavage is afforded by the means of observation possessed by the Staff of the Geological Survey of Great Britain and Ireland. If the operations of such a survey were conducted with the science and skill necessary for carrying out such an important work, nothing would be wanting to complete the data requisite for the solution of many interesting problems in physical geology. But it is much to be feared, that so long as geologists remain in the half-educated condition with which they are at present contented, that such an amount of intelligent observation is not to be looked for; and that those who wish for satisfactory foundations of facts on which to build their theories, must even be content to take their hammers in their hands, and having strapped on their knapsacks, to seek in the field of nature the facts for themselves' (p. 421).
4. *Prince Albert*: Queen Victoria's consort, Prince Albert (1819–61), second son of the Duke of Saxe-Coburg-Gotha who had married Victoria in 1840. One of the principal projectors of the Great Exhibition in 1851, he strongly supported the reform of British science and technical education.
5. *ex termini necessitate*: without exception (Latin).
6. *the old Stoical Doctrine*: the teaching that a sage would display judgement without bias and emotion and always act for the greatest good. Haughton clearly had a poor opinion of the Department of Science and Art and of the Geological Survey.
7. *Scientific Popery*: Although the doctrine of papal infallibility was not promulgated until 1870, Protestants (like Haughton) frequently likened Roman Catholicism to a

dictatorship. When governments dictated or controlled scientific work, this became equivalent to popery in Haughton's view.

8. *official puffer*: an advocate or spokesman.
9. *a contrâ*: on the contrary (Latin).

To Auguste Arthur De La Rive[1] 26 October 1856 1289

26th, Oct, 1856.

My dear Sir

During my recent visit to the Continent my attention was perpetually attracted to the beautiful colour of the water in many rivers. Down the Stelvio pass,[2] for example, on the Italian side there is a stream which shines sometimes like a fluorescent solution of sulphate of quinine.[3] I am told that the colour of the Rhone as it leaves the lake of Geneva (which unhappily I have not yet seen) is a deep and beautiful blue.[4] Would you kindly inform me whether any experiments have been made with a view of determining the true cause of this colour?[5] I do not think it can be referred to the natural colour of the water, as I am sure a layer of distilled water of equal depth would not produce some of the effects that I have seen in the rivers of the Tyrol.

During my holiday I spent many days upon the glaciers with my friend Mr Huxley. We paid much attention to the structure of the ice, a theory of which has been propounded by Prof Forbes. I cannot agree with this theory,[6] and when I have time to put my observations together I will tell you more particularly the grounds of my dissent.

Believe me dear Sir | Most sincerely Yours | John Tyndall

RI MS JT/1/339
LT Transcript Only

---

1. The letter lacks an addressee, but was clearly addressed to de la Rive from the answer he made from Venice on 7 November 1856, for which see volume 6.
2. *Stelvio pass*: The highest mountain pass in northern Italy and in the southern Tyrol at an elevation of 9,045 ft. Tyndall and Frankland had been through the pass between 3 and 4 September 1856 (RI MS JT/13c/898).
3. *sulphate of quinine*: Quinine, an alkaloid prepared from the bark of the cinchona tree, had been used to treat malaria since the seventeenth century. It was commercially purified and sold as quinine sulphate whose fluorescent properties were first noted by John Herschel in 1845.
4. *A deep and beautiful blue*: Tyndall saw this for himself in July 1857 and attributed it to the refraction of light by suspended particles derived from rocks ground down by glaciers. He made no reference to any observers who had previously tried to explain the phenomenon

despite having been supplied these by de la Rive. See J. Tyndall, *Glaciers of the Alps* (London: J. Murray, 1860), pp. 33–34, 261–62, and RI MS JT/2/13c/952–54. For de la Rive's reply on 7 November 1856, see vol. 6.

5. *the true cause of this colour*: See note 3. Although probably not suggesting the blue colour was due to fluorescence, it can be caused by organic matter in the waters.
6. *I cannot agree with this theory*: Forbes's viscous theory of glacial structure and motion. See letter 1274, n. 13 and Introduction to this volume.

From Wilhelm Josef Grailich[1] 26 October 1856 1290

Wien 26 Oct. 1856.

Hochgeehrter Herr!

Wenn Sie in dem Strudel grossartigen Lebens, in das Sie nun wieder zurückversetzt sein dürften, einen Augenblick Musse gewinnen sollten, so erlauben Sie dass ich ihn für eine Reminiscenz Ihres Wiener Aufenthaltes in Anspruch nehme. Bei Ihrer vielseitigen Anregung durch die Eindrücke einer weiteren Reise, bei dem Ueberblicke der Erlebnisse in Frankreich, Italien und Deutschland kann die Zeit unserer Naturforscherversammlung freilich nur als eine Episode neben vielen anderen in Ihrem Gedächtnisse Platz finden und wenn ich auch mit gutem Rechte hinzufügen mag dass Sie gewiss nirgends herzlicher willkommen waren als in unseren Kreisen, so konnte es Ihnen doch nirgends an freundlichem Entgegenkommen fehlen, wo Menschen wohnen die für offene Männlichkeit und geistreiche Persönlichkeit Sinn und Verständniss tragen. Uns, und speciell mir, bildet Ihre und Ihrer Freunde Ankunft in Wien, allerdings den Mittelpunct um den sich alle die frohen Erinnerungen der Septembertage gruppiren: und wenn ich etwas bedaure, so ist es der Umstand nicht mehr Zeit gefunden zu haben um über die denkwürdigen Untersuchungen mit Ihnen zu sprechen, welche sich an Ihren Namen knüpfen. Ich habe seitdem Veranlassung gefunden Ihr Memoire vom Jan. 1855 „on the nature of the Force by which Bodies are repelled etc." zu studiren und ich habe nicht bald etwas mit grösserem Vergnügen vom Anfang bis zu Ende gelesen—der klare Vortrag, die meisterhafte Handhabung der inductiven Methode, die Sicherheit mit welcher die entscheidenden Thatsachen frei von jeder theoretischen Ansicht geprüft und festgestellt werden reiht [reihen] diese Arbeit würdig an die Untersuchungen über die magnetischen Verhältnisse comprimirter Massen. Ich habe H. Regierungsrath v. Ettingshausen auf dieselbe aufmerksam gemacht und zweifle nicht dass er mit gleichem Beifalle Ihrer Darstellung folgen und Ihre Versuchsweise im physical. Institute einführen wird, da nicht geleugnet werden kann dass die Thatsachen des Magnetismus in keiner andern Weise vollständiger und schlagender dargelegt werden können. Ihr Ausspruch über die Ampère-Webersche Ansicht scheint mir allerdings zu streng; obschon ich von Haus aus Optiker und mehr in den Theorien dieses Theiles der Physik bewandert bin so erinnere ich mich doch lebhaft des Eindrucks, den das Studium von Ampère's Arbeiten auf mich machte: die Art wie er, wenn auch nicht immer auf dem kürzesten analytischen Wege, Analyse und Experiment verknüpft, jeden Grundsatz der Theorie scharfsinnig durch den Versuch läutert und feststellt und endlich ohne Spitzfindigkeit mit Nothwendigkeit zur Annahme von Molecularströmen geführt wird, hat in mir stets ein Gefühl der Sicherheit geweckt wie es künstliche und gesuchte Theorien nicht zu erzeugen vermögen. Die Thatsache, die Sie p. 39 gegen das Corollarium der Weberschen Deduction aufführen, enthält freilich eine grosse Schwierigkeit und ich bin begierig zu erfahren was unser Landsmann darauf wird erwiedern können. — Auch den Satz Plückers, welchen Sie in der Note p. 2 anführen habe ich nicht ganz wie Sie verstanden; „unabhängig von der magnetischen oder diamagnetischen Beschaffenheit der Substanz" habe ich so aufgefasst: „die Substanz mag m. oder d. sein, immer äussert sich die Wirkung in der Richtg[2] der Axe wie eine Resultirende, und es mag wohl erlaubt sein die Erscheinungen nach diesen Resultirenden zu betrachten, wie wir die Erscheinungen von Massenanziehung auf den Mittelpunct der Massen u.s.w. beziehen." Doch ich bin in diesen Dingen immer noch blosser Zuschauer und muss um Ihre Nachsicht bitten, dass ich es versuche meine Ansichten gegen einen Meister in diesem Fache zu äussern. —

Beiliegend bin ich so frei Ihnen meine Arbeit über die Reflexion und Brechung des Lichts an Zwillingsflächen optisch einaxiger Krystalle zu übersenden. Wenn sie auch nicht den Anspruch machen darf sich den theoretischen Arbeiten, welche über die Prinzipien der Undulationstheorie Licht verbreiten, anzureihen, so dürfte doch manche Thatsache zu welcher sie leitet, von Interesse sein; so z.B. dass es in Zwillingskrystallen eine Ebene gibt in welcher alle einfallenden extraordinären Strahlen gebrochen werden, ohne Reflexion zu erleiden: für die gewöhnlichen Substanzen gibt es nur einen einzigen Einfallswinkel (den Polarisationswinkel, insofern die Substanz nicht elliptisch polarisirt) unter welchem diess statt findet;—der Umstand dass ordinäre Strahlen reflectirt werden, ohne Brechung zu erleiden, u.s.f. Da ich alles durch Beobachtung festgestellt habe, so kann ich nun auf Grund dieser Rechnungsresultate zur weiteren Aufklärung mancher sonderbarer Reflexionsphenomene, welche Zwillingskrystalle zeigen, fortschreiten. Ich habe einen gedrängten Auszug des beiliegenden Memoires (Zusammenstell[un]g der Resultate) in Poggendorff veröffentlicht. —Wenn Sie das zweite beiliegende Exemplar H. Prof. Faraday, dem Manne den ich unter allen lebenden Naturforschern am höchsten verehre, mittheilen und ihm in meinem Namen meine Hochachtung ausdrücken wollen, werden Sie mich sehr verbinden.

Und nun erlauben Sie dass ich noch einmal auf die lustige Zeit Ihres hiesigen Aufenthaltes zurückkomme. Sie begreifen, dass während meines ganzen Schreibens das Bild der holden Gemalin Ihres werthen Freundes Hooker nicht von meiner Seite wich; wie befindet sie sich, wie kam sie zurück, wie gedenkt sie ihres Aufenthaltes in den staubigen, windigen, eckigen Strassen Wiens? Stürmisch, wie diese Zeit an mir vorüberrauschte, hat sie einen halb mährchenhaften Eindruck in mir zurückgelassen, und die „fairy queen" dieses Herbstmährchens ist Mistress Hooker. Ihre heitere, unbefangene Weise, die Anmuth mit welcher sie im fremden Lande die Zauber der Heimath zu schaffen und einem vielgeplagten Redacteur manche sonnige Stunde in sein wolkiges Dasein zu bringen wusste, wird in unseren Kreisen nicht vergessen werden und mancher fröhliche Toast wird noch auf die Erinnerung an ihre liebliche Erscheinung ausgebracht werden. Wenn alle englischen Damen ihr gleichen, so müsste die Völkerwanderung wieder beginnen und die Damen des Continents hätten Anlass zur bittersten Eifersucht. Wenigstens wäre es dann erklärlich warum nach einander Römer, Angelsachsen, Dänen und Normannen das ferne Eiland zum Zielpunkte ihrer Wanderlust machten. Wenn Sie sie sehen, oder ihr schreiben so bitte ich meine hochachtungsvollsten Grüsse zu melden, wie ich auch ihren verehrten Gemahl aufs herzlichste grüsse. Auch H. Prof. Frankland bitte ich, mich gelegentlich zu empfehlen, und vor allem seien Sie gebeten, Ihr Wohlwollen zu bewahren

Ihrem | aufrichtig ergebenen | DJGrailich

Vienna 26 Oct. 1856.

Esteemed Sir!

If you should gain a moment of leisure in the whirl of wonderful life to which you should have returned by now, then allow me to claim for it a reminiscence of your stay in Vienna. With the varied inspiration you will have received through the impressions of your extended journey and with the perspective of your experiences in France, Italy and Germany, the time of our Meeting of Natural Scientists[3] can admittedly find a place in your memory only as one episode among many others; and if I may also quite rightly add that you were certainly nowhere more warmly welcomed than in our circles, then you could nowhere have lacked friendly good will wherever people reside who bear a sense of and an understanding for open manliness and intellectually stimulating personality. To us, and particularly to me, your arrival and that of your friends in Vienna certainly form the centre round which all the happy memories of those September days are grouped: and if there is anything that I regret, then it is the circumstance that I did not find more time to speak with you about the memorable investigations which are connected with your name. I have since found occasion to study your memoir[4] from Jan. 1855 'On the nature of the Force by which Bodies are repelled etc.' and I have not read in a long time anything from beginning to end with greater pleasure—the clear exposition, the masterful handling of the inductive method, the confidence with which decisive facts are tested and determined independently of any theoretical opinion, all rank this work worthily among the investigations into the magnetic conditions of compressed masses. I have made Councillor v. Ettingshausen aware of it and I do not doubt that he will follow your description with the same approval and will introduce your way of conducting experiments in the physical institute,[5] as it cannot be denied that the facts of magnetism cannot be expounded more completely and more convincingly in any other way. However, your statement on the Ampère-Weberian view[6] seems too severe to me; although I am inherently an optician and better versed in the theories of this branch of physics, I nevertheless vividly remember the that the study of Ampère's works made on me: the manner in which, even if not always following the shortest analytical path, he combines analysis and experiment, astutely refines and establishes every theoretical principle by experiment, and is finally led, without sophistry and of necessity, to the acceptance of molecular streams, has always awakened in me a feeling of certainty, in a way that artifical and searched-for theories cannot engender. The fact that you present on p. 39 against the corollary of the Weber's deduction[7] presents, admittedly, a great difficulty and I am curious to learn what our compatriot will be able offer in reply to it. —Also, I did not understand Plücker's sentence, which you quote in the Note on p. 2, entirely as you did; 'regardless of the magnetic or the diamagnetic composition of the substance', I interpreted thus: 'the substance may be m. or d.,[8] the effect will always express itself in the direction of the axis as a <u>resultant</u>, and it may well be permissible to consider the phenomena according to these resultants, in the way in which we relate the phenomena of attraction of masses to the centre of the masses, and so forth'. But I am only a mere onlooker in these things and have to beg your indulgence for trying to express my views in opposition to a master in this subject.

I take the liberty to send you enclosed my work on the reflection and refraction of light on twin surfaces of optically uniaxial crystals.[9] Even if it may not aspire to be included among the theoretical works which shed light on the principles of wave theory, then many a fact to which it leads might nevertheless be of interest; for instance, that there is a <u>plane</u> in twin crystals in which all incident extraordinary rays are refracted without experiencing reflection: for ordinary substances there is only a single angle of incidence (the angle of polarisation, in so far as the substance is not elliptically polarised) at which this takes place;—the circumstance that ordinary rays are reflected without experiencing refraction, and so forth. As I have determined all of this through observation, I can proceed to further clarify, based on the results of these calculations, a good many strange reflection phenomena which twin crystals display. I have published in Poggendorff[10] a brief excerpt of the enclosed memoire (compilation of results). —If you would convey the second enclosed copy to Prof. Faraday, the man whom I admire the most among all living natural scientists, and express my deep respect to him on my behalf, I shall be very obliged to you.

And now permit me to come back once again to the merry time of your stay here. You will understand that while writing all of this, the image of your dear friend Hooker's fair wife never left my side; how is she, how did she make her way back, how does she remember her stay in the dusty, windy, angular streets of Vienna? As tempestuously as those days rushed past me, it left a half fairy-tale impression in me, and the 'fairy queen' of this autumn fairytale is Mistress Hooker. Her cheerful, natural way, the grace with which she knew how to create the charms of her home country in a foreign land and to bring many a sunny hour to a much tormented editor's cloudy existence[11] will not be forgotten in our circles and many a cheerful toast will still be raised to the memory of her lovely appearance. If, then the great migration of peoples[12] would have to begin again and the ladies of the continent would have cause for the most bitter jealousy. At least, it would then be undertsandable why, one after the other, Romans, Anglo-Saxons, Danes, and Normans made the distant island the destination of their wanderlust. If you see her, or write to her, then I ask you to convey my most respectful regards, just as I give her esteemed spouse my warmest regards also. I also ask you to give my compliments to Prof. Frankland when you have the occasion, and above all, may I ask that you continue to favour

Your | truly obedient servant | DJGrailich[13]

RI MS JT/1/G/12

1. *Grailich*: Wilhelm Josef Grailich (1829–59), Austrian physicist and mineralogist. A pupil and assistant of Andreus von Ettingshausen at the Vienna Polytechnic, he was elected Professor of Physics at the University of Vienna in 1857. His research was largely on crystal optics, especially fluorescence. During the meeting of the Deutscher Naturforscher und Ärtze Versammlung in Vienna in September 1856, he had acted as Secretary to the Physics Division and befriended Tyndall.
2. *Richtg*: abbreviation for Richtung = direction (German).
3. *our Meeting of Natural Scientists*: The meeting of the Gesellschaft Deutscher Naturforscher und Ärtze Versammlung that Tyndall had attended in Vienna.
4. *your memoir*: Tyndall's Bakerian Lecture, 'On the Nature of the Force by which Bodies are Repelled from the Poles of a Magnet', *Phil. Trans.*, 145 (1855), pp. 1–51. See letter 1031, n. 2.
5. *the physical institute*: The Physical Institute of the University of Vienna had been founded by Ettingshausen in 1850 and provided seminars and a laboratory course.

6. *the Ampère-Weberian view*: In 1826 Ampère had shown that the force between two parallel conductors (whether straight or coiled) depended on the inverse square of the distance between them. In the case of coiled wire, he showed that mathematically the angular displacements would cancel out. Weber explained these results as caused by the presence of positively–and negatively–charged particles whose force of interaction depended upon their relative velocities and accelerations.
7. *the corollary of the Weberian deduction*: The penultimate section of his printed Bakerian Lecture (note 4) was entitled: 'On M. Weber's Theory of Diamagnetic Polarity, and Ampère's Theory of Molecular Currents', pp. 38–40. At p. 39 Tyndall stated: 'The theory, notwithstanding its great beauty, is so extremely artificial, that I imagine the general conviction of its truth cannot be very strong; but there is one conclusion flowing from it which appears to me to be in direct opposition to experimental facts. The conclusion is, *that the magnetism of two iron particles in the line of magnetization is increased by the reciprocal action; but that, on the contrary, the diamagnetism of two bismuth particles lying in this direction is diminished by their reciprocal action.*'
8. *the substance may be m or d*: magnetic or diamagnetic.
9. *optically uniaxial crystals*: W. J. Grailich, 'Bewegung des Lichtes in optisch–einaxigen Zwillingskrystallen' [The motion of light in crystals with a single optical axis], *Wien, Sitz. Ber.,* 11 (1854), pp. 46–87; 12 (1854), pp. 230–63.
10. *in Poggendorff*: J. Grailich, 'Ueber die Brechung und Reflexion des Lichts an Zwillingsflächen optisch–einaxiger Krystalle' [On the refraction and reflection of light on twin surfaces of optically uniaxial crystals], *Poggend. Annal.* 98 (1856), pp. 203–14.
11. *tormented editor's cloudy existence*: Grailich was referring to his exhausting duties as secretary of the physics section of papers read at the Vienna conference.
12. *the great migration of peoples*: Grailich's light-hearted remark was a reference to the fact that during the nineteenth century some five million German-speaking Europeans emigrated to England and the United States of America.
13. *DJGrailich*: Grailich's first name was Wilhelm, but the 'D' here probably stood for Dr.

From Henry Moseley 28 October 1856 1291

Olveston | 28 Oct. 1856.

My dear Sir,

The Rev. Stephen Hawtrey[1] who is the Mathematical Master at Eton[2] is desirous to consult you on a course of instruction in Experimental science for the Etonians, which he purposes should extend over a period of two or three years. I have taken the liberty of giving him a letter of introduction to you; being convinced that the subject will commend itself to you as of no slight importance to the interests of education and of science.

Will you now permit me to consult you on a private matter. I have two boys[3] who are at a private classical school with a view to their being sent to a public school.

One of them (the elder) is acquitting himself well in Latin and Greek which appear to be tolerably well adapted to the education of his mind.

Of the younger (not quite 11 years old) his master reports that he is singularly without a talent for the acquisition of language, having no verbal memory—showing in such matters a vagrant mind, &c. &c. He reports moreover (what I also know) that he is far from being without intelligence in common things and is very clever in Arithmetic. His is in short (I doubt not) a mind of that class which cannot be educated by means of language but is capable of being educated through science of some kind or other. Now can you recommend me a school where this would be done; and where he would have for his associates boys of his own class? Does Earlswood[4]—the name I think of a school in Hampshire which you are acquainted with—satisfy these conditions?

Yours truly | H. Moseley.

RI MS JT/3/899
LT Transcript Only

1. *The Rev. Stephen Hawtrey*: Stephen Hawtrey (1808–86), a cousin of Edward Craven Hawtrey, the Provost (headmaster) of Eton College (1834–42), had been appointed Mathematical Master there before 1845. He combined this position with the headmastership of St Mark's School, Windsor and serving as Vicar of Trinity Church, Windsor. Tyndall subsequently gave a number of lectures at Eton, but experienced disciplinary problems (RI MS JT/2/13c/1035–37).
2. *Eton*: Eton College was founded by Henry VI in 1440 as a boys' boarding school at Eton, a village close to Windsor, in Berkshire.
3. *I have two boys*: Moseley's sons were Henry Nottidge Moseley (1844–91) who became a distinguished naturalist after education at Harrow School and Exeter College, Oxford. His son, Henry Moseley (1887–1914) became a distinguished physicist. The other son has not been identified and was not recorded in either the 1851 or 1861 Census.
4. *Earlswood*: Moseley meant Queenwood College. Given Tyndall's reservations about the headmaster, George Edmondson, he is unlikely to have recommended it as suited to the younger son.

BIOGRAPHICAL REGISTER

This register contains the names and biographical details of people who are mentioned three or more times in the letters included in this volume. Biographical information about people mentioned once or twice is contained in the notes appended to the appropriate letter(s).

Agassiz, Louis (1807–73), had studied in Germany and France before being appointed to a position at the University of Neuchâtel. In 1847 he moved to the chair of zoology and geology at Harvard and founded the Museum of Comparative Zoology. He was a key figure in developing the study of the Earth's history, proposed the theory of ice ages, and made a close study of glaciers in the Alps. In 1840 he published his important two-volume *Études sur les Glaciers* (Neuchâtel: Gent et Gassmann), which helped establish the scientific study of glaciers.

Airy, George Biddell (1801–92), was educated at Cambridge, graduating in 1823. He was appointed Lucasian professor of mathematics three years later, and in 1828 transferred to the Plumian Chair of astronomy and directorship of the Cambridge University observatory. In 1835 he was appointed Astronomer Royal—a position he held until 1881—and set about improving the instrumentation and research activities of the Royal Observatory, Greenwich. He vigorously pursued research on several topics, including optical diffraction, calculated the Earth's mean density, and analyzed the inequalities in the motions of the Earth and Venus.

Ampère, André-Marie (1775–1836), served as professor of experimental physics at the Collège de France, to which he was appointed in 1824. Like Faraday, he was intrigued by the discovery in 1820 by Hans Christian Øersted that a compass needle is influenced by a neighboring current-carrying wire. He subsequently directed his research to understanding the relationship between electricity and magnetism, developing his own law of the force between parallel current-carrying wires. Ampère's researches and theories were familiar to Tyndall.

Anderson, Charles (1790–1866), served in the Royal Artillery, achieving the rank of sergeant. In 1827 he was first employed at the RI to assist Faraday in the project (on behalf of the Board of Longitude) of improving optical glass. With the termination of that project in 1830 Anderson became Faraday's laboratory assistant and also assisted him in his RI lectures.

Arago, François (1786–1853), had been a staunch French Republican and eminent physicist who made a number of contributions especially to the science of optics and championed Fresnel's wave theory of light. His innovations in the field of electromagnetism included Arago's disc.

Ashburton, Lord: see Baring, William Bingham.

Babbage, Charles (1791–1871), was a Cambridge-educated mathematician who held the Lucasian Professorship in mathematics at Cambridge from 1828 to 1839, but never delivered a lecture. A highly controversial character, he contributed to a wide range of subjects, including mathematics, astronomy, cryptography, the productivity of manufacturing, and, most famously, the theory of computing. He designed and tried to implement the first automatic calculating engines—the difference engine and the analytical engine.

Baring, William Bingham, 2nd Baron Ashburton (1799–1864), was educated at Oxford and served as MP for various constituencies between 1826 and 1848. Initially a Whig, from 1837 he supported Peel and the Tories, serving as joint secretary of the Board of Control and Paymaster General; he was appointed a privy counsellor in 1845. His first wife, Harriet (1805–57), was a socialite and hostess who attracted many literary and scientific figures to their houses in Piccadilly and Alresford in Hampshire.

Barlow, John (1798–1869), studied at Trinity College, Cambridge, and was ordained in 1823. He was subsequently appointed curate at Uckfield (Sussex), rector of Little Bowden (Northamptonshire), and from 1854 until his death he was chaplain-in-ordinary at Kensington Palace. From 1842 to 1860 he was the hardworking secretary of the RI and was close to both Faraday and Tyndall. Having married the daughter of an MP who made his fortune in India, he lived near the RI and entertained many of its wealthy and aristocratic supporters as well as fellow scientists. Tyndall sometimes dined with the Barlows. Barlow was, however, in the habit of dozing during RI lectures; thus on 6 March 1855 Tyndall recorded that Barlow was "asleep in the front seat" (Journal, RI JT 2/13c/732).

Barnard, Jane (1832–1911), Faraday's niece, lived at the RI and assisted the aging Faradays. Like her uncle and aunt she was a member of the Sandemanian church.

Bevington, James Buckingham (1804–92), was a Quaker leather merchant and father of a pupil (Geoffrey) at Queenwood College. The Bevingtons lived at Wandsworth, close to Clapham Common, and owned Neckinger Mills in Bermondsey.

Biot, Jean-Baptiste (1774–1862), was a product of the École Polytechnique and later professor of physics at the Collège de France. He conducted experiments on balloon flights and also made a close study of meteorites. His main research was on the polarization of light, for which he was awarded the RS's Rumford Medal in 1840.

Brewster, David (1781–1868), was a Scottish man of science who wrote on a wide range of scientific topics but contributed primarily to optics. He was also known as a controversialist— for example, he was locked in a bitter dispute with Charles Wheatstone over the invention of the stereoscope. Tyndall encountered him at meetings of the BAAS and they both served as jurors at the 1855 Exposition Universelle.

Bunsen, Robert (1811–99), studied at the University of Göttingen, where he lectured from 1833 to 1836. He then held positions at the Polytechnikum at Kassel, the University of Marburg (1839–51), the University of Breslau, and, from 1852, the University of Heidelberg. Among his many innovations in chemistry was his discovery of cacodyl (a compound containing arsenic). He made extensive use of both electrolysis and spectrum analysis to analyze chemical substances and developed the cell (battery) that bears his name. Frankland was working with Bunsen in 1848, when Tyndall traveled to Marburg to study with Bunsen. Tyndall later described Bunsen as "a master of the language of experiment[. . . . His] lectures were full of matter. [. . . He] was a man of fine presence, tall, handsome, courteous, and without a trace of affectation or pedantry" (Tyndall, *New Fragments* [London: Longmans, Green & Co., 1892], 238).

Carlyle, Jane Welsh (1801–66), was the wife of Thomas Carlyle and a literary figure in her own right.

Carlyle, Thomas (1795–1881), the Scottish moralist, historian, and essayist, was one of the major literary influences on Tyndall, who avidly read his publications. After reading *Past and Present* (1843) Tyndall commented that Carlyle "must be a true hero. My feelings towards him are those of worship 'transcendental wonder' as he defines it" (Journal, 18 July 1847, RI JT 2/13a/231).

Chapman, John (1821–94), had a varied early career, including a period studying medicine in Paris, but in the mid-1840s he settled in London and purchased a publishing company. In 1851 he became the editor and proprietor of the *Westminster Review*, which was supported by philosophical radicals, such as George Eliot, John Stuart Mill, and Harriet Martineau. Many of the scientific articles were written by Thomas Henry Huxley, who was joined by Tyndall in 1853.

Coxe, James (1810–78), was educated at the Universities of Edinburgh, Göttingen, Heidelberg, and Paris, and initially practiced medicine in Edinburgh. In 1855 he was appointed to the Royal Commission that examined the condition of the lunatic asylums in Scotland and, following the 1857 act establishing the General Board of Lunacy for Scotland, served as one of its two paid commissioners. He was knighted in 1863 for his services.

Coxe, Mary Anne (née Cumming, 1806–75), was the wife of James Coxe.

Davy, Humphry (1778–1829), had worked at the Pneumatic Institution in Bristol before gaining employment at the RI in 1801. There he achieved fame as a lecturer and for his chemical researches that isolated such elements as sodium, potassium, and chlorine. He is often attributed—probably wrongly—with the invention of the miners' safety lamp. In addition to being the foremost scientist at the RI, Davy served as president of the RS from 1820 to 1827.

de la Rive, Auguste (1801–73), served as professor of natural philosophy at the Academy of Geneva and worked principally on electricity, although he also investigated the specific heats of gases and calculated the temperature of the Earth's crust. He published on electrical discharges through rarefied gases and on the aurora borealis. His principal book was *Traité d'électricité théoretique et appliquée* (3 vols., Paris: J.-B. Baillière et H. Baillière, 1854–58).

de la Rue, Warren (1815–89), was head of a large firm of stationers in London. His scientific interests were broad and included innovative research in the fields of photography, astronomy, and chemistry.

Debus, Heinrich (1824–1915), studied chemistry in Marburg (1845–48), where he worked as Bunsen's assistant and where he met Tyndall, Frankland,

and Hirst. Moving to England in 1851, he taught chemistry at Queenwood College and was for a time a colleague of Tyndall's. Later he taught at Clifton College (Bristol), Guy's Hospital, and the Royal Naval College, Greenwich. His research was principally on organic chemistry, including the oxidization of ethyl alcohol to form glyoxal ($C_2H_2O_2$) and glyoxylic acid ($C_2H_2O_3$).

Despretz, César (1791–1863), was a Catholic chemist and physicist who taught at the Faculté des Sciences in Paris. His research covered a wide range of topics, but in the early 1850s he was primarily concerned with the conduction of heat and the functioning of the voltaic pile. He also published textbooks in both physics and chemistry.

Dove, Heinrich Wilhelm (1803–79), was educated at the University of Berlin and later taught at Berlin's Friedrich-Wilhelms-Universität. In his research he focused on meteorology and was awarded the RS's Copley Medal in 1853 for his contributions to understanding the distribution of heat in the atmosphere.

Drewry, Henry Runciman (1801–86), was a clerk at the War Department who was promoted to chief clerk in February 1855.

du Bois-Reymond, Emil (1818–96), was an up-and-coming German anatomist and physiologist from Berlin who focused on the emerging field of electrophysiology and challenged the views propounded by Matteucci. Tyndall, who first met him Germany, described him as "a fine fellow—athletic, with a neck like an ox and a chest like Hercules, his face ruddy brown as if he spent his whole time in the fields" (Journal, 28 April 1851, RI JT 2/13b/539). He made several trips to England during the early 1850s and delivered a course of lectures on electrophysiology at the RI in the spring of 1855.

Duboscq, Louis Jules (1817–86), was a French inventor and instrument maker known for producing high-quality optical instruments. Tyndall presumably met him at the Exposition Universelle in Paris and subsequently ordered a number of optical instruments from him.

Edmundson, Anne (née Singleton, 1799–1863), was the daughter of William Singleton, George Edmondson's reading master at Ackworth, and sister of Josiah Singleton, one of the masters at Queenwood whom Edmondson dismissed. Tyndall disliked her and objected to her interference in the school's affairs.

Edmondson, George (1798–1863), was an Ackworth-trained Quaker educationalist who had spent seven years in Russia, at the invitation of Alexander

I, helping to improve the country's agriculture. After returning to England he opened a school at Blackburn in 1825, then one near Preston, before becoming the head teacher at Queenwood College when it opened in 1847. Under his influence science and technological subjects were emphasized. Although he hired Tyndall, Hirst, Debus, and Frankland, he was considered to be weak, quarrelsome, and dominated by his wife. "No man neglects his duty more systematically, [. . .] or who sucks more pleasure out of his meat and drink," commented Tyndall (Journal, 12 April 1855, RI JT 2/13c/921).

Emerson, Ralph Waldo (1803–82), was an American and the leading transcendentalist poet and essayist. Tyndall admired him greatly, avidly read his books, and heard him lecture in Halifax in January 1848. Particularly during the 1840s and 1850s Emerson exerted a major influence on Tyndall; as he told Hirst, "In Emerson you behold one of the noblest souls that ever was struck in clay" (vol. 2, letter 0342).

Ettingshausen, Andreas von (1796–1878), had held positions in mathematics and physics at the University of Vienna since 1817 and was appointed director of its physics institute in 1852. He introduced advanced courses in mathematical physics and in 1837 he presented at a scientific meeting his improved design for electromagnetic machines.

Faraday, Michael (1791–1867), was an eminent natural philosopher and the Fullerian professor of chemistry at the RI and Superintendent of the House from 1852. His extensive research focused on electricity and encompassed electrochemistry, electromagnetism, diamagnetism, and magneto-optics (the Faraday effect). He was Tyndall's main patron and helped him obtain the position of professor of natural philosophy at the RI in 1853. Despite significant differences between their views on religion and on the nature of magnetism, they worked closely together. Tyndall greatly respected Faraday and on one occasion commented to Hirst that "he and I are still like father and son" (letter 1170). Tyndall's *Faraday as a Discoverer* (London: Longmans, Green & Co, 1868) was his tribute to his mentor.

Faraday, Sarah (née Barnard, 1800–79), lived with her husband in their rooms at the RI. Tyndall frequently met her and commented in his journal, "She is very motherly to me, and Faraday says that were it not for her own feebleness she must become quite a mother to me, and take care of me" (2 March 1855, RI JT 2/13c/730).

Finch, John (1784–1859), was an iron merchant and temperance reformer from Liverpool who helped Robert Owen set up the socialist community at

Harmony Hall (Queenwood) in 1839 and was its first acting governor. He remained at Queenwood after it became a school in 1847 and served as one of its governors.

Forbes, David (1828–76), the brother of Edward Forbes, studied chemistry and then worked as a metallurgist, including ten years in Norway helping to develop that country's mining industry. He wrote a number of papers on geology, mineralogy, and chemistry, for which he was admitted to the RS in 1858 with Tyndall's support.

Forbes, Edward (1815–54), was a highly productive naturalist who published about one hundred papers and several substantive books—covering a wide range of geological and zoological topics—during his short life. In his early years he traveled extensively and in 1843 he was appointed to the poorly remunerated chair of botany at King's College London, but had to take on other work that affected his health. Just a few months before his death he was appointed the highly cherished professorship of natural history at the University of Edinburgh.

Forbes, James David (1809–68), studied the broad curriculum at Edinburgh University and then specialized in law before being admitted to the Faculty of Advocates. Becoming increasingly interested in science, he secured the chair of natural philosophy at Edinburgh in 1833. While much of his research was on physics, he also studied the behavior of glaciers, undertook field trips to Switzerland and Norway, and published extensively on the topic. His theory that a glacier is a viscous body was challenged by Tyndall in his "Observations on Glaciers," *Roy. Inst. Proc.* 2 (1854–58): 320–27.

Foucault, Léon (1819–68), was famed for his use of a long pendulum to demonstrate the rotation of the earth. The experiment was first carried out in January 1851 and seen by Tyndall during his visit to the 1855 Exposition Universelle. Foucault also contributed to optics (including a method for measuring the speed of light) and discovered eddy currents (induced electric currents in metals).

Fox, Luther Owen (1812–79), the surgeon in Broughton, had been educated at University College London and Saint Bartholomew's Hospital and then obtained his licentiate of the Royal College of Surgeons in 1833 and his fellowship in 1852. He attended the pupils and staff at Queenwood College and was on close terms with Tyndall, who often visited and dined with him. Six of Fox's eight sons attended Queenwood.

Francis, William (1817–1904), was the illegitimate son of the printer and publisher Richard Taylor, with whom he entered into partnership ("Taylor & Francis") in 1852. He had learned the printing trade in Germany in the 1830s before training as a chemist with Liebig in Giessen, where he took his doctorate in 1842. He made many translations for his father's *Scientific Memoirs* and *Philosophical Magazine* and became the managing editor of the latter in 1851. He first met Tyndall in October 1850 and became a close friend and facilitated his entry into the scientific community in the 1850s.

Frankland, Edward (1825–99), worked initially as assistant to the chemist Lyon Playfair and then taught science at Queenwood College, where he met Tyndall. He and Tyndall then commenced studies at the University of Marburg, but in 1848 Frankland proceeded to Liebig's laboratory at the University of Giessen. Returning to England, he taught briefly at the Putney College of Engineering before being appointed to the chair of chemistry at Owens College, Manchester, in 1851; later he became professor at the Royal College of Chemistry. One of Tyndall's closest friends, he was elected to the RS in 1853 (with Tyndall's support) and was awarded the Royal Medal in 1857. In his early research Frankland isolated a number of new organometallic compounds, thus helping to establish the field of organic chemistry.

Frankland, Sophie (née Fick, 1821–74), married Edward Frankland in 1851. Her brother was the German physiologist Adolf Eugen Fick.

Galbraith, Joseph Allen (1819–90), served from 1854 to 1870 as professor of natural and experimental philosophy at Trinity College Dublin. With Samuel Haughton he wrote a series of textbooks on physics and mathematics including *Manual of Natural Philosophy* (Dublin: Hodges & Smith, 1850) and *Manual of Optics* (London: Longman, Brown, Green & Longmans, 1854).

Gale, Sarah (c. 1801–67), was probably the widowed innkeeper of the Tally Ho in Broughton. Her husband, Thomas Gale, had died in 1841. She does not appear to have been the mother of George Gale (b. 1836), the Queenwood College pupil who had clashed with Haas in 1855.

Gassiot, John Peter (1797–1877), a wine merchant by trade, pursued research on electricity at his home at Clapham Common. He published papers on the functioning of the voltaic pile (battery) and was a founding member of the London Electrical Society in 1837. Gassiot was a prominent member of the London scientific community.

Goethe, Johann Wolfgang von (1749–1832), was the outstanding German literary figure of the late eighteenth and early nineteenth centuries. His extensive writings included his theory of colors and a study of the structure and function of plants. During his time in Marburg Tyndall became familiar with some of Goethe's works. In late 1855 Tyndall read G. H. Lewes's recently published *The Life and Works of Goethe.*

Grove, William Robert (1811–96), was a Welsh lawyer who developed an improved form of cell (battery) using zinc and platinum electrodes in different acids separated by a porous ceramic jar. He also developed the first fuel cell, which used hydrogen and oxygen to generate electricity. A keen supporter of the RI, he presented his ideas about the correlation of forces (which includes notions of energy conservation) at a course of lectures at the RI in 1846. He was professor of experimental philosophy at the LI (1841-46).

Haas, John (b. 1828), was a teacher of modern languages and drawing at Queenwood College, having joined the school in 1848. He was a German-speaking Swiss who had trained at the famous Pestalozzian school at Hofwyl. He, Debus, and Tyndall often spoke in German to preserve their conversation from other teachers; he had also improved Tyndall's French. In 1856 he married another Swiss, Eliza Wegmuller. On the death of Edmondson in 1863 he returned to his birthplace, Morat (Murton), where he directed a school.

Harris, William Snow (1791–1867), was an Edinburgh-trained physician. He lived in Plymouth, took a keen interest in the emerging science of electricity, and developed lightning conductors for ships.

Haughton, Samuel (1821–97), was an Irish mathematician, geologist, anatomist, and physiologist who published a series of scientific and mathematical textbooks with his Dublin colleague, Joseph Allen Galbraith. He was professor of geology and mineralogy at Trinity College Dublin (1851–81), where he and Galbraith taught the growing number of students who were taking public examinations to enter the Indian Civil Service or the Royal Artillery and Royal Engineers. He took holy orders in 1847. Tyndall supported his admission to the RS in 1858.

Hirst, Anna (née Martin, 1831–57), was the Irish wife of Thomas Archer Hirst. They had met in Marburg, in 1852, when she was acting as companion to her sister, Mary Simpson (née Martin), the wife of the Irish chemist

Maxwell Simpson, who was studying in Bunsen's laboratory. Her brother was John Martin. She married Hirst at her family home in County Down, Ireland, on 28 December 1854. Following a honeymoon in Paris, she joined her husband at Vine Cottage in Broughton. She soon began to display symptoms of tuberculosis, leading Hirst to resign from Queenwood College in June 1856 to nurse her in the south of France. She died in Paris on 1 July 1857 and was buried in the cemetery at Montmartre.

Hirst, Thomas Archer (1830–1892), was a Yorkshire mathematician and educator. His friendship with Tyndall had begun in Halifax in 1845 when they were both working for the surveyor Richard Carter. Encouraged by Tyndall, he studied at the universities of Marburg, Göttingen, and Berlin between 1849 and 1852. He returned to England with a doctorate to replace Tyndall as a mathematics teacher at Queenwood College in 1853. Like Tyndall, he did not get on with the headmaster, George Edmondson. He resigned in June 1856 to look after his sick wife and stayed on the Continent until 1860. He was Tyndall's closest and lifelong friend, as reflected in their huge correspondence.

Hofmann, August Wilhelm (1818–92), was a German organic chemist who had studied with Justus Liebig at the University of Giessen. On the recommendation of Liebig and Prince Albert he was appointed director of the newly established Royal College of Chemistry in Oxford Street in 1845. In 1853 the college was incorporated into the Metropolitan School of Science Applied to Mining and the Arts, which the British government had founded in 1851 as the Government School of Mines and of Science Applied to the Arts. His work on the "ammonia type" kickstarted the British dyestuffs industry.

Hooker, Joseph Dalton (1817–1911), was a British botanist who specialized in taxonomy and the geographical distribution of plants. He was the younger son of Sir William Jackson Hooker, the director of the Royal Botanic Gardens at Kew. After studying medicine at the University of Glasgow, he became a naval surgeon and spent the years 1839–43 on a geomagnetic survey of the southern oceans. From 1847 to 1851 he explored and collected botanical specimens in India. He had been a close friend of Darwin's since 1844. In June 1855 Hooker joined his father as deputy director of Kew Gardens. Hooker was first introduced to Tyndall by Huxley in November 1855 and he became one of Tyndall's closest friends. Together with Huxley, in August 1856 Hooker helped Tyndall with his first studies of Alpine glaciers.

Hooker, Frances Harriet (née Henslow, 1825–74), was the daughter of the Cambridge botanist John Steven Henslow, who had been Darwin's teacher.

She married Joseph Hooker in August 1851 and together they had four sons and two daughters. She was fluent in French and German and in addition to her many private translations for her husband she published a translation of Emmanuel le Maout, *A General System of Botany* (London: Longman, 1873). She was much admired by the Austrian physicist Josef Grailich.

Hughes, Edward (1819–59), was an Irish educator born in Enniskillen and headmaster of the Royal Naval Lower School at Greenwich Hospital (a charitable institution founded in 1694 for the maintenance and education of the orphans of deceased seamen). There he taught English, mathematics, and navigation. In response to the need for suitable textbooks for elementary schools sponsored by the Committee of Council on Education, beginning in 1854 he edited and published a series of *Reading-Lesson-Books* containing essays on science, literature, geography, and economics to which Tyndall contributed.

Huxley, Thomas Henry (1825–95), was a British zoologist, educator, and science popularizer. Following medical training at Charing Cross Hospital in London, he acted as a ship's surgeon (1846–50). After eking a living by writing for Chapman's *Westminster Review,* in 1854 he was appointed lecturer on natural history at the Royal School of Mines, and paleontologist to the Geological Survey in 1855. In the same year he joined Tyndall (whom he first met in 1851 when traveling to the Ipswich meeting of the BAAS) as Fullerian Professor of Natural History at the RI and lecturer at Saint Thomas's Hospital. While in Australia he met Henrietta (Nettie) Anne Heathorn, whom he married in July 1855 after an eight-year engagement.

Jones, Henry Bence (1813–73), was a British physician and medical chemist. He was educated at Harrow and Trinity College, Cambridge, before studying medicine at Saint George's Hospital, London. Attracted by the potential applications of chemistry to clinical medicine, he studied organic chemistry with Liebig at Giessen (1841–43) before assuming a career as a successful private physician and physician at Saint George's Hospital. He published important research on protein chemistry and made an abstract in English of du Bois-Reymond's work on electrophysiology (1852) that annoyed Matteucci. Though not elected secretary of the RI until 1860, as a manager since 1853 and close friend of Faraday's, he supported Tyndall's appointment at the RI in 1853. He frequently entertained Tyndall at both his London home and Folkestone, where he kept a summer house. In May 1842 he married a cousin, Lady Millicent Acheson, daughter of the 2nd Earl of Gosford, with whom he had seven children.

Lee, Stanlake (1815–96), was rector of the church at Broughton in Hampshire, which was under the patronage of the Barings. Lee read classics at Queen's College, Oxford, where he graduated in 1840. He took holy orders in 1841 and served as rector at Broughton from 1842 until his death. He published a number of sermons and frequently attended functions at Queenwood College.

Le Verrier, Urbain Jean Joseph (1811–77), was a French astronomer. A graduate of the École Polytechnique (1831), he began his career there as *répétiteur d'astronomie*. In August 1846 he predicted the position of the then unknown planet Neptune. In 1854 he succeeded Arago as director of the Paris Observatory.

Lyell, Charles (1797–1875), was a Scottish geologist and lawyer, whose *Principles of Geology* (3 vols., 1830–33) promulgated the uniformitarian view of the processes of geological change. Educated at Exeter College, Oxford, he studied law at Lincoln's Inn. He served as professor of geology at King's College, London (1831–33), after which he lived on private means. He married Mary Elizabeth Horner in July 1832. A prominent figure in the Geological Society, he served as president 1835–37 and 1849–51. He was knighted in 1848.

Magnus, Heinrich Gustav (1802–70), was a German chemist and physicist and consummate experimentalist. A pupil of Mitscherlich in Berlin and of Berzelius in Stockholm, he was professor of technology and physics at the University of Berlin from 1834 until his death. In Berlin he developed a superbly equipped physics laboratory at his home, to which he invited young German physicists to work, including Tyndall in 1851. Although his early research was primarily in chemistry, by the 1850s he had begun to investigate a range of physical problems, as his friendly letters to Tyndall demonstrate.

Martin, John (1812–75), was an Irish nationalist ("Honest John Martin") and the brother of Hirst's wife, Anna. He had been convicted of treason in 1848 and transported to Van Dieman's Land (Tasmania) in 1849. He was granted release in 1854 on condition that he live in exile and chose to live in Paris. It was there that Hirst first met him when on his honeymoon. Following a full pardon in May 1856, he returned to Ireland, where he continued to engage in the cause for home rule.

Matteucci, Carlo (1811–68), was an Italian chemist, physicist, physiologist, and politician who had been in correspondence with Faraday since 1833. After graduating in physics at the University of Bologna in 1828 he undertook further studies in Paris before being appointed professor of physics at the University of Pisa in 1840. His pioneering research in the field of electrophysiology led him into a vexatious dispute with du Bois-Reymond that is reflected in his correspondence with Tyndall on diamagnetism. He had married an Englishwoman, Robinia Elizabeth Young, in September 1846.

Maule, Fox: see Panmure, Lord.

Moore, Julia (1803–1904), was one of the three daughters of the surgeon James Carrick Moore. Her sister Harriet Jane Moore was a talented watercolorist, and her brother, another John Carrick Moore, was a field geologist and a member of the RI since 1836. The wealthy Moores had close connections with the RI and frequently entertained Tyndall at their Wimbledon home. The family impressed Tyndall "strongly with the feeling of true human worth" (RI JT 2/13c/933).

Moseley, Henry (1801–72), was an English mathematician and educationalist. Following schooling at a naval school in Portsmouth he studied mathematics at Saint John's College, Cambridge, where he graduated seventh wrangler (1826) and later received his MA (1836). He took holy orders in 1827, and during a curacy in Somerset he published a book on hydrodynamics (1830) that led to his appointment as professor of natural and experimental philosophy at King's College, London (1831–44). He then became a school inspector for the Committee of Council of Education, where he did important work in introducing science teaching into elementary schools and teacher training colleges. Despite his appointment as resident canon at Bristol Cathedral in 1853 and as vicar of Olveston in Gloucestershire, he was closely involved in arranging examinations for admission to the Royal Military Academy at Woolwich in 1855 following the reform of military education at the time of the Crimean War.

Owen, Richard (1804–92), was an English anatomist and paleontologist who served as Hunterian Professor of Comparative Anatomy at the Royal College of Surgeons (1836–56). He was then appointed superintendent of the natural history collections at the British Museum. Tyndall first met him in 1852 but was not formally introduced until February 1856, when he was greatly impressed "by his vast scientific acquirements . . . elevated by his love for and cultivation of poetry" (RI JT 2/13c/727).

Panmure, Lord, (1801–74), was a Scottish career politician. Born Fox Maule, he became the second Baron Panmure on the death of his father, William Ramsay Maule, in 1852. After serving in the army between 1820 and 1832, he was elected as MP for various Perthshire constituencies (1835–52), and acted as undersecretary of state (1835–41), and secretary at war (1846–52; 1855–58), following which he succeeded to the earldom of Dalhousie in Scotland. Although Tyndall never met him, Tyndall was obliged to apologize to him for criticizing the way marks for science questions were allocated in examinations for recruits to the Royal Military Academy.

Pare, William (1805–73), was a Birmingham tobacconist who became an Owenite and leading member of the cooperative movement. He was a governor of Robert Owen's original Harmony Hall commune (1842–44) and remained a landlord of its successor, Queenwood College, while acting as manager to an ironworks outside Dublin. He was elected to the Statistical Society in 1855 for his analyses of projected railway lines in Britain and abroad. Tyndall met him several times while teaching at Queenwood.

Payne, Thomas (Tom), was a cousin of Tyndall's. It is unclear whether this was on his mother's or father's side. He had worked with Tyndall as a civil assistant in the Irish Ordnance Survey in 1839 and then emigrated to Melbourne (c. 1840) to make his fortune. He returned to Ireland via Palestine in 1854 and spent some time with Tyndall in London in December 1854 and January 1855 in search of a wife (RI JT 2/13c/704, 707, 712). He appears to have returned to Australia and married an Irish woman (Rosa M.), since a daughter was born there in 1858. By 1861 he was registered as a landed proprietor living in Paddington, London, where he frequently entertained Tyndall (1861 Census).

Plücker, Julius (1801–68), was a German mathematician and experimental physicist. He was professor of physics at the University of Bonn, 1847–68. He first established a reputation for his work on analytical geometry, but under the influence of Faraday's research he began experimental work in electricity and magnetism in the 1840s. His claim, from 1848 onward, that crystals displayed magnetic effects was challenged by Tyndall and, to Faraday's embarrassment, led to ill feeling and controversy.

Pollock, Juliet (née Creed, 1819–99), was the daughter of the vicar of Corse, Gloucestershire, and wife of William Frederick Pollock, barrister and author. They married in 1844 and had three sons. She and her husband were staunch supporters of the RI and frequently entertained Tyndall to dinner. On such occasions he often played with their children and displayed special affection for young Walter Herries Pollock, a future editor of the *Saturday Review*.

Pridie [Priddie/Priddy], Robert [Roby/Rob] Hall (b. 1832), was a Yorkshireman whom Hirst and Tyndall befriended when they joined the Mutual Improvement Society in Halifax. The son of the Reverend James Pridie, an independent minister in Halifax, Pridie and another member of the Mutual Improvement Society in Halifax, Francis Booth, were mentored by Hirst (just as Hirst was mentored by Tyndall). When Booth died of tuberculosis in March 1853, Hirst financially supported his widow, and Pridie, who was by then a bank clerk in Halifax, arranged the payments. Tyndall took over these payments while Hirst was in Europe in 1856.

Regnault, Henri Victor (1810–78), was a French chemist, physicist, and photographer. A pupil of Liebig's at the University of Giessen, he was professor of physics at the Collège de France, 1841–54, and director of the famous porcelain factory at Sèvres, 1854–70. He suffered a serious accident in 1856, to the great concern of Faraday.

Riess [Rieß], Peter Theophil (1804–83), was one of the many German physicists who befriended Tyndall during his studies in Berlin in 1851 and 1853. He became professor of physics at the Academie der Wissenschaften in Berlin (1842–83) and in 1842 he was the first Jew to be elected to the Royal Prussian Academy of Sciences. His studies of static electricity were described in *Die Lehre von den Reibungselektrizität* (2 vols., Berlin: August Hirschwald, 1853), following which he studied diamagnetism. Tyndall arranged for the translation of several of his papers to appear in *Phil. Mag.*

Sabine, Edward (1788–1883), was an Anglo-Irish physicist who organized the "magnetic crusade" of worldwide magnetic readings. A graduate of the Royal Military Academy at Greenwich, he reached the rank of general in the 1850s. He was closely connected with the administration of the RS and was its treasurer between 1850 and 1861. He was important in promoting Tyndall in the early part of his career.

Steinheil, Karl August (1801–70), was a German physicist and astronomer who was professor of mathematics and physics at the University of Munich, 1832–70. In 1854 he opened an optical workshop in Munich that became renowned for its instruments and telescopes for astronomy and telegraphy.

Stokes, George Gabriel (1819–1903), was an English mathematician and physicist. He had graduated senior wrangler and Smith's Prizeman at Cambridge in 1841 and served as Lucasian Professor of mathematics at Cambridge, 1849–1903. He also augmented his income as a lecturer at the Royal School of Mines (1854–60). As secretary of the RS (1854–85), he was responsible for refereeing papers for *Phil. Trans.*—a task for which he frequently sought Tyndall's help. He also mentored Tyndall's early contributions to *Phil. Trans.*

Tennyson, Alfred (1809–92), was a Lincolnshire poet educated at Trinity College, Cambridge. He was appointed poet laureate in succession to Wordsworth in 1850, the same year that he published *In Memoriam.* In 1853 he moved to Farringford on the Isle of Wight, from where he published *Maud* in 1855. Like most nineteenth-century men of science, Tyndall greatly admired Tennyson's poetry, though he did not converse with him until 1858.

Thomson, William (1824–1907), was a distinguished Irish-Scottish mathematician, physicist, and inventor. He was educated at the University of Glasgow and Peterhouse College, Cambridge, where he was second wrangler and Smith's Prizeman in 1841. He was professor of natural philosophy in Glasgow from 1846, although he paid frequent visits to London. He had first met Tyndall at the Edinburgh meeting of the BAAS in 1850, and from the start the two physicists had a strained relationship.

Tyndale, Hector (1821–80), was an American businessman and man of culture who was Tyndall's cousin. He was the son of Tyndall's father's brother, Robinson Tyndall, who had emigrated to Philadelphia from Leighlinbridge around 1820 and founded a glass and ceramics import business. Hector inherited the business and first met his physicist cousin in 1854 on one of his frequent business trips to Europe. They immediately struck up an intellectual rapport and began an occasional correspondence. He became involved in Republican politics in 1856 but had no sympathy for the abolitionist cause espoused by John Brown. His mother, Sarah, a Quaker, introduced him to Walt Whitman in 1856.

Tyndall, Sarah (née Macassey [Macasey/McAssey], 1793–1867), was Tyndall's mother. She had been widowed in March 1847. She remained in the family cottage at Leighlinbridge, Ireland, supported financially by Tyndall. Although the three letters to her that have survived for the period 1855–56 show detachment and aloofness, Tyndall later told Hirst of his deep affection for her rare qualities.

Wartmann, Elie-François (1817–86), was a Swiss physicist and archaeologist who worked on colorblindness, synesthesia, magnetochemistry, and electrodynamics. He was the son of the Geneva astronomer Louis-François Wartmann and the first French-speaking physicist to draw attention to Ohm's law.

In 1852 Tyndall had translated one of Wartmann's papers on the effect of an electric current on plant growth for *Phil. Mag.* In July 1855 Wartmann served as a juror at the Exposition Universelle in Paris, where he spent a great deal of time in Tyndall's company.

Weber, Wilhelm Eduard (1804–91), was a German physicist noted for his work on electrodynamics. He was a graduate of the University of Halle, where he began a distinguished career as professor of physics in 1828, the year he first met Gauss. From 1831 until his resignation in political protest in 1837, he served as professor of physics at the University of Göttingen, where he collaborated with Gauss on magnetic and telegraphic researches. In 1843 he was called to the chair of physics at the University of Leipzig, where he engaged in research on electrodynamics. He returned to the chair at Göttingen in 1848 and resumed his partnership with Gauss. Tyndall arranged for several of his papers to be translated in *Phil. Mag.*

Wertheim, Theodor (1820–64), was an Austrian chemist best known for his study of the chemistry of garlic. He had studied in Berlin with Mitscherlich and with Redtenbacher in Prague before serving as professor of chemistry at the University of Pest from 1853 until 1860. Tyndall met him in Paris in 1855 when both were serving as jurors at the Exposition Universelle.

Wheatstone, Charles (1802–75), was an English physicist, inventor, and businessman. He had been a musical instrument maker before becoming professor of experimental philosophy at King's College, London (1834–75), where he was a reluctant lecturer. There he developed the electric telegraph with William Fothergill Cooke, and a number of optical instruments, including the stereoscope. The latter led him into controversy with David Brewster. He married Emma West in 1847 and had five children by 1855. The family settled in Hammersmith, where Wheatstone frequently entertained Tyndall, whom he had first met in June 1852.

Whitehouse, Mary A. (b. 1815), was Tyndall's landlady at 9 Chadwell Street, Islington, where Tyndall lodged from 18 June 1852 until 18 August 1855. Her husband, Henry, was a "chaser and embosser." Tyndall recorded in his journal: "Tomorrow I quit a house where I have experienced great attention and where everything possible was done to promote my comfort" (RI JT 2/13c/782).

Wright, Richard Sears (1829–92), was an English mathematician and teacher who taught engineering and surveying at Queenwood College from 1851. According to Hirst, who encouraged his development as a geometer, he suffered from epilepsy. He married in July 1853. Tyndall went on holiday with the Wrights on the Isle of Wight in 1855 and attended the funeral of their daughter, Maud, in January 1857. He left Queenwood in 1863 and opened his own school in Southsea.

Wrottesley, John (1798–1867), was an English astronomer and member of parliamentary commissions. He was a son of John Wrottesley, a Staffordshire landowner, farmer, and MP who was made 1st Baron Wrottesley in 1838. He was educated at Westminster School and Christ College, Oxford (BA 1819) and trained as a lawyer. He was one of the founders of the Royal Astronomical Society in 1820 and its secretary (1831–41). On inheriting his father's title as 2nd Baron Wrottesley, he re-erected his Blackheath observatory at Wrottesley Hall at Tettenhall. He succeeded Lord Rosse as president of the RS in 1854, but resigned in 1857.

Wynne, Anne (née Osborne, b. 1808), was the daughter of Sir Daniel Toler Osborne of Beechwood, Tipperary, who married George Wynne in 1834. The couple had a daughter and three sons, the second-eldest being Edward Wynne. During the 1850s Tyndall was a frequent visitor to the Wynnes' home at Harrow.

Wynne, Edward ("Eddie") (b. 1838), was a son of Colonel George Wynne. After Wynne served a brief shipbuilding apprenticeship with John Scott Russell, Tyndall arranged his tuition with Hirst at Queenwood College so that he could obtain a commission in the Royal Engineers. After failing in 1856, he entered Woolwich in 1857, following which he served in Canada.

Wynne, George (1804–90), was an Irish member of the Royal Engineers (commissioned as lieutenant in 1826) who had met Tyndall while seconded to the Irish Ordnance Survey, 1839–43. After further military service he worked for the Board of Trade, serving as government inspector of railways from 1847 to 1858, before spending the remainder of his life in military service, rising to general in 1877. Tyndall and Wynne had a high regard for each other and Tyndall wrote an appreciation of Wynne: "The Late General Wynne," *Times*, 12 July 1890, 11.

INDEX

The index refers to the letters using their serial numbers. The letter 'n' denotes a footnote reference and appears if an item is not otherwise mentioned in the letter to which the footnote pertains. A number in **bold** signifies a letter to or from a correspondent.

Abel-Rémusat, Jean-Pierre, 1187n
Abernethy, John, 1198
Académie des Sciences, 1105
Addams, Robert, 1273, 1278
Aggasiz, Louis, 1273, 1274, 1278, 1283
Airy, George Biddell, 1037, 1052, **1128**, **1129**, **1202**, 1203, **1204**
Akademie der Wissenschaften zu Berlin, 1216
Albert, Prince of Wales, 1208, 1288
Alciabiades, 1241
Alfred, Duke of Saxa Coburg, 1208
Alps, 1266, 1271, 1273, 1274, 1276, 1280, 1281, 1283
Amici, Giovanni Battista, 1109
Ampère, André-Marie, 1105, 1139, 1142, 1290
Anderson, Charles, 1105, **1145**
Annalen der Physik und Chemie, 1084, 1116, 1210, 1286
Ansted, David Thomas, 1170
Arago, Dominique, 1277
Arago, François, 1105, 1162
  Arago's disk, 1277, 1286
Archimedes, 1042
Ashburton, Lady, 1169, 1170, 1187
Ashburton, Lord. *See* Baring, William Bingham
Asher, Adolphus, 1084
Athenaeum Club, 1159, 1219
Atkinson, Edmund, 1157, 1216, 1272
Auer, Johann, 1280
Aytoun, William Edmondstoune, 1231

Babbage, Charles, 1159, 1209, **1254**, **1265**
Babbage, Henry Provost, **1209**
Babington, Charles Cardale, 1206
Bakerian Lecture, 1031, 1034, 1036, 1047n, 1052, 1076, 1118, 1119, 1123, 1128n, 1139, 1140, 1153, 1168, 1172, 1203, 1216, 1290
Baker, William, **1165**
Ballin, Hermine, **1026**, 1031
Barbezat, Edouard, 1124
Baring, Luisa, 1187
Baring, Lydia, 1187
Baring, William Bingham, 2nd Baron Ashburton, 1036, 1089, **1169**, 1170, 1183, 1187, **1189**, 1223
Barlow, Cecilia (née Law), 1032
Barlow, John, 1032, 1037, 1123
Barnard, Edward, 1031n
Barnard, Jane, 1101, 1105, 1132, 1134, 1138, 1259, 1273, 1276
Becquerel, Edmond, 1042, 1105, 1109
Beek, Albert van, 1045
Bennett, Mary, 1187
Bennett, William, 1187
Berlin Academy of Sciences, 1100
Best, Grace, 1242
Best, Samuel, 1132, 1242
Bevington, Geoffrey, 1166, 1192
Bevington, James Buckingham, 1035, 1047, 1152, 1154, 1163, 1164, **1166**, 1170, **1182**, **1192**, **1195**, 1196, 1235
Bevington, Mary, 1166, 1182, 1192
Billet, Félix, 1125
Biot, Jean-Baptiste, 1105, 1109, 1152
Blackwell, John Kenyon, 1042n
Blaikley, Alexander, 1208

boiler explosions, 1045, 1047, 1048, 1216, 1230, 1244
Bonaparte, Charles Lucien, 2nd Prince of Canino and Musignano, 1105, 1109
Bonaparte, Joseph-Charles-Paul, 1109
Bonaparte, Louis-Napoléon, 1109
Booth, Francis, 1267, 1283
Brayley, Edward William, 1043
Bremer, Fredrika, 1190
Brewster, David, 1109, 1124, 1137, 1138, **1191**, 1223
British Association for the Advancement of Science, Edinburgh Meeting (1850) of, 1053; Liverpool meeting (1854) of, 1052; Glasgow meeting (1855) of, 1091, 1119, 1126, 1132, 1134, 1135, 1136, 1137, 1138, 1144, 1149
Brodie, Benjamin Collins, 1157
Brookfield, Jane, 1079, 1187
Brookfield, Magdalene, 1187
Brookfield, William, 1079n
Brookfield, William Henry, 1187
Browning, Robert, 1187
Brown, Samuel, 1143
Bruce, James, 23rd Earl of Elgin, 1089
Brummel, George Bryan, 1187
Brunner, Jean, 1124
Buckle, Henry Thomas, 1088
Buff, Heinrich, **1142**
Bunsen, Robert, 1049, 1075, 1087, 1140, 1153, 1232
Byron, Lord, 1066n

Campbell, George John Douglas, 8th Duke of Argyll, 1135
Cardwell, Edward, 1089
Carlyle, Jane Welsh, 1236, 1274
Carlyle, Thomas, 1089, 1118, 1170, 1187, 1190, 1274
Carter, John, 1124
Carter, Richard, 1040
Cartier, George-Étienne, 1074
Cauchy, Augustin-Louis, 1109
Cayley, Arthur, 1088, 1203
Chapman, John, 1097, 1162, 1163, 1274
Chateaubriand, François-René, 1101
Cherriman, John Bradford, 1074
Childrey, Joshua, 1200
Christison, Robert, 1206
Church, A., 1142
Churchill, John, 1st Duke of Marlborough, 1187
Cincinnatus, Lucius Quinctius, 1118
Claudet, Antoine, 1208, 1256, 1264
Clerk, George, 1109
Clunn, Thomas, 1033n
Cole, Henry, 1187
collodion balloons, 1046, 1049, 1073
Combe, Andrew, 1066
Combe, George, 1066n
Committee of Industrial Instruction, 1207
Comptes Rendus de l'Académie des Sciences, 1100
Cooper, William White, 1044n, 1203n
Coriolanus, Gaius Marcius, 1241
Coxe, James, 1051, 1077, 1099, 1140, 1205, 1231
Coxe, May Anne (née Cumming), 1052, **1066**, **1077**, 1099, **1205**, **1231**
Crelle, August Leopold, 1281
Crimea, 1077, 1096, 1099, 1140, 1144, 1187
Crystal Palace, 1237. *See also* Great Exhibition of the Works of All Nations
Cumming, Charles, 1242
Cumming, William Fullerton, 1087

Dallas, George Mifflin, 1229
Daniell, John Frederic, 1216
Darwin, Charles, 1033
Daubeny, Charles, 1282
Davy, Humphry, 1030, 1045
Debus, Heinrich, 1041, 1050, **1065**, 1089, 1094, 1097, 1102, 1109, 1118, 1132n, 1140, **1157**, 1164, **1200**, 1226, 1233, 1241, 1242, 1243, 1245, **1269**, 1272, 1281, 1283, **1287**
de la Beche, Henry, 1071
de la Rive, Auguste Arthur, 1100, 1105, 1109, 1129, 1138, 1139, **1172**, 1210, 1253, 1279, **1289**
de la Rive, Lucien, 1279
de la Rue, Warren, 1085, **1257**
Denison, Evelyn, 1109
Department of Science and Art, 1207
Despretz, César, 1029, 1109, **1159**, **1160**, **1219**, 1223, **1240**
Deville, Henri Étienne Sainte-Clair, 1279
diamagnetism, 1029, 1039, 1042, 1053, 1054, 1061, 1082, 1119, 1137, 1139, 1142, 1149, 1151, 1152, 1154, 1158, 1161, 1172, 1192, 1232, 1261, 1277. *See also* magnetism
Dickinson, Aspden P., 1094
Dickinson, John Charles, 1094
Dirichlet, Peter Gustav Lejeune, 1100

Dix, Dorothea Lynde, 1077
Dollfus-Ausset, Daniel, 1273
d'Orléans, Henri Eugène Philippe Louis, duc d'Aumale, 1089
Dove, Heinrich Wilhelm, 1084, 1124, 1125, 1210
Dressner, Carl, 1087
Drewry, Henry Runciman, 1214, 1222
Drummond, Emily, 1047n, 1089
Drummond, Maria, 1047n
Drummond, Mary, 1047n
Dub, Julius, 1116
du Bois-Reymond, Emil, 1066, **1067**, 1068, 1070, 1072, 1080, 1081, 1084, 1085, 1089, 1091, 1100, 1123, 1171, 1246, **1249**, 1263, 1286
du Bois-Reymond, Jeannette (née Claude), 1067
Duboscq, César, 1031n, 1109
Duboscq, Louis Jules, **1131**, 1132, **1167**, 1180
Duboscq's lamp, 1031
Dunker, Wilhelm, 1087

Edmondson, Anne (née Singleton), 1077n, 1086
Edmondson, George, 1035, 1036, 1050, 1086, 1089, 1094, 1114, 1194, 1220, 1233, 1241, 1242, 1243, 1245, 1255, 1272, 1281, 1287
Edmondson, George William, 1086, 1245
Edmondson, Jane, 1245
education, 1243
Edward, Albert, Prince of Wales, 1208
Ehrenberg, Christian Gottfried, 1071
electricity, 1062, 1070n
electricity, animal, 1089, 1263
electric telegraph, 1093n
electro-physiology, 1066, 1228, 1286
Ellice, Edward, 1198
Elwent, Mr., 1232
Emerson, Ralph Waldo, 1066n, 1089, 1187, 1190, 1287
Eton College, 1291
Ettingshausen, Andreas von, 1109, 1280, 1290
Euclid, 1223
Eugenie, Empress of France, 1070
examinations, 1094, 1098, 1107, 1108, 1109, 1110, 1111, 1112, 1113, 1118, 1121, 1122, 1127, 1130, 1148, 1177, 1188, 1206, 1213, 1217, 1218, 1252, 1259, 1285
Exposition Universelle. *See* Paris Universal Exhibition
Eyre and Spottiswoode printing company, 1030

Fairbairn, William, 1048
Faraday, Michael, **1030**, 1031, 1036, 1037, 1039, 1040, **1042**, 1052, 1053, **1054**, 1057, 1058, 1064, **1069**, 1070n, 1076, 1081, 1082, 1084, 1089, 1091, 1092, 1099, 1100, **1101**, **1105**, 1116n, 1118, 1129, **1132**, **1133**, **1134**, 1137, **1138**, 1139, 1142, **1144**, 1151, 1152, 1154, 1159, 1161, 1162, 1163, 1164, 1166, 1168, 1170, 1172, 1174, 1202, 1203, 1208, 1216, **1223**, **1224**, 1227, 1229, 1230, **1234**, 1238, 1239, 1249, 1250, **1259**, 1268, **1273**, **1276**, **1278**, 1279
Faraday, Sarah (née Barnard), 1101, 1105, 1132, 1134, 1138, 1223, 1259, 1273, 1276, 1278n
Feilitzsche, F. C. O. von, 1076, 1082, 1116, 1168
Fick, Adolph, 1180, 1282
Fick, Emilie (née von Cölln), 1180
Finch, John, 1086, 1089
Fitton, William Henry, 1096
Fitzroy, Robert, 1150
Forbes, David, 1033, 1037
Forbes, Edward, 1037, 1143, 1153n
Forbes, James David, 1142, 1272, 1273, 1274, 1280, 1283, 1285, 1286, 1289
Foucault, Léon, 1052, 1105, 1109
Fox, Luther Owen, 1092n, 1211, 1272
Francis, William, 1047, 1050, 1052, 1055, 1063, 1084, 1085, 1089, 1099, 1109, 1132, 1140, 1157, 1171, 1178, 1238, 1259, 1262, 1266, 1267, 1272
Frankenheim, Moritz Ludwig, 1282
Frankland, Edward, **1031**, **1046**, **1049**, **1062**, **1073**, **1093**, 1132, 1136n, 1137, 1138, 1145, 1147, **1149**, **1153**, 1180, 1189, 1223, 1230, **1266**, **1268**, 1270, **1271**, 1272, 1276, 1280, **1282**
Frankland, Sophie (née Fick), 1026, 1153, 1268, 1271
Frederick II, King of Prussa, 1187
Froment, Paul-Gustave, 1124, 1277

Galbraith, Joseph Allen, **1212**, 1215, 1217, 1218, **1251**, 1255
Gale, George, 1198

Gale, Mr., 1035
Gale, Sarah, 1198
Gassiot, John Peter, 1070n, 1081, 1084, 1159
Gaugain, Jean Mothée, 1100
Gauss, Carl Friedrich, 1100, 1202
Genth, Frederick Augustus, 1140
Geological Club, 1033, 1034, 1037
Geological Society, 1033, 1037
geology, 1150, 1245, 1257, 1259, 1288. *See also* slaty cleavage
Gerland, B. Wilhelm, 1153n
Gerling, Christian Ludwig, 1280
Gerling, Mathilde, 1180
Gesellschaft Deutscher Naturforsher und Ärzte (Society of German Natural Scientists and Physicians), 1087, 1216, 1274, 1280, 1290
Ginty, Margaret (née Roberts), **1059**, 1064
Ginty, William Gilbert, 1059
Girard, Heinrich, 1087
glaciers, 1258, 1262, 1266, 1270, 1273, 1274, 1276, 1280, 1283, 1285, 1286, 1289
Glaisher, Cecilia Louisa, 1120n
Glaisher, James, **1120**
Gmelin, Leopold, 1054
Goëthe, Johann Wolfgang von, 1170, 1179, 1187, 1190, 1195, 1255, 1287
Goldsmid, Aaron Asher, 1085
Goldsmid, Isaac Lyon, 1085n
Goldsmith, Oliver, 1065
Gooch, Daniel, 1051
Graham, Thomas, 1170
Grailich, Wilhelm Josef, 1282, **1290**
Graves, Charles, 1121
Great Exhibition of the Works of All Nations (1851), 1237, 1031, 1140
Green, George, 1087
Grey, George, **1207**
Grieve, Thomas, 1227
Grosvenor, Eleanor, 1229
Grove, William Robert, 1036, 1159, 1161, 1162, 1172, 1203, 1219
Guthrie, Frederick, 1153

Haas, John, 1086, 1157, 1164, 1179, 1233, 1242, 1243, 1245
Hall, Anna Maria, 1153n
Hamilton, William Rowan, 1137
Hankel, Wilhelm Gottlieb, 1116
Hants and Wilts Educational Society, 1132
Hardy, Mr., 1051
Hare, Robert, 1170
Harrington, George Fellows, 1154
Harris, William Snow, **1210**, **1239**, 1259
Haughton, Samuel, 1238, 1241, 1243, 1251, 1255, 1259, **1288**
Hawker, Charles, **1104**
Hawtrey, Stephen, 1291
health, 1070, 1090, 1099, 1133, 1154, 1245, 1255, 1270
heat, 1055, 1228
Heber, Reginald, 1101n
Heine, Heinrich, 1140, 1190, 1274
Helmholtz, Hermann Ludwig Ferdinand von, 1091
Henfrey, Arthur, 1206
Henry IV, King of Navarre and King of France, 1281
Herries, John Charles, 1079
Herries, Maria Julia, 1250
Herschel, John Frederick William, **1258**, **1261**, **1262**
Hill, Henry, 1035, 1195
Hirst, Anna (née Martin), 1027, 1028, 1035, 1036, 1040, 1047, 1068, 1083, 1094, 1211, 1221, 1226, 1233, 1235, 1241, 1242, 1243, 1245, 1249, 1259, 1267, 1280, 1281
Hirst, John Henry, 1094, 1218, 1272
Hirst, Thomas Archer, **1027**, **1028**, **1029**, **1035**, **1036**, **1037**, **1039**, **1040**, **1047**, **1050**, 1052, **1055**, **1056**, **1058**, **1064**, **1068**, **1070**, **1072**, **1078**, **1080**, **1081**, **1083**, **1085**, **1086**, **1089**, **1090**, **1092**, **1094**, **1095**, **1097**, 1099, **1102**, **1109**, **1112**, **1113**, **1114**, **1115**, **1117**, **1137**, **1151**, **1152**, **1154**, 1157, **1162**, **1163**, **1164**, 1166, **1170**, **1174**, **1179**, **1183**, **1184**, **1185**, **1187**, **1190**, **1194**, 1195, **1196**, **1197**, **1198**, **1203**, **1208**, **1211**, 1216, **1218**, **1220**, **1221**, **1225**, **1226**, **1227**, **1229**, **1233**, **1235**, **1236**, **1238**, **1241**, **1242**, **1243**, **1245**, 1247, 1249, **1255**, 1259, **1267**, **1272**, **1280**, **1281**, **1283**, **1284**, 1287
Hofmann, August Wilhelm, 1089, 1229
homeopathy, 1187
Hooker, Frances Harriet (née Henslow), 1280, 1282, 1283, 1290
Hooker, Joseph Dalton, 1170, 1266, 1271, 1273, 1282, 1283
Hooker, William, 1109
Hopkins, Evan, 1154
Hopkins, William, 1170

Horace, 1041
Hornblower, Jemima Hanbury, 1026, 1031
Hueter, Viktor, **1041**
Hughes, Edward, **1141**, **1146**, **1147**, 1154n, 1183
Humboldt, Alexander von, 1100, 1200
Hunt, Robert, 1071
Huxley, George, 1188
Huxley, Henrietta (née Heathorn), **1188**, 1273, 1283
Huxley, Mary ('Polly') (née MacIntyre), 1188
Huxley, Thomas Henry, **1043**, **1044**, **1071**, 1085, **1106**, 1148n, 1198, 1226, 1266, 1271, 1272, 1273, 1274, 1280, 1283, 1289

Iffland, Johannes, 1041
Illustrated London News, 1187
induction coil, 1229
Institut de France, 1109
instruments, scientific, 1038, 1131, 1139, 1155, 1156, 1167, 1186, 1191, 1229. *See also* Ruhmkorff's coil; stereoscope

Jacobi, Moritz von, 1116
Jennings, Thomas, 1153
Jewsbury, Geraldine Endsor, **1237**, 1274
Jones, Henry Bence, 1031, 1055, 1164, 1167, 1170, 1201, 1230, 1249, 1259, 1278
Jones, Olivia M., **1201**
Joule, James Prescott, 1082, 1091, 1116, 1153, 1181

Kelland, Philip, 1135, 1137
Kingsley, Charles, 1187
Kirchhoff, Gustav, 1232
Klettenberg, Susanna Katherina von, 1200
Knoblauch, Hermann, 1171, **1180**
Knoblauch, Karl Friedrich, 1180
Knochenhauer, Karl Wilhelm, 1100
Kohlrausch, Rudolph, 1216, **1232**

laboratories, 1244
Labouchere, Henry, 1109
Lacey, Henry, 1092n
Lallemand, Claude François, 1221, 1226
Langberg, Christian, 1240
Latham, Robert Gordon, 1088
Leader, 1170
Lee, Elizabeth, 1242
Lee, Stanlake, 1241, 1242
Leidenfrost, Johann, 1142
Leland, Charles Godfrey, 1140
Lenoir, George André, 1062, 1093
Lenz, Heinrich Friedrich Emil, 1116
Lequesne, Philip Winter, 1220
Lerebours, Nicolas, 1191
Lerebours, Noel, 1191n
Le Verrier, Urbain Jean Joseph, 1055, 1109, 1125
Lewes, George Henry, 1170, 1187, 1195, 1200
Leyden battery, 1186
Leyser, Georg Moritz Ludwig, 1139, 1286
Liebig, Justus von, 1137, 1153
Lloyd, Humphrey, 1038
London Institution, 1028, 1043, 1081, 1102, 1183
Longfellow, Henry Wadsworth, 1187, 1188, 1229
Luscinus, Gaius Fabricus, 1118
Luther, Martin, 1066
Lyell, Charles, **1033**, **1034**, 1036n, 1037, **1096**, 1192, 1216

Macaulay, Thomas Babington, 1187
Mackay, Charles, 1118
Mackrell, George, 1142
Maggi, Pietro Giuseppe, 1091
magnetism, 1037, 1038, 1039, 1042, 1052, 1053, 1054, 1057, 1058, 1060, 1061, 1091, 1116, 1119, 1129, 1138, 1144, 1151, 1152, 1154, 1155, 1168, 1174, 1176, 1202, 1277, 1290. *See also* diamagnetism
Magnus, Anna, 1084
Magnus, Bertha (née Humblot), 1084
Magnus, Christine, 1084
Magnus, Heinrich Gustav, 1075, **1084**, 1125, 1165n, 1168, 1171, **1216**, 1218, **1230**, **1244**, 1276, 1278, 1279
Magnus, Paul Gustav, 1084
Mallet, Robert, 1208
Malone, Thomas, 1240
Marchand, R. F., 1054
Martin, John, 1027, 1036, 1052, 1095, 1102, 1109, 1112, 1226, 1227, 1233, 1281
Martin, Robert, 1027n
Martin, Samuel, 1027n
Masson, Victor, 1123
masturbation, 1220, 1225n, 1243
mathematics, 1087
Mathieu, Claude-Louis, 1124
Matteucci, Carlo, **1123**, **1161**, 1164, 1168,

1172, 1197, 1208, **1228**, **1253**, 1259, 1263, **1277**, 1279, **1286**
Matteucci, Robinia Elizabeth (née Young), 1123n
Maule, Fox, Second Baron Panmure, 1104, 1214, 1217, 1218, 1222, 1225, 1226, 1245
McCoy, Frederick, 1148
mechanical rotation, 1265
Melloni, Macedonio, 1100
Merz, Ludwig, 1191
Merz, Siegmund, 1191
Methodism, 1037
Miller, William Allen, 1216
Miller, William Hallows, 1076, 1170
Millikin, Richard Alfred, 1153n
Moigno, François-Napoléon-Marie, 1027, 1109
Moncel, Théodose du, 1109, 1167
Monsell, William, 1107, 1110
Monteith, Mr., 1205
Moore, Julia, **1246**, **1248**, **1252**, **1260**
Moran, John, 1027n
Moseley, Henry, **1098**, 1102, 1104, **1107**, **1108**, **1110**, **1111**, **1121**, **1122**, **1127**, **1130**, **1148**, **1177**, **1206**, 1212, **1215**, 1217, 1218, **1285**, **1291**
Moseley, Henry, Jr., 1291
Moseley, Henry Nottidge, 1291
Müller, Johann Heinrich Jakob, 1116
Murchison, Roderick Impey, 1170, 1234
Museum of Practical Geology, 1234
Muspratt, Edmund Knowles, 1153
Muspratt, James Sheridan, 1153

Napleon III, Emperor of France, 1070
National Review, 1182
Newman, John, 1145
Newman, John Henry, 1154n
Nero, 1140
Newton, Isaac, 1131, 1154
Nicholas I, Czar of Russia, 1052
Nichol, John Pringle, **1135**
Nicholson, John Helm, 1138n
Noad, Henry Minchin, 1240
Nobili, Leopoldo, 1186
Nuovo Cimento, 1123, 1286

O'Brien, William Smith, 1052n
Ohm, Georg Simon, 1139
optics, 1131, 1191
Owen, Richard, 1044n, 1047, 1192
Owen, Robert, 1085n

pachytrop, 1062
Pare, John, 1086, 1089
Pare, William, 1080, 1081
Paris Universal Exhibition (1855), 1064, 1109, 1112, 1118, 1124, 1125, 1140, 1180, 1191, 1207
Parsons, William, Third Earl of Rosse, 1191
Payne, Thomas (Tom), 1028, 1052, 1099, 1118
Peele, Robert, 1089
Peel, Frederick, 1089
Peltier, Jean Charles Athanase, 1155, 1259
Pepys, William Haseldine, 1045
Percy, Algernon, 4th Duke of Northumberland, 1092, 1229
Percy, John, 1085
Phillips, John, 1135
Philosophical Club, 1085, 1170
Philosophical Magazine, 1035, 1039, 1042, 1047, 1049, 1052, 1053, 1057, 1060, 1061, 1091, 1095, 1100, 1147, 1158, 1159, 1160, 1168, 1171, 1172, 1174, 1175, 1176, 1178, 1179, 1203, 1208, 1210, 1232, 1257, 1262, 1286, 1288
Philosophical Transactions of the Royal Society of London, 1091, 1181, 1228, 1263, 1264
Physical Institute of the University of Vienna, 1290
physics, 1117, 1125, 1132, 1140, 1142, 1217, 1230, 1232, 1244, 1232, 1244, 1286, 1290
Piria, Raffaele, 1123
Plateau, Joseph Antoine Ferdinand, 1042
Playfair, George, 1229
Playfair, Lyon, 1170
Plücker, Julius, 1082, 1172, 1174, 1208, 1230, 1280
Poe, Edgar Allan, 1052n, 1063n
Poggendorff, Johann Christian, 1084, 1168, 1180
Poisson, Siméon Denis, 1139
Pollock, Juliet (née Creed), **1032**, **1079**, **1250**
Pollock, Walter Herries, 1079, 1250
Pollock, William Fredrick, 1079n
Powell, Baden, 1154
Pridie [Priddie/Priddy], Robert [Roby/Rob] Hall, 1267, 1272, 1283
Proceedings of the Royal Society of London, 1091

Queenwood College, 1034, 1047, 1050, 1080, 1085, 1086, 1089, 1094, 1114, 1132, 1133, 1157, 1179, 1190, 1194, 1196, 1220, 1221, 1229, 1233, 1235, 1241, 1242, 1243, 1245, 1246, 1281, 1284, 1287, 1291
Queenwood Mutual Improvement Society, 1050, 1080
Queenwood Observer, 1208
Quintus Icilius, Gustav von, 1171

race, 1099
Ramsay, Andrew Crombie, 1043n, 1044, **1143**
Rauch, Christian Daniel, 1187
Red Lion Club, 1149, 1153
Regnault, Henri Victor, 1105, 1109, 1244, 1273, 1283
Reichenbach, Carl, 1153, 1282
Reich, Ferdinand, 1197
reputation, 1179
Richard I, King of England, 1102
Riess, Fanny (née Gütterbock), 1168
Riess, Peter Theophil, 1055, 1068, 1084, **1100**, **1168**, **1171**, 1172, 1210, 1259
Robinson, Thomas Romney, **1119**
Rogers, Henry, 1154n
Roscoe, Henry Enfield, 1208
Rose, Gustav, 1084, 1282
Rose, Heinrich, 1168
Royal Institution, 1030, 1031, 1034, 1036, 1040, 1046, 1047, 1092, 1156, 1192, 1245
Royal Military Academy, 1098, 1104, 1107, 1109
Royal Society, 1049, 1057, 1067, 1114, 1153, 1208, 1229; Council of, 1076; government grant fund of the, 1091, 1153, 1170, 1172, 1173, 1179, 1199, 1239, 1256, 1259
Rubini, Giovanni Baptista, 1231
Rufus, William, King William II, 1132
Ruhmkorff, Heinrich Daniel, 1125
Ruhmkorff's coil, 1080, 1083, 1109, 1125n, 1156, 1226, 1227
Russell, Harriette (née Osborne), 1196
Russell, John, 1192
Russell, John Scott, 1193, 1194, 1196, 1198
Russell, William James, 1153

Sabine, Edward, **1038**, **1074**, **1075**, 1077, 1124, **1173**, 1203, 1228, 1277, 1285
Sabine, Elizabeth Juliana (née Leeves), 1038, 1277
Sandemanian sect, 1026
Savage, Sarah, 1278
Schachner, Mr., 1079
Scheerer, T., 1054
Schlotten, Karl, 1180
Schrötter, Anton, 1278
Schweins, Ferdinand, 1087
sciences, natural, 1122n, 1130, 1217, 1218
Scipio Africanus, Publius Cornelius, 1179
Scoresby, William, 1037
Scott, Walter, 1187
Scrope, George Julius Poulett, 1261
Secretan, Marc, 1191n
Seebeck, Thomas Johann, 1142, 1155n
Séguier, Armand, 1124
Shakespeare, William, 1182
Sharpey, William, 1173, **1263**
Shaw, George Bernard, 1198n
Sheffield Literary and Philosophical Society, 1275
Siemens, Werner, 1091
Simon, Isaac, 1084n
slaty cleavage, 1150, 1172, 1234, 1259, 1261, 1275, 1280
Smith, Dr., 1281
Smyth, Charles Piazzi, 1135
snow crystals, 1120
Somerset House, 1219
Sorby, Henry Clifton, **1150**, 1172, **1275**
Soret, Jacques-Louis, 1172
Spedding, James, 1187
Spenser, Edmund, 1028
spiritualism, 1170
Spottiswoode, William, 1281
Stegmann, Friedrich Ludwig, **1087**
Steinheil, Karl August, 1109, 1124, 1125, 1191
Stephens, Walter, 1094
stereoscope, 1109, 1264
Sterling, John, 1187
Stevelly, John, 1137, 1138
Stokes, George Gabriel, 1061, **1076**, **1082**, 1085, **1116**, 1131, **1181**, 1203, 1232, **1256**, 1259, **1264**
Stokes, William, 1259
Stuken, Anna Christina, 1041
Sylvester, James Joseph, **1088**, **1265**

table-turning, 1170
Taylor, Francis, 1220, 1221
Taylor, Frank, 1220, 1221

Taylor, Richard, 1063
Taylor, Sarah, 1063
Temple, Henry John, Third Viscount Palmerston, 1192
Tennent, Edith Laetitia, 1229
Tennent, Emerson, 1229
Tennent, Sarah, 1229n
Tennent, William Emerson, 1229
Tennyson, Alfred, 1089, 1164, 1187, 1229, 1250, 1274
Thackeray, William Makepeace, 1229
Thénard, Louis Jacques, 1105, 1109
thermoelectric pile, 1186
Thomson, William, **1053**, 1054, **1057**, **1060**, **1061**, 1064, 1076, 1082, **1091**, 1137, 1138, **1155**, **1156**, **1158**, 1172, **1175**, **1176**, **1178**, **1186**, **1199**, **1203**, 1216, 1232, 1272
Thornhill, William, 1065
Tidmarsh, John Francis, 1281
Times, the, 1052, 1213, 1214, 1218, 1220
Tirel, Walter, 1132
Toronto, University of, 1074, 1075, 1078, 1080, 1081
translation, 1171
Trevelyan, Arthur, 1142, 1180
Tribe, Walter Harry, 1245
Tufnell, C. J., 1153n
Tyndale, Hector, 1028, **1052**, **1063**, **1099**, 1112, **1118**, **1140**, **1274**
Tyndale, Sharon, 1052
Tyndall, Sarah (née Macassey [Macasey/McAssey]), **1103**, **1136**, **1270**

unifilar, 1038

Verdet, Emile, 1068
Vincent, Benjamin, 1278

Walker, Charles Vincent, 1279
Wanklyn, James Alfred, 1153
War Department, 1214, 1217, 1218, 1222, 1223, 1225
Wartmann, Elie-François, 1105, 1109, 1112, 1124, **1125**
Watson, Joseph John William, 1031
Weber, Wilhelm Eduard, 1057, **1139**, 1142, 1154, 1158, 1168, 1171, 1216n, 1232, 1290
Weld, Charles, 1089, 1091
Wellesley, Henry, 1st Earl of Cowley, 1109
Wells, William Charles, 1261
Welsh, John, 1038
Wertheim, Theodor, 1109, 1124
Westmacott, James Sherwood, 1187
Westminster Review, 1092, 1094, 1095, 1097, 1162, 1163, 1274
Weyrausch, Kaspar, 1180
Weyrausch, Mathilde Wilhelmine Amoene, 1180
Whately, Edward W., 1205
Whately, Teckla (née Coxe), 1205
Wheatstone, Charles, 1091, 1055, 1109, 1159, 1170, 1208, 1223, 1230
Whewell, William, 1137, 1138
Whitehouse, Mary A., 1109, 1112, 1113
Whitelaw, George, 1030
Whiteside, James, 1229
Whitman, Walt, 1274
Wiedmann, Gustav Heinrich, 1240
William I, King of England, 1102
Williams, George, 1220
Williamson, Alexander William, 1057, 1058, 1064n
Williamson, William Crawford, 1268
Willis, Robert, 1109
Wordsworth, William, 1187
Wright, Fanny, 1242
Wright, Richard Sears, 1086, 1114, 1117n, 1152, 1157, 1164, 1184, 1233, 1242, 1243, 1245, 1255, 1267, 1272, 1281, 1283, 1287
Wright, William, 1063n
Wrottesley, John, 1036, 1037, 1170, 1173, 1203
Wynne, Anne (née Osborne), 1077
Wynne, Edward ("Eddie"), 1077, 1193, 1196, 1197, 1198, 1203, 1211, 1233, 1251, 1267, 1281
Wynne, George, 1044, **1045**, 1047, **1048**, **1051**, 1072, 1077, **1193**, 1194, 1203, 1208, 1211, 1251

Young, James, 1153
Young, Robert, 1123
Young, Thomas, 1182

Zammineer, Friedrich, 1142